COMPUTATIONAL METHODS IN SUBSURFACE FLOW

COMPUTATIONAL METHODS IN SUBSURFACE FLOW

PETER S. HUYAKORN

GeoTrans, Inc.
Reston, Virginia

GEORGE F. PINDER

Department of Civil Engineering
Princeton University
Princeton, New Jersey

1983

ACADEMIC PRESS

A Subsidiary of Harcourt Brace Jovanovich, Publishers

New York London
Paris San Diego San Francisco São Paulo Sydney Tokyo Toronto

ACADEMIC PRESS, INC.
111 Fifth Avenue, New York, New York 10003

United Kingdom Edition published by
ACADEMIC PRESS, INC. (LONDON) LTD.
24/28 Oval Road, London NW1 7DX

Library of Congress Cataloging in Publication Data

Huyakorn, P. S.
Computational methods in subsurface flow.

Includes index.
1. Groundwater flow--Mathematics. 2. Finite element method. I. Pinder, George Francis, Date . II. Title.
TC176.H89 1983 551.49 82-24456
ISBN 0-12-363480-6

PRINTED IN THE UNITED STATES OF AMERICA

83 84 85 86 9 8 7 6 5 4 3 2 1

To Saroach and Pongpen,
Wendy and Justin

Contents

4. Finite Element Simulation of Isothermal Flow in Porous Media

5. Finite Element Simulation of Solute and Energy Transport in Porous Media

6. Finite Element Simulation of Fluid Flow and Deformation in Unfractured and Fractured Porous Media

7. Alternative Finite Element Techniques and Applications

8. The Finite Difference Method

9. Finite Difference Simulation of Single and Multiphase Isothermal Fluid Flow and Solute Transport

10. Alternative Finite Difference Methods

Preface

Although the use of numerical analysis in the simulation of various subsurface phenomena is well documented in a number of journals and monographs, there has been no concerted effort to organize this material under one cover. In undertaking this task we have attempted to consider the application of all of the commonly encountered numerical techniques to subsurface problems. Among the problems considered in this book are groundwater flow and contaminant transport, moisture movement in variably saturated soils, land subsidence and similar flow and deformation processes in soil and rock mechanics, and oil and geothermal reservoir engineering. Because of the breadth of subject matter, we have provided only a generic overview. Details of specific formulations, particularly those relating to thermodynamically complex systems, are not dealt with herein. Rather, the reader should consult the original papers referenced at the end of each chapter.

This book is intended for senior undergraduate and graduate students in geoscience and engineering and also for professional groundwater hydrologists, engineers, and research scientists who want to solve or model subsurface problems using numerical techniques. We assume that the reader has some knowledge of calculus and matrix algebra and a basic understanding of subsurface phenomena. The content of the book is partitioned with respect to numerical procedures and physical systems. Thus, the reader may approach a simulation problem in one of two ways. Either the numerical method or the physical problem can be employed to ferret out the relevant theoretical development. To accommodate those readers who would like to strengthen their backgrounds in either numerical analysis or porous-medium physics, we have provided a brief introduction into the fundamental concepts in each of these topical areas. At the same time, however, we have tried to provide sufficient mathematical-physical depth to allow application of the methodology to field problems.

Chapter 1 of the book presents introductory material on partial differential equations and outlines various solution approaches that have been applied to subsurface flow. A brief literature review on each numerical technique and its applications is also included to give readers a historical overview.

Chapters 2 and 3 present the fundamental theory of the finite element method. These chapters should be read by those who are not familiar with the basic concepts and techniques for formulating finite element equations and interpolation (basis or shape) functions. Chapters 4–6 provide details on how the finite element approach, known as the Galerkin finite element method, can be used to solve a wide range of subsurface problems. The readers may elect to read different parts of these chapters depending on the type of problems in which they are interested. In Chapter 4, attention is focused on isothermal fluid flow in porous medium reservoirs. The subjects treated range from simple problems of saturated groundwater flow to more complex ones of moisture movement and multiphase flow in petroleum reservoirs. In Chapter 5, more difficult problems of solute transport and energy transport are treated. Various types of practical situations are discussed. These include contaminant transport in single-phase flow and energy transport in single- and two-phase flow. Chapter 6 deals with another specialized topic, that of fluid flow and mechanical deformation of conventional and fractured porous media. Various finite element approximations of fluid flow and solid displacement equations are presented. Included in this chapter is the treatment of material nonlinearities and discrete-fracture as well as double porosity conceptual idealizations of fractured systems.

Chapter 7 is devoted to other alternative finite element techniques that have been used to solve subsurface flow problems. Included in the presentation are the point and subdomain collocation techniques and the increasingly popular boundary element technique. The chapter is intended for those who want to gain an insight into different ways of formulating piecewise approximations of partial differential equations in subsurface problems.

Chapters 8 and 9 are devoted to finite difference techniques and their applications to single- and multiphase flow and solute transport. In Chapter 8, a simple and concise way to derive finite difference formulas is presented. The material is intended for readers who are not familiar with the basic theory of finite differences. In Chapter 9, several types of difference approximations and solution schemes are described for single- and multiphase flow problems in reservoir engineering. Wherever possible, we attempt to employ standard finite difference notations commonly adopted in most references on the subject.

The final chapter of the book is devoted to other alternative numerical methods that are based on combinations of the standard finite difference approach and classical mathematics. Included in the discussion are semianalytical finite difference techniques and their applications to unsaturated flow and free-surface flow problems, and the method of characteristics and its application to single- and two-phase flow and solute transport problems.

Although we have attempted to provide a broad coverage of various numerical methods for solving subsurface problems, we have not been able to include information pertaining to field applications of numerical models. Interested readers can obtain such information from many original references, some of which are cited in this book.

The successful completion of the manuscript is made possible by the efforts of those who gave generously of their time in reading and editing the early drafts of the book, and those who helped proofread the final copy. Particular recognition is due to R. H. Page, M. B. Allen, D. E. Dougherty, V. V. Nguyen, L. M. Abriola, M. A. Celia, and B. H. Lester, who reviewed the drafts of the manuscript and helped verify the type setting. We also wish to thank Dorothy Hannigan, who overcame our handwriting to provide a beautifully typed manuscript. Thanks are also extended to M. Sayabovorn and S. Saejong for their help in the preparation of the manuscript.

1

Introduction

1.1 Purpose and Scope

The objective of this book is to provide an introduction to the various numerical approaches to the simulation of subsurface flow phenomena. The book examines a wide range of practical problems, including groundwater flow, unsaturated flow, flow in fractured media, solute and energy transport, geothermal reservoir simulation, oil and gas reservoir simulation, and land subsidence. The major emphasis is placed on the finite element and finite difference methods that have proved to be powerful tools for solving differential equations usually intractable analytically. Other numerical techniques considered to be variants of the finite element and the finite difference methods are also presented. These include the collocation method, the boundary element method, and the method of characteristics.

The material of the book is divided into four basic parts. The first part is devoted to fundamental concepts in numerical methods. Having developed the basic theory, we next demonstrate in the second part how finite elements can be applied to a broad range of problems in groundwater hydrology and reservoir engineering. Alternative techniques related to finite elements and applicable to these problems are also included. The third and fourth parts deal with the finite difference theory and applications of this and other associated techniques to some of the problems for which the finite element method may not be optimal. In progressing from topic to topic, adequate background material concerning the physical concepts of subsurface phenomena is provided for readers who may be unfamiliar with this subject.

In the following sections of this chapter, we provide introductory material on differential equations and outline various solution approaches that have been applied to subsurface flow.

1.2 Introduction to Partial Differential Equations

1.2.1 Differential Equations and Operators

By a differential equation, we mean an equation that relates two or more variables in terms of derivatives or differentials. If such an equation involves only two variables, one of which is dependent on the other, it is called an "ordinary differential equation" (ODE). On the other hand, if the equation involves more than two variables and is written in terms of partial derivatives of one variable with respect to the remaining variables, it is called a "partial differential equation" (PDE).

An nth-order ordinary differential equation can be written in the form

$$F(x, u, du/dx, d^2u/dx^2, \ldots, d^nu/dx^n) = 0, \tag{1.2.1.1}$$

where F is a specified function, u the dependent variable that is an unknown function of the independent variable x, and n the order of the highest derivative.

Equation (1.2.1.1) is said to be linear when it is first degree in u and derivatives of u, and quasi-linear when it is first degree in the highest derivative of u. A linear nth-order equation can be written in the form

$$a_0u + a_1(du/dx) + \cdots + a_n(d^nu/dx^n) = f(x),$$

or

$$\sum_{k=0}^{n} a_k(d^ku/dx^k) = f(x), \tag{1.2.1.2}$$

where the coefficients a_k are either constants or functions of x.

If $f(x) = 0$, the equation is said to be homogeneous; otherwise it is said to be inhomogeneous. In a manner similar to an ordinary differential equation, an nth-order partial differential equation relating three variables x, y and u can be written in the form

$$F\left(x, y, u, \frac{\partial u}{\partial x}, \frac{\partial u}{\partial y}, \frac{\partial^2 u}{\partial x^2}, \frac{\partial^2 u}{\partial x\,\partial y}, \frac{\partial^2 u}{\partial y^2}, \ldots, \frac{\partial^n u}{\partial x^n}, \ldots, \frac{\partial^n u}{\partial y^n}\right) = 0. \tag{1.2.1.3}$$

Just as in the case of an ordinary differential equation, Eq. (1.2.1.3) is said to be linear when it is first degree in u and partial derivatives of u, and quasi-linear when it is first degree in the highest partial derivatives of u.

In this book, we shall be mainly concerned with second-order linear and nonlinear partial differential equations. A linear second-order partial differential equation can be written in the form

$$a\frac{\partial^2 u}{\partial x^2} + b\frac{\partial^2 u}{\partial x\,\partial y} + c\frac{\partial^2 u}{\partial y^2} + d\frac{\partial u}{\partial x} + e\frac{\partial u}{\partial y} + gu = f(x, y), \tag{1.2.1.4}$$

where the coefficients a, b, c, d, e, and g are constants or functions of x and y.

Irrespective of whether a given differential equation is a partial or an ordinary differential equation, we define its solution as a function u that reduces the equation to an identity upon substitution of u and its appropriate derivatives into that equation. A general solution of a linear differential equation is defined as the collection of all solutions of that equation.

Linear Operator

Frequently it is convenient to write differential equations in the abbreviated form

$$L(u) = f, \tag{1.2.1.5}$$

where f is a known function of the independent variable(s), and L is called a "differential operator." For an nth-order linear differential equation, the operator L is given by

$$L = \sum_{k=0}^{n} a_k \frac{d^k}{dx^k}. \tag{1.2.1.6}$$

For a linear second-order partial differential equation, L is given by

$$L = a\frac{\partial^2}{\partial x^2} + b\frac{\partial^2}{\partial x\,\partial y} + c\frac{\partial^2}{\partial y^2} + d\frac{\partial}{\partial x} + e\frac{\partial}{\partial y} + g. \tag{1.2.1.7}$$

An operator L is said to be a "linear operator" if for any functions u_1, u_2 and any constants c_1, c_2 the following relation holds:

$$L(c_1u_1 + c_2u_2) = c_1L(u_1) + c_2L(u_2). \tag{1.2.1.8}$$

It can be readily shown that the operators given by Eqs. (1.2.1.6) and (1.2.1.7) are linear operators. Indeed, the operators of linear differential equations are always linear. This means that the principle of superposition is applicable. In other words, if $u_1, u_2, \ldots, u_k$ are solutions of the equation $L(u) = f$, then their linear combination $u = c_1u_1 + c_2u_2 + \cdots + c_ku_k$ is also a solution.

Self-adjoint Operator

An operator L is said to be self-adjoint if L is identical to its own adjoint L^*. The adjoint L^* can be obtained via integration by parts or Green's theorem. As an illustration, let L be defined in a domain D, and let u and v be two functions in the field of definition of L. Then the inner product

$$(Lu, v) = \int_D (Lu)v\, dD$$

can be transformed via Green's theorem (or integration by parts) into

$$(Lu, v) = (u, L^*v) + \text{boundary integral terms.}$$

For a self-adjoint operator, $L = L^*$.

A differential equation containing a self-adjoint operator is said to be a self-adjoint equation. As an example, consider a linear problem governed by

$$Lu = f \qquad \text{in} \quad \text{domain } D$$

and

$$B_p u = g_p \qquad \text{on} \quad \text{boundary } \Gamma,$$

where L and B_p are linear operators, and f and g_p are functions defined in D and on Γ, respectively. Suppose that $L = d^4/dx^4$ and D corresponds to $0 < x < 1$. Thus

$$\begin{aligned}
(Lu, v) &= \int_0^1 \left(\frac{d^4u}{dx^4}\right) v\, dx = \frac{d^3u}{dx^3} v \Big|_0^1 - \int_0^1 \frac{d^3u}{dx^3}\frac{dv}{dx}\, dx \\
&= \frac{d^3u}{dx^3} v \Big|_0^1 - \frac{d^2u}{dx^2}\frac{dv}{dx}\Big|_0^1 + \int_0^1 \frac{d^2u}{dx^2}\frac{d^2v}{dx^2}\, dx \\
&= \frac{d^3u}{dx^3} v \Big|_0^1 - \frac{d^2u}{dx^2}\frac{dv}{dx}\Big|_0^1 + \frac{du}{dx}\frac{d^2v}{dx^2}\Big|_0^1 - \int_0^1 \frac{du}{dx}\frac{d^3v}{dx^3}\, dx \\
&= \frac{d^3u}{dx^3} v \Big|_0^1 - \frac{d^2u}{dx^2}\frac{dv}{dx}\Big|_0^1 + \frac{du}{dx}\frac{d^2v}{dx^2}\Big|_0^1 - u\frac{d^3v}{dx^3}\Big|_0^1 + \int_0^1 u\frac{d^4v}{dx^4}\, dx.
\end{aligned}$$

This can be put in the form

$$(Lu, v) = (u, Lv) + B_r(u, v)\Big|_0^1 + B_r^*(u, v)\Big|_0^1,$$

where

$$B_r(u, v) = -u\frac{d^3v}{dx^3} + \frac{du}{dx}\frac{d^2v}{dx^2} = \sum_{r=0}^{1} (-1)^{r+1} F_r u G_r v,$$

$$B_r^*(u, v) = -\frac{d^2u}{dx^2}\frac{dv}{dx} + \frac{d^3u}{dx^3} v = \sum_{r=2}^{3} (-1)^{r+1} G_r^* u F_r^* v,$$

in which F_r, F_r^*, G_r, and G_r^* are boundary operators.

The boundary conditions of the type $F_r u = g_r$ and $G_r^* u = g_r^*$ are called the essential (Dirichlet) and natural (Neumann) boundary conditions, respectively. Note that for the 2nth-order equation $d^{2n}u/dx^{2n} = f$, we have $L = L^*$, $F_r(r = 0, 1, \ldots, n-1) = F_r^*\ (r = 2n-1, 2n-2, \ldots, n)$, and $G_r^*(r = 2n-1, 2n-2, \ldots, n) = G_r(r = 0, 1, \ldots, n-1)$. The symmetric property of self-adjoint operators proves to be very useful in their finite element solutions. The coefficient matrix obtained from the finite element approximation of a self-adjoint equation is always symmetric.

1.2.2 Classification of Physical Problems

The ordinary or partial differential equation alone is not sufficient to describe a specific physical problem. This is because a general solution of an nth-order ODE involves n independent arbitrary constants, and a general solution of an nth-order PDE involves n independent arbitrary functions. In order to define uniquely a given physical problem, the values of the constants or the forms of the functions in the general solution must be specified. This can be achieved via the use of additional information arising from the physical situation. The additional information is presented in the form of auxiliary conditions associated with the physical problem. These auxiliary conditions are referred to more specifically as "initial and boundary conditions." We can classify physical problems based on their auxiliary conditions.

Initial Value Problems

These are time dependent problems governed by an ODE or by a special type of PDE. If a given problem is governed by an nth-order ODE, complete description of this problem then requires the solution of the governing equation

$$F(t, u, du/dt, d^2u/dt^2, \ldots, d^nu/dt^n) = 0 \tag{1.2.2.1a}$$

in a time domain $t_0 \leqslant t \leqslant \infty$, subject to the initial conditions

$$u(t_0) = u_0, \qquad d^ku(t_0)/dt^k = u_0^k \qquad (k = 1, 2, \ldots, n-1), \tag{1.2.2.1b}$$

where $u_0, u_0^1, \ldots, u_0^{n-1}$ are initial values of the function u and its $(n-1)^{\text{th}}$ derivatives at the initial time t_0.

If a given problem is governed by a first-order PDE, the complete description of this problem then requires the solution of the equation

$$F(x, t, u, \partial u/\partial x, \partial u/\partial t) = 0 \qquad \text{in} \quad \text{region} \ -\infty < x < \infty, \quad 0 \leqslant t \leqslant \infty \tag{1.2.2.2a}$$

in a prescribed region of the (x, t) space, subject to the initial condition

$$u(x, 0) = u_0(x). \tag{1.2.2.2b}$$

Such a problem is commonly referred to as a Cauchy initial value problem to distinguish it from the classical initial value problem associated with an ODE.

Boundary Value Problems

These are problems governed by ordinary or partial differential equations in which the independent variables are spatial coordinates. If a given problem is governed by an nth-order ODE, a complete description of this problem requires solution of the equation

$$F(x, u, du/dx, d^2u/dx^2, \ldots, d^nu/dx^n) = 0 \qquad (1.2.2.3a)$$

in a prescribed space region $x_1 \leqslant x \leqslant x_2$, subject to the boundary conditions

$$g_k[x_1, u(x_1), du(x_1)/dx, \ldots, d^{n-1}u(x_1)/dx^{n-1}] = 0, \qquad (1.2.2.3b)$$

$$g_l[x_2, u(x_2), du(x_2)/dx, \ldots, d^{n-1}u(x_2)/dx^{n-1}] = 0, \qquad (1.2.2.3c)$$

where

$$k = 1, 2, \ldots, M, \qquad l = 1, 2, \ldots, N, \qquad M + N = n.$$

If a given problem is governed by an nth-order PDE involving the coordinates x and y, a complete description of this problem then requires the solution of the equation

$$F(x, y, u, \partial u/\partial x, \partial u/\partial y, \ldots, \partial^n u/\partial x^n, \ldots, \partial^n u/\partial y^n) = 0 \qquad (1.2.2.4a)$$

in a closed region R of (x, y) space, subject to the boundary condition

$$g_k(x, y, u, \partial u/\partial x, \partial u/\partial y, \ldots, \partial^{n-1}u/\partial x^{n-1}, \ldots, \partial^{n-1}u/\partial y^{n-1}) = 0, \qquad k = 1, 2, \ldots, m \leqslant n-1, \qquad (1.2.2.4b)$$

on the boundary Γ of the region R.

Initial Boundary Value Problems

This type of problem is governed by partial differential equations in which the independent variables are spatial coordinates and time. If a given problem is governed by an nth-order PDE involving the coordinates x, y and time t, a complete description of this problem then requires the solution of the governing equation

$$F\left(x, y, t, \frac{\partial u}{\partial x}, \frac{\partial u}{\partial y}, \frac{\partial u}{\partial t}, \ldots, \frac{\partial^n u}{\partial x^n}, \ldots, \frac{\partial^n u}{\partial y^n}, \ldots, \frac{\partial^n u}{\partial t^n}\right) = 0 \qquad (1.2.2.5a)$$

in the interior of space region R during time interval $t_0 \leqslant t \leqslant \infty$, subject to the initial conditions

$$u(x, y, t_0) = u_0(x, y), \tag{1.2.2.5b}$$

$$\frac{d^k u}{dt^k}(x, y, t_0) = u_0^k(x, y), \qquad k = 1, \dots, n-1.$$

in the region R, and the boundary conditions

$$g_k\left(x, y, t, \frac{\partial u}{\partial x}, \frac{\partial u}{\partial y}, \dots, \frac{\partial^{n-1} u}{\partial x^{n-1}}, \dots, \frac{\partial^{n-1} u}{\partial y^{n-1}}\right) = 0, \tag{1.2.2.5c}$$

$$k = 1, 2, \dots, m \leqslant n-1, \qquad t > t_0$$

on the boundary Γ of the region R. Although we have assumed that the order of the highest-order derivative is the same in time and space, this is not generally the case, and the preceding discussion may be modified accordingly.

1.2.3 Classification of Partial Differential Equations

All partial differential equations of the form $L(u) = f$ can be classified as elliptic, parabolic, or hyperbolic. For a second-order linear or quasi-linear equation in two independent variables x and y, the classification is based on analogy with the terminology associated with conic sections. The equation is first written in the form

$$a\frac{\partial^2 u}{\partial x^2} + 2b\frac{\partial^2 u}{\partial x\, \partial y} + c\frac{\partial^2 u}{\partial y^2} = F\left(x, y, u, \frac{\partial u}{\partial x}, \frac{\partial u}{\partial y}\right), \tag{1.2.3.1}$$

where a, b, and c are functions of x and y only and the equation is linear if F is linear.

(1) If $b^2 - ac > 0$, Eq. (1.2.3.1) is said to be a hyperbolic equation. Via the coordinate transformation $\xi = \xi(x, y)$, $\eta = \eta(x, y)$, it can be shown that the hyperbolic equation is reducible to the canonical form

$$\frac{\partial^2 u}{\partial \xi^2} - \frac{\partial^2 u}{\partial \eta^2} = G\left(\xi, \eta, u, \frac{\partial u}{\partial \xi}, \frac{\partial u}{\partial \eta}\right). \tag{1.2.3.2}$$

(2) If $b^2 - ac = 0$, Eq. (1.2.3.1) is said to be a parabolic equation and is reducible to the following canonical form:

$$\frac{\partial^2 u}{\partial \eta^2} = G\left(\xi, \eta, u, \frac{\partial u}{\partial \xi}, \frac{\partial u}{\partial \eta}\right). \tag{1.2.3.3}$$

(3) If $b^2 - ac < 0$, Eq. (1.2.3.1) is said to be an elliptic equation

and is reducible to the following canonical form:

$$\frac{\partial^2 u}{\partial \xi^2} + \frac{\partial^2 u}{\partial \eta^2} = G\left(\xi, \eta, u, \frac{\partial u}{\partial \xi}, \frac{\partial u}{\partial \eta}\right). \tag{1.2.3.4}$$

This classification criterion can be generalized to linear and quasi-linear partial differential equations involving n independent variables x_1, x_2, x_3, ..., x_n. Such equations can be written in the form

$$a_{ij} \frac{\partial^2 u}{\partial x_i \, \partial x_j} = F\left(x_1, x_2, \ldots, x_n, u, \frac{\partial u}{\partial x_1}, \ldots, \frac{\partial u}{\partial x_n}\right), \tag{1.2.3.5}$$

where summation notation is assumed for repeated indices ($i, j = 1, 2, \ldots, n$) and the coefficients a_{ij} are functions of x_1, x_2, ..., x_n. To classify Eq. (1.2.3.5), we look for the eigenvalues of a_{ij}. In other words, we set

$$\begin{vmatrix} a_{11}-\lambda & a_{12} & \cdots & a_{1n} \\ a_{21} & a_{22}-\lambda & \cdots & a_{2n} \\ \vdots & \vdots & \vdots & \vdots \\ a_{n1} & & \cdots & a_{nn}-\lambda \end{vmatrix} = 0. \tag{1.2.3.6}$$

The left-hand side of Eq. (1.2.3.6) is a polynomial of degree n. Thus, there exist n roots (or n eigenvalues) for λ. If all the roots are positive, we say that Eq. (1.2.3.5) is elliptic. If some of the roots are zero, Eq. (1.2.3.5) is parabolic. If some of the roots are positive and some are negative, Eq. (1.2.3.5) is hyperbolic. Hyperbolic and parabolic equations usually occur in time dependent (initial boundary value) problems, whereas elliptic equations occur in equilibrium or steady state (boundary value) problems. For parabolic and hyperbolic equations the solution domains are usually open, whereas for elliptic equations the solution domains are defined by closed boundaries.

1.3 Partial Differential Equations in Subsurface Flow

Subsurface flow problems are usually governed by second-order partial differential equations. Let us now consider a few of the important equations we are likely to encounter.

1.3.1 Single-Phase Liquid Flow in Porous Media

For a situation of two-dimensional flow in a homogeneous and isotropic porous medium, the governing equation can be written in the form

$$A\left(\frac{\partial^2 u}{\partial x^2} + \frac{\partial^2 u}{\partial y^2}\right) = B \frac{\partial u}{\partial t}, \tag{1.3.1.1}$$

where u is an unknown function of position and A and B are parametric coefficients of the porous medium.

In groundwater hydrology, the dependent variable u usually corresponds to the piezometric (or hydraulic) head and A and B are coefficients of transmissivity and storage of the confined water-bearing formation (confined aquifer). In petroleum engineering, u usually corresponds to the fluid pressure and A and B are the mobility and compressibility factors of the reservoir formation.

Using the previous criterion for classifying second-order partial differential equations, it can be readily shown that Eq. (1.3.1.1) is parabolic. When the flow reaches a steady state condition, Eq. (1.3.1.1) reduces to

$$\partial^2 u/\partial x^2 + \partial^2 u/\partial y^2 = 0, \tag{1.3.1.2}$$

which is the well-known (elliptic) Laplace equation.

1.3.2 *Single-Phase Ideal Gas Flow in Porous Media*

For a situation of two-dimensional flow in a homogeneous and isotropic porous medium under isothermal conditions, the governing equation can be written in the form

$$A[\partial(p\, \partial p/\partial x)\, \partial x + \partial(p\, \partial p/\partial y)/\partial y] = \phi\, \partial p/\partial t, \tag{1.3.1.3}$$

where p is the absolute pressure, A the mobility factor, and ϕ the effective porosity of the reservoir formation.

Equation (1.3.1.3) is a nonlinear parabolic equation. When the flow reaches a steady-state condition, the equation reduces to

$$\partial(p\, \partial p/\partial x)/\partial x + \partial(p\, \partial p/\partial y)/\partial y = 0. \tag{1.3.1.4}$$

Equation (1.3.1.4) is a nonlinear elliptic equation that can be converted to the linear Laplace equation

$$\partial^2 u/\partial x^2 + \partial^2 u/\partial y^2 = 0, \tag{1.3.1.5a}$$

where

$$u = p^2. \tag{1.3.1.5b}$$

1.3.3 *Solute Transport in Porous Media*

For a two-dimensional problem of density-independent solute transport in a homogeneous and isotropic porous medium, the governing equation takes the form

$$D(\partial^2 C/\partial x^2 + \partial^2 C/\partial y^2) - v_x\, \partial C/\partial x - v_y\, \partial C/\partial y = \partial C/\partial t, \tag{1.3.1.6}$$

where C is the concentration of the solute, D the dispersion coefficient of the porous medium, and v_x and v_y the seepage velocity components in the x and y directions, respectively.

When a steady-state condition is reached, Eq. (1.3.1.6) reduces to

$$D(\partial^2 C/\partial x^2 + \partial^2 C/\partial y^2) - v_x\,\partial C/\partial x - v_y\,\partial C/\partial y = 0. \quad (1.3.1.7)$$

It can be shown that Eq. (1.3.1.6) is parabolic and Eq. (1.3.1.7) is elliptic.

1.4 Solution Methodology

Returning to the discussion of the boundary and initial boundary value problems, we see that these problems reduce to finding the unknown function u that satisfies the governing PDE (or ODE as the case may be) together with the associated boundary and initial conditions. There are many alternative methods for solving such problems. These methods can be classified as analytical and numerical. Some of the important ones are

(1) analytical methods:
 (a) separation of variables,
 (b) similarity solutions,
 (c) complex variable techniques,
 (d) Fourier and Laplace transformations,
 (e) Green's functions,
 (f) regular and singular perturbations,
 (g) power series;
(2) numerical methods:
 (a) finite difference methods,
 (b) Galerkin or variational finite element methods,
 (c) collocation methods,
 (d) method of charcteristics,
 (e) boundary element methods.

For a number of problems involving linear or quasi-linear equations and regions of simple (or regular) geometry, it is possible to obtain exact solutions by analytical methods. This is often accomplished by a separation of variables or by applying a transformation that makes the variables separable and leads to a similarity solution. Occasionally, the complex-variable technique, the Green's function approach, or a Laplace or Fourier transformation (or both) of the differential equation leads to an exact solution.

For nonlinear problems with regions of regular geometry, very few analytical solutions exist, and these are usually approximate solutions obtained by using perturbation or power series methods. Because regular and singular perturbation methods are applicable primarily when the nonlinear terms in the equation are small in comparison to the linear terms, their usefulness is limited. The power series method is powerful and has been employed with some success, but because the method requires evaluation of a coefficient for each term in the series, it is relatively tedious. Also, it is difficult, if not impossible, to demonstrate that the power series converges to the correct solution.

Problems involving regions of irregular geometry are generally intractable analytically. For such problems, the use of numerical methods to obtain approximate solutions is advantageous. There are currently five widely used numerical methods. These methods are closely related. In several cases the finite difference, finite element, and collocation methods yield the same approximation. The method of characteristics is a variant of the finite difference method and is particularly suitable for solving hyperbolic equations. Finally, the boundary element method, a variant of the conventional finite element method, is especially useful in the solution of elliptic equations for which Green's functions exist. In this volume, we are concerned almost exclusively with numerical techniques.

1.5 Computational Methods in Subsurface Flow

1.5.1 Introduction

It is interesting to contemplate the factors that have come together to create the simulation capability we enjoy today. One is tempted to attribute current technology to advances in numerical analysis. Yet numerical analysis, albeit somewhat primitive, has been around since the time of the Babylonians [see Kelly (1967)], although we have seen significant developments in simulation methodology only since World War II.

It seems the modern-day computer is at least in part responsible for this flurry of activity. Whereas the conceptual underpinnings of today's computers were introduced in the nineteenth century with the work of Babbage in 1834 and Kelvin in 1876, it appears to have been the extraordinary advances in the electronics industry that have catapulted mathematical simulation into its current position of prominence. It is indeed phenomenal how rapidly computational capability has grown. Consider, for example, the description of the IBM 709 presented by Forsythe and Wasow (1960). They reported that this system executed addition in 24 μsec (floating point addition took at least 84 μsec) and multiplication in from 24 to 240 μsec and that it had a high-speed memory of up to 32,768 words. Projecting into the sixties, they forecast machine memories 50 to 100 times as large and arithmetic calculation speeds up to 100 times as fast. Further advances, they perceived, would come slowly because of technological constraints. Compare their projections with the capabilities of today's IBM 3033, which performs a multiplication–addition in 1 μsec and boasts core storage of 3 million words—remarkably accurate forecasting!

The promise and eventual reality of enhanced computer power kindled the interest of scientists and engineers interested in applying numerical methods to practical problems. Thus, there evolved in parallel with computer technology an increased interest in more efficient and flexible numerical methods. In the ensuing paragraphs we present a brief overview of the role of several of these numerical schemes in the simulation of subsurface phenomena.

1.5.2 The Finite Difference Method

The most popular numerical approach to the simulation of large reservoir systems is the method of finite differences. There are several practical reasons for this popularity. Finite difference methods are conceptually straightforward. The fundamental concepts are readily understood and do not require advanced training in applied mathematics. Moreover, due to their extensive history, they boast a firm theoretical foundation. In addition, because of the form and algebraic simplicity of the equations arising from difference approximations, several clever algorithms have been developed for their solution.

The use of finite differences in subsurface simulation began in earnest with the development of the alternating direction algorithms of Peaceman and Rachford in 1955. The oil industry recognized early the importance of numerical simulation in the forecast of reservoir performance. In the period following the classic paper by Peaceman and Rachford (1955) there was an astounding growth in the numerical methodologies proposed and the physical systems considered. Early work focused on single-phase reservoir simulation (Bruce *et al.*, 1953). This was followed by "black oil" reservoir simulation, which involved two components, a nonvolatile component (black oil) and a volatile gas soluble in the oil phase. Coats *et al.* (1967), Douglas *et al.* (1959), Sheldon *et al.* (1960), and Stone and Garder (1961) are important references in this topical area.

More recently, models have been developed for forecasting reservoir behavior under tertiary recovery operations. Many of these simulators are thermodynamically, chemically, and hydrodynamically very sophisticated. Consider, for example, the work of Coats (1980) in predicting miscible flood performance, Gottfried (1965) and Shutler (1970) in thermal oil recovery simulation, and Bondor *et al.* (1972) in polymer flood behavior forecasting. For problems of this complexity and degree of nonlinearity the simplicity and computational efficiency of finite difference methods has a justifiable attraction. Three excellent monographs on the application of finite difference theory to reservoir engineering have recently been published and the interested reader is referred to these publications for technical details (Aziz and Settari, 1979; Crichlow, 1977; Peaceman, 1977).

The first application of finite difference theory to groundwater flow appears to have been made by Remson *et al.* (1965). This work and those that followed shortly thereafter (Freeze and Witherspoon, 1966; Bittinger *et al.*, 1967; Pinder and Bredehoeft, 1968) focused on a description of the hydrodynamics of single-phase fluid flow. Thus numerical simulators represented a natural extension of the electric analog network models that were widely used for groundwater flow forecasting at that time.

At about the same time as these groundwater models were being developed, there was a parallel effort under way in soil physics. The unsaturated flow equations, which are similar to those encountered in multiphase oil reservoir simulation, appear to have been of interest to a number of researchers during the late fifties. Papers published at that time include Ashcroft *et al.* (1962), Hanks and Bowers (1962), Philip (1957), Rubin and Steinhardt (1963), and Whisler and Klute (1965). The ability to accommodate spatially variable material properties made the numerical approach attractive, and the finite difference approach, being the only well-established numerical methodology available at the time, was the obvious choice. In a recent review article Haverkamp *et al.* (1977) compared six of the most frequently used numerical schemes and reported significant differences in computational efficiency.

The majority of saturated–unsaturated groundwater flow simulators [e.g., Cooley (1971), Freeze (1971)] assume an immobile air phase. This obviously reduces the number of dependent variables and the associated solution time. Brutsaert (1973) and Green *et al.* (1970) developed simulators that relaxed this assumption. In their approach the air and water phase equations were solved simultaneously using the same basic methodology employed in the oil industry for two-phase flow. Their results were compared against field information.

There is a rather extensive literature on the use of finite difference methods for the solution of the mass transport equation. The governing equation is applicable to oil, groundwater, and unsaturated flow simulation. One of the earliest attempts to solve the transport problem was made by Peaceman and Rachford (1962), who employed standard finite difference methods in an oil reservoir simulation. Stone and Brian (1963) presented a scheme that circumvented some of the well-known numerical difficulties encountered by Peaceman and Rachford.

Multidimensional groundwater transport models have been developed by Shamir and Harleman (1967) and Oster *et al.* (1970). Density dependent formulations were published by Fried and Ungemach (1971) and Henry and Hilleke (1972). A transport simulator based on a polygonal finite difference net was employed by Orlob and Woods (1967) and appears to represent the first large-scale application of transport modeling in regional groundwater simulation.

The simulation of mass transport in saturated–unsaturated media is of more recent vintage. The first work appears to have been that of Tanji *et al.* (1967). A number of important papers followed; a relatively recent contribution is provided by Wierenga (1977). The majority of papers now appearing in the literature in this topical area seem to have abandoned finite difference methods in search of more accurate simulation tools. These alternative possibilities follow.

We have yet to consider the use of finite difference methods in non-isothermal flow. Probably the most exotic problem to be considered along these lines is the simulation of geothermal reservoir behavior. The majority of geothermal reservoir models are formulated using finite differences. It is difficult to establish with certainty the geothermal reservoir model development chronology. Certainly among the earliest papers were those of Brownell *et al.* (1975), Faust and Mercer (1976), Lasseter *et al.* (1975), Thomas and Pierson (1976), and Toronyi (1974). A closely related yet distinctly different simulator was presented by Coats (1974) to forecast oil reservoir behavior subject to steamflooding. Because of the highly nonlinear nature of the geothermal equations and their considerable mathematical complexity, it is not surprising that finite difference methods remain today the generally accepted numerical approach for this class of problems. For a review of the available literature on geothermal simulation see Pinder (1979).

1.5.3 The Finite Element Method

The application of the finite element method to subsurface problems followed two distinctly different paths. In the petroleum industry, Galerkin's method, the theoretical foundation upon which most modern finite element models are built, was used almost exclusively with rectangular elements. Enhanced accuracy was achieved using higher-degree polynomial basis functions, occasionally augmented with analytical solutions. In subsurface hydrology and subsidence, on the other hand, the tendency was to use triangular or isoparametric irregular quadrilateral elements. The resulting dichotomy of approach has led to two quite different bodies of literature.

The Galerkin finite element concepts appear to have been introduced into the oil reservoir engineering literature via the classic paper of Price *et al.* (1968). Although the authors' objective was to overcome problems generally associated with finite difference solutions to the convection-dominated transport equation, it was evident that the approach had much greater potential appeal. The method was subsequently applied to a two-phase flow waterflooding problem by Douglas *et al.* (1969). McMichael and Thomas (1973) extended this concept to two-dimensional two-phase flow. Whereas solutions of high accuracy were generally achieved, it was also conceded that the Galerkin finite element schemes were, for the most part, noncompetitive with finite difference algorithms because of their computational inefficiency. Recent work has sought to change this prognosis. The alternating direction procedures introduced by Douglas *et al.* (1969) certainly hold promise. Moreover, the use of Lobatto quadrature for evaluation of the coefficient matrices has recently been recommended by Young (1981). This approach was introduced earlier in a more general context by Gray and van Genuchten (1978). Hybridization of ideas along these lines may soon lead to highly efficient two-phase flow algorithms.

In groundwater hydrology, finite element theory has been employed using triangular and isoparametric elements. In contrast to work in the petroleum industry, most, but not all, groundwater simulations use first-degree Lagrange polynomials as basis functions. The classic papers by Javandel and Witherspoon (1968) and Zienkiewicz *et al.* (1966) appear to be the first two publications describing the use of triangular finite element theory in porous media flow. Shortly thereafter Neuman and Witherspoon (1970, 1971) and Taylor and Brown (1967) demonstrated the application of this methodology to the analysis of free surface Darcy groundwater flow. Their work was later extended by Huyakorn (1973) and Volker (1969) to deal with non-Darcy free surface flow. The finite element method is particularly suited to this application because of the simplicity of deforming the mesh and updating the element matrix coefficients to accommodate the changing geometry of the solution domain and the changing element properties due to nonlinear material behavior. This advantage is even more evident when the method is applied to the simulation of land subsidence. In this instance, the finite element grid deforms as the land subsides [Sanhu and Wilson (1969), Safai and Pinder (1980)]. In addition, it is straightforward to extend the finite element stress analysis to deal with the elastoplastic material properties and anisotropic behavior frequently encountered in practice.

The unsaturated flow problem that exhibits a dynamic air phase does not appear to have been solved using finite elements. The problem solved by McMichael and Thomas (1973), however, is conceptually very similar. The single-equation static air phase was considered by Neuman (1973). He found that in solving these nonlinear equations an enhanced solution was obtained when the time matrix was modified to resemble a finite difference formulation on an irregular subspace.

The original work on Galerkin's method by Price *et al.* (1968) was motivated by the unsatisfactory finite difference solutions obtained for the convective–diffusive transport equation. Aziz *et al.* (1968) were also investigating this approach to solving the heat transport equation at about the same time. In fact, this work involved a considerably more challenging natural convection problem in higher dimensions. Both of the preceding investigations employed the Galerkin approach in the sense adopted by the petroleum industry. Guymon *et al.* (1970), on the other hand, approached the finite element simulation of mass and energy transport using a formulation analogous to that employed in engineering mechanics. Pinder (1973) attempted to utilize the best features of both approaches in his finite element model of contaminant transport in groundwater flow. His work was later extended to simulate density dependent mass transport in two and three dimensions (Segol and Pinder, 1976; Taylor and Huyakorn, 1978). Today the Galerkin finite element approach is widely used in both the petroleum industry and hydrologic community for the solution of multidimensional transport in both single- and multiphase systems [see, for example, Duguid and Reeves (1976), Guvanasen and Volker (1981), Segol (1977), and Young (1981)].

The finite element method has also been used for the simulation of nonisothermal flow in geothermal reservoirs. Papers dealing with this subject include those by Huyakorn and Pinder (1977), Mercer *et al.* (1975), Mercer and Faust (1975), and Voss (1978). Other nonisothermal applications include the simulation of coupled moisture and thermal transport in unsaturated soils by Guymon and Berg (1977) and, recently, the simulation of seasonal thermal energy storage by Huyakorn *et al.* (1982) and Voss *et al.* (1980).

The simulation of pressure propagation and mass and energy transport in fractured reservoirs has received increased attention in recent years. There are two avenues of approach to the problem. One requires identification and mathematical definition of the geometry of each fracture in the porous medium. The second assumes the fractures and porous blocks represent two overlapping continua. Finite element methods have been used to solve the equations arising from both models. Clearly, the flexibility inherent in the finite element approach is particularly useful in the discrete fracture model. Early work along these lines was conducted by Wilson and Witherspoon (1970) and Noorishad *et al.* (1971). There is currently considerable activity in this area. Several researchers have extended the earlier work to investigate the transient coupling effect of fluid flow and solid deformation and to understand the nonlinear mechanical behavior of fractures or joints. These include Ayatollahi (1978), Goodman and Dubois (1972), and Hilber and Taylor (1976). A finite element formulation was also used by Bibby (1981), Duguid and Abel (1974), and O'Neill (1977) to simulate the double continuum-based fractured porous media equations. In this case, the use of finite elements was not an essential aspect of the mathematical simulator as it was in the discrete model formulation.

The simulation of mass transport with chemical reactions is an exceedingly challenging problem. It is only in recent years that any serious attempt has been made to subject this problem to quantitative analysis. Although early work in this area employed finite difference approximations (Lai and Jurinak, 1971), the more recent calculations have used either Galerkin theory (Rubin and James, 1973) or orthogonal collocation (Nguyen *et al.*, 1981). In these problems the enhanced accuracy of the finite element Galerkin methodology appears to offset the somewhat increased numerical complexity.

1.5.4 The Boundary Element Method

The boundary element method (BEM) is a relative newcomer in applied numerical analysis. One of the early papers in this area was written by Jaswon and Ponter (1963), who applied the method to a problem in solid mechanics. Since that time, the approach has spread into many areas of engineering analysis. In subsurface hydrology, the boundary element method has been championed by Liggett (1977) and his colleagues. The method has been used most effectively for solving the problem of groundwater flow with a free surface. For problems of elliptic type, particularly those with moving boundaries, the method holds promise for reducing computational effort through a reduction in problem dimensionality. One also achieves a concomitant savings in input data preparation. When parabolic problems or problems involving variable coefficients are encountered, the methodology loses much of its appeal. Although the number of equations to be solved may decrease, the dimensionality is generally not reduced. Finally, one should recognize that the BEM matrix is full, whereas other numerical procedures generally result in sparse and banded matrices with their associated advantages in computer storage and computational efficiency. For a concise yet comprehensive discussion of the application of BEM in engineering, the reader is referred to Brebbia and Walker (1980).

1.5.5 The Collocation Method

Although the collocation method has been used since the 1930s for the solution of engineering problems [see Finlayson (1972)], it is only during the last decade that the technique has been employed over subspaces in a manner analogous to the finite element method. The collocation approach, when applied to a discretized mesh, provides the advantage of the finite element method with additional attractive features. Collocation methods do not require integration procedures in the formulation of the approximating equations. Moreover, the resulting matrix equation exhibits a coefficient structure that may be amenable to efficient solution using modern methods in matrix algebra. An apparent disadvantage to collocation methods is the predisposition to use basis functions of higher continuity. The most commonly used member of this family of functions is the Hermite cubic polynomial. The Lagrange polynomials can also be used although they do not exhibit first-derivative continuity across an element.

For reasons that are not at all clear, the collocation method has not received a great deal of attention in the study of subsurface problems. Chang and Finlayson (1977) have presented results for both Hermite and Lagrange polynomials for second-order operators. Houstis *et al.* (1979) and Mercer and Faust (1976) have compared the performance of collocation and other numerical schemes and report mixed results. In the oil industry collocation has received relatively little attention although there is a comprehensive monograph on the subject (Douglas and Dupont, 1974). Sincovec (1977) presents the collocation methodology and demonstrates its suitability for solving the transport equation. He also describes difficulties to be expected in solving the multiphase flow equations. These problems had also been indicated earlier by Mercer and Faust (1976) but were overcome recently by Allen and Pinder (1983). An important contribution to the use of collocation methods in oil reservoir simulation was made by Bangia *et al.* (1978). They recognized the underlying numerical similarity in the Galerkin and collocation techniques and used this insight to develop an alternating direction collocation procedure for single-phase flow.

Collocation methods appear to have been first applied to groundwater flow by Pinder *et al.* (1978). In this formulation, collocation was performed on irregular subspaces. The detailed theory is found in Frind and Pinder (1979). Later this approach was extended to consider the transport equation (Celia and Pinder, 1980). The relative merits of the collocation approach as compared to more standard techniques for the simulation of subsurface flow have yet to be definitively established for field applications. The next few years should prove particularly interesting in this regard.

1.5.6 Method of Characteristics

The "method of characteristics" was developed to simulate convection-dominated transport. Recognizing that the convection-dominated transport equation behaves as a first-order rather than second-order partial differential equation, Garder *et al.* (1964) introduced the method of characteristics as an appropriate solution methodology. This approach to solving partial differential equations is known to be effective for equations of hyperbolic type.

Although the method of characteristics does not appear to have been widely used in the oil industry, where it was developed, it did find extensive application in groundwater hydrology. Pinder and Cooper (1970) and Reddell and Sunada (1970) used this method to solve the density-dependent transport equations. Later, the same scheme was widely used to simulate the movement of contaminants in the subsurface [see, for example, Bredehoeft and Pinder (1973)].

The practical difficulties associated with the utilization of the method of characteristics are evident to anyone who attempts to simulate a field situation. The method involves tracking mathematical particles as they move along characteristic curves. The programming sophistication required to generate an efficient yet generally applicable code is considerable. Moreover, the method does not have the same degree of mathematical rigor as that identified with other solution procedures. In spite of these drawbacks, the method of characteristics remains the technology of choice for simulating many convection-dominated two-dimensional groundwater transport problems.

It is evident that each of the five generally recognized subsurface simulation methodologies has its own advantages and disadvantages. The finite difference approach is simple in concept and implementation; the finite element approach provides a flexible mesh geometry and can be more accurate for certain operators; the collocation method can provide both enhanced accuracy and flexible mesh geometry; the boundary element method can reduce the dimensionality of problems involving certain operators; finally, the method of characteristics can be effectively employed to simulate convection-dominated transport problems.

References

Allen, M. B., and Pinder, G. F. (1983). Collocation Simulation of Multiphase Porous-Medium Flow, *Soc. Pet. Eng. J.* **23,** 135–142.

Ashcroft, G., Marsh, D. D., Evans, D. D., and Boersma, L. (1962). Numerical method for solving the diffusion equation: I. Horizontal flow in semi-infinite media. *Soil Sci. Soc. Am. Proc.* **26,** 522–525.

Ayatollahi, M. S. (1978). Stress and Flow in Fractured Porous Media. Ph.D. thesis, Department of Material Science and Mineral Engineering, University of California, Berkeley.

Aziz, K., and Settari, A. (1979). "Petroleum Reservoir Simulation." Applied Sci. Publ., Essex, England.

Aziz, K., Holst, P. H., and Kanra, P. S. (1968). Natural convection in porous media. *Pet. Soc. C.I.M.* Paper 6813, presented at 19th Annual Technical Meeting, Calgary, Alberta, Canada.

Bangia, V. K., Bennett, C., Reynolds, A., Raghavan, R., and Thomas, G. (1978). Alternating direction collocation methods for simulating reservoir performance. Paper SPE 7414, presented at 53rd Annual Technical Conference and Exhibition of Soc. Pet. Eng. AIME, Houston, Texas.

Bibby, R. (1981). Mass transport in dual-porosity media. *Water Resour. Res.* **17** (4), 1075–1081.

Bittinger, M. W., Duke, H. R., and Longenbaugh, R. A. (1967). Mathematical simulations for better aquifer management. Pub. no. 72 IASH, pp. 509–519.

Bondor, P. L., Hirasaki, G. J., and Tham, M. J. (1972). Mathematical simulation of polymer flooding in complex reservoir. *Soc. Petrol. Eng. J.* **12,** 369–382.

Brebbia, C. A., and Walker, S. (1980). "Boundary Element Techniques in Engineering." Butterworths, London.

Bredehoeft, J. D., and Pinder, G. F. (1973). Mass transport in flowing groundwater. *Water Resour. Res.* **9,** 192–210.

Brownell, D. H., Jr., Garg, S. K., and Pritchett, J. W. (1975). Computer simulation of geothermal reservoirs. Paper SPE 5381, presented at the 45th Annual California Regional Meeting of the Soc. of Pet. Eng., AIME, Ventura, California.

Bruce, G. H., Peaceman, D. W., and Rachford, H. H. Jr. (1953). Calculation of unsteady-state gas flow through porous media. *Petrol. Trans. AIME* **198,** 74–92.

Brutsaert, W. (1973). Numerical solutions of multiphase well flow. *ASCE* J. Hydraul. Div. **99** (HY1), 1981–2001.

Celia, M. A., and Pinder, G. F. (1980). Alternating direction collocation solution to the transport equation. *In* Proceedings of the 3rd International Conference on Finite Elements in Water Resources," S. Y. Wang, C. V. Alsonso, C. A. Brebbia, W. G. Gray, and G. F. Pinder, eds., pp. 3.36–3.48. University of Mississippi.

Chang, P. W., and Finlayson, B. A. (1977). Orthogonal collocation on finite elements for elliptic equations. *In* "Advances in Computer Methods for Partial Differential Equations III," R. Vichnevetsky, ed., 79–86. IMACS, New Brunswick, New Jersey.

Coats, K. H. (1974). Simulation of steamflooding with distillation and solution gas. Paper SPE 5015, presented at the 19th Annual Fall Meeting of Soc. Pet. Eng. AIME, Houston, Texas.

Coats, K. H. (1980). An equation of state compositional model. *Soc. Pet. Eng. J.* **20,** 363–376.

Coats, K. H., Nielsen, R. L., Terhune, M. H., and Weber, A. G. (1967). Simulation of three-dimensional, two-phase flow in oil and gas reservoirs. *Soc. Pet. Eng. J.* **7,** 377–388.

Cooley, R. L. (1971). A finite difference method for unsteady flow in variably saturated porous media: Application to a single pumping well. *Water Resour. Res.* **7** (6), 1607–1625.

Crichlow, H. B. (1977). "Modern Reservoir Engineering—A Simulation Approach." Prentice-Hall, Englewood Cliffs, New Jersey.

Douglas, J., Jr., and Dupont, T. (1974). "Collocation Methods for Parabolic Equations in a Single Space Variable." Springer-Verlag, Berlin and New York.

Douglas, J., Jr., Peaceman, D. W., and Rachford, H. H. Jr. (1959). A method for calculating multi-dimensional immiscible displacement. *Trans. Soc. Pet. Eng. AIME* **216,** 297–306.
Douglas, J., Jr., Dupont, T., and Rachford, H. H. Jr. (1969). The application of variational methods to waterflooding problems. *J. Can. Pet. Technol.* **8,** 79–85.
Duguid, J. O., and Abel, J. F. (1974). Finite element galerkin method for analysis of flow in fractured porous media. *In* "Finite Element Methods in Flow Problems," J. T. Oden *et al.,* eds., UAH Press, Huntsville, Alabama.
Duguid, J. O., and Reeves, M. (1976). Material Transport through Porous Media: A Finite Element Galerkin Model. ORNL–4928, Oak Ridge Nat. Lab., Oak Ridge, Tennessee.
Faust, C. R., and Mercer, J. W. (1976). An analysis of finite-difference and finite element techniques for geothermal reservoir simulation. Paper SPE5742, presented at the Soc. Pet. Eng. Symposium on Numerical Simulation of Reservoir Performance, Los Angeles.
Finlayson, B. A. (1972). "The Method of Weighted Residuals and Variational Principles." Academic Press, New York.
Forsythe, G. E., and Wasow, W. R. (1960). "Finite-Difference Methods for Partial Differential Equations." Wiley, New York.
Freeze, R. A. (1971). "Three-dimensional, transient, saturated–unsaturated flow in a groundwater basin. *Water Resour. Res.* **7** (2), 347–366.
Freeze, R. A., and Witherspoon, P. A. (1966). Theoretical analysis of regional groundwater flow; 1. Analytical and numerical solutions to the mathematical model. *Water Resour. Res.* **2,** 641–656.
Fried, J. J., and Ungemach, P. O. (1971). A dispersion model for a quantitative study of a ground water pollution by salt. *Water Res.* **5,** 491–495.
Frind, E. O., and Pinder, G. F. (1979). A collocation finite element method for potential problems in irregular domains. *Int. J. Num. Methods Eng.* **14,** 681–701.
Garder, A. O., Peaceman, D. W., and Pozzi, A. L. (1964). Numerical calculation of multidimensional miscible displacement by the method of characteristics. *Soc. Pet. Eng. J.* **4,** 26–36.
Goodman, R. E., and Dubois, J. (1972). "Duplication and dilatancy of jointed rocks. *ASCE J. Soil Mech. Found. Div.* **98** (SM4), 399–422.
Gottfried, B. S. (1965). A mathematical model of thermal oil recovery in linear systems. Paper SPE 1117, presented at Soc. Pet. Eng. Production Research Symposium, Tulsa, Oklahoma.
Gray, W. G., and van Genuchten, M. T. (1978). Economical alternatives to Gaussian quadrature and isoparametric quadrilaterals. *Int. J. Num. Methods Eng.* **12,** 1478–1484.
Green, D. W., Dabiri, H., and Weinaug, C. F. (1970). Numerical modelling of unsaturated groundwater flow and comparison to a field experiment. *Water Resour. Res.* **6,** 862–874.
Gupta, S. K., Tanji, K. K., and Luthin, J. N. (1975). "A three-dimensional finite element groundwater model. Contribution No. 152, Water Resources Research Center University of California, Davis.
Guvanasen, V., and Volker, R. E. (1981). Simulating Mass Transport in Unconfined Aquifers. *ASCE J. Hydraul. Div.* **107** (HY4), 461–477.
Guymon, G. L., and Berg, R. L. (1977). Galerkin finite element analog of coupled moisture and heat transport in algid soils. *In* "Proceedings of the First International Conference on Finite Elements in Water Resources," 3.107–3.114. Pentech Press, Plymouth, England.
Guymon, G. L., Scott, V. H., and Herrmann, L. R. (1970). A general numerical solution of the two-dimensional diffusion–convection equation by the finite element method. *Water Resour. Res.* **6,** 1611–1617.
Hanks, R. J., and Bowers, S. A. (1962). Numerical solution of the moisture flow equation for infiltration into layered soils. *Soil Sci. Soc. Am. Proc.* **26,** 530–534.
Haverkamp, R., Vauclin, M., Touma, J., Wierenga, P. J., and Vachaud, G. (1977). A

comparison of numerical simulation models for one-dimensional infiltration. *Soil Sci. Soc. Am. J.* **41,** 285–294.

Henry, H. R., and Hilleke, J. B. (1972). Exploration of multiphase fluid flow in a saline aquifer system affected by geothermal heating. Final Report to U.S. Geol. Survey by the Bureau of Engineering Research, University of Alabama.

Hilber, H. M., and Taylor, R. L. (1976). A finite element model of fluid flow in systems of deformable fractured rock. Report No. UC SESM 76–5, University of California, Berkeley.

Houstis, E. N., Mitchell, W. F., and Papatheodorou, T. S. (1979). A C^1-collocation method for mildly nonlinear elliptic equations on general 2-D domains. *In* "Advances in Computer Methods for Partial Differential Equations III," R. Vichnevetsky and R. S. Stepleman, eds., IMACS, New Brunswick, New Jersey.

Huyakorn, P. S. (1973). Finite element solution of two-regime flow toward wells. Water Research Lab., Report No. 137, University of New South Wales, Australia.

Huyakorn, P. S., and Pinder, G. F. (1977). A pressure-enthalpy finite element model for simulating hydrothermal reservoirs. *In* "Advances in Computer Methods for Partial Differential Equations II," R. Vichnevetsky, ed. IMACS, New Brunswick, New Jersey.

Huyakorn, P. S., and Taylor, C. (1977). Finite element models for coupled groundwater flow and convective dispersion. *In* "Proceedings of the First International Conference on Finite Elements in Water Resources." 1.131–1.151. Pentech Press, Plymouth, England.

Huyakorn, P. S., Dougherty, D. E., and Faust, C. R. (1982). Numerical simulation of thermal energy storage problems. *Proc. Tenth IMACS World Congr. Syst. Simul. Sci. Comput.,* **2,** 296–299.

Jaswon, M. A., and Ponter, A. R. (1963). An integral equation solution of the torsion problem. *Proc. R. Soc. London, Ser. A* **273,** 237–246.

Javandel, I., and Witherspoon, P. A. (1968). Application of the finite element method to transient flow in porous media. *Soc. Pet. Eng. J.* **8,** 241–252.

Kelly, L. G. (1967). "Handbook of Numerical Methods and Applications." Addison-Wesley, Reading, Massachusetts.

Lai, S. H., and Jurinak, J. J. (1971). Numerical approximation of cation exchange in miscible displacement through soil columns. *Soil Sci. Soc. Amer. Proc.* **35,** 894–899.

Lasseter, T. J., Witherspoon, P. A., and Lippmann, M. J. (1975). The numerical simulation of heat and mass transfer in multidimensional two-phase geothermal reservoirs. *Proc. 2nd U.N. Symp. Dev. Use Geotherm. Res.* **3,** 1715–1723.

Liggett, J. A. (1977). Location of free surface in porous media. *ASCE J. Hydraul. Div.,* **103** (HY4), 353–365.

McMichael, C. L., and Thomas, G. W. (1973). Reservoir simulation by Galerkin's method. *Soc. Pet. Eng. J.* **13** (3), 125–138.

Mercer, J. W., and Faust, C. R. (1975). Simulation of water- and vapour-dominated hydrothermal reservoirs. SPE 5520 presented at the 50th Annu. Fall Meet. Soc. Pet. Eng. AIME, Dallas, Texas.

Mercer, J. W., and Faust, C. R. (1976). The application of finite-element techniques to immiscible flow in porous media. *In* "Finite Elements in Water Resources," W. G. Gray, G. F. Pinder and C. A. Brebbia, eds., 1.21–1.57. Pentech Press, Plymouth, England.

Mercer, J. W., Pinder, G. F., and Donaldson, I. G. (1975). A Galerkin-finite element analysis of the hydrothermal system at Wairokei, New Zealand. *J. Geophys. Res.* **80** (17), 2608–2621.

Neuman, S. P. (1973). Saturated–unsaturated seepage by finite elements. *ASCE J. Hydrol. Div.* **99** (HY12), 2233–2251.

Neuman, S. P., and Witherspoon, P. A. (1970). Finite element method of analyzing steady seepage with a free surface. *Water Resour. Res.* **6,** 889–897.

Neuman, S. P., and Witherspoon, P. A. (1971). Analysis of nonsteady flow with a free surface using the finite element method. *Water Resour. Res.* **7,** 611–623.

Nguyen, V. V., Pinder, G. F., Gray, W. G., and Botha, J. F. (1981). Numerical simulation of uranium in-situ mining. Water Resources Program Report 81–WR–10, Princeton University, Princeton, New Jersey.

Noorishad, J., Witherspoon, P. A., and Brekke, T. L. (1971). A method for coupled stress and flow analysis of fractured rock masses. Geotechnical Publication No. 71–6, University of California, Berkeley.

O'Neill, K. (1977). The Transient Three Dimensional Transport of Liquid and Heat in Fractured Porous Media. Ph.D. thesis, Princeton University, Princeton, New Jersey.

Orlob, G. T., and Woods, P. C. (1967). Water-quality management in irrigation systems. *ASCE J. Irrig. Drain. Div.* **93,** 49–66.

Oster, C. A., Sonnichsen, J. C., and Jaske, P. T. (1970). Numerical solution of the convective diffusion equation. *Water Resour. Res.* **6,** 1746–1752.

Peaceman, D. W. (1977). "Fundamentals of Numerical Reservoir Simulation." Elsevier, Amsterdam.

Peaceman, D. W., and Rachford, H. H., Jr. (1955). The numerical solution of parabolic and elliptic differential equations. *J. Soc. Ind. Appl. Math.* **3,** 28–41.

Peaceman, D. W., and Rachford, H. H., Jr. (1962). Numerical calculation of multidimensional miscible displacement. *Soc. Pet. Eng. J.* **2,** 327–339.

Philip, J. R. (1957). The theory of infiltration: 1. The infiltration equation and its solution. *Soil Sci.* **83,** 345–357.

Pinder, G. F. (1973). A Galerkin-finite element simulation of groundwater contamination on Long Island, New York. *Water Resour. Res.* **9,** 1657–1669.

Pinder, G. F., Frind, E. O., and Celia, M. A. (1978). Groundwater flow simulation using collocation finite elements. *In* "Proceedings of the 2nd International Conference on Finite Elements in Water Resources," C. A. Brebbia, W. G. Gray and G. F. Pinder, eds., 1.171–1.185. Pentech Press, Plymouth, England.

Pinder, G. F. (1979). State-of-the-art review of geothermal reservoir modelling. LBL-9093, GSRMP-5, UC-66a, NTIS, University of California at Berkeley.

Pinder, G. F., and Bredehoeft, J. D. (1968). Application of the digital computer for aquifer evaluation. *Water Resour. Res.* **4,** 1069–1093.

Pinder, G. F., and Cooper, H. H., Jr. (1970). A numerical technique for calculating the transient position of the saltwater front. *Water Resour. Res.* **6,** 875–882.

Price, H. S., Cavendish, J. C., and Varga, R. A. (1968). Numerical methods of higher order accuracy for diffusion convection equations. *Soc. Pet. Eng. J.*, 293–303.

Reddell, D. L., and Sunada, D. K. (1970). Numerical simulation of dispersion in groundwater aquifers. Hydrol. Paper 41, Colorado State University, Fort Collins, Colorado.

Remson, I., Appel, C. A., and Webster, R. A. (1965). Groundwater models solved by digital computer. *ASCE J. Hydraul. Div.* **91** (HY3), 133–147.

Rubin, J., and James, R. V. (1973). Dispersion-affected transport of reacting solutes in saturated porous media: Galerkin method applied to equilibrium-controlled exchange in unidirectional steady water flow. *Water Resour. Res.* **9,** 1332–1356.

Rubin, J., and Steinhardt, R. (1963). Soil water relations during rain infiltration: 1. Theory. *Soil Sci. Soc. Am. Proc.* **27,** 246–251.

Safai, N. M., and Pinder, G. F. (1979). Vertical and horizontal land deformation in a desaturating porous medium. *Adv. Water Resour.* **2,** 19–25.

Safai, N. M., and Pinder, G. F. (1980). Vertical and horizontal land deformation due to fluid withdrawal. *Int. J. Num. Anal. Methods Geomech.* **4,** 131–142.

Sanhu, R. S., and Wilson, E. L. (1969). Finite element analysis of land subsidence. *Land Subsidence Proc.* **IAHS2,** 393–400.

Segol, G. (1977). A three-dimensional Galerkin-finite element model for the analysis of contaminant transport in saturated–unsaturated porous media. *In* "Proceedings of the First International Conference on Finite Elements in Water Resources," 2.123–2.145. Pentech Press, Plymouth, England.

Segol, G., and Pinder, G. F. (1976). Transient simulation of saltwater intrusion in southeastern Florida. *Water Resour. Res.* **12,** 62–70.

Settari, A., Price, H. S., and Dupont, T. (1977). Development and application of variational methods for simulation of miscible displacement in porous media. *Soc. Pet. Eng. J.* **17,** 228–246.

Shamir, U. Y., and Harleman, D. F. (1967). Dispersion in layered porous media. *ASCE J. Hydraul. Div.* **93,** 237–260.

Sheldon, J. W., Harris, C. D., and Bavly, D. (1960). A method for general reservoir behavior simulation on digital computers. Paper 1521-G, Soc. Pet. Eng. 35th Annual Fall Meeting, Denver, Colorado.

Shutler, N. D. (1970). Numerical, three-phase simulation of the linear steamflood process. *Trans. Soc. Pet. Eng. AIME* **246,** 232–246.

Sincovec, R. F. (1977). Generalized collocation methods for time-dependent, nonlinear boundary-value problems. *Soc. Pet. Eng. J.* 345–352.

Stone, H. L., and Brian, P. T. L. (1963). Numerical solution of convective transport problems. AIChE J. **9,** 681–688.

Stone, H. L., and Garder, A. O., Jr. (1961). "Analysis of gas-cap or dissolved-gas drive reservoirs", *Soc. Pet. Eng. J.* 92–104.

Tanji, K. K., Dutt, G. R., Paul, J. L., and Doneen, L. D. (1967). II. A computer method for predicting salt concentrations in soils at variable moisture contents. *Hilgardia* **38,** 307–318.

Taylor, C., and Huyakorn, P. S. (1978). Three-dimensional groundwater flow and solute transport. *In* "Finite Elements in Fluids," R. Gallagher, ed., vol. 3, chap.17, Wiley, New York.

Taylor, R. L., and Brown, C. B. (1967). Darcy flow solutions with a free surface. *Proc. Am. Soc. Civ. Eng.* **93** (HY2), 25–33.

Thomas, L. K., and Pierson, R. (1976). Three dimensional geothermal reservoir simulation. Paper SPE6104, presented at the 51st Annual Fall Technical Conference and Exhibition of the Soc. of Pet. Eng. AIME, New Orleans.

Toronyi, R. M. (1974). Two-Phase, Two Dimensional Simulation of a Geothermal Reservoir and the Wellbore System. Ph.D. Dissertation, Pennsylvania State University.

Volker, R. E. (1969). Nonlinear flow through porous media by finite elements. *ASCE J. Hydraul. Div.* **95** (HY6), 2093–2114.

Voss, C. I. (1978). Finite Element Simulation of Multipurpose Geothermal Reservoirs. Ph.D. dissertation, Dept. of Civil Engineering, Princeton University, Princeton, New Jersey.

Voss, C. I., Kinnmark, I., and Hyden, H. (1980). Simulation of low temperature seasonal energy storage in a shallow glaciofluvial aquifer. Report No. 8011, National Swedish Board for Energy Source Development.

Whisler, F. D., and Klute, A. (1965). The numerical analysis of infiltration, considering hysteresis, into a vertical soil column at equilibrium under gravity. *Soil Sci. Soc. Am. Proc.* **29,** 489–494.

Wierenga, P. J. (1977). Solute distribution profiles computed with steady state and transient water movement models. *Soil Sci. Soc. Am. J.* **41,** 1050–1055.

Wilson, C. R., and Witherspoon, P. A. (1970). An investigation of laminar flow in fractured porous rocks. No. 70-6, Dept. of Civil Engineering, University of California, Berkeley.

Young, L. C. (1981). A finite element method for reservoir simulation. *Soc. Pet. Eng. J.* **21,** 115–128.

Zienkiewicz, O. C., Meyer, P., and Cheung, Y. K. (1966). Solution of anisotropic seepage problems by finite elements. *Proc. Am. Soc. Civ. Eng.* **92** (EM1), 111–120.

2

The Finite Element Method

2.1 General

The finite element method is a numerical method for solving differential equations by means of "piecewise approximation." As distinct from the finite difference method, which regards the solution region as an array of grid points, the finite element method envisions the region as being made up of many small interconnected subregions called "finite elements." Such elements, which generally take simple shapes (e.g. triangular, quadrilateral and rectangular) are then assembled in various ways to represent a solution domain of arbitrary geometry.

The origin of the finite element method can be separately traced to applied mathematicians, physicists, and engineers. The first attempt to solve elliptical partial differential equations using a piecewise approximation was made by Courant (1943). The first important application of such a technique to a problem in structural mechanics was by Turner *et al.* (1956). The name "finite element method" was suggested in a later paper by Clough (1960), who used finite elements to perform plane stress analysis. Since then, there has been a rapid development and the finite element method is now generally recognized as a standard tool for solving a wide range of physical problems. For a more comprehensive review of the history of finite elements, the reader is referred to Heubner (1975).

We develop an understanding of the essential concepts of finite element analysis without undue mathematical complexity by the consideration of simple discrete systems. Having achieved this goal, more mathematically abstract formulations are presented later in this chapter.

2.2 Basic Concepts in Finite Element Analysis

The finite element analysis of a physical problem can be described as follows:

(1) The physical system is subdivided into a series of finite elements that are connected at a discrete number of nodal points; this process is

often called "discretization." Typical subdivisions of a pipe flow network and a continuum are shown in Figs. 2.1a and 2.1b, respectively. Each element is identified by its element number and the lines connecting the nodal points situated on the element boundary.

(2) A matrix expression is developed to relate the nodal variables of each element. The resulting matrix is commonly referred to as an "element matrix." For a discrete problem, the element matrix relation can often be established via direct physical reasoning. For a continuum problem, the element matrix expression must be obtained via a more general mathematical formulation that normally makes use of either a variational or weighted residual method.

(3) The element matrices are combined or assembled to form a set of algebraic equations that describes the entire global system. The coefficient matrix of this final set of equations is called the "global matrix." The assembly procedure is performed in such a way that certain compatibility conditions are satisfied at each node shared by different elements.

(4) Prescribed boundary conditions are incorporated into the assembled or global matrix equation.

(5) The resulting set of simultaneous algebraic equations is solved. Here, many different solution algorithms can be employed. Among the widely used algorithms are the Gauss elimination and Choleski decomposition algorithms that take into account the banded and symmetric feature of the coefficient matrix [see Bathe and Wilson (1976)].

The fifth step is the final step if we only want to obtain the nodal values of the unknown function. Additional computation must be made if other quantities involving derivatives of the function are to be derived from the nodal values.

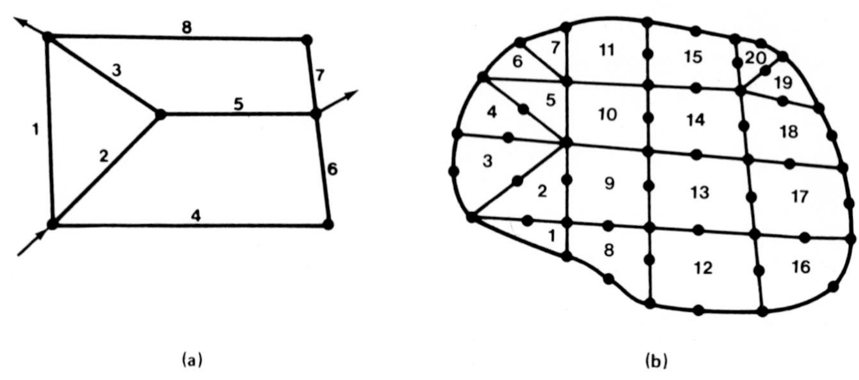

Fig. 2.1. Typical finite element discretization: (a) one-dimentional elements; (b) two-dimensional elements.

2.3 Solution of Discrete Problems

We now employ the preceding procedure to analyze two discrete systems.

2.3.1 One Dependent Variable

Consider the steady state flow of water through a porous medium consisting of three zones of different hydraulic conductivities as shown in Fig. 2.2. Water is injected at a unit rate into one end of the medium while the hydraulic head is maintained at a constant value equal to five at the other end. We use the finite element method to determine the hydraulic head distribution in the porous medium as follows:

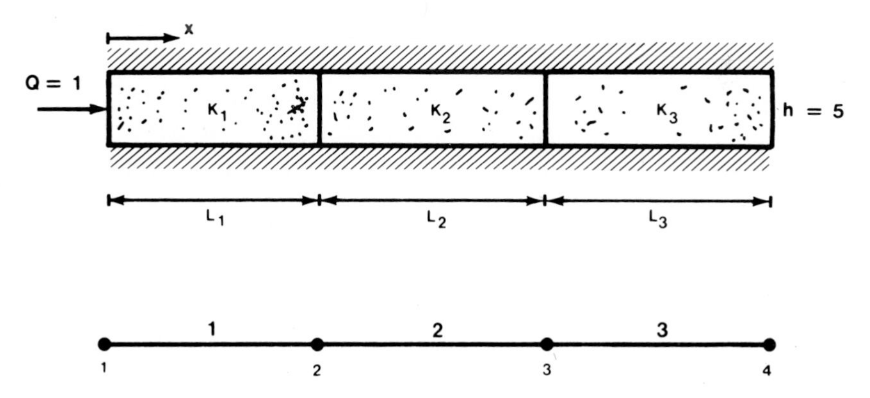

Fig. 2.2. One-dimensional steady flow through a composite porous medium.

Step 1 Discretize the porous medium. We first subdivide the system into three elements as shown in Fig. 2.2. The elements are identified as follows:

Element number	*Nodal connections*	
	Node 1	*Node 2*
1	1	2
2	2	3
3	3	4

Step 2 Determine the element characteristics. We consider a typical isolated element (e) as illustrated in Fig. 2.3. The values of the flow rate Q and the hydraulic head h are defined at each node. For convenience, we assign a positive sign to the flow rate entering the element. The

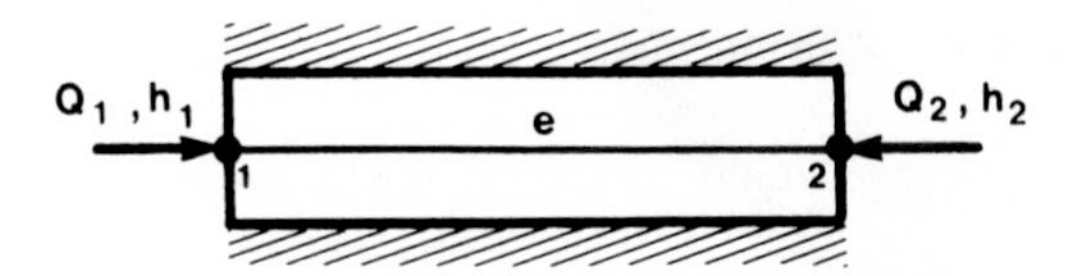

Fig. 2.3. Typical element of the porous medium flow system.

unknown function for this simple problem is the hydraulic head. The relation between the flow rate and the hydraulic head is given by Darcy's law. The Darcy equation for each node can be written as

$$Q_1 = \frac{KA}{L}(h_1 - h_2), \tag{2.3.1.1a}$$

$$Q_2 = \frac{KA}{L}(h_2 - h_1), \tag{2.3.1.1b}$$

where K, A, and L denote the hydraulic conductivity, cross-sectional area, and length of the element, respectively.

Equations (2.3.1.1a) and (2.3.1.1b) may be written in matrix form as

$$\begin{bmatrix} \dfrac{KA}{L} & \dfrac{-KA}{L} \\ \dfrac{-KA}{L} & \dfrac{KA}{L} \end{bmatrix}^e \begin{Bmatrix} h_1 \\ h_2 \end{Bmatrix}^e = \begin{Bmatrix} Q_1 \\ Q_2 \end{Bmatrix}^e, \tag{2.3.1.2}$$

or

$$[C]^e\{h\}^e = \{Q\}^e, \tag{2.3.1.3}$$

where the superscript e is used to indicate that the matrix equation is written at the element level, and

$$c_{11} = \left(\frac{KA}{L}\right)^e, \quad c_{12} = -\left(\frac{KA}{L}\right)^e, \quad c_{21} = c_{12}, \quad c_{22} = \left(\frac{KA}{L}\right)^e.$$

It should be noted that the element matrix $[C]^e$ is a symmetric but singular matrix; i.e., its determinant is zero.

Step 3 Assemble the element equations. After assembly of all element equations, the global set of equations is obtained and this can be written as

$$\begin{bmatrix} c_{11} & c_{12} & c_{13} & c_{14} \\ c_{21} & c_{22} & c_{23} & c_{24} \\ c_{31} & c_{32} & c_{33} & c_{34} \\ c_{41} & c_{42} & c_{43} & c_{44} \end{bmatrix} \begin{Bmatrix} h_1 \\ h_2 \\ h_3 \\ h_4 \end{Bmatrix} = \begin{Bmatrix} Q_1 \\ Q_2 \\ Q_3 \\ Q_4 \end{Bmatrix}, \tag{2.3.1.4}$$

or, equivalently,

$$[C]\{h\} = \{Q\}, \tag{2.3.1.5}$$

where $[C]$ is the "global matrix."

The assembly process is performed in the following manner: First, we rewrite the element matrix equation in terms of the "global numbering scheme." Thus, for element 1,

$$\begin{bmatrix} c_{11}^{(1)} & c_{12}^{(1)} & 0 & 0 \\ c_{21}^{(1)} & c_{22}^{(1)} & 0 & 0 \\ 0 & 0 & 0 & 0 \\ 0 & 0 & 0 & 0 \end{bmatrix} \begin{Bmatrix} h_1 \\ h_2 \\ h_3 \\ h_4 \end{Bmatrix} = \begin{Bmatrix} Q_1^{(1)} \\ Q_2^{(1)} \\ 0 \\ 0 \end{Bmatrix}. \tag{2.3.1.6}$$

For element 2,

$$\begin{bmatrix} 0 & 0 & 0 & 0 \\ 0 & c_{11}^{(2)} & c_{12}^{(2)} & 0 \\ 0 & c_{21}^{(2)} & c_{22}^{(2)} & 0 \\ 0 & 0 & 0 & 0 \end{bmatrix} \begin{Bmatrix} h_1 \\ h_2 \\ h_3 \\ h_4 \end{Bmatrix} = \begin{Bmatrix} 0 \\ Q_1^{(2)} \\ Q_2^{(2)} \\ 0 \end{Bmatrix}. \tag{2.3.1.7}$$

For element 3,

$$\begin{bmatrix} 0 & 0 & 0 & 0 \\ 0 & 0 & 0 & 0 \\ 0 & 0 & c_{11}^{(3)} & c_{12}^{(3)} \\ 0 & 0 & c_{21}^{(3)} & c_{22}^{(3)} \end{bmatrix} \begin{Bmatrix} h_1 \\ h_2 \\ h_3 \\ h_4 \end{Bmatrix} = \begin{Bmatrix} 0 \\ 0 \\ Q_1^{(3)} \\ Q_2^{(3)} \end{Bmatrix}. \tag{2.3.1.8}$$

The global matrix equation is now obtained by adding contributions from all elements. Combination of Eqs. (2.3.1.6)–(2.3.1.8) gives

Node no.	1	2	3	4
1	$c_{11}^{(1)}$	$c_{12}^{(1)}$	0	0
2	$c_{21}^{(1)}$	$c_{22}^{(1)} + c_{11}^{(2)}$	$c_{12}^{(2)}$	0
3	0	$c_{21}^{(2)}$	$c_{22}^{(2)} + c_{11}^{(3)}$	$c_{12}^{(3)}$
4	0	0	$c_{21}^{(3)}$	$c_{22}^{(3)}$

$$\begin{Bmatrix} h_1 \\ h_2 \\ h_3 \\ h_4 \end{Bmatrix} = \begin{Bmatrix} Q_1^{(1)} \\ Q_2^{(1)} + Q_1^{(2)} \\ Q_2^{(2)} + Q_1^{(3)} \\ Q_2^{(3)} \end{Bmatrix} \tag{2.3.1.9}$$

Equation (2.3.1.9) can be written in the form

$$[C]\{h\} = \{Q\}, \tag{2.3.1.10}$$

where

$$\{h\} = \begin{Bmatrix} h_1 \\ h_2 \\ h_3 \\ h_4 \end{Bmatrix}, \qquad \{Q\} = \begin{Bmatrix} Q_1 \\ Q_2 \\ Q_3 \\ Q_4 \end{Bmatrix} = \begin{Bmatrix} Q_1^{(1)} \\ Q_2^{(1)} + Q_1^{(2)} \\ Q_2^{(2)} + Q_1^{(3)} \\ Q_2^{(3)} \end{Bmatrix},$$

and

$$[C] = \begin{bmatrix} c_{11} & c_{12} & c_{13} & c_{14} \\ c_{21} & c_{22} & c_{23} & c_{24} \\ c_{31} & c_{32} & c_{33} & c_{34} \\ c_{41} & c_{42} & c_{43} & c_{44} \end{bmatrix}.$$

The values of the matrix elements c_{ij} are obtained directly from Eq. (2.3.1.9). Note that $c_{ij} = c_{ji}$ and that Eq. (2.3.1.10) is just Eq. (2.3.1.5) presented earlier.

We now enforce the continuity of flow at the internal nodes, 2 and 3. At node 2,

$$Q_2^{(1)} + Q_1^{(2)} = 0. \tag{2.3.1.11}$$

At node 3,

$$Q_2^{(2)} + Q_1^{(3)} = 0. \tag{2.3.1.12}$$

Thus, the right-hand-side vector $\{Q\}$ becomes

$$\{Q\} = \begin{Bmatrix} Q_1^{(1)} \\ 0 \\ 0 \\ Q_2^{(3)} \end{Bmatrix}. \tag{2.3.1.13}$$

Step 4 Incorporate prescribed boundary conditions. There are two boundary conditions: at node 1, $Q_1 = 1$, and at node 4, $h_4 = 5$. The first boundary condition is inserted simply by setting $Q_1^{(1)} = Q_1 = 1$. Thus, we obtain

$$\begin{bmatrix} c_{11} & c_{12} & c_{13} & c_{14} \\ c_{21} & c_{22} & c_{23} & c_{24} \\ c_{31} & c_{32} & c_{33} & c_{34} \\ c_{41} & c_{42} & c_{43} & c_{44} \end{bmatrix} \begin{Bmatrix} h_1 \\ h_2 \\ h_3 \\ h_4 \end{Bmatrix} = \begin{Bmatrix} 1 \\ 0 \\ 0 \\ Q_4 \end{Bmatrix}. \tag{2.3.1.14}$$

The second boundary condition ($h_4 = 5$) must be inserted carefully if the symmetry of $[C]$ is to be preserved. One way of accomplishing this is to write Eq. (2.3.14) in the form

$$\begin{bmatrix} c_{11} & c_{12} & c_{13} & 0 \\ c_{21} & c_{22} & c_{23} & 0 \\ c_{31} & c_{32} & c_{33} & 0 \\ 0 & 0 & 0 & 1 \end{bmatrix} \begin{Bmatrix} h_1 \\ h_2 \\ h_3 \\ h_4 \end{Bmatrix} = \begin{Bmatrix} 1 - 5c_{14} \\ -5c_{24} \\ -5c_{34} \\ 5 \end{Bmatrix}. \tag{2.3.1.15}$$

Step 5 Solve the final set of algebraic equations. This final set is represented by the matrix equation (2.3.1.15). The solution can be obtained using one of many standard elimination schemes.

2.3.2 Two Dependent Variables

The purpose of this example is to demonstrate how the procedure presented in the first example can be extended to handle a discrete system having more than one dependent variable (degree of freedom) at each node. We consider a simple pin-jointed truss composed of two bars as shown in Fig. 2.4. This may appear to be a rather unusual example to consider in light of the general thrust of this text, but we shall, in fact, encounter the displacement equations once again in the discussion of porous medium deformation in Chapter 6. (The reader who is not interested in the solution of this type of problem can skip this subsection.) The unknown dependent variables in the present problem are the horizontal and vertical displacements u_x and u_y. The truss is assumed to be subjected to horizontal and vertical forces of 5 units each at nodal point 2. To find the force–displacement characteristic of this truss, we perform the following steps:

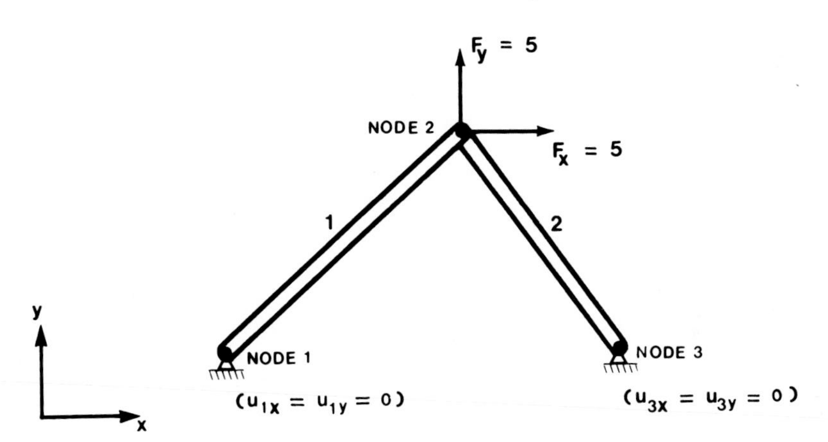

Fig. 2.4. A simple pin-jointed truss.

Step 1 We discretize the given structure into two elements which are identified as follows:

Element number	*Nodal connections*	
1	1	2
2	2	3

Step 2 Once again consider a typical isolated element (*e*) illustrated in Fig. 2.5. Let (x', y') denote a local coordinate system and let $F_{ix'}$,

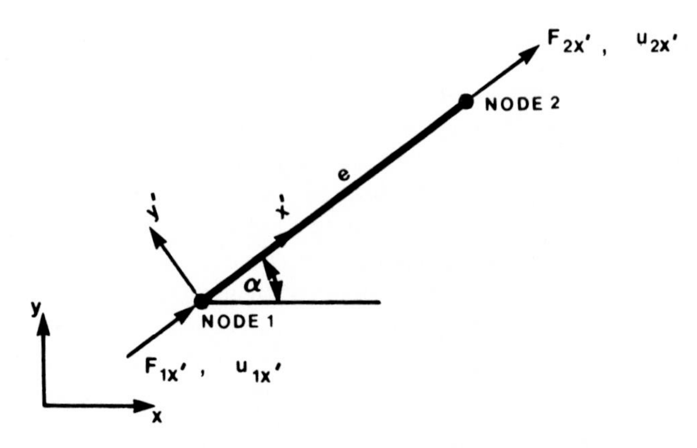

Fig. 2.5. A typical finite element with the nodal variables in the local coordinate system.

$u_{ix'}$ denote, respectively, the axial force and the displacement along the x' axis at node i. The force–displacement relation is given by Hooke's law. Thus, for nodes 1 and 2 we obtain

$$F_{1x'} = \left(\frac{EA}{L}\right)(u_{1x'} - u_{2x'}), \tag{2.3.2.1}$$

$$F_{2x'} = \left(\frac{EA}{L}\right)(u_{2x'} - u_{1x'}), \tag{2.3.2.2}$$

where E is Young's modulus, A the cross-sectional area, and L the rod length.

These two equations can be written in matrix form as

$$\begin{Bmatrix} F_{1x'} \\ F_{2x'} \end{Bmatrix} = \left(\frac{EA}{L}\right)\begin{bmatrix} 1 & -1 \\ -1 & 1 \end{bmatrix}\begin{Bmatrix} u_{1x'} \\ u_{2x'} \end{Bmatrix}, \tag{2.3.2.3a}$$

or (2.3.2.3b)

$$\{F'\} = [\mathrm{K}']\{u'\},$$

where $[K']$ is referred to as the "element stiffness matrix." The matrix equation (2.3.2.3a) is written with respect to the local coordinate axes (x', y'). It is necessary to transform it into a matrix equation written with respect to the global coordinate axes (x, y). In the global axes, we have two components of force and displacement at each node (Fig. 2.6). Listing the forces acting on the nodes, we have

$$\{F\} = \begin{Bmatrix} F_{1x} \\ F_{1y} \\ F_{2x} \\ F_{2y} \end{Bmatrix}. \tag{2.3.2.4a}$$

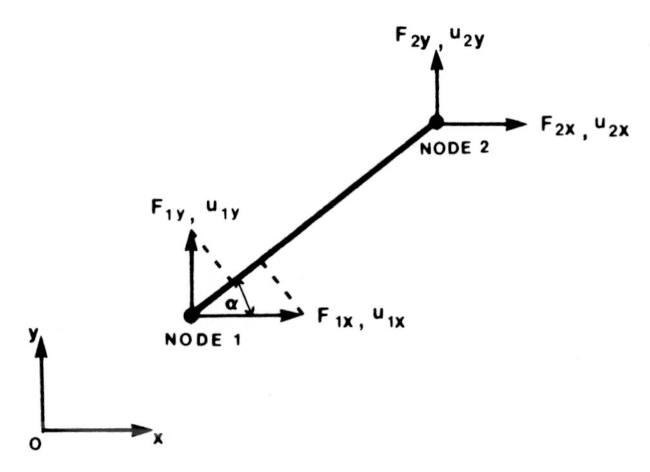

Fig. 2.6. A typical finite element with the nodal variables in the global coordinate system.

Similarly, for the corresponding displacements

$$\{u\} = \begin{Bmatrix} u_{1x} \\ u_{1y} \\ u_{2x} \\ u_{2y} \end{Bmatrix}. \qquad (2.3.2.4b)$$

By resolving the forces and displacements at each node along the local x' axis, we obtain the following transformation for the displacements:

$$\begin{Bmatrix} u_{1x'} \\ u_{2x'} \end{Bmatrix} = \begin{bmatrix} \cos\alpha & \sin\alpha & 0 & 0 \\ 0 & 0 & \cos\alpha & \sin\alpha \end{bmatrix} \begin{Bmatrix} u_{1x} \\ u_{1y} \\ u_{2x} \\ u_{2y} \end{Bmatrix}, \qquad (2.3.2.5a)$$

or

$$\{u'\} = [T]\{u\}, \qquad (2.3.2.5b)$$

and for the forces

$$\{F'\} = [T]\{F\}, \qquad (2.3.2.5c)$$

where $[T]$ is called the coordinate transformation matrix. It can be easily shown that $[T]$ is an orthogonal matrix, i.e.,

$$[T][T]^t = [I] \qquad (2.3.2.6)$$

where $[T]^t$ denotes the transpose of $[T]$ and $[I]$ is a 2×2 identity matrix. Substituting Eqs. (2.3.2.5b) and (2.3.2.5c) into (2.3.2.3b), we obtain

$$[T]\{F\} = [K'][T]\{u\}. \qquad (2.3.2.7)$$

Premultiplication of Eq. (2.3.2.7) by $[T]^t$ and use of (2.3.2.6) gives

$$\{F\} = [T]^t[K'][T]\{u\}, \tag{2.3.2.8a}$$

or

$$\{F\} = [K]\{u\}, \tag{2.3.2.8b}$$

where

$$[K] = [T]^t[K'][T]. \tag{2.3.2.8c}$$

Equation (2.3.2.8b) is the required element matrix equation. It can be seen that matrix $[K]$ is the element matrix which is simply obtained by performing the "congruent transformation" on matrix $[K']$, i.e., (2.3.2.8c). The matrix $[K]$ is evaluated as

$$[K] = \frac{AE}{L}\begin{bmatrix} \cos^2\alpha & \cos\alpha\sin\alpha & -\cos^2\alpha & -\cos\alpha\sin\alpha \\ \cos\alpha\sin\alpha & \sin^2\alpha & -\cos\alpha\sin\alpha & -\sin^2\alpha \\ -\cos^2\alpha & -\cos\alpha\sin\alpha & \cos^2\alpha & \cos\alpha\sin\alpha \\ -\cos\alpha\sin\alpha & -\sin^2\alpha & \cos\alpha\sin\alpha & \sin^2\alpha \end{bmatrix}. \tag{2.3.2.8d}$$

As in the first example, note again that the element matrix $[K]$ is symmetric. We see that the end product of step 2 is Eq. (2.3.2.8b), which can be put in the form

$$\{F\}^e = [K]^e\{u\}^e, \tag{2.3.2.9}$$

where, once again, the superscript e is used to indicate that the matrix equation is written at the element level.

The procedure for generating $[K]^e$ for this two degree of freedom problem can be summarized as follows:

(i) generate $[K']$, the element stiffness matrix with respect to the local axes, using Hooke's law;
(ii) derive $[T]$, the coordinate transformation matrix; and
(iii) obtain $[K]^e$ from the congruent transformation

$$[K]^e = [T]^t[K'][T].$$

Step 3 Assemble the element equations. Again the assembly process is performed by rewriting the element matrix equations in terms of the global numbering scheme. If we enforce the conditions, $u_{ix}^{(e)} = u_{ix}$, $u_{iy}^{(e)} = u_{iy}$ (i = 1, 2, 3 nodal indices), the matrix equation for element 1 can be written as

Unknown or degree of freedom no. 1 2 3 4 5 6

$$\begin{array}{c|} 1 \\ 2 \\ 3 \\ 4 \\ 5 \\ 6 \end{array}\begin{bmatrix} k_{11}^{(1)} & k_{12}^{(1)} & k_{13}^{(1)} & k_{14}^{(1)} & 0 & 0 \\ k_{21}^{(1)} & k_{22}^{(1)} & k_{23}^{(1)} & k_{24}^{(1)} & 0 & 0 \\ k_{31}^{(1)} & k_{32}^{(1)} & k_{33}^{(1)} & k_{34}^{(1)} & 0 & 0 \\ k_{41}^{(1)} & k_{42}^{(1)} & k_{43}^{(1)} & k_{44}^{(1)} & 0 & 0 \\ 0 & 0 & 0 & 0 & 0 & 0 \\ 0 & 0 & 0 & 0 & 0 & 0 \end{bmatrix} \begin{Bmatrix} u_{1x} \\ u_{1y} \\ u_{2x} \\ u_{2y} \\ u_{3x} \\ u_{3y} \end{Bmatrix} = \begin{Bmatrix} F_{1x}^{(1)} \\ F_{1y}^{(1)} \\ F_{2x}^{(1)} \\ F_{2y}^{(1)} \\ 0 \\ 0 \end{Bmatrix}. \quad (2.3.2.10a)$$

Similarly, for element 2, we obtain

$$\begin{bmatrix} 0 & 0 & 0 & 0 & 0 & 0 \\ 0 & 0 & 0 & 0 & 0 & 0 \\ 0 & 0 & k_{11}^{(2)} & k_{12}^{(2)} & k_{13}^{(2)} & k_{14}^{(2)} \\ 0 & 0 & k_{21}^{(2)} & k_{22}^{(2)} & k_{23}^{(2)} & k_{24}^{(2)} \\ 0 & 0 & k_{31}^{(2)} & k_{32}^{(2)} & k_{33}^{(2)} & k_{34}^{(2)} \\ 0 & 0 & k_{41}^{(2)} & k_{42}^{(2)} & k_{43}^{(2)} & k_{44}^{(2)} \end{bmatrix} \begin{Bmatrix} u_{1x} \\ u_{1y} \\ u_{2x} \\ u_{2y} \\ u_{3x} \\ u_{3y} \end{Bmatrix} = \begin{Bmatrix} 0 \\ 0 \\ F_{2x}^{(2)} \\ F_{2y}^{(2)} \\ F_{3x}^{(2)} \\ F_{3y}^{(2)} \end{Bmatrix}. \quad (2.3.2.10b)$$

The two-element matrix equations are then combined to obtain the global matrix equation

$$[K]\{\mathscr{U}\} = \{\mathscr{F}\}, \quad (2.3.2.11)$$

where

$$\{\mathscr{U}\} = \begin{Bmatrix} u_{1x} \\ u_{1y} \\ u_{2x} \\ u_{2y} \\ u_{3x} \\ u_{3y} \end{Bmatrix}, \qquad \{\mathscr{F}\} = \begin{Bmatrix} F_{1x}^{(1)} \\ F_{1y}^{(1)} \\ F_{2x}^{(1)} + F_{2x}^{(2)} \\ F_{2y}^{(1)} + F_{2y}^{(2)} \\ F_{3x}^{(2)} \\ F_{3y}^{(2)} \end{Bmatrix}, \quad (2.3.2.12a)$$

and the matrix $[K]$ is obtained from

$$[K] = \sum_{e=1}^{2} [K]^e. \quad (2.3.2.12b)$$

Finally, we enforce the equilibrium conditions at node 2, which is shared by both elements of the truss. In so doing, the right-hand-side vector $\{\mathscr{F}\}$ is modified to

$$\{\mathscr{F}\} = \begin{Bmatrix} F_{1x}^{(1)} \\ F_{1y}^{(1)} \\ 5 \\ 5 \\ F_{3x}^{(2)} \\ F_{3y}^{(2)} \end{Bmatrix}. \quad (2.3.2.12c)$$

Steps 4 and 5 The remaining two steps are performed in a similar manner to that described in Example 1.

2.4 Solution of Steady-State Continuum Problems

The finite element procedure described in the foregoing sections can be readily extended to deal with continuum problems. The extension involves the establishment of a more general way to formulate the element matrix equations. This can be accomplished via the use of either a variational or a weighted residual approach. Early applications of the finite element method were formulated using variational techniques. Recently, the situation has changed. The so-called Galerkin weighted residual approach has gained increasing popularity owing to its generality in application, particularly to non-self-adjoint problems (see Chapter 1 for definition of self-adjoint differential operators), which cannot be cast in variational form. We now present a brief description of both the variational and the Galerkin approaches.

2.4.1 The Variational Approach

In this approach, we replace the problem of finding a solution of the governing differential equation by an equivalent variational problem that consists of finding an unknown function that extremizes (makes stationary) a certain integral quantity, subject to the prescribed boundary conditions. Such an integral is called a "functional" because it is a function of the unknown function. The two problem statements are equivalent in that an exact solution of one is also the solution of the other. This equivalence is demonstrated using the calculus of variations. It shows that the necessary conditions for extremization of the functional are indeed the governing equation and boundary conditions of the same physical problem. We now illustrate how the finite element method, formulated using a variational principle, can be used to obtain an approximate solution.

Let Ω denote a functional that is to be extremized over a one-dimensional region R. A general expression for this functional is given by

$$\Omega = \int_R F(x, u, du/dx, d^2u/dx^2, \ldots)\, dx. \tag{2.4.1.1}$$

Usually, the integrand F can be obtained via the "Euler–Lagrange differential equation," which represents the necessary condition for stationarity of the functional. The details of this procedure, however, are not dealt with in this book; the interested reader is referred to Weinstock (1952).

To solve the variational problem using the finite element method, we discretize the region R into a series of m finite elements and let the subregion of a typical element e be denoted R^e. Within this subregion, we approximate the unknown function u by a trial function $\hat{u}$, which is given by

$$\hat{u} = \sum_{I=1}^{n^e} N_I(x) u_I, \tag{2.4.1.2}$$

where $N_I(x)$ are linearly independent functions selected a priori and referred to as "basis (or interpolation or shape) functions," u_I are values of u at the nodes of the element, and n^e is the number of nodes assigned to element e.

Next, we assume that the total functional is equal to the sum of all element contributions, that is,

$$\Omega = \sum_{e=1}^{m} \Omega^e, \tag{2.4.1.3}$$

where

$$\Omega^e = \int_{R^e} F\left(x, \hat{u}, \frac{d\hat{u}}{dx}, \frac{d^2\hat{u}}{dx^2}, \dots\right) dx \tag{2.4.1.4}$$

and the summation is taken over all m elements.

It is apparent from Eq. (2.4.1.4) that Ω^e is a function of the nodal parameters of element e. Upon substitution of Eq. (2.4.1.2) into (2.4.1.3) and enforcing nodal compatibility, the total functional Ω becomes a function

$$\Omega = \Omega(u_1, u_2, \dots, u_n), \tag{2.4.1.5}$$

where n is the total number of nodes in the finite element network.

For Ω to be extremized, the necessary conditions are

$$\partial\Omega/\partial u_I = 0, \qquad I = 1, 2, \dots, n. \tag{2.4.1.6}$$

Substituting Eq. (2.4.1.3) into (2.4.1.6), we obtain

$$\partial\Omega/\partial u_I = \sum_{e=1}^{m} \partial\Omega^e/\partial u_I = 0, \qquad I = 1, 2, \dots, n. \tag{2.4.1.7}$$

Equation (2.4.1.7) implies that

$$\partial\Omega^e/\partial u_I = Q_I^e, \qquad I = 1, 2, \dots, n_e, \tag{2.4.1.8a}$$

where Q is a dummy variable. Equation (2.4.1.8a) can be written in matrix form as

$$\{\partial\Omega^e/\partial u\} = \{Q\}^e, \tag{2.4.1.8b}$$

where

$$\{\partial\Omega^e/\partial u\} = \begin{Bmatrix} \partial\Omega^e/\partial u_1 \\ \vdots \\ \partial\Omega^e/\partial u_{ne} \end{Bmatrix}.$$

This equation represents a set of n^e algebraic equations that characterize the behavior of element e. The assumption that we can represent the total functional as the sum of the functionals for all individual elements provides the basis for formulating element equations from a variational principle. (This assumption is valid, provided the basis functions satisfy certain continuity, and completeness conditions to be discussed.) If the functional Ω is a quadratic function of u and its derivatives, then Ω^e is also quadratic and Eq. (2.4.1.8b) for an element e can always be written as

$$\{\partial\Omega^e/\partial u\} = [C]^e\{u\}^e - \{F\}^e = \{Q\}^e, \tag{2.4.1.9}$$

where $[C]^e$ is the element coefficient matrix, $\{u\}^e$ the column vector containing nodal values of function u, and $\{F\}^e$ the column vector of resultant nodal actions.

After assembly of all elements, the global matrix equation is obtained. This takes the form

$$[C]\{u\} - \{F\} = \{0\} \quad \text{or} \quad [C]\{u\} = \{F\}, \tag{2.4.1.10}$$

in which

$$[C] = \sum_{e=1}^{m} [C]^e, \quad \{F\} = \sum_{e=1}^{m} \{F\}^e, \quad \text{and} \quad \sum_{e=1}^{m} \{Q\}^e = \{0\}. \tag{2.4.1.11}$$

The preceding finite element approximation may be regarded as a variant of the well-known classical Rayleigh–Ritz procedure for obtaining an approximate solution of the variational problem. According to this procedure, we assume that the unknown function u can be approximated over the entire region R by a trial solution of the form

$$\hat{u} = \sum_{I=1}^{n} C_I\psi_I(x), \tag{2.4.1.12}$$

where $\psi_I(x)$ are linearly independent basis functions, C_I unknown parameters to be determined subsequently, and n the number of terms in the finite series.

The basis functions $\psi_I(x)$ are selected in such a way that $\hat{u}$ satisfies the essential boundary conditions regardless of the choice of parameters C_I. When substituting Eq. (2.4.1.12) into the functional to be extremized, we obtain, after performing integration, the following expression:

$$\Omega = \Omega(C_1, C_2, \ldots, C_n). \tag{2.4.1.13}$$

For extremization of Ω, it is necessary that

$$\partial\Omega/\partial C_I = 0, \qquad I = 1, 2, \ldots, n, \tag{2.4.1.14}$$

which results in a set of n algebraic equations in n unknowns, C_1, C_2, ..., C_n. It is apparent that the distinction between the Rayleigh–Ritz and the finite element methods lies in the definition of basis functions. In the Rayleigh–Ritz method, the basis functions are defined in the entire region R, whereas in the finite element method these functions are defined piecewise (element by element).

The basis functions for the Rayleigh–Ritz method are required to satisfy the essential boundary conditions of the problem, whereas the piecewise basis functions for the finite element method need to satisfy certain continuity and completeness conditions to be discussed later. Because the Rayleigh–Ritz method uses functions defined over the entire region, it can be used only for regions of relatively simple geometric configuration. In the finite element method, the same geometric limitations exist, but only for each element. Because these simply shaped elements can be assembled to represent exceedingly complex geometries, the finite element method is a far more versatile tool than the classical Rayleigh–Ritz method.

2.4.2 *The Galerkin Approach*

Although variational methods provide a convenient approach for deriving the element matrix equation, it is not the only approach available. Frequently, we encounter practical problems for which the classical functionals cannot be derived, or, in other words, the variational principles do not exist. For these cases we have to employ a more general approach in formulating the element matrix equations. The Galerkin method is one such approach and has been widely used. It is a special case of the method of weighted residuals (MWR), which can be described as follows.

Consider a continuum problem governed by the differential equation

$$L(u) - f = 0 \tag{2.4.2.1}$$

in the region R enclosed by the boundary B. To obtain an approximate solution, the method is applied in three steps. The first step is to approximate the unknown function u by a trial function of the form

$$\hat{u} = \sum_{I=1}^{n} N_I C_I, \tag{2.4.2.2}$$

where N_I are linearly independent basis functions defined over the entire solution domain and C_I the unknown parameters to be determined subsequently. It is a common practice to select the n basis functions in such a way that all essential boundary conditions are satisfied.

Because the trial function $\hat{u}$ is only an approximation, it is not likely to satisfy Eq. (2.4.2.1) exactly. Substitution of $\hat{u}$ in Eq. (2.4.2.1) thus results in an error or residual:

$$\varepsilon = L(\hat{u}) - f. \tag{2.4.2.3}$$

The method of weighted residuals seeks to determine the unknowns C_I in such a way that the error is minimal in some specified sense. This is accomplished by forming a weighted integral of ε over the entire solution domain and then setting this integral (weighted residual) to zero. The second step of the procedure thus consists of selecting n linearly independent "weighting" functions W_I and requiring that

$$\int_R W_I \varepsilon \, dR = \int_R W_I(L(\hat{u}) - f)\, dR = 0 \qquad \text{for} \quad I = 1, 2, \ldots, n. \tag{2.4.2.4}$$

Once we specify the functional form of the weighting functions, we can employ Eq. (2.4.2.2) to represent $\hat{u}$ and combine this information with Eq. (2.4.2.4) to provide a set of simultaneous equations in the n unknowns C_I, $I = 1, 2, \ldots, n$. The final step is to solve these equations for C_I and hence obtain an approximate representation of the unknown function u via the use of Eq. (2.4.2.2). Various classical weighted residual methods can now be generated, depending on the choice of the weighting functions. The methods commonly employed are as follows [see also Finlayson (1972)].

(a) THE POINT COLLOCATION METHOD In this approach one specifies a set of points x_I in the solution domain, denoted as collocation points, and then selects the weighting functions to be Dirac delta functions, defined as

$$W_I = \delta(x - x_I), \qquad I = 1, 2, \ldots, n. \tag{2.4.2.5}$$

The Dirac delta functions have the attractive property that

$$\int_R \varepsilon\delta(x - x_I)\, dR = \varepsilon(x_I) = 0, \qquad I = 1, 2, \ldots, n.$$

Thus, the procedure consists of simply evaluting the residuals at the collocation points and therefore involves a minimum of computational effort.

(b) The Subdomain Collocation Method In this method, the solution domain is divided into a number of subdomains, and the weighting functions are chosen such that

$$W_I = \begin{cases} 1, & x \in R_I \\ 0, & x \notin R_I \end{cases} \qquad I = 1, 2, \ldots, n, \tag{2.4.2.6}$$

where the R_I refer to the specified subdomains.

(c) The Galerkin Method In the Galerkin method, the weighting functions are chosen to be identical to the basis functions, i.e., $W_I = N_I$. Thus, the weighted residual equation becomes

$$\int_R N_I \varepsilon \, dR = 0, \tag{2.4.2.7a}$$

or

$$\int_R N_I(L(\hat{u}) - f) \, dR = 0, \qquad I = 1, 2, \ldots, n. \tag{2.4.2.7b}$$

Of the three weighted residuals methods described, the Galerkin method is the one most naturally suited to finite element applications. For this reason, it has been widely accepted as the methodology for generating the finite element equations.

In the Galerkin finite element formulation, we proceed as follows:

Step 1 The domain is once again subdivided into a series of elements and the unknown function is represented over a typical element subregion by the trial function

$$\hat{u} = \sum_{I=1}^{n^e} N_I u_I, \tag{2.4.2.8}$$

where N_I are the piecewise defined basis functions for element e, u_I the unknown nodal values, and n_e the number of nodes on the element.

Step 2 Next, we note that the integral on the left side of Eq. (2.4.2.7) is equal to the summation of the integrations performed over the element subregions. This permits us to write the equations governing the behavior of a typical element as

$$\int_{R^e} N_I[L(\hat{u}) - f] \, dR = Q_I^e, \qquad I = 1, 2, \ldots, n^e. \tag{2.4.2.9}$$

At this stage, the integral obtained via the Galerkin criterion contains higher-order differentials than the variational functional. This is undesirable because a higher continuity requirement would have to be imposed on the element basis functions (the higher the order of continuity, the narrower our choice of functions becomes). Fortunately, in most cases, we can overcome this difficulty by applying Green's theorem (integration by parts) to the higher-order terms in the integral expression of Eq. (2.4.2.9). The order of the integrand is thereby reduced, and this enables us to use interpolating functions with a lower-order interelement continuity requirement.

Step 3 Having the required element equations, one must now assemble all element equations into a global matrix equation and incorporate the boundary conditions.

2.4.3 *Derivation of Linear Basis Functions*

Requirements for Basis Functions

The procedures for formulating the element equations via the variational and weighted residual approaches rely on the assumption that the integral over the entire solution region is equal to the summation of the integrals performed over element subregions. To ensure that this assumption is valid and that our approximate solution converges to the correct solution as we refine the finite element mesh, the interpolating (basis or shape) functions must satisfy certain requirements. These requirements are as follows:

(1) At element interfaces, the unknown function u and any of its derivatives up to one order less than the highest derivative appearing in the functional or the weighted residual integral must be continuous. This is called the continuity requirement. Thus, suppose the integrand in the element equation contains up to $(m + 1)$th derivatives of function u, then at the element interfaces we must have continuity in the mth derivative of u. This is called the C^m-continuity requirement.

(2) The trial function u and its derivatives must be able to represent any constant values of u and its derivatives appearing in the functional (or the weighted residual integral) as, in the limit, the element size is reduced to zero. This is known as the "completeness" requirement.

Linear Basis Functions

The element basis functions commonly employed take the form of polynomials. In this section, we describe a direct procedure for deriving the linear basis functions of the one-dimensional and triangular elements.

One-Dimensional Elements

Consider the typical element shown in Fig. 2.7a. Let the trial function $\hat{u}(x)$ be represented within this element by the linear polynomial

$$\hat{u} = a_1 + a_2 x, \tag{2.4.3.1}$$

where a_1 and a_2 are constants. These constants can be evaluated by setting $\hat{u}(x_1) = u_1$ and $\hat{u}(x_2) = u_2$, where x_1 and x_2 define the nodal

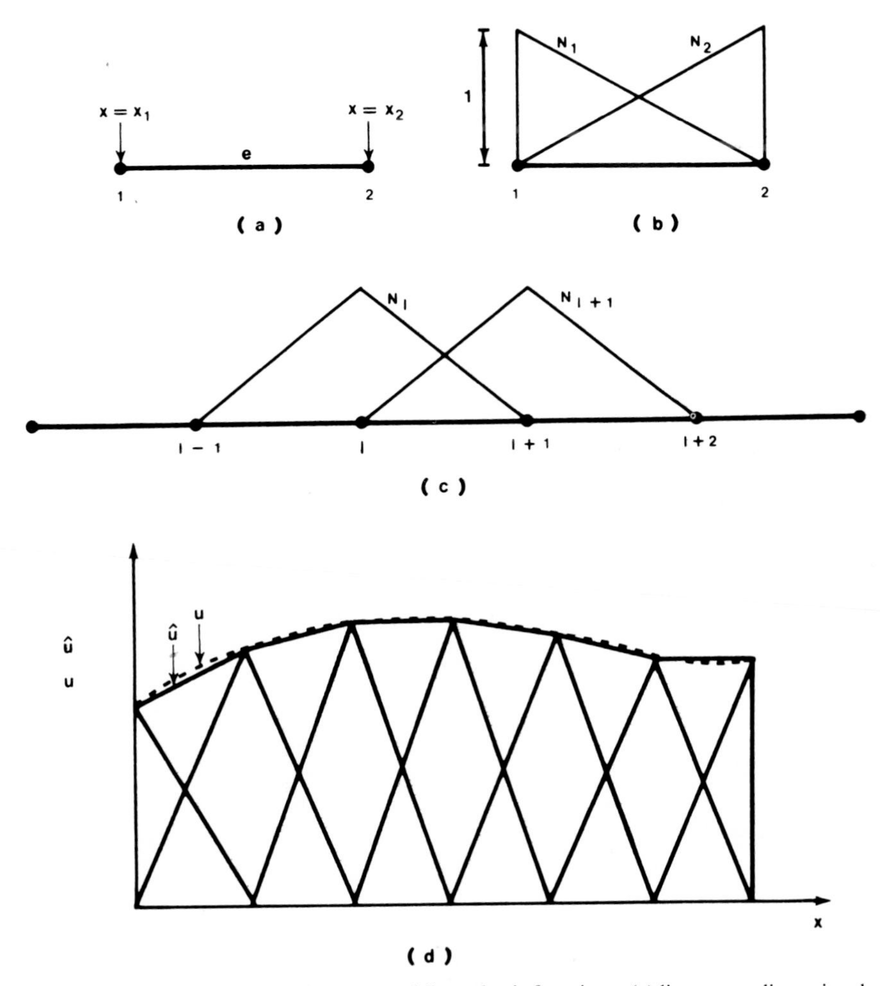

Fig. 2.7. One-dimensional elements and linear basis functions: (a) linear one-dimensional element; (b) element basis functions; (c) global basis functions for nodes I and $I + 1$; (d) approximation of u by trial function $\hat{u}$.

points of the element. Applying these two constraints yields the matrix equation

$$\begin{Bmatrix} u_1 \\ u_2 \end{Bmatrix} = \begin{bmatrix} 1 & x_1 \\ 1 & x_2 \end{bmatrix} \begin{Bmatrix} a_1 \\ a_2 \end{Bmatrix}. \tag{2.4.3.2}$$

Solving for the unknowns a_1 and a_2, we obtain

$$\begin{Bmatrix} a_1 \\ a_2 \end{Bmatrix} = \frac{1}{L} \begin{bmatrix} x_2 & -x_1 \\ -1 & 1 \end{bmatrix} \begin{Bmatrix} u_1 \\ u_2 \end{Bmatrix}, \tag{2.4.3.3}$$

where $L = x_2 - x_1$. The function $\hat{u}(x)$ may now be written as

$$\hat{u}(x) = \frac{1}{L}(x_2 u_1 - x_1 u_2) + \frac{1}{L}(u_2 - u_1)x. \tag{2.4.3.4a}$$

Equation (2.4.3.4a) is rearranged to give

$$\hat{u}(x) = \frac{1}{L}(x_2 - x)u_1 + \frac{1}{L}(x - x_1)u_2. \tag{2.4.3.4b}$$

Comparison of Eqs. (2.4.3.4b) and (2.4.2.8) shows that the element shape functions are given by

$$N_1^e = \frac{1}{L}(x_2 - x), \qquad N_2^e = \frac{1}{L}(x - x_1) \qquad \text{for} \quad x_1 \leqslant x \leqslant x_2 \tag{2.4.3.5}$$

in which the superscript e is introduced to denote the fact that these functions belong to element e. A sketch of the functions N_1^e and N_2^e is depicted in Fig. 2.7b. It becomes apparent, when we assemble the elements, that the global basis function associated with each internal node I of the mesh is given by

$$N_I = \begin{cases} (x - x_{I-1})/(x_I - x_{I-1}) & \text{for} \quad x_{I-1} \leqslant x \leqslant x_I \\ (x_{I+1} - x)/(x_{I+1} - x_I) & \text{for} \quad x_I \leqslant x \leqslant x_{I+1}. \end{cases} \tag{2.4.3.6}$$

As illustrated in Fig. 2.7c, N_I takes the shape of a hat. For this reason, it is called the "chapeau" basis function. Knowing the expressions for the basis functions N_I of all nodes in the network, we can determine the distribution of the trial function $\hat{u}$ over the entire region. A typical plot of $\hat{u}$ is shown in Fig. 2.7d. It can be seen that $\hat{u}$ is merely a "piecewise" linear approximation of the unknown function u.

Triangular Elements

To obtain the basis functions for triangular elements, we simply extend the preceding procedure. Consider the typical triangular element with the nodes numbered in the counterclockwise direction, as shown in Fig.

2.8. Let the trial function be represented within this element by the linear polynomial

$$\hat{u}(x, y) = a_1 + a_2x + a_3y, \tag{2.4.3.7}$$

where a_1 and a_2 and a_3 are constants. These constants are determined by writing

$$\begin{Bmatrix} u_1 \\ u_2 \\ u_3 \end{Bmatrix} = \begin{bmatrix} 1 & x_1 & y_1 \\ 1 & x_2 & y_2 \\ 1 & x_3 & y_3 \end{bmatrix} \begin{Bmatrix} a_1 \\ a_2 \\ a_3 \end{Bmatrix}. \tag{2.4.3.8a}$$

Solving for a_1, a_2, and a_3, and substituting these into Eq. (2.4.3.7), we obtain

$$\hat{u}(x, y) = (1/2A)[(\alpha_1 + \beta_1 x + \gamma_1 y)u_1 + (\alpha_2 + \beta_2 x + \gamma_2 y)u_2 + (\alpha_3 + \beta_3 x + \gamma_3 y)u_3], \tag{2.4.3.8b}$$

where

$$\alpha_1 = x_2y_3 - x_3y_2, \qquad \beta_1 = y_2 - y_3, \qquad \gamma_1 = x_3 - x_2,$$

$$\alpha_2 = x_3y_1 - x_1y_3, \qquad \beta_2 = y_3 - y_1, \qquad \gamma_2 = x_1 - x_3,$$

$$\alpha_3 = x_1y_2 - x_2y_1, \qquad \beta_3 = y_1 - y_2, \qquad \gamma_3 = x_2 - x_1,$$

and

$$A = \frac{1}{2}\begin{vmatrix} 1 & x_1 & y_1 \\ 1 & x_2 & y_2 \\ 1 & x_3 & y_3 \end{vmatrix} = \text{area of the triangular element.}$$

From Eq. (2.4.3.8) we deduce that the basis functions for the triangular element are given by

$$N_I = (1/2A)(\alpha_I + \beta_I x + \gamma_I y) \qquad \text{for} \quad I = 1, 2, 3. \tag{2.4.3.9}$$

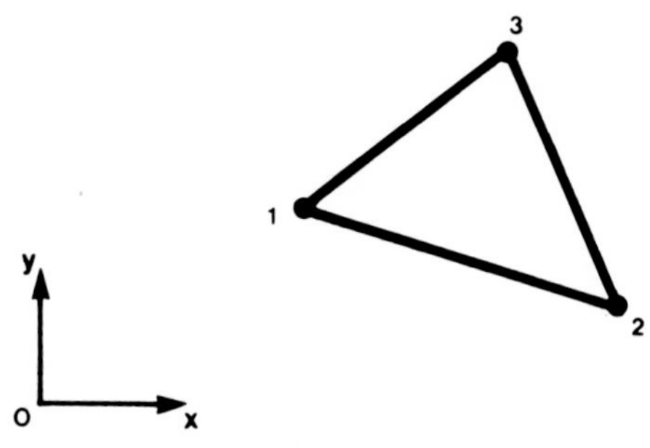

Fig. 2.8. A typical triangular element with counterclockwise node numbering.

To illustrate the application of the variational and Galerkin methods in formulating the element equations, we present the following examples:

Example 1 Consider a simple problem of one-dimensional flow in a uniform porous medium 10 units long. The governing equation for this problem is given by

$$K\, d^2p/dx^2 = 0, \qquad 0 \leqslant x \leqslant 10, \tag{2.4.3.10}$$

where the coefficient K is assumed to be unity. Suppose also that the boundary conditions are the Dirichlet boundary conditions given by

$$p = \begin{cases} 0 & \text{at} \quad x = 0 \qquad (2.4.3.11a) \\ 1 & \text{at} \quad x = 1. \qquad (2.4.3.11b) \end{cases}$$

Variational Approach

We first formulate this problem via the variational approach. Using the Euler–Lagrange equation, the functional for this problem can be shown to take the form

$$\Omega = \int_0^{10} \frac{K}{2}\left(\frac{dp}{dx}\right)^2 dx. \tag{2.4.3.12}$$

We seek the function p that minimizes this functional. The following steps are now taken to derive the element matrix equation.

Step 1 We discretize the domain into m one-dimensional elements and write the trial function for the typical element in the form

$$\hat{p}(x) = N_J(x)p_J \qquad (J = 1, 2), \tag{2.4.3.13}$$

in which J represents the node numbers of the element, and the summation convention is used on the repeated subscripts.

Step 2 Since the integral in Eq. (2.4.3.12) contains only first-order derivatives of p, we only require C^0 continuity across the element interfaces. This requirement is met by the linear basis functions, which also satisfy the completeness criterion. Thus, we may write

$$\Omega = \sum_{e=1}^{m} \Omega^e, \tag{2.4.3.14}$$

where

$$\Omega^e = \int_{x_1}^{x_2} \frac{K}{2}\left(\frac{d\hat{p}}{dx}\right)^2 dx. \tag{2.4.3.15}$$

From Eqs. (2.4.3.15) and (2.4.3.13) we obtain

$$\Omega^e = \int_{x_1}^{x_2} \frac{K}{2}\left(\frac{dN_I}{dx}p_I\right)\left(\frac{dN_J}{dx}p_J\right) dx. \tag{2.4.3.16}$$

Performing the differentiation of Ω^e with respect to the nodal values gives

$$\frac{\partial \Omega^e}{\partial p_I} = \int_{x_1}^{x_2} K \frac{dN_I}{dx}\frac{dN_J}{dx} p_J \, dx, \qquad I = 1, 2, \tag{2.4.3.17}$$

which can be written in matrix form as

$$[C]^e\{p\}^e = \{\partial\Omega/\partial p\}^e, \tag{2.4.3.18}$$

where

$$[C]^e = \int_{x_1}^{x_2} K \begin{bmatrix} \dfrac{dN_1}{dx}\cdot\dfrac{dN_1}{dx} & \dfrac{dN_1}{dx}\cdot\dfrac{dN_2}{dx} \\ \\ \dfrac{dN_2}{dx}\cdot\dfrac{dN_1}{dx} & \dfrac{dN_2}{dx}\cdot\dfrac{dN_2}{dx} \end{bmatrix} dx,$$

$$\{p\}^e = \begin{Bmatrix} p_1 \\ p_2 \end{Bmatrix}, \qquad \{\partial\Omega/\partial p\}^e = \begin{Bmatrix} \partial\Omega^e/\partial p_1 \\ \partial\Omega^e/\partial p_2 \end{Bmatrix}.$$

Note that the element matrix $[C]^e$ is symmetric. On evaluating the derivatives of the shape functions in (2.4.3.5), we obtain

$$[C]^e = \frac{K}{L}\begin{bmatrix} 1 & -1 \\ -1 & 1 \end{bmatrix}. \tag{2.4.3.19}$$

The matrix equation (2.4.3.18) describes the characteristics of a particular finite element. Once it is obtained, the procedure described in Section 2.3.1 for assembling the elements and incorporating boundary conditions can be employed.

Galerkin Approach

Subdivide the region into m elements, as in the variational approach, and proceed as follows.

Step 1 Construct the trial function and use the Galerkin criterion to form the weighted residual integral.

The trial function $\hat{p}$ is written in terms of the global basis functions and nodal values as follows:

$$\hat{p} = N_J p_J, \qquad J = 1, 2, \ldots, n, \tag{2.4.3.20}$$

where N_J denotes the global basis function at node J, n is the total number of nodes in the finite element network, and the nodal summation convention is again employed on repeated subscripts. By substituting $\hat{p}$ into Eq. (2.4.3.10), we obtain a residual ε, which can be expressed as

$$\varepsilon = K\frac{d^2\hat{p}}{dx^2}. \tag{2.4.3.21}$$

Step 2 Apply the Galerkin criterion and construct the weighted residual integral of the form

$$\int_0^{10} N_I \varepsilon \, dx = 0,$$

or

$$\int_0^{10} N_I K \frac{d^2\hat{p}}{dx^2}\, dx = 0, \qquad I = 1, 2, \ldots, n. \tag{2.4.3.22}$$

Applications of Green's theorem (integration by parts) to Eq. (2.4.3.22) yields

$$-\int_0^{10} K\frac{dN_I}{dx}\frac{d\hat{p}}{dx}\, dx + K\frac{d\hat{p}}{dx}N_I\bigg|_0^{10} = 0, \qquad I = 1, 2, \ldots, n. \tag{2.4.3.23}$$

Because the Dirichlet boundary conditions (prescribed values of p) at $x = 0$ and $x = 10$ are to be imposed separately on the global matrix equation finally obtained, the boundary equations may be dropped. Thus we obtain for the remaining internal nodes

$$\int_0^{10} K\frac{dN_I}{dx}\frac{d\hat{p}}{dx} = 0, \qquad I = 2, 3, \ldots, n - 1. \tag{2.4.3.24}$$

Substitution of Eq. (2.4.3.20) into (2.4.3.24) yields

$$\int_0^{10} K\frac{dN_I}{dx}\frac{dN_J}{dx}p_J\, dx = 0, \qquad I = 2, 3, \ldots, n - 1. \tag{2.4.3.25}$$

Because the basis functions satisfy the continuity and completeness requirement, Eq. (2.4.3.25) can be written in the form

$$\int_0^{10} K\frac{dN_I}{dx}\frac{dN_J}{dx}p_J\, dx = \sum_{e=1}^{m}\left[\int_{x_1}^{x_2} K\frac{dN_I}{dx}\frac{dN_J}{dx}p_J\, dx\right] = 0$$

$$\text{for} \quad I = 2, 3, \ldots, n - 1,$$

where x_1 and x_2 are the boundary coordinates of element e. From Eq.

(2.4.3.25) we can readily extract the element equation. Switching from the global to local node numbers, we obtain for a typical element e

$$C_{IJ}^e p_J^e = Q_I^e \qquad \text{for} \quad I = 1, 2, \tag{2.4.3.26}$$

where

$$C_{IJ}^e = \int_{x1}^{x2} K \frac{dN_I}{dx} \frac{dN_J}{dx} \, dx, \qquad \sum_{e=1}^{m} Q_I^e = 0.$$

In matrix notation, Eq. (2.4.3.26) becomes the required element matrix equation

$$[C]^e\{p\}^e = \{Q\}^e, \tag{2.4.3.27}$$

where

$$[C]^e = \int_{x1}^{x2} K \begin{bmatrix} \frac{dN_1}{dx} \cdot \frac{dN_1}{dx} & \frac{dN_1}{dx} \cdot \frac{dN_2}{dx} \\ \frac{dN_2}{dx} \cdot \frac{dN_1}{dx} & \frac{dN_2}{dx} \cdot \frac{dN_2}{dx} \end{bmatrix} dx,$$

$$\{p\}^e = \begin{Bmatrix} p_1 \\ p_2 \end{Bmatrix}, \qquad \{Q\}^e = \begin{Bmatrix} Q_1 \\ Q_2 \end{Bmatrix}.$$

It should be noted that the element matrix obtained from the Galerkin approach is identical to that from the variational approach. Thus, the remaining steps of the finite element solution need no further elaboration.

Example 2 As a second example, we consider the problem of two-dimensional, horizontal flow in an isotropic reservoir formation (Fig. 2.9). The governing equation in the flow region R can be written in the form

$$\frac{\partial}{\partial x_1}\left(T \frac{\partial p}{\partial x_1}\right) + \frac{\partial}{\partial x_2}\left(T \frac{\partial p}{\partial x_2}\right) = 0 \quad \text{in} \quad R, \tag{2.4.3.28}$$

where p is the fluid pressure and T is the transmissivity of the porous medium.

On the boundary portion B_1 we assume the Dirichlet boundary condition

$$p = \bar{p}(x, y), \tag{2.4.3.29a}$$

and on boundary portion B_2 we assume the boundary condition

$$T \frac{\partial p}{\partial x_1} n_1 + T \frac{\partial p}{\partial x_2} n_2 = q, \tag{2.4.3.29b}$$

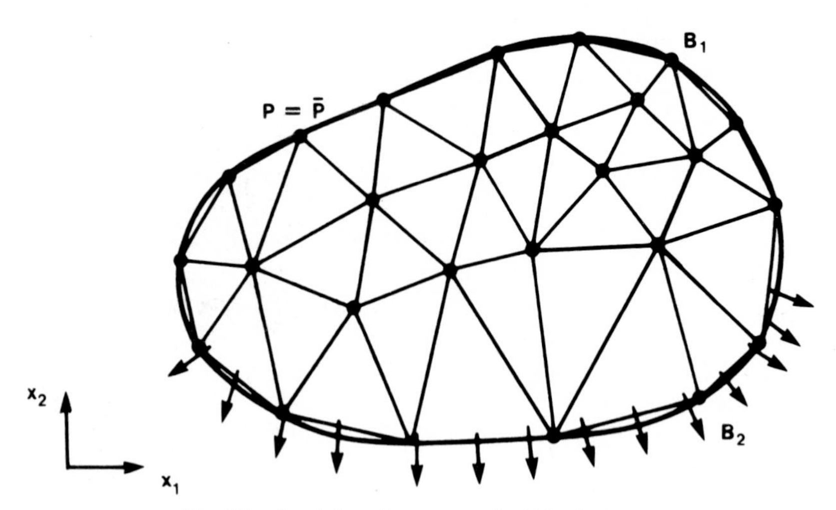

Fig. 2.9. Areal flow in a reservoir. $T\,(\partial p/\partial x_i)n_i = q$.

where $\bar{p}$ is a prescribed pressure distribution on the boundary B_1, n_1 and n_2 are the components of the outward unit normal vector on the boundary B_2, and q is the prescribed flux distribution.

Variational Approach

We first use the Euler–Lagrange equation to obtain the functional of the form

$$\Omega = \int_R \frac{T}{2}\left[\frac{\partial p}{\partial x_i}\frac{\partial p}{\partial x_i}\right] dR - \int_{B_2} qp\, dB, \qquad (2.4.3.30)$$

where the summation convention is implied on the coordinate subscript i, B_2 is the flow boundary, and it is understood that the function p is prescribed as $\bar{p}$ on B_1.

Step 1 The flow region R is divided into m triangular elements. Within each element, the trial function is assumed to take the form

$$\hat{p}(x_1, x_2) = N_J(x_1, x_2)p_J, \qquad J = 1, 2, 3, \qquad (2.4.3.31)$$

where N_J denotes element basis functions, J the nodal subscript that ranges from 1 to 3, and the summation convention is once again implied on the repeated subscript J. Here we use capital letters to denote nodal subscripts and small letters to denote coordinate subscripts.

Step 2 We then replace the total functional Ω by

$$\Omega = \sum_{e=1}^{m} \Omega^e, \qquad (2.4.3.32)$$

where

$$\Omega^e = \int_{R^e} \frac{T}{2}\left[\frac{\partial \hat{p}}{\partial x_i}\frac{\partial \hat{p}}{\partial x_i}\right] dR - \int_{B_2^e} q\hat{p}\, dB. \tag{2.4.3.33}$$

From Eqs. (2.4.3.31) and (2.4.3.33) we obtain

$$\Omega^e = \int_{R^e} \frac{T}{2}\left[\frac{\partial N_I}{\partial x_i} p_I \frac{\partial N_J}{\partial x_i} p_J\right] dR - \int_{B_2} qN_I p_I\, dB. \tag{2.4.3.34}$$

Differentiating Eq. (2.4.3.34) with respect to the nodal values gives

$$\frac{\partial \Omega^e}{\partial p_I} = \int_{R^e} T \frac{\partial N_I}{\partial x_i}\frac{\partial N_J}{\partial x_i} p_J\, dR - \int_{B_2^e} qN_I\, dB \qquad \text{for} \quad I = 1, 2, 3. \tag{2.4.3.35}$$

Equation (2.4.3.35) can be written in matrix form as

$$[C]^e\{p\}^e = \{F\}^e + \{\partial\Omega/\partial p\}^e, \tag{2.4.3.36}$$

where

$$[C]^e = \begin{bmatrix} C_{11}^e & C_{12}^e & C_{13}^e \\ C_{21}^e & C_{22}^e & C_{23}^e \\ C_{31}^e & C_{32}^e & C_{33}^e \end{bmatrix}, \qquad \{F\}^e = \begin{Bmatrix} F_1^e \\ F_2^e \\ F_3^e \end{Bmatrix},$$

$$C_{IJ}^e = \int_{R^e} T \frac{\partial N_I}{\partial x_i}\frac{\partial N_J}{\partial x_i} dR, \qquad F_I^e = \int_{B_2^e} qN_I\, dB.$$

Equation (2.4.3.36) describes the behavior of a triangular element of the porous medium domain. The components of vector $\{F\}^e$ may be interpreted as nodal fluxes on the boundary portion B_2^e of an exterior element whose boundary portion B_2^e is subject to the boundary condition given by Eq. (2.4.3.29b). It follows that for the "interior" elements which lie inside the flow domain, the vector $\{F\}^e$ is equal to zero. The matrix $[C]^e$ is referred to as the element seepage matrix and is evaluated using (2.4.3.9) as

$$N_I = (1/2A)(\alpha_I + \beta_I x_1 + \gamma_I x_2), \qquad I = 1, 2, 3,$$

$$\partial N_I/\partial x_1 = \beta_I/2A, \qquad \partial N_I/\partial x_2 = \gamma_I/2A,$$

$$C_{IJ}^e = \int_{R^e} T\left[\frac{\partial N_I}{\partial x_1}\frac{\partial N_J}{\partial x_1} + \frac{\partial N_I}{\partial x_2}\frac{\partial N_J}{\partial x_2}\right] dR$$

$$= \int_{R^e} T\left[\frac{(\beta_I\beta_J + \gamma_I\gamma_J)}{4A^2}\right] dR,$$

Performing the preceding integration, one obtains

$$C_{IJ}^e = \frac{T}{4A}(\beta_I\beta_J + \gamma_I\gamma_J),$$

or, in matrix form,

$$[C]^e = \frac{T}{4A}\begin{bmatrix} \beta_1\beta_1 & \beta_1\beta_2 & \beta_1\beta_3 \\ & \beta_2\beta_2 & \beta_2\beta_3 \\ \text{sym} & & \beta_3\beta_3 \end{bmatrix} + \frac{T}{4A}\begin{bmatrix} \gamma_1\gamma_1 & \gamma_1\gamma_2 & \gamma_1\gamma_3 \\ & \gamma_2\gamma_2 & \gamma_2\gamma_3 \\ \text{sym} & & \gamma_3\gamma_3 \end{bmatrix},$$

where sym means symmetric.

If it is assumed, as in Fig. 2.10, that nodes 1 and 2 lie on B_2^e, the F_I^e corresponds to F_1^e and F_2^e, and these terms can be evaluated as, for example,

$$F_1^e = \int_0^L qN_1\, dS. \tag{2.4.3.37}$$

Now along the local coordinate S, the linear functions q and N_1 are given by

$$q = q_1 + \frac{S}{L}(q_2 - q_1), \qquad N_1 = 1 - \frac{S}{L}, \tag{2.4.3.38}$$

which upon substitution into (2.4.3.37) yields

$$F_1^e = \int_0^L \left[q_1 + \frac{S}{L}\left(q_2 - q_1\right)\right]\left[1 - \frac{S}{L}\right] dS = \frac{L}{2}\left[q_1 + \frac{q_2 - q_1}{3}\right].$$

Similarly, it can be shown that

$$F_2^e = \frac{L}{2}\left[q_2 + \frac{q_1 - q_2}{3}\right].$$

Thus, the column vector $\{F\}^e$ in Eq. (2.4.3.36) is given by

$$\{F\}^e = \frac{L}{2}\begin{Bmatrix} (2q_1 + q_2)/3 \\ (q_1 + 2q_2)/3 \\ 0 \end{Bmatrix}. \tag{2.4.3.39}$$

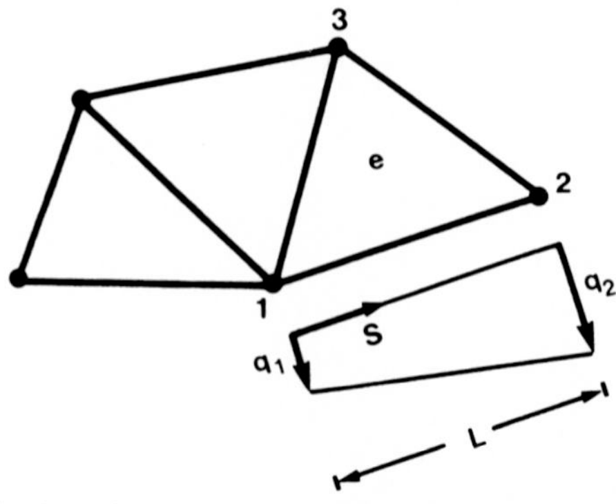

Fig. 2.10. Evaluation of the flux term across the exterior element boundary.

Galerkin Approach

Step 1 Let us first represent the trial function $\hat{p}$ in terms of the global basis functions and nodal values as

$$\hat{p} = N_J p_J, \qquad J = 1, 2, \ldots, n. \tag{2.4.3.40}$$

Thus we construct a weighted residual equation of the form

$$\int_R N_I \left[\frac{\partial}{\partial x_1}\left(T \frac{\partial \hat{p}}{\partial x_1}\right) + \frac{\partial}{\partial x_2}\left(T \frac{\partial \hat{p}}{\partial x_2}\right)\right] dR = 0, \tag{2.4.3.41a}$$

or

$$\int_R N_I \left[\frac{\partial}{\partial x_i}\left(T \frac{\partial \hat{p}}{\partial x_i}\right)\right] dR = 0, \qquad I = 1, 2, \ldots, n, \tag{2.4.3.41b}$$

where the repeated subscripts imply coordinate summation.

Step 2 Next, we apply Green's theorem to Eq. (2.4.3.41b) and obtain

$$\int_B N_I T \frac{\partial \hat{p}}{\partial x_i} n_i \, dB - \int_R T \frac{\partial N_I}{\partial x_i} \frac{\partial \hat{p}}{\partial x_i} dR = 0 \qquad \text{for} \quad I = 1, 2, \ldots, n, \tag{2.4.3.42}$$

where B is the entire boundary of the flow region. It is understood that for the nodes that are on boundary portion B_1, Eq. (2.4.3.42) is to be replaced by the appropriate Dirichlet (prescribed pressure) boundary conditions introduced by modifying the global matrix, as described in Section 2.3.1.

Without affecting the end result of the matrix formulation process, Eq. (2.4.3.42) can now be replaced by the following equation, which incorporates the prescribed flux condition on boundary portion B_2:

$$\int_R T \frac{\partial N_I}{\partial x_i} \frac{\partial N_J}{\partial x_i} p_J \, dR - \int_{B_2} N_I q \, dB = 0 \qquad \text{for} \quad I = 1, 2, \ldots, n. \tag{2.4.3.43}$$

Equation (2.4.3.43) can now be expressed as a sum of elemental contributions:

$$\sum_{e=1}^{m} \left[\int_{R^e} T \frac{\partial N_I}{\partial x_i} \frac{\partial N_J}{\partial x_i} p_J \, dR - \int_{B_2^e} N_I q \, dB \right] = 0 \qquad \text{for} \quad I = 1, 2, \ldots, n, \tag{2.4.3.44}$$

where N_I now represents the element basis functions.

From Eq. (2.4.3.44) we can extract the element equation of the form

$$Q_I^e = \int_{R^e} T \frac{\partial N_I}{\partial x_i} \frac{\partial N_J}{\partial x_i} p_J \, dR - \int_{B_2^e} N_I q \, dB \qquad \text{for} \quad I = 1, 2, 3. \qquad (2.4.3.45)$$

At this stage, it becomes apparent that the result of the Galerkin formulation is identical to that of the variational formulation; that is, Eq. (2.4.3.45) is fundamentally the same as Eq. (2.4.3.35) with Q_I^e corresponding to $\partial\Omega^e/\partial p_I$. Thus, the remaining steps of the Galerkin approach for this problem need no further elaboration.

2.5 Solution of Time Dependent Continuum Problems

2.5.1 Formulation of Finite Element Equations

The steady-state flow equations presented in the two previous examples are elliptic equations. When transient flow is considered, the governing equation becomes parabolic, and it is necessary to perform temporal as well as spatial discretizations. The approximation in the time domain can be obtained using either finite differences or Galerkin finite elements in time.

Consider the one-dimensional parabolic equation

$$\frac{\partial}{\partial x}\left(\kappa \frac{\partial p}{\partial x}\right) = \alpha \frac{\partial p}{\partial t}, \qquad 0 \le x \le 10, \qquad (2.5.1.1)$$

with the following initial and boundary conditions

$$p(x, 0) = p_0, \qquad (2.5.1.2a)$$

$$p(0, t) = 0, \qquad (2.5.1.2b)$$

$$p(10, t) = 0. \qquad (2.5.1.2c)$$

For this particular example, we only present the Galerkin finite element approach. Because p is a function of both space and time, we represent it by the trial function

$$\hat{p}(x, t) = N_J(x) p_J(t), \qquad J = 1, 2, \ldots, n. \qquad (2.5.1.3)$$

If we perform the steps described in the first steady state flow example, we obtain the following global equation:

$$\int_0^{10} \kappa \frac{dN_I}{dx} \frac{dN_J}{dx} \, dx + \int_0^{10} \alpha N_I N_J \frac{dp_J}{dt} \, dx = 0, \qquad (2.5.1.4a)$$

or

$$C_{IJ} p_J + M_{IJ} \frac{dp_J}{dt} = 0 \qquad \text{for} \quad I = 2, 3, \ldots, n - 1, \qquad (2.5.1.4b)$$

where

$$C_{IJ} = \int_0^{10} \kappa \frac{dN_I}{dx}\frac{dN_J}{dx}\,dx,$$

$$M_{IJ} = \int_0^{10} \alpha N_I N_J\,dx.$$

It should be noted that, except for the existence of the time derivative term, Eq. (2.5.1.4) is identical to (2.4.3.25).

Temporal Approximation via Finite Differences

Equation (2.5.1.4) represents a system of first-order differential equations. To solve this system numerically, we subdivide the period of analysis into a number of equal time increments as shown in Fig. 2.11a. To solve for the unknown nodal values at time level $k + 1$, we write the global equation (2.5.1.4b) at time level $k + \theta$ as

$$C_{IJ}p_J^{k+\theta} + M_{IJ}(dp_J/dt)^{k+\theta} = 0, \tag{2.5.1.5}$$

where θ is a time-weighting factor, $0 \leqslant \theta \leqslant 1$ and the superscript is used to denote the time level.

Next, we replace the time derivative by the finite difference approximation

$$(dp_J/dt)^{k+\theta} = (p_J^{k+1} - p_J^k)/\Delta t. \tag{2.5.1.6}$$

We then assume that at the old time level k the nodal values p_J^k are known and that over the time increment Δt defined by the old and the

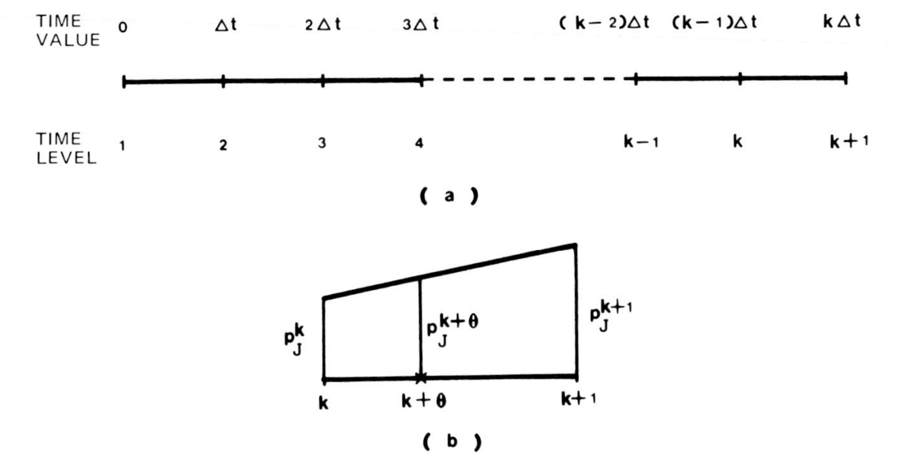

Fig. 2.11. Temporal approximation: (a) discretization, (b) evaluation of $p_J^{k+\theta}$.

current time levels, k and $k + 1$, the nodal values vary linearly as depicted in Fig. 2.11b. Thus, it follows that the nodal values at time level $k + \theta$ are given by

$$p_J^{k+\theta} = (1 - \theta)p_J^k + \theta p_J^{k+1}. \qquad (2.5.1.7)$$

Substituting Eqs. (2.5.1.6) and (2.5.1.7) into (2.5.1.5), we obtain

$$C_{IJ}[\theta p_J^{k+1} + (1 - \theta)p_J^k] + (M_{IJ}/\Delta t)(p_J^{k+1} - p_J^k) = 0. \qquad (2.5.1.8)$$

Equation (2.5.1.8) corresponds to a system of algebraic equations. From it, various time-stepping schemes can be obtained, depending on the chosen value of θ.

EXPLICIT SCHEME This scheme is obtained by substituting $\theta = 0$ into Eq. (2.5.1.8) to give

$$[M_{IJ}/\Delta t]p_J^{k+1} = [(M_{IJ}/\Delta t) - C_{IJ}]p_J^k. \qquad (2.5.1.9)$$

FULLY IMPLICIT SCHEME This scheme is obtained by substituting $\theta = 1$ into Eq. (2.5.1.8). After rearranging the terms so that the unknowns appear on the right-hand side and the knowns on the left-hand side, we obtain

$$[C_{IJ} + (M_{IJ}/\Delta t)]p_J^{k+1} = (M_{IJ}/\Delta t)p_J^k. \qquad (2.5.1.10)$$

CRANK–NICOLSON IMPLICIT SCHEME This scheme is obtained by substituting $\theta = \frac{1}{2}$ into Eq. (2.5.1.8) as

$$[C_{IJ}/2 + (M_{IJ}/\Delta t)]p_J^{k+1} = [(M_{IJ}/\Delta t) - (C_{IJ}/2)]p_J^k. \qquad (2.5.1.11)$$

Whichever time-stepping scheme is employed, the solution of the transient problem starts by using the initial conditions of the global equations to represent values at the old time level. Boundary conditions are then treated in the same manner as for the steady state problem. Once these steps are accomplished, recursive solutions for p_J^{k+1} can be obtained in terms of the known values p_J^k at the old time level.

Of the three time-stepping schemes, the explicit scheme may, in certain instances, require less computational effort, but, as will be shown later, this scheme is only "conditionally" stable. (In other words, unless the value of Δt is less than a certain magnitude, the solution obtained from the explicit scheme will have an uncontrollable exponential error growth.) On the other hand, both the fully implicit and the Crank–Nicolson schemes, are "unconditionally" stable.

Temporal Approximation via Weighted Residuals

An alternative way to obtain the temporal approximation is via the use of the weighted residual method. In this approach, we consider the particular time increment of time levels k and $k + 1$. Within this time increment $p_J(t)$, the nodal value at time t may be approximated by a trial function $\hat{p}_J(t)$ of the form

$$\hat{p}_J(t) = N^k(t)p_J^k + N^{k+1}(t)p_J^{k+1}, \qquad (2.5.1.12)$$

where $N^k(t)$ denotes the linear interpolation function associated with the kth time level. Using the local time coordinate ξ, defined in Fig. 2.12, Eq. (2.5.1.12) becomes

$$\hat{p}_J(\xi) = N^k(\xi)p_J^k + N^{k+1}(\xi)p_J^{k+1}, \qquad (2.5.1.13a)$$

where $N^k(\xi)$ and $N^{k+1}(\xi)$ are given by

$$N^k(\xi) = 1 - \xi, \qquad N^{k+1}(\xi) = \xi, \qquad (2.5.1.13b)$$

$$\frac{d}{d\xi}N^k(\xi) = -1, \qquad \frac{d}{d\xi}N^{k+1}(\xi) = 1. \qquad (2.5.1.13c)$$

Equation (2.5.1.4b) is now rewritten as

$$C_{IJ}p_J + M_{IJ}\,dp_J/dt = 0. \qquad (2.5.1.14)$$

To solve (2.5.1.14), we form the weighted residual integral of the form

$$\int_{t_k}^{t_{k+1}} w_j[C_{IJ}\hat{p}_J + M_{IJ}\,d\hat{p}_J/dt]\,dt = 0, \qquad (2.5.1.15)$$

where w_j is the weighting function.

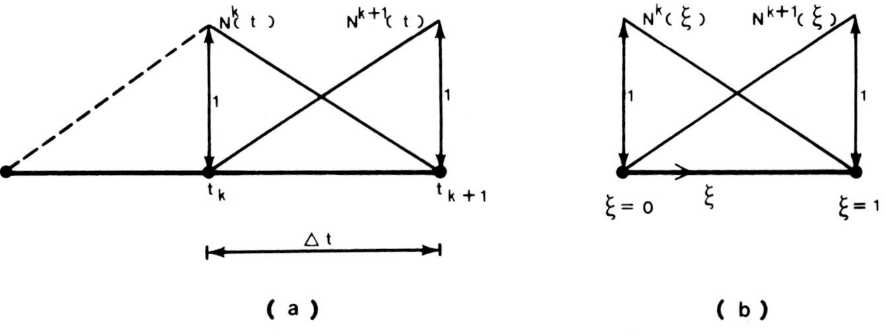

Fig. 2.12. Interpolation functions for time levels k and $k+1$: (a) global time coordinate; (b) local time coordinate: $N^k(\xi) = 1 - \xi$; $N^{k+1}(\xi) = \xi$; $\xi = 1 + (t - t_{k+1})/\Delta t$; $d\xi = dt/\Delta t$.

In terms of the local coordinate ξ, Eq. (2.5.1.15) becomes

$$\int_0^1 w_j \left[C_{IJ}\hat{p}_J + \frac{M_{IJ}}{\Delta t}\frac{d\hat{p}_J}{d\xi} \right] \Delta t \, d\xi = 0. \qquad (2.5.1.16)$$

Substitution of Eq. (2.5.1.13) into (2.5.1.16) yields

$$\int_0^1 w_j \left[C_{IJ}((1 - \xi)p_J^k + \xi p_J^{k+1}) + \frac{M_{IJ}}{\Delta t}(p_J^{k+1} - p_J^k) \right] d\xi = 0, \qquad (2.5.1.17)$$

which can be expressed in the same form as the weighted average finite difference equation (2.5.1.8). Thus, Eq. (2.5.1.17) becomes

$$C_{IJ}[\theta p_J^{k+1} + (1 - \theta)p_J^k] + \frac{M_{IJ}}{\Delta t}(p_J^{k+1} + p_J^k) = 0 \qquad (2.5.1.18a)$$

provided the weighting factor θ is defined as

$$\theta = \int_0^1 w_j \xi \, d\xi \Big/ \int_0^1 w_j \, d\xi. \qquad (2.5.1.18b)$$

As before, various time-stepping schemes can be obtained depending on the value of θ which, in turn, is governed by the choice of the weighting function w_j. Figure 2.13 shows typical choices of w_j and the resulting values of θ. The first three, (a)–(c), correspond to collocation schemes

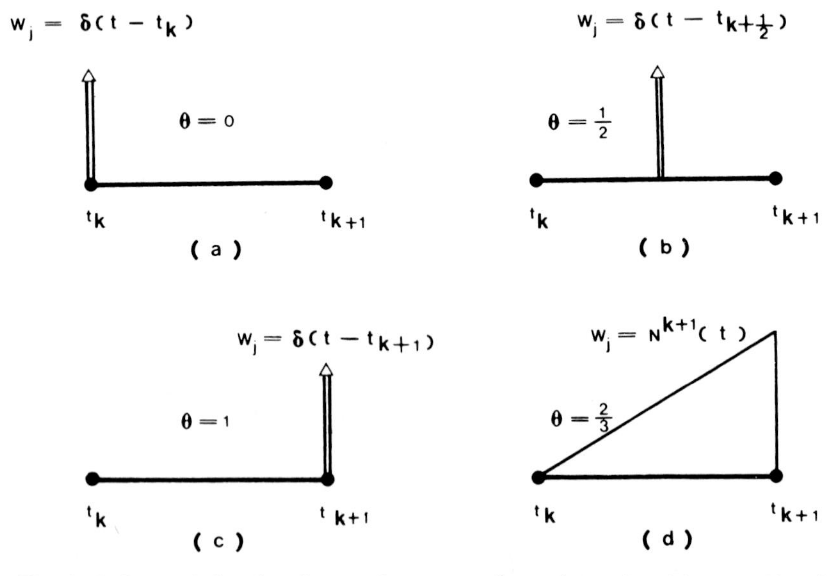

Fig. 2.13. Interpolation functions and corresponding values of θ: (a) $w_j = \delta(t - t_k)$; (b) $w_j = \delta(t - t_{k+1/2})$; (c) $w_j = \delta(t - t_{k+1})$; (d) $w_j = N^{k+1}(t)$.

with Dirac delta functions applied at levels k, $k + 1/2$, and $k + 1$, respectively, and these yield the explicit, the Crank–Nicolson, and fully implicit time-stepping formulas, respectively. The remaining weighting function is just the basis function at time level $k + 1$. Thus, this corresponds to the use of the Galerkin criterion and leads to the implicit time-stepping scheme with $\theta = 2/3$.

2.5.2 *Stability Analysis*

In solving the transient problem using time-stepping methods, it is important to ensure that the time-stepping process is computationally stable. Because computers can store only a finite number of digits to represent each real number, there exist round-off errors whose cumulative effect is the difference between the computed and the exact solutions of the finite element matrix equation. A computationally stable scheme must be able to control the growth of the cumulative round-off error. To develop a mathematical criterion for testing stability, it is convenient to express the previously developed time-stepping schemes in a general matrix form:

$$[A]\{p\}^{k+1} = [B]\{p\}^k, \tag{2.5.2.1}$$

where $\{p\}^k$ denotes the column vector whose elements are the nodal values p_J^k. Equation (2.5.2.1) can be written as

$$\{p\}^{k+1} = [G]\{p\}^k, \tag{2.5.2.2}$$

where $[G]$ is commonly referred to as the "amplification" matrix and $[G] = [A]^{-1}[B]$, provided $[A]$ is nonsingular. We now define the error vector at time level k as

$$\{E\}^k = \{P\}^k - \{p\}^k, \tag{2.5.2.3}$$

where $\{P\}^k$ denotes the exact solution at time level k. Since Eq. (2.5.2.2) is also satisfied by $\{P\}^k$, it follows that

$$\{P\}^{k+1} = [G\}\{P\}^k. \tag{2.5.2.4}$$

Subtraction of Eq. (2.5.2.2) from (2.5.2.4) and use of Eq. (2.5.2.3) leads to

$$\{E\}^{k+1} = [G]\{E\}^k. \tag{2.5.2.5}$$

Taking compatible matrix and vector norms and employing the Schwartz inequality, one obtains

$$\|E\|^{k+1} \leqslant \|G\| \cdot \|E\|^k. \tag{2.5.2.6}$$

It is apparent from Eq. (2.5.2.6) that the error will not grow with increasing value of k if and only if

$$\|G\| \leqslant 1. \tag{2.5.2.7}$$

Thus, the stability of the numerical solution can be assured if the norm of the amplification matrix is made smaller than unity. For a symmetric matrix $[G]$ the appropriate norm to use is the spectral norm $\|G\|_2$. Thus, the inequality (2.5.2.7) becomes

$$\|G\|_2 \leqslant 1. \tag{2.5.2.8}$$

Since $\|G\|_2$ is defined as

$$\|G\|_2 = \max_I |\lambda_I|, \tag{2.5.2.9}$$

where λ_I denotes eigenvalues of $[G]$, the stability criterion becomes

$$\max_I |\lambda_I| \leqslant 1. \tag{2.5.2.10}$$

EXPLICIT SCHEME For the explicit scheme, i.e., (2.5.1.9),

$$[G] = [M]^{-1}([M] - \Delta t\,[C]) = [I] - \Delta t\,[M]^{-1}[C], \tag{2.5.2.11a}$$

where $[I]$ is the unit matrix. Equation (2.5.2.11a) can be written as

$$[I] = [G] + \Delta t\,[M]^{-1}[C], \tag{2.5.2.11b}$$

Taking matrix norms on both sides, we obtain

$$1 \leqslant \|G\|_2 + \Delta t\,\|M^{-1}\|_2 \cdot \|C\|_2,$$

or

$$\|G\|_2 \geqslant \|I\|_2 - \Delta t\,\|M^{-1}\|_2 \cdot \|C\|_2. \tag{2.5.2.11c}$$

For stability, we must have

$$\|G\|_2 \leqslant 1,$$

and hence

$$1 - \Delta t\,\|M^{-1}\|_2 \cdot \|C\|_2 \leqslant 1. \tag{2.5.2.12}$$

It can be seen that there exists a bound for Δt outside of which the inequality in (2.5.2.12) is not valid. For this reason, the explicit scheme is said to be "conditionally stable."

FULLY IMPLICIT SCHEME For the fully implicit scheme, i.e., (2.5.1.10),

$$[G] = [C_{IJ} + (M_{IJ}/\Delta t)]^{-1}[M_{IJ}/\Delta t]. \tag{2.5.2.13}$$

It can be shown that for all values of Δt, we have (see Chung, 1979)

$$\|G\|_2 \leq \|I\|_2 = 1. \tag{2.5.2.14}$$

Crank–Nicolson Scheme For the Crank–Nicolson scheme, i.e., (2.5.1.11), one obtains

$$[G] = [C_{IJ}/2 + (M_{IJ}/\Delta t)]^{-1}[(M_{IJ}/\Delta t) - (C_{IJ}/2)]. \tag{2.5.2.15}$$

Once again, it can be shown that

$$\|G\|_2 \leq 1$$

for any value of Δt.

One-Dimensional Example For illustrative purposes, we consider once again the one-dimensional transient flow example described in Section 2.5.1. To simplify the global matrix computation, we divide the solution domain into only three elements; each has nodal spacing of Δx. We also assume that the coefficients κ and α of the governing equation are constant.

The local element matrices $[C]^e$ and $[M]^e$ can be obtained simply as

$$[C]^e = \kappa \int_0^{\Delta x} \begin{bmatrix} \dfrac{dN_1}{dx}\dfrac{dN_1}{dx} & \dfrac{dN_1}{dx}\dfrac{dN_2}{dx} \\ \\ \dfrac{dN_2}{dx}\dfrac{dN_1}{dx} & \dfrac{dN_2}{dx}\dfrac{dN_2}{dx} \end{bmatrix} dx = \frac{\kappa}{\Delta x}\begin{bmatrix} 1 & -1 \\ -1 & 1 \end{bmatrix},$$

$$[M]^e = \alpha \int_0^{\Delta x} \begin{bmatrix} N_1N_1 & N_1N_2 \\ N_2N_1 & N_2N_2 \end{bmatrix} dx = \frac{\Delta x}{6}\begin{bmatrix} 2 & 1 \\ 1 & 2 \end{bmatrix}.$$

By assembling the three elements and incorporating the Dirichlet boundary conditions $p_1 = p_4 = 0$ at the two end nodes, the resulting global matrix equation can be reduced to

$$\frac{\kappa}{\Delta x}\begin{bmatrix} 2 & -1 \\ -1 & 2 \end{bmatrix}\begin{Bmatrix} p_2 \\ p_3 \end{Bmatrix} + \frac{\alpha\,\Delta x}{6}\begin{bmatrix} 4 & 1 \\ 1 & 4 \end{bmatrix}\begin{Bmatrix} dp_2/dt \\ dp_3/dt \end{Bmatrix} = \begin{Bmatrix} 0 \\ 0 \end{Bmatrix},$$

from which we obtain

$$[C] = \frac{\kappa}{\Delta x}\begin{bmatrix} 2 & -1 \\ -1 & 2 \end{bmatrix}, \qquad [M] = \frac{\alpha\,\Delta x}{6}\begin{bmatrix} 4 & 1 \\ 1 & 4 \end{bmatrix}.$$

To obtain the stability criterion for the explicit scheme, we compute $[G]$ from

$$[G] = [I] - \Delta t\,[M]^{-1}[C].$$

Thus,

$$[G] = \begin{bmatrix} g_{11} & g_{12} \\ g_{21} & g_{22} \end{bmatrix},$$

where

$$g_{11} = g_{22} = \frac{18\kappa\,\Delta t}{5(\Delta x)^2}, \qquad g_{12} = g_{21} = -\frac{12\kappa\,\Delta t}{5\alpha(\Delta x)^2}.$$

The eigenvalues of $[G]$ obtained are

$$\lambda_1 = \frac{6\kappa\,\Delta t}{\alpha(\Delta x)^2} - 1, \qquad \lambda_2 = \frac{6\kappa\,\Delta t}{5\alpha(\Delta x)^2} - 1.$$

Hence,

$$\max_J |\lambda_J| = |\lambda_1| \leqslant 1, \qquad J = 1, 2,$$

which leads to the following stability condition:

$$-1 \leqslant \frac{6\kappa\,\Delta t}{\alpha(\Delta x)^2} - 1 \leqslant 1,$$

or

$$\Delta t \leqslant \alpha(\Delta x)^2/3\kappa.$$

In this simple example, we are able to obtain an analytical representation for the stability constraint. For more complicated problems, however, it is not generally possible to obtain the eigenvalues explicitly in terms of Δx and Δt. The eigenvalues must be computed with the overall mesh parameters embedded into matrices $[C]$ and $[M]$.

In this chapter, we have developed the fundamental theory and application of the finite element method. For additional reading, the reader can refer to Cook (1974), Hinton and Owen (1979), Heubner (1975), Norrie and De Vries (1978), and Segerlind (1976).

References

Bathe, K. J., and Wilson, E. L. (1976). "Numerical Methods in Finite Element Analysis." Prentice-Hall, Englewood Cliffs, New Jersey.

Chung, T. J. (1979). "Finite Element Analysis in Fluid Dynamics." McGraw-Hill, New York.

Clough, R. W. (1960). The finite element method in plane stress analysis. *ASCE J. Struc. Div. Proc. 2nd Conf. Electronic Computation,* 345–378.

Cook, R. D. (1974). "Concepts and Applications of Finite Element Analysis." Wiley, New York.

Courant, R. (1943). Variational methods for the solution of problems of equilibrium and vibration. *Bull. Amer. Math. Soc.* **49,** 1–43.
Finlayson, B. A. (1972). "The Method of Weighted Residuals and Variational Principles with Application in Fluid Mechanics, Heat and Mass Transfer." Academic Press, New York.
Huebner, K. H. (1975). "Finite Element Method for Engineers." Wiley, New York.
Hinton, E., and Owen, D. R. J. (1979). "An Introduction to the Finite Element Computations," chapters 1–4. Pineridge Press, Swansea, United Kingdom.
Norrie, D. H., and De Vries, G. (1978). "An Introduction to Finite Element Methods." Academic Press, New York.
Segerlind, L. J. (1976). "Applied Finite Element Analysis." Wiley, New York.
Turner, M., Clough, R., Martin, H., and Topp, L. (1956). Stiffness and deflection analysis of complex structures. *J. Aeronaut. Sci.* **23,** 805–823.
Weinstock, R. (1952). "Calculus of variations with applications to physics and engineering." McGraw-Hill, New York.

3

Element Families and Interpolation Functions

3.1 General

In Chapter 2, we presented the derivation of the finite element equations for simple continuum problems involving one unknown function. It is apparent that after the unknown function has been approximated within each element in terms of appropriate basis functions and nodal parameters, the derivation of the element matrix equation follows a standard procedure. Although we elected to use the simplest one-dimensional and triangular elements in the examples heretofore given, it should be obvious by now that other higher-order functions could be chosen. A question then arises as to which type of element is most appropriate for a particular problem. Unfortunately, there is no clear-cut answer to this question. The optimal element varies generally from problem to problem. The selection of particular elements is thus very much dependent on the experience and judgment of the analyst. In making the selection, consideration should be given to factors such as the geometry of the global domain, the degree of accuracy required for the solution, and the complexity and cost of element matrix computation.

In this chapter, we present a systematic approach to the construction of shape functions for various families of one-, two-, and three-dimensional elements. Because the prime motivation of the finite element approach is to represent accurately a solution domain of arbitrary shape through an assemblage of simple element shapes, the elements commonly employed are geometrically simple. Typical one-, two-, and three-dimensional elements are shown in Fig. 3.1. A one-dimensional element is just a line segment. A two-dimensional element may be a triangle or a quadrilateral, and a three-dimensional element may be a tetrahedron, a hexahedron, or a triangular prism. In the special case where the global domain exhibits axial symmetry, we can employ a three-dimensional element generated by revolving a plane triangular or quadrilateral element around the axis of symmetry. This will be demonstrated in Chapter 4.

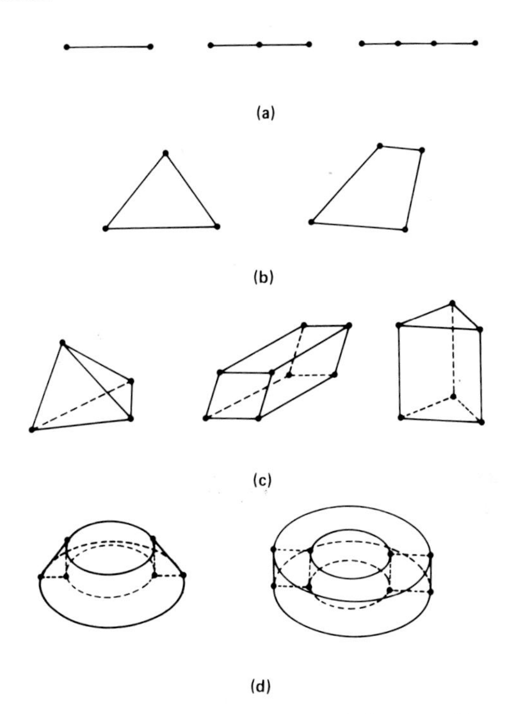

Fig. 3.1. Typical finite elements: (a) one-dimensional elements; (b) two-dimensional elements; (c) three-dimensional elements; (d) three-dimensional ring elements.

The number of nodes in a particular element depends upon the type of nodal parameters, the degree of the interpolation function, and the continuity required by the method of approximation. Generally, the interpolation functions employed are polynomials of various degrees. Although not frequently used, other types of functions, e.g., trigonometric functions, could also serve as interpolation functions. Polynomials are popular because, among other things, they are simple functions to manipulate mathematically.

3.2 Polynomial Series

3.2.1 Construction of Complete Series

In general, the distribution of an unknown variable within a finite element can be represented by a polynomial whose coefficients are "generalized parameters" directly related to nodal coordinates and nodal values of the element. Once this polynomial is obtained, the expressions for the element shape functions can be easily derived.

ONE-DIMENSIONAL CASE In one dimension, a complete mth-order polynomial may be written as

$$P_m(x) = \sum_{i=0}^{m} a_i x^i, \tag{3.2.1.1}$$

where the number of terms in the series equals $m + 1$.

For $m = 1$, $P_1(x) = a_0 + a_1 x$; for $m = 2$, $P_2(x) = a_0 + a_1 x + a_2 x^2$, etc.

TWO-DIMENSIONAL CASE In two dimensions, a complete mth-order polynomial may be written as

$$P_m(x, y) = \sum_{k=1}^{n} a_k x^i y^j, \qquad i + j \leq m, \tag{3.2.1.2}$$

where n is the number of terms in the series and i and j are nonnegative integer exponents of x and y, respectively. The total number of terms of the polynomial is given by

$$n = \tfrac{1}{2}(m + 1)(m + 2).$$

The values of i and j are related to the value of k as follows:

$$k = \tfrac{1}{2}(i + j)(i + j + 1) + j + 1.$$

For $m = 1$, $P_1(x, y) = a_1 + a_2 x + a_3 y$; for $m = 2$, $P_2(x, y) = a_1 + a_2 x + a_3 y + a_4 x^2 + a_5 xy + a_6 y^2$, etc.

A convenient way to illustrate a complete two-dimensional polynomial is by means of the so-called "Pascal triangle" scheme wherein the terms in the series are placed in a triangular array in ascending order, i.e.,

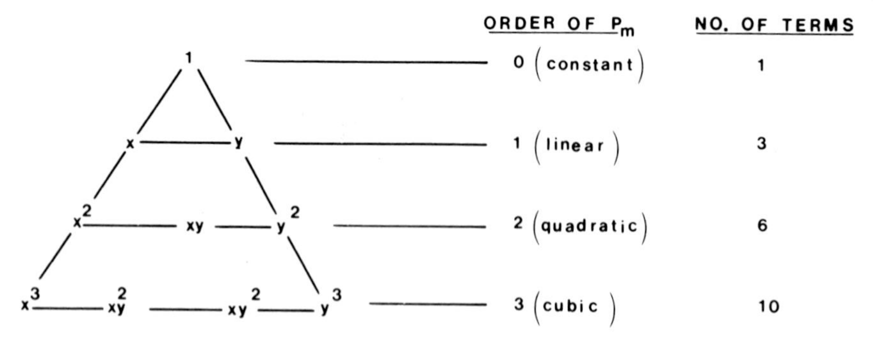

THREE-DIMENSIONAL CASE In three dimensions, a complete mth-order polynomial takes the form

$$P_m(x, y, z) = \sum_{l=1}^{n} a_l x^i y^j z^k; \qquad i + j + k \leq m, \tag{3.2.1.3}$$

where the total number of terms n is given by

$$n = \tfrac{1}{6}(m + 1)(m + 2)(m + 3).$$

For $m = 1$,

$$P_1(x, y, z) = a_1 + a_2x + a_3y + a_4z;$$

for $m = 2$,

$$P_2(x, y, z) = a_1 + a_2x + a_3y + a_4z + a_5x^2 + a_6xy + a_7xz + a_8y^2 + a_9yz + a_{10}z^2.$$

In a manner similar to the two-dimensional case, we can illustrate a complete three-dimensional polynomial by placing the terms at different planar levels of a tetrahedron as illustrated.

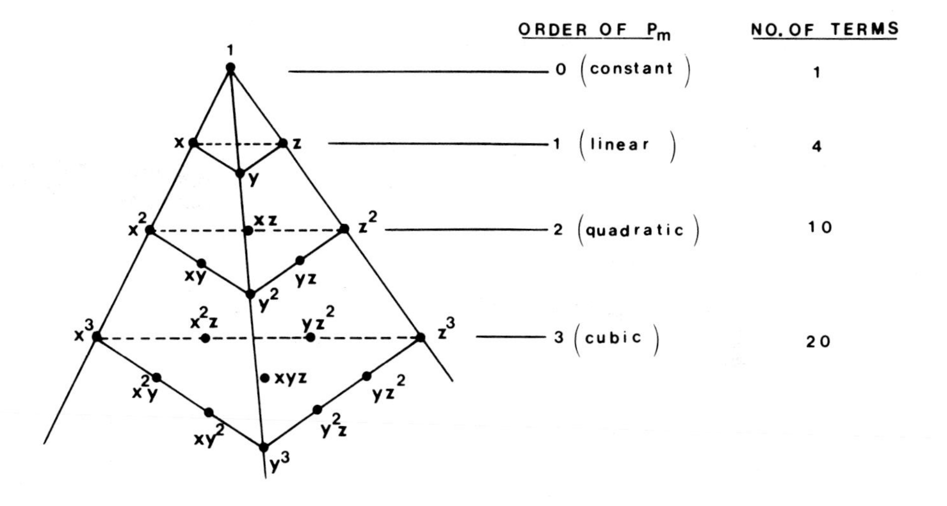

3.2.2 *Truncation of Polynomial Series*

Frequently we choose to use a "truncated" polynomial expansion to represent the distribution of the unknown variable within a finite element. The question then arises as to which terms of the original polynomial series should be omitted. To answer this question we need to reexamine the compatibility and completeness requirements discussed in Chapter 2. Such requirements are essential to ensure continuity of the unknown variable and convergence to the correct solution as the mesh is refined. The completeness requirement presents no problem for elements with straight edges, because it is readily met by choosing a series that includes all constant and linear terms. The compatibility requirement is met in the case of a first- or second-order operator if the unknown function is continuous at the element interfaces. (This type of continuity is commonly referred to as C^0 continuity.)

In addition to satisfying the two essential requirements, the representation of the unknown variable within an element must be invariant with respect to any transformation from one Cartesian system to another. Polynomials that display such an invariance property are said to possess "geometric isotropy." It can be shown that any complete polynomial of degree m has geometric isotropy. There is a simple guideline that permits us to choose the appropriate polynomial terms such that the resulting truncated series possesses the desired geometric isotropy (Heubner, 1975). According to this guideline, the truncated terms in the polynomial series should occur in "symmetric" pairs. For example, suppose we wish to construct a cubic polynomial for a two-dimensional element that has eight nodal variables assigned to it. In this case, a complete cubic polynomial contains 10 terms,

$$P(x, y) = a_1 + a_2x + a_3y + a_4x^2 + a_5xy + a_6y^2 + a_7x^3 + a_8x^2y + a_9xy^2 + a_{10}y^3.$$

We may drop only terms that occur in symmetric pairs. These are $(a_7x^3, a_{10}y^3)$ and (a_8x^2y, a_9xy^2). Thus the resulting 8-term cubic polynomial having geometric isotropy would be

$$P(x, y) = a_1 + a_2x + a_3y + a_4x^2 + a_5xy + a_6y^2 + a_8x^2y + a_9xy^2,$$

or

$$P(x, y) = a_1 + a_2x + a_3y + a_4x^2 + a_5xy + a_6y^2 + a_7x^3 + a_{10}y^3.$$

We can readily use this idea to construct other truncated (or incomplete) polynomial expansions. This may be done conveniently via the use of the Pascal triangle and the tetrahedron array described previously.

3.2.3 Direct Method for Deriving Shape Functions

We have shown how an unknown variable can be represented within a finite element by a polynomial series whose coefficients are generalized parameters. The number of such parameters is often chosen to be equal to the total number of degrees of freedom associated with the particular element. The evaluation of the generalized parameters in terms of the nodal values and coordinates is accomplished by evaluating the polynomial series at each nodal degree of freedom. The general procedure may be described as follows:

First we represent the distribution of the variable u by a trial function of the form

$$\hat{u} = \{P\}^T\{a\}. \tag{3.2.3.1}$$

For a two-dimensional case involving a second-degree polynomial, $\{P\}^T$ and $\{a\}$ are given by

$$\{a\} = \begin{Bmatrix} a_1 \\ \vdots \\ a_6 \end{Bmatrix}, \tag{3.2.3.2}$$

$$\{P\}^T = \{1, x, y, x^2, xy, y^2\}. \tag{3.2.3.3}$$

Next, we evaluate the trial function $\hat{u}$ at each nodal degree of freedom and obtain

$$\{u\} = [G]\{a\}, \tag{3.2.3.4}$$

where

$$[G] = \begin{bmatrix} 1 & x_1 & y_1 & x_1^2 & x_1y_1 & y_1^2 \\ 1 & x_2 & y_2 & x_2^2 & x_2y_2 & y_2^2 \\ 1 & x_3 & y_3 & x_3^2 & x_3y_3 & y_3^2 \\ 1 & x_4 & y_4 & x_4^2 & x_4y_4 & y_4^2 \\ 1 & x_5 & y_5 & x_5^2 & x_5y_5 & y_5^2 \\ 1 & x_6 & y_6 & x_6^2 & x_6y_6 & y_6^2 \end{bmatrix}. \tag{3.2.3.5}$$

From Eq. (3.2.3.4) we obtain

$$\{a\} = [G]^{-1}\{u\}. \tag{3.2.3.6}$$

Substitution of Eq. (3.2.3.6) into (3.2.3.1) yields

$$\hat{u} = \{P\}^T[G]^{-1}\{u\} = \{N\}^T\{u\}, \tag{3.2.3.7}$$

in which $\{N\}^T$ is a row vector of shape functions. It follows that

$$\{N\}^T = \{P\}^T[G]^{-1}. \tag{3.2.3.8}$$

This procedure is straightforward and simple to apply. However, it should be noted that in a few cases, certain element geometries can lead to a singular matrix $[G]$. These cases occur when the nodal arrangement is such that one degree of freedom is dependent on another or a combination of the terms in the polynomial series yields a shape function that gives zero values at all nodal points. Such a "null" shape function represents a degeneracy of rank one of the $[G]$ matrix. Another disadvantage of the direct method just described is the effort required to compute $[G]^{-1}$ when it exists. For a large number of elements with many degrees of freedoms, the computational cost can be high. In view of this disadvantage, it is preferable to obtain the shape functions by other means. One way to accomplish this is through the use of interpolation procedures that rely on the concept of natural coordinate systems. In the next section, we discuss the use of such procedures to obtain the shape functions for different types of elements.

3.3 One-Dimensional Elements

3.3.1 Lagrange Elements

We begin with the use of Lagrange interpolation to obtain one-dimensional shape functions. Consider the typical one-dimensional element consisting of $m + 1$ nodal points as shown in Fig. 3.2a. The locations of these nodal points are specified by their coordinate values x_1, x_2, ..., x_{m+1}. We wish to approximate the unknown function u by a trial function written in the form

$$\hat{u}(x) = \sum_{I=1}^{m+1} N_I(x) u_I, \tag{3.3.1.1}$$

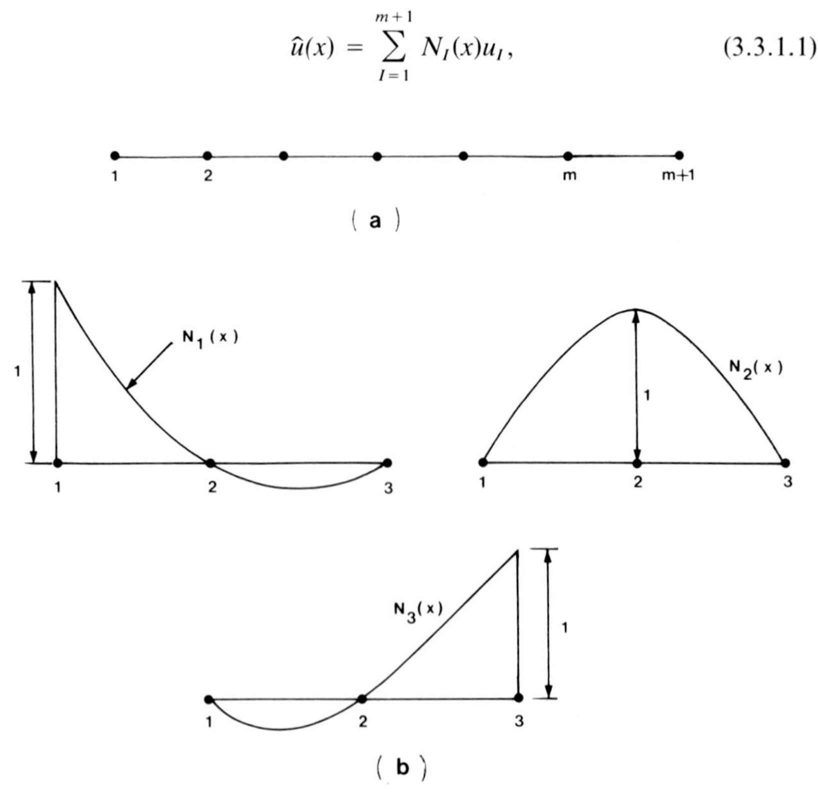

Fig. 3.2. Interpolation using one-dimensional Lagrange polynomials.

where N_I are the shape functions having the following property:

$$N_I = \begin{cases} 1 & \text{at} \quad x = x_I \\ 0 & \text{at} \quad x = x_J, \quad J \neq I. \end{cases}$$

Such shape functions are indeed Lagrange polynomials of degree m. Thus, they can be obtained using Lagrange's formula for interpolation as follows:

$$N_I = \prod_{\substack{J=1 \\ J \neq I}}^{m+1} \left[\frac{x - x_J}{x_I - x_J} \right]$$
$$= \frac{(x - x_1)(x - x_2) \cdots (x - x_{I-1})(x - x_{I+1}) \cdots (x - x_{m+1})}{(x_I - x_1)(x_I - x_2) \cdots (x_I - x_{I-1})(x_I - x_{I+1}) \cdots (x_I - x_{m+1})}. \quad (3.3.1.2)$$

As an example, we consider a three-node element shown in Fig. 3.2b. In this case, $m = 2$ and N_1, N_2, and N_3 are given by

$$N_1 = \frac{(x - x_2)(x - x_3)}{(x_1 - x_2)(x_1 - x_3)},$$

$$N_2 = \frac{(x - x_1)(x - x_3)}{(x_2 - x_1)(x_2 - x_3)}, \qquad N_3 = \frac{(x - x_1)(x - x_2)}{(x_3 - x_2)(x_3 - x_1)}.$$

It can be easily shown that Lagrange interpolation functions satisfy the requirement of continuity of the unknown function across element interfaces. As mentioned earlier, such a continuity requirement is referred to as the "C^0 continuity" requirement.

The above Lagrange formula has been written in terms of global coordinates. Another way of writing this formula is by means of "natural coordinates." Such coordinates represent a local coordinate system that is related to the global system in such a way that the natural coordinates take on values of ± 1 or zero at the two end points of the line element. Consider the line segment shown in Fig. 3.3. Let the distances from point P to point 1 and $m + 1$ be denoted by L_2 and L_1, respectively.

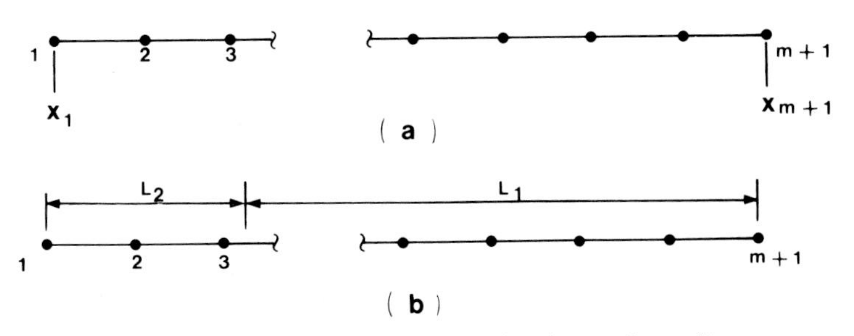

Fig. 3.3. Representation of a line segment in global and natural coordinate systems: (a) global coordinate representation; (b) natural coordinate representation.

Further, let us nondimensionalize L_1 and L_2 so that

$$L_1 + L_2 = 1. \tag{3.3.1.3}$$

It is apparent that if we move point P to the left until it corresponds to point 1, $L_1 = 1$ and $L_2 = 0$. Similarly, if we move point P to the right until it corresponds to point $m + 1$, $L_1 = 0$ and $L_2 = 1$. By inspection, we can write the function that relates L_1, L_2 to the global coordinate x in the form

$$x = L_1 x_1 + L_2 x_{m+1}. \tag{3.3.1.4}$$

Collecting Eqs. (3.3.1.3) and (3.3.1.4), we obtain

$$\begin{Bmatrix} 1 \\ x \end{Bmatrix} = \begin{bmatrix} 1 & 1 \\ x_1 & x_{m+1} \end{bmatrix} \begin{Bmatrix} L_1 \\ L_2 \end{Bmatrix}. \tag{3.3.1.5}$$

Inversion of the matrix in Eq. (3.3.1.5) gives

$$\begin{Bmatrix} L_1 \\ L_2 \end{Bmatrix} = \frac{1}{x_{m+1} - x_1} \begin{bmatrix} x_{m+1} & -1 \\ -x_1 & 1 \end{bmatrix} \begin{Bmatrix} 1 \\ x \end{Bmatrix}. \tag{3.3.1.6}$$

Thus, using Eq. (3.3.1.5) we can transform the original element subregion into the corresponding subregion in the natural coordinate system as shown in Fig. 3.3b.

It can be shown that in terms of the natural coordinates the Lagrange shape function for node I takes the form

$$N_I(L_1, L_2) = \Psi_p(L_1)\Phi_q(L_2), \tag{3.3.1.7}$$

where $\Psi_p(L_1)$ and $\Phi_q(L_2)$ are functions of L_1 and L_2, respectively. These functions are given by

$$\Psi_p(L_1) = \begin{cases} \displaystyle\prod_{k=1}^{p} \left(\frac{mL_1 - k + 1}{k}\right) & \text{for} \quad k \geqslant 1 \\ 1 & \text{for} \quad k = 0, \end{cases} \tag{3.3.1.8a}$$

$$\Phi_q(L_2) = \begin{cases} \displaystyle\prod_{k=1}^{q} \left(\frac{mL_2 - k + 1}{k}\right) & \text{for} \quad k \geqslant 1 \\ 1 & \text{for} \quad k = 0, \end{cases} \tag{3.3.1.8b}$$

where p denotes the number of nodes that lie on the right of node I and q denotes the number of nodes that lie on the left of node I.

As an example, we now show how the quadratic shape functions ($m = 2$) for the typical three-node element are obtained. First, we write

N_1, N_2, and N_3 in the form

$$N_1 = \Psi_2(L_1)\Phi_0(L_2), \tag{3.3.1.9a}$$

$$N_2 = \Psi_1(L_1)\Phi_1(L_2), \tag{3.3.1.9b}$$

$$N_3 = \Psi_0(L_1)\Phi_2(L_2). \tag{3.3.1.9c}$$

Now, from Eq. (3.3.1.8), $\Psi_0(L_1) = 1$, $\Psi_1(L_1) = 2L_1$, $\Psi_2(L_1) = L_1(2L_1 - 1)$, $\Phi_0(L_2) = 1$; the values of Ψ and Φ for the nodes from left to right are $\Phi_1(L_2) = 2L_2$ and $\Phi_2(L_2) = L_2(2L_2 - 1)$. Substitution of these into Eqs. (3.3.1.9a)–(3.3.1.9c) yields

$$N_1 = L_1(2L_1 - 1), \tag{3.3.1.10a}$$

$$N_2 = 4L_1L_2, \tag{3.3.1.10b}$$

$$N_3 = L_2(2L_2 - 1). \tag{3.3.1.10c}$$

The main advantage in using the natural coordinate instead of the x coordinate is the ease of integration in the natural coordinate system.

When the shape functions are expressed in terms of L_1 and L_2, the computation of element matrices involves evaluation of the integral terms that can be written in the form

$$\int_0^l L_1^a L_2^b \, dx,$$

where l is the total length of the finite element and a and b are integer powers. This integral can be transformed as

$$\int_0^l L_1^a L_2^b \, dx = l \int_0^1 L_1^a (1 - L_1)^b \, dL_1. \tag{3.3.1.11}$$

The integral on the right side of Eq. (3.3.1.11) is the standard beta function. It is given by

$$\int_0^1 L_1^a (1 - L_1)^b \, dL_1 = \frac{a!b!}{(a + b + 1)!}. \tag{3.3.1.12a}$$

Substituting Eq. (3.3.1.11) into (3.3.1.12a), we obtain

$$\int_0^l L_1^a L_2^b \, dx = \frac{la!b!}{(a + b + 1)!}, \tag{3.3.1.12b}$$

which is readily evaluated. We shall see in subsequent sections how the natural coordinate system can be readily extended to two and three dimensions.

3.3.2 Hermite Elements

Although problems in subsurface flow generally involve the solution for an unknown function u, there may be occasions when it is desirable to introduce the first derivative of u as an additional nodal parameter. If u and its first derivatives are to be specified at the nodes, the shape functions must then be chosen such that continuity of the first derivative of u exists at the element interfaces. (This type of continuity is commonly referred to as C^1 continuity). In this section, we demonstrate how the desired shape functions can be constructed using Hermite polynomials. In general, an nth-order Hermite polynomial in x is a polynomial of degree $2n + 1$ and may be denoted $H^n(x)$. Thus, a first-order Hermite is cubic in x. Hermite polynomials are useful as interpolation functions because their values and the values of their derivatives up to order n are either unity or zero at the end points of the closed interval $[x_1, x_2]$. This property can be represented symbolically if we assign two subscripts and write $H^n_{mI}(x)$, where m denotes the order of the derivative and I refers to either node 1 or node 2 of a line element. The symbolic representation is thus given by

$$\frac{d^k H^n_{mI}(x_J)}{dx^k} = \begin{cases} \delta_{IJ} & \text{for} \quad k = m, \quad m = 0, 1, \ldots, n \\ 0 & \text{for} \quad k \neq m, \quad m = 0, 1, \ldots, n. \end{cases} \tag{3.3.2.1}$$

To derive the expression for the first-order Hermite polynomial H^1_{mI}, we consider the typical element with two degrees of freedom each at nodes 1 and 2 as shown in Fig. 3.4. Let us define the local coordinate ξ as

$$\xi = 2[x - \tfrac{1}{2}(x_1 + x_2)]/L_e; \qquad L_e = x_2 - x_1. \tag{3.3.2.2}$$

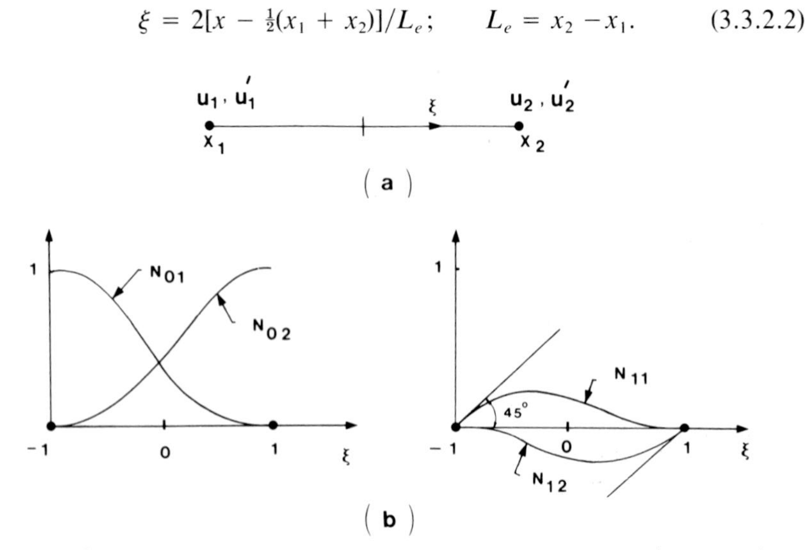

Fig. 3.4. Cubic Hermite element and its shape functions; $\xi = 2\,[x - \tfrac{1}{2}(x_1 + x_2)]/L_e$.

We seek to construct the trial function that takes the form

$$\hat{u} = H^1_{01}u_1 + H^1_{02}u_2 + H^1_{11}u'_1 + H^1_{12}u'_2$$
$$= H^1_{0I}u_I + H^1_{1I}u'_I = N_{0I}u_I + N_{1I}u'_I, \qquad (3.3.2.3)$$

where $H^1_{0I} = N_{0I}$ and $H^1_{1I} = N_{1I}$, u_I and u'_I denote, respectively, the values of u and du/dx at node I, N_{0I} and N_{1I} are shape functions associated with u_I and u'_I, and the summation convention is used on repeated subscripts. To derive the expression for the function N_{01}, we first write it in the form

$$N_{01} = a_1 + a_2\xi + a_3\xi^2 + a_4\xi^3. \qquad (3.3.2.4)$$

To determine a_1, a_2, a_3, and a_4, we use these four conditions: $N_{01} = 1$ at node 1; $N_{01} = 0$ at node 2, $dN_{01}/dx = 0$ at node 1; and $dN_{01}/dx = 0$ at node 2. Thus,

$$1 = a_1 - a_2 + a_3 - a_4, \qquad N_{01} = 1, \quad \xi = -1,$$
$$0 = a_1 + a_2 + a_3 + a_4, \qquad N_{01} = 0, \quad \xi = 1,$$
$$0 = (2a_2 - 4a_3 + 6a_4)/L_e, \qquad dN_{01}/dx = 0, \quad \xi = -1,$$
$$0 = (2a_2 + 4a_3 + 6a_4)/L_e, \qquad dN_{01}/dx = 0, \quad \xi = 1.$$

The solution of these equations gives

$$a_1 = \tfrac{1}{2}, \qquad a_2 = -\tfrac{3}{4}, \qquad a_3 = 0, \qquad a_4 = \tfrac{1}{4}.$$

Thus, N_{01} is given by

$$N_{01} = \tfrac{1}{4}(2 - 3\xi + \xi^3) = \tfrac{1}{4}(\xi - 1)^2(\xi + 2). \qquad (3.3.2.5a)$$

In a similar manner, we obtain the remaining shape functions:

$$N_{02} = -\tfrac{1}{4}(\xi + 1)^2(\xi - 2), \qquad (3.3.2.5b)$$
$$N_{11} = \tfrac{1}{8}(\xi + 1)(\xi - 1)^2L_e, \qquad (3.3.2.5c)$$
$$N_{12} = \tfrac{1}{8}(\xi + 1)^2(\xi - 1)L_e. \qquad (3.3.2.5d)$$

The above interpolation functions are plotted in Fig. 3.4.

3.4 Two-Dimensional Elements

In this section, we present families of triangular and quadrilateral elements that can be applied to problems requiring C^0 continuity of the function u. For such problems we usually choose the nodal values of the unknown function to be the degrees of freedom (or unknowns) of the element. To ensure interelement continuity, we require that the number of nodes (and hence the number of nodal values) along a side of the element be just sufficient to determine uniquely the variation of u along that side. For example, if u is assigned to have a quadratic representation, then three values of u must be specified along each element side.

3.4.1 Triangular Elements

Basis functions for the family of triangular elements can be simply derived using natural (area) coordinates. Consider the triangular element illustrated in Fig. 3.5a. Each side of this element is identified by the

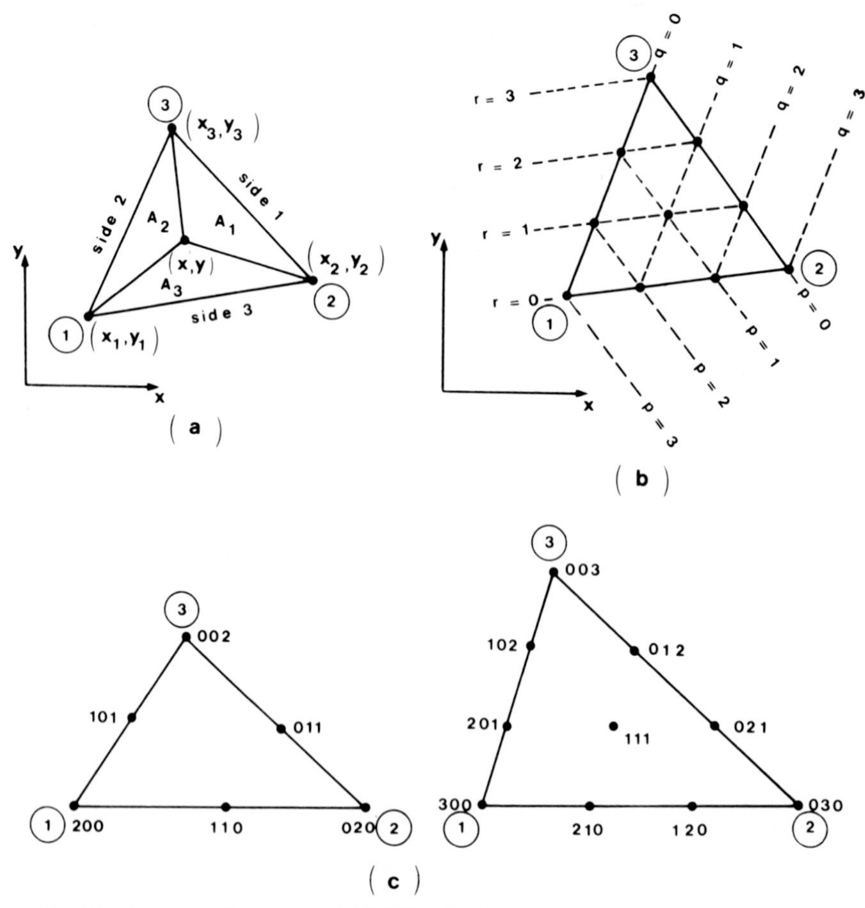

Fig. 3.5. Area coordinates and node designation for triangles: (a) area coordinate system; (b) coordinate designation; (c) node designations for quadratic and cubic triangles.

opposite vertex. Thus, side 1 is opposite to vertex 1, etc. We now identify a point within the triangle by specifying its local coordinates (L_1, L_2, L_3). These coordinates are defined as follows:

$$L_1 = A_1/A, \qquad L_2 = A_2/A, \qquad L_3 = A_3/A, \tag{3.4.1.1}$$

where A is the total area of the triangle and A_i ($i = 1, 2, 3$) denotes the areas of the subtriangle i. Because the sum of A_1, A_2, and A_3 must equal A, it follows that

$$L_1 + L_2 + L_3 = 1. \tag{3.4.1.2}$$

In addition to satisfying Eq. (3.4.1.2) the coordinate L_i possesses the property that $L_i = 1$ at node i and $L_i = 0$ at the remaining nodes. In accordance with the concepts outlined in Section 3.3.1, the coordinates L_1, L_2, L_3 are natural coordinates of a triangular region. Furthermore, these coordinates fulfill the requirements of shape functions. Thus, they can be used to describe the Cartesian coordinates, x and y, of an arbitrary point within the triangle (see Fig. 3.5) as follows:

$$x = L_1x_1 + L_2x_2 + L_3x_3, \tag{3.4.1.3}$$

$$y = L_1y_1 + L_2y_2 + L_3y_3. \tag{3.4.1.4}$$

Collecting Eqs. (3.4.1.2)–(3.4.1.4), one obtains

$$\begin{Bmatrix} 1 \\ x \\ y \end{Bmatrix} = \begin{bmatrix} 1 & 1 & 1 \\ x_1 & x_2 & x_3 \\ y_1 & y_2 & y_3 \end{bmatrix} \begin{Bmatrix} L_1 \\ L_2 \\ L_3 \end{Bmatrix}. \tag{3.4.1.5}$$

Inversion of the coefficient matrix leads to

$$L_i = (1/2A)\,(\alpha_i + \beta_i x + \gamma_i y) \qquad i = 1, 2, 3, \tag{3.4.1.6}$$

where

$$\alpha_1 = x_2y_3 - x_3y_2, \qquad \beta_1 = y_2 - y_3, \qquad \gamma_1 = x_3 - x_2,$$

and the remaining coefficients are obtained by a cyclic permutation of subscripts.

To construct the shape functions for higher-order elements, we first establish a particular scheme for numbering the nodal points of such elements. Via this scheme, the nodes are given the three-digit label pqr, where p, q, and r are integers satisfying the relation (see Gallagher, 1975)

$$p + q + r = m, \tag{3.4.1.7}$$

in which m is the degree of the interpolation polynomial for the particular triangle.

As illustrated in Fig. 3.5b, the integers p, q, and r denote the position of a particular node along sides 1, 2, and 3 of the triangle, respectively. Fig. 3.5c shows this node numbering scheme for typical quadratic and cubic triangles. It should be noted that the vertices of the triangle also carry the designations 1, 2, and 3 for the purpose of identifying the area coordinates.

We write $N_{pqr}(L_1, L_2, L_3)$ to denote the shape function for node pqr as a function of the coordinates (L_1, L_2, L_3). This function may be expressed by the formula

$$N_{pqr}(L_1, L_2, L_3) = \Phi_p(L_1)\Phi_q(L_2)\Phi_r(L_3), \tag{3.4.1.8}$$

where

$$\Phi_p(L_1) = \begin{cases} \prod_{k=1}^{P} \dfrac{mL_1 - k + 1}{k} & \text{for} \quad p \geqslant 1 \\ 1 & \text{for} \quad p = 0, \end{cases} \tag{3.4.1.9}$$

m is the degree of the polynomial, and the remaining functions $\Phi_q(L_2)$ and $\Phi_r(L_3)$ are given by the same formula with appropriate adjustment of the subscripts.

As an example, consider the development of the shape functions of the quadratic triangle shown in Fig. 3.5c:

$$N_{200} = \Phi_2(L_1)\Phi_0(L_2)\Phi_0(L_3) = L_1(2L_1 - 1).$$

Similarly, the remaining shape functions are derived as

$$N_{020} = L_2(2L_2 - 1), \qquad N_{002} = L_3(2L_3 - 1),$$

$$N_{110} = 4L_1L_2, \qquad N_{011} = 4L_2L_3, \qquad N_{101} = 4L_3L_1.$$

Once all the shape functions are determined, the trial function for the element can be obtained from

$$\begin{aligned} \hat{u} = {} & N_{002}u_{002} + N_{200}u_{200} + N_{020}u_{020} + N_{101}u_{101} \\ & + N_{110}u_{110} + N_{011}u_{011}. \end{aligned} \tag{3.4.1.10}$$

To facilitate the formulation of the element matrix equation, we simply rewrite Eq. (3.4.1.10) using our standard nodal summation convention

$$\hat{u} = N_I u_I, \qquad I = 1, 2, \ldots n_e, \tag{3.4.1.11}$$

where n_e corresponds to the number of nodes on the element and a single subscript I is used to denote the node number. Thus, for each value of I we have to identify the corresponding value of pqr from Fig. 3.5c.

In forming the element coefficient matrix, there are two distinct operations that must be performed. The first operation is differentiation of the shape functions with respect to the Cartesian coordinates. This can be achieved by using the chain rule of differentiation. Before differentiating, we write the functions N_I in terms of L_1 and L_2 using the relation $L_3 = 1 - L_1 - L_2$. Thus $\partial N_I/\partial x$ is given by

$$\frac{\partial N_I}{\partial x} = \frac{\partial N_I}{\partial L_1}\frac{\partial L_1}{\partial x} + \frac{\partial N_I}{\partial L_2}\frac{\partial L_2}{\partial x}. \tag{3.4.1.12}$$

As an example, suppose N_I corresponds to the shape function N_{110} of the quadratic triangle. In this case, N_I is given by

$$N_I = 4L_1L_2. \tag{3.4.1.13}$$

Thus,

$$\frac{N_I}{\partial x} = 4L_2\frac{\partial L_1}{\partial x} + 4L_1\frac{\partial L_2}{\partial x}. \tag{3.4.1.14}$$

Now from (3.4.1.6)

$$\frac{\partial L_1}{\partial x} = \frac{\beta_1}{2A}, \qquad \frac{\partial L_2}{\partial x} = \frac{\beta_2}{2A}. \tag{3.4.1.15}$$

Substitution of (3.4.1.15) into (3.4.1.14) gives

$$\partial N_I/\partial x = 2(\beta_1 L_2 + \beta_2 L_1)/A. \tag{3.4.1.16}$$

The second operation that has to be performed is integration over the triangle. The integral generally takes the form

$$\int_{R_e} L_1^a L_2^b L_3^c \, dR = \int_A L_1^a L_2^b L_3^c \, dA.$$

It can be shown in several different ways that integrals of this form are readily evaluated using the following convenient relation (Zienkiewicz, 1977):

$$\int_A L_1^a L_2^b L_3^c \, dA = 2Aa!b!c!/(a + b + c + 2)!, \tag{3.4.1.17}$$

where a, b, and c are integer constants.

From the above development, it is apparent that the use of area coordinates to formulate basis functions permits simple evaluation of the integrals occurring in the formulation of the element matrix.

3.4.2 Quadrilateral Elements

In this section, we deal with two families of quadrilateral (four-sided) elements. The first is referred to as the "Lagrange family" and the second we call the "serendipity family."

Lagrange Elements

The Lagrange family is so called because the basis functions for the elements in this family can be derived simply by taking the tensor product of one-dimensional Lagrange polynomials. The simplest element in the Lagrange family is the four-node quadrilateral element shown in Fig. 3.6. Before deriving the shape functions for this element, we first introduce the local coordinates (ξ, n) and call them "isoparametric" coordinates. In this isoparametric system, the original quadrilateral subregion reduces to a square whose corners are located at $\xi = \pm 1$, $\eta = \pm 1$. The transformation between the global and local coordinates can be written in the form

$$x = \sum_{I=1}^{4} N_I(\xi, \eta) x_I, \tag{3.4.2.1}$$

$$y = \sum_{I=1}^{4} N_I(\xi, \eta) y_I, \tag{3.4.2.2}$$

where $N_I(\xi, \eta)$ is now defined as some as yet unspecified function of coordinates (ξ, η) and is associated with node I of the element. If N_I is to satisfy the condition we normally impose on basis functions, it must

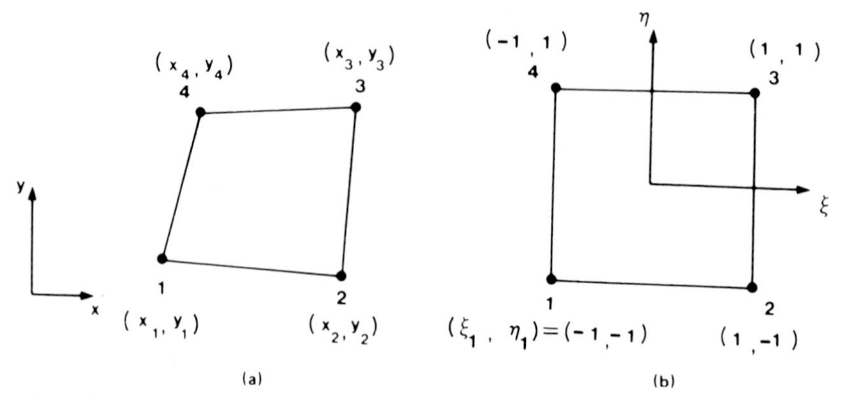

Fig. 3.6. Representation of bilinear element in global and isoparametric local coordinates: (a) global coordinates; (b) isoparametric coordinates.

possess the following properties:

$$N_I = \begin{cases} 1 & \text{at node } I \\ 0 & \text{at the remaining nodes,} \end{cases}$$

$$\sum_{I=1}^{n_e} N_I = 1.$$

One possible choice is

$$N_I(\xi, \eta) = \Psi_I(\xi)\Phi_I(\eta), \tag{3.4.2.3}$$

where $\Psi_I(\xi)$ and $\Phi_I(\eta)$ are the first-degree Lagrange polynomials for node I. Thus, we obtain for node 1,

$$\begin{aligned} N_1(\xi, \eta) &= \left(\frac{\xi - \xi_2}{\xi_1 - \xi_2}\right)\left(\frac{\eta - \eta_4}{\eta_1 - \eta_4}\right) \\ &= \left(\frac{\xi - 1}{-1 - 1}\right)\left(\frac{\eta - 1}{-1 - 1}\right) = \frac{1}{4}(\xi - 1)(\eta - 1). \end{aligned} \tag{3.4.2.4a}$$

Similarly, the remaining shape functions are

$$N_2(\xi, \eta) = \tfrac{1}{4}(1 + \xi)(1 - \eta), \tag{3.4.2.4b}$$

$$N_3(\xi, \eta) = \tfrac{1}{4}(1 + \xi)(1 + \eta), \tag{3.4.2.4c}$$

$$N_4(\xi, \eta) = \tfrac{1}{4}(1 - \xi)(1 + \eta). \tag{3.4.2.4d}$$

These expressions for N_1, N_2, N_3, and N_4 can be put in the form

$$N_I(\xi, \eta) = \tfrac{1}{4}(1 + \xi\xi_I)(1 + \eta\eta_I), \qquad I = 1, \ldots, 4. \tag{3.4.2.4e}$$

Because the four shape functions are formed by taking the tensor product of linear Lagrange polynomials, they are often referred to as "bilinear" shape functions.

Higher-order Lagrange elements can be formulated in the same way as linear elements. With higher-order interpolation it is possible to use elements with curved boundaries. Figure 3.7 shows the typical biquadratic and bicubic elements in global and isoparametric local coordinate systems. The general coordinate transformation relation for these elements may be written in the form

$$x = \sum_{I=1}^{n_e} N_I(\xi, \eta)x_I, \tag{3.4.2.5a}$$

$$y = \sum_{I=1}^{n_e} N_I(\xi, \eta)y_I, \tag{3.4.2.5b}$$

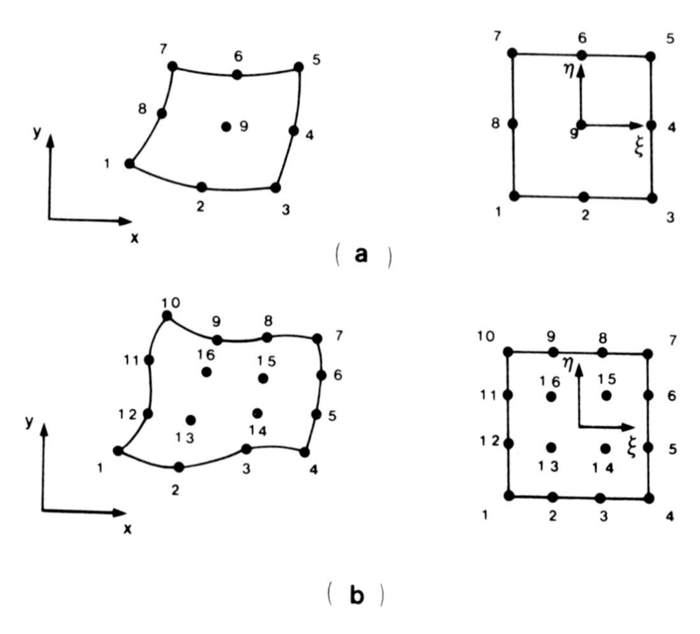

Fig. 3.7. Elements of Lagrange family: (a) biquadratic; (b) bicubic.

where n_e is once again the number of nodes on a particular element. Regardless of the value of n_e, the expressions for the shape functions can be derived using Eq. (3.4.2.3) with appropriate degree of the Lagrange polynomials. Often, the use of curved-sided elements permits a better approximation of curved boundaries. However, it should be noted that the higher-order Lagrange elements contain a substantial number of internal nodes, and this may not be desirable. For this reason, we also consider the second family of quadrilateral elements. This family is referred to as the serendipity family.

Serendipity Elements

The elements in this family contain only exterior nodes. Figure 3.8 shows typical linear, quadratic, and cubic elements. The shape functions for these elements can be derived formally by inspection. The resulting expressions are (Ergatoudis *et al.*, 1968)

(1) *Linear element*

$$N_I(\xi, \eta) = \tfrac{1}{4}(1 + \xi\xi_I)(1 + \eta\eta_I);$$

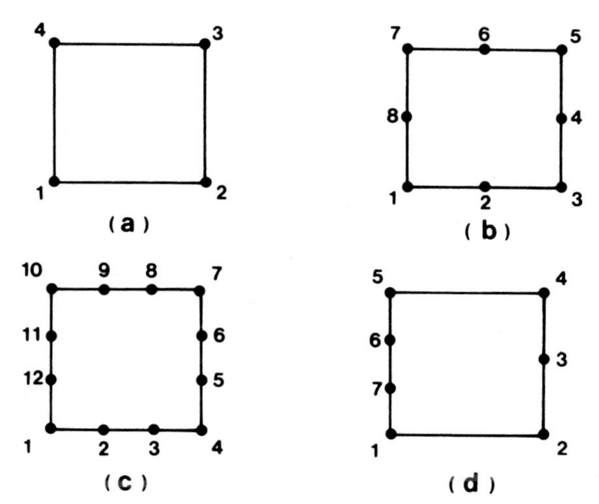

Fig. 3.8. Elements of serendipity family defined in local coordinates: (a) linear element; (b) quadratic element; (c) cubic element; (d) mixed order element.

(2) *Quadratic element*

$$N_I(\xi, \eta) = \begin{cases} \frac{1}{4}(1 + \xi\xi_I)(1 + \eta\eta_I)(\xi\xi_I + \eta\eta_I - 1) \\ \qquad \text{for corner nodes,} \\ \frac{1}{2}(1 - \xi^2)(1 + \eta\eta_I) \\ \qquad \text{for midside nodes at } \xi_I = 0, \eta_I = \pm 1, \\ \frac{1}{2}(1 + \xi\xi_I)(1 - \eta^2) \\ \qquad \text{for midside nodes at } \xi_I = \pm 1, \eta = 0; \end{cases}$$

(3) *Cubic element*

$$N_I(\xi, \eta) = \begin{cases} \frac{1}{32}(1 + \xi\xi_I)(1 + \eta\eta_I)[9(\xi^2 + \eta^2) - 10] \\ \qquad \text{for corner nodes,} \\ \frac{9}{32}(1 + \xi\xi_I)(1 - \eta^2)(1 + 9\eta\eta_I) \\ \qquad \text{for nodes at } \xi_I = \pm 1, \eta_I = \pm\frac{1}{3}, \\ \frac{9}{32}(1 + \eta\eta_I)(1 - \xi^2)(1 + 9\xi\xi_I) \\ \qquad \text{for nodes at } \xi_I = \pm\frac{1}{3}, \eta_I = \pm 1. \end{cases}$$

In designing a two-dimensional mesh, there may be occasions when we want to use quadrilateral elements of different kinds. To allow the transition from, say, linear elements to higher-order elements, a transition (or mixed order) element is required. Such a transition element is illustrated in Fig. 3.8d. The shape functions for this particular element can be obtained as follows:

For side nodes, the shape functions are the same as those given previously for the quadratic and cubic elements.

For corner nodes, the shape functions $N_I(\xi, \eta)$ may be written in the form

$$N_I(\xi, \eta) = \psi_I(\xi, \eta)[\Phi_I(\xi) + \Theta_I(\eta)],$$

where

$$\psi_I(\xi, \eta) = \tfrac{1}{4}[(1 + \xi\xi_I)(1 + \eta\eta_I)].$$

The functions $\Phi_I(\xi)$ and $\Theta_I(\eta)$ depend on the order of the element sides and are given as follows (Pinder and Gray, 1977):

(1) For a linear side, $\Phi_I = \Theta_I = \frac{1}{2}$.
(2) For a quadratic side,

$$\Phi_I(\xi) = \xi\xi_I - \tfrac{1}{2}, \qquad \Theta_I(\eta) = \eta\eta_I - \tfrac{1}{2}.$$

(3) For a cubic side,

$$\Phi_I(\xi) = \tfrac{9}{8}\xi^2 - \tfrac{5}{8}, \qquad \Theta_I(\eta) = \tfrac{9}{8}\eta^2 - \tfrac{5}{8}.$$

Having obtained the shape functions for various types of element in the Lagrange and serendipity families, we are now left with the task of evaluating the coefficients of the element equations by carrying out the integrations appearing in them. The integrals involved for second-order operators are typically

$$\int_{R^e} \left(\frac{\partial N_I}{\partial x}\frac{\partial N_J}{\partial x} + \frac{\partial N_I}{\partial y}\frac{\partial N_J}{\partial y} \right) dx\, dy,$$

$$\int_{R^e} \left(N_I \frac{\partial N_J}{\partial x} + N_I \frac{\partial N_J}{\partial y} \right) dx\, dy, \qquad \text{and} \qquad \int_{R^e} N_I N_J \, dx\, dy.$$

To evaluate these integrals, it is necessary to express $\partial N_I/\partial x$ and $\partial N_I/\partial y$ in terms of ξ and η. This can be achieved using the chain rule of differentiation as follows:

$$\frac{\partial N_I}{\partial \xi} = \frac{\partial N_I}{\partial x}\frac{\partial x}{\partial \xi} + \frac{\partial N_I}{\partial y}\frac{\partial y}{\partial \xi}, \tag{3.4.2.6a}$$

$$\frac{\partial N_I}{\partial \eta} = \frac{\partial N_I}{\partial x}\frac{\partial x}{\partial \eta} + \frac{\partial N_I}{\partial y}\frac{\partial y}{\partial \eta}, \tag{3.4.2.6b}$$

or, in matrix form

$$\begin{Bmatrix} \dfrac{\partial N_I}{\partial \xi} \\[2ex] \dfrac{\partial N_I}{\partial \eta} \end{Bmatrix} = \begin{bmatrix} \dfrac{\partial x}{\partial \xi} & \dfrac{\partial y}{\partial \xi} \\[2ex] \dfrac{\partial x}{\partial \eta} & \dfrac{\partial y}{\partial \eta} \end{bmatrix} \begin{Bmatrix} \dfrac{\partial N_I}{\partial x} \\[2ex] \dfrac{\partial N_I}{\partial y} \end{Bmatrix} = [J] \begin{Bmatrix} \dfrac{\partial N_I}{\partial x} \\[2ex] \dfrac{\partial N_I}{\partial y} \end{Bmatrix}. \tag{3.4.2.7}$$

For isoparametric elements the Jacobian matrix $[J]$ is

$$[J] = \begin{bmatrix} \dfrac{\partial x}{\partial \xi} & \dfrac{\partial y}{\partial \xi} \\ \\ \dfrac{\partial x}{\partial \eta} & \dfrac{\partial y}{\partial \eta} \end{bmatrix} = \begin{bmatrix} \dfrac{\partial N_I}{\partial \xi} x_I & \dfrac{\partial N_I}{\partial \xi} y_I \\ \\ \dfrac{\partial N_I}{\partial \eta} x_I & \dfrac{\partial N_I}{\partial \eta} y_I \end{bmatrix}; \tag{3.4.2.8}$$

where the repeated indices indicate summation over the range from 1 to n_e. To obtain the required derivatives, we must invert the Jacobian matrix. Thus,

$$\begin{Bmatrix} \dfrac{\partial N_I}{\partial x} \\ \\ \dfrac{\partial N_I}{\partial y} \end{Bmatrix} = [J]^{-1} \begin{Bmatrix} \dfrac{\partial N_I}{\partial \xi} \\ \\ \dfrac{\partial N_I}{\partial \eta} \end{Bmatrix}. \tag{3.4.2.9}$$

In addition to transforming the derivative from (x, y) to (ξ, η), the differential area must be changed using the relation

$$\int_R F(x, y)\, dx\, dy = \int_{R(\xi,\eta)} F(x(\xi, \eta), y(\xi, \eta))|J|\, d\xi\, d\eta, \tag{3.4.2.10}$$

where $|J|$ is the determinant of the Jacobian coordinate transformation matrix $[J]$. The transformations in Eqs. (3.4.2.9) and (3.4.2.10) are based on the assumption that $[J]^{-1}$ exists, that is, $|J| \neq 0$. We can test the validity of this assumption by checking the sign of $|J|$. If its sign does not change over the element subregion then our assumption is valid and the transformation from the (x, y) to (ξ, η) plane is acceptable.

After the preceding operations have been carried out, the integrals reduce to the form

$$\int_{-1}^{1} \int_{-1}^{1} f(\xi, \eta)\, d\xi\, d\eta,$$

where f is a function of ξ and η owing to the preceding transformations and the fact that the original bases were defined in local (ξ, η) coordinates.

The form of function f may be complicated. However, this does not present any difficulty because the integrals can be easily evaluated numerically using quadrature formulas; these will be presented in Section 3.6.

3.4.3 Hermite Elements

The shape functions for Hermite quadrilateral elements can be constructed by taking the tensor product of one-dimensional Hermite poly-

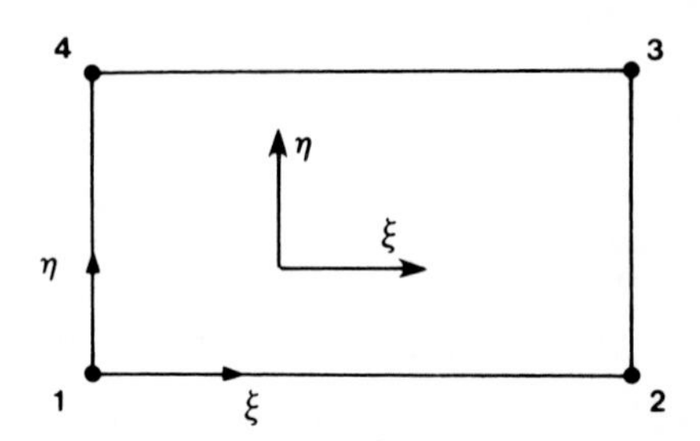

Fig. 3.9. Bicubic Hermite element. u_l, $\left(\frac{\partial u}{\partial \xi}\right)_l$, $\left(\frac{\partial u}{\partial \eta}\right)_l$, $\left(\frac{\partial^2 u}{\partial \xi\, \partial \eta}\right)_l$ are specified at node l, where $l = 1, 2, 3, 4$.

nomials [see Van Genuchten *et al.* (1977)]. For the bicubic element shown in Fig. 3.9, we obtain 16 shape functions. Adopting the previous double subscript notation we can denote these shape functions as N_{1I}, N_{2I}, N_{3I}, and N_{4I}, where I refers to the nodes of the element, and N_{1I}, N_{2I}, N_{3I} and N_{4I} are identified with the nodal parameters u_I, $(\partial u/\partial \xi)_I$, $(\partial u/\partial \eta)_I$, and $(\partial^2 u/\partial \xi\, \partial \eta)_I$, respectively. Thus, the trial function for a typical element is written

$$\hat{u}(x, y) = u_I N_{1I} + \left(\frac{\partial u}{\partial \xi}\right)_I N_{2I} + \left(\frac{\partial u}{\partial \eta}\right)_I N_{3I} + \left(\frac{\partial^2 u}{\partial \xi\, \partial \eta}\right)_I N_{4I} \qquad I = 1, \ldots, 4.$$

Note that the mixed second derivative is required as a nodal parameter because of the combination of Hermite polynomials that arise in the product.

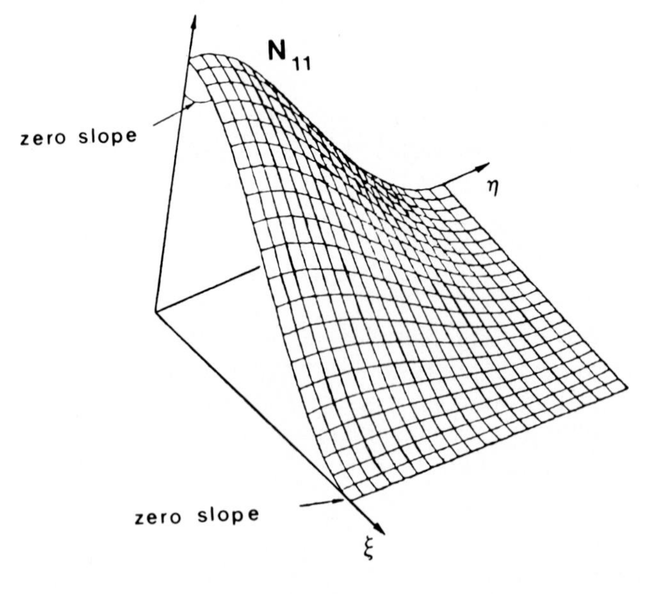

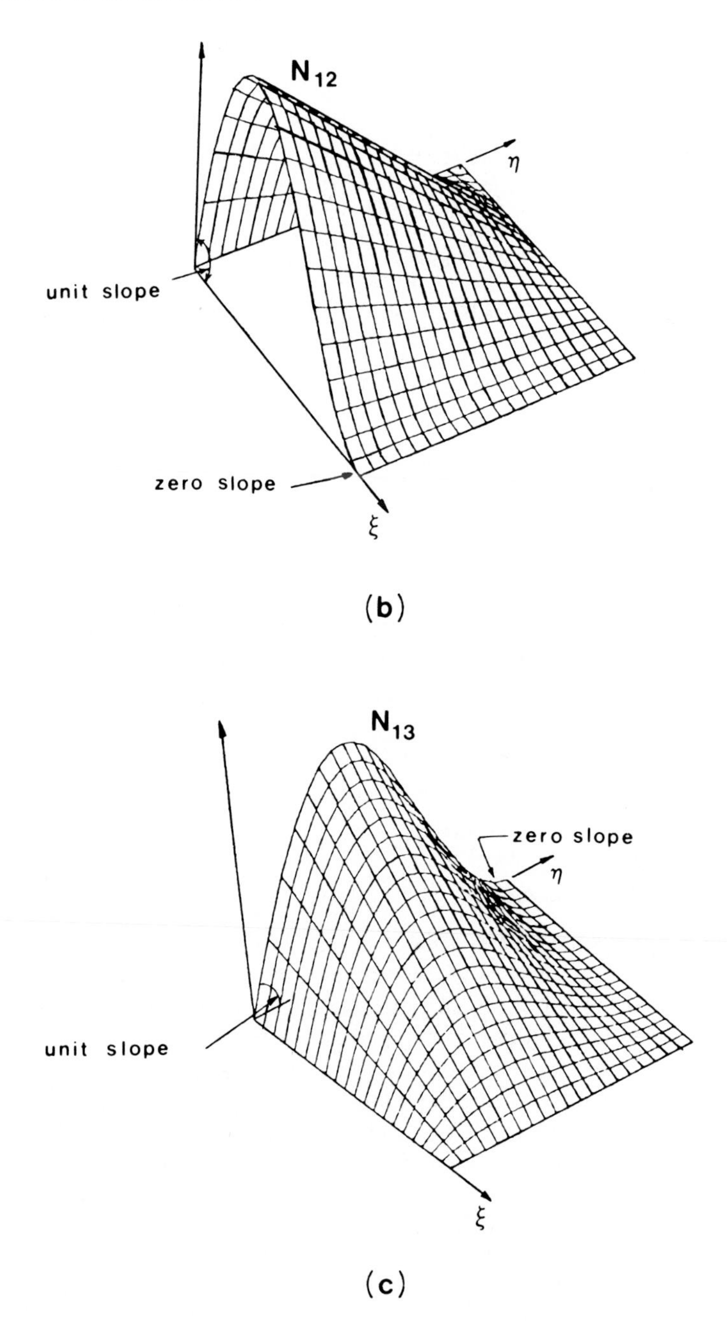

Fig. 3.10. Bicubic Hermite interpolation functions: (a) N_{11}; (b) N_{12}; (c) N_{13}. (After Pinder and Gray, 1977.)

The 16 shape functions may be expressed as

$$
\begin{aligned}
N_{11} &= H_{01}(\xi)H_{01}(\eta), & N_{12} &= H_{02}(\xi)H_{01}(\eta),\\
N_{13} &= H_{01}(\xi)H_{02}(\eta), & N_{14} &= H_{02}(\xi)H_{02}(\eta),\\
N_{21} &= H_{11}(\xi)H_{01}(\eta), & N_{22} &= H_{12}(\xi)H_{01}(\eta),\\
N_{23} &= H_{11}(\xi)H_{02}(\eta), & N_{24} &= H_{12}(\xi)H_{02}(\eta),\\
N_{31} &= H_{01}(\xi)H_{11}(\eta), & N_{32} &= H_{02}(\xi)H_{11}(\eta),\\
N_{33} &= H_{01}(\xi)H_{12}(\eta), & N_{34} &= H_{02}(\xi)H_{12}(\eta),\\
N_{41} &= H_{11}(\xi)H_{11}(\eta), & N_{42} &= H_{12}(\xi)H_{11}(\eta),\\
N_{43} &= H_{11}(\xi)H_{12}(\eta), & N_{44} &= H_{12}(\xi)H_{12}(\eta).
\end{aligned}
$$

The behavior of the typical shape functions N_{11}, N_{12}, and N_{13} is illustrated in Fig. 3.10. By the above procedure for deriving the two-dimensional basis functions, we can achieve C^1 continuity only if the element is rectangular in shape. For a nonrectangular quadrilateral element, we require the specification of the two additional second derivatives $\partial^2 u/\partial x^2$ and $\partial^2 u/\partial y^2$ at each corner node to achieve C^1 continuity. Detailed discussion is given by Fellipa (1966).

3.5 Three-Dimensional Elements

3.5.1 Tetrahedral Elements

The formulation of basis functions for three-dimensional elements is a direct extension of the two-dimensional case. The simplest three-dimensional element is the four-node tetrahedron. Higher-order elements can be formed by adding more nodes to the basic tetrahedron. As might be expected, the tetrahedral family shown in Fig. 3.11 exhibits properties similar to those of the triangular family. Complete nth order polynomials in three dimensions are obtained by placing $\frac{1}{6}(n + 1)(n + 2)(n + 3)$ nodes on the tetrahedron and arranging these nodes as illustrated in Fig. 3.11.

The shape functions can be derived by using volume coordinates (Fig. 3.12), which we define as

$$L_i = V_i/V, \qquad i = 1, 2, 3, 4, \tag{3.5.1.1}$$

where V is the entire volume of the tetrahedron and V_i is the volume defined by point P and the face opposite to vertex i. The volume coordinates and the Cartesian coordinates are related by

$$\begin{Bmatrix} x \\ y \\ z \\ 1 \end{Bmatrix} = \begin{bmatrix} x_1 & x_2 & x_3 & x_4 \\ y_1 & y_2 & y_3 & y_4 \\ z_1 & z_2 & z_3 & z_4 \\ 1 & 1 & 1 & 1 \end{bmatrix} \begin{Bmatrix} L_1 \\ L_2 \\ L_3 \\ L_4 \end{Bmatrix}. \tag{3.5.1.2}$$

Inversion of the coefficient matrix leads to

$$L_i = (1/6V)(a_i + b_i x + c_i y + d_i z), \qquad i = 1, 2, 3, 4, \tag{3.5.1.3}$$

where

$$a_1 = \begin{vmatrix} x_2 & y_2 & z_2 \\ x_3 & y_3 & z_3 \\ x_1 & y_4 & z_4 \end{vmatrix}, \qquad b_1 = -\begin{vmatrix} 1 & y_2 & z_2 \\ 1 & y_3 & z_3 \\ 1 & y_4 & z_4 \end{vmatrix},$$

$$c_1 = \begin{vmatrix} x_2 & 1 & z_2 \\ x_3 & 1 & z_3 \\ x_4 & 1 & z_4 \end{vmatrix}, \qquad d_1 = \begin{vmatrix} x_2 & y_2 & 1 \\ x_3 & y_3 & 1 \\ x_4 & y_4 & 1 \end{vmatrix},$$

$$V = \tfrac{1}{6}\begin{vmatrix} 1 & x_1 & y_1 & z_1 \\ 1 & x_2 & y_2 & z_2 \\ 1 & x_3 & y_3 & z_3 \\ 1 & x_4 & y_4 & z_4 \end{vmatrix}$$

and the remaining coefficients are obtained by a cyclic permutation of the subscripts. As might be expected, the shape functions for the linear tetrahedral element are given by

$$N_I = L_I, \qquad I = 1, 2, 3, 4. \tag{3.5.1.4}$$

For higher-order elements, we apply a formula similar to Eq. (3.4.1.8), which was used for the triangular family. Thus, using a four-digit nodal subscript, we can write

$$N_{\mathrm{pqrs}}(L_1, L_2, L_3, L_4) = \Phi_p(L_1)\Phi_q(L_2)\Phi_r(L_3)\Phi_s(L_4), \tag{3.5.1.5}$$

where

$$\Phi_p(L_1) = \begin{cases} \prod_{k=1}^{p} (mL_1 - k + 1)/k & \text{for} \quad p \geqslant 1 \\ 1 & \text{for} \quad p = 0, \end{cases} \tag{3.5.1.6}$$

m is the order of the polynomial, and the remaining functions $\Phi_q(L_2)$, $\Phi_r(L_3)$, and $\Phi_s(L_4)$ are given by the same formula, with appropriate adjustment of the subscripts. Application of Eqs. (3.5.1.5) and (3.5.1.6) to quadratic and cubic tetrahedra yields the following:

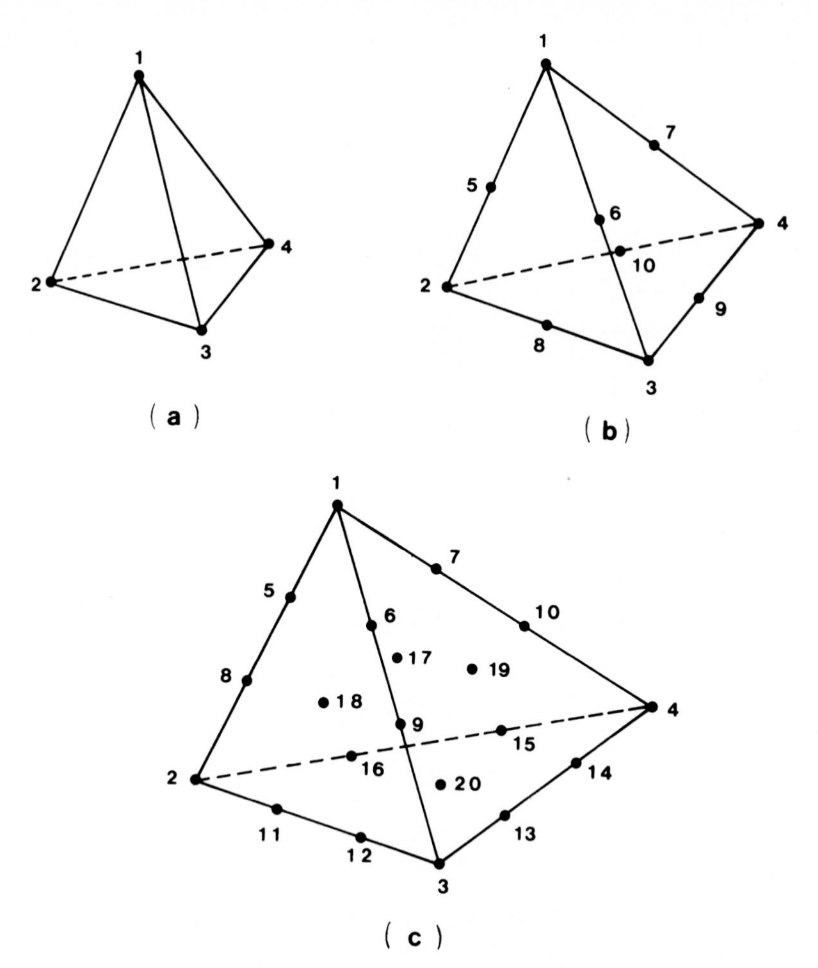

Fig. 3.11. Tetrahedral elements: (a) linear; (b) quadradic; (c) cubic.

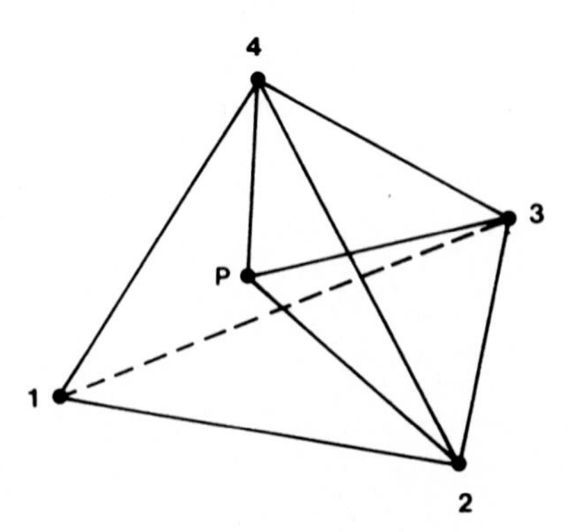

Fig. 3.12. Volume coordinates.

(1) *Quadratic element*

(a) Corner nodes:

$$N_I = (2L_I - 1)L_I, \qquad I = 1, 2, 3, 4;$$

(b) Midside nodes:

$$N_5 = 4L_1L_2, \qquad N_6 = 4L_1L_3, \qquad \text{etc.};$$

(2) *Cubic element*

(a) Corner nodes:

$$N_I = \tfrac{1}{2}(3L_I - 1)(3L_I - 2)L_I \qquad I = 1, 2, 3, 4;$$

(b) Side nodes:

$$N_5 = \tfrac{9}{2}L_1L_2(3L_1 - 2), \qquad \text{etc.};$$

(c) Midface nodes:

$$N_{17} = 27L_1L_2L_4, \qquad N_{18} = 27L_1L_2L_3, \qquad \text{etc.}$$

The integration formula in volume coordinates that corresponds to the triangular element is (Zienkiewicz, 1977)

$$\int_V L_1^a L_2^b L_3^c L_4^d \, dV = 6Va!b!c!d!/(a + b + c + d)!. \qquad (3.5.1.7)$$

3.5.2 *Hexahedral Elements*

The two-dimensional quadrilateral elements may be extended to three-dimensional hexahedral elements. For elements in the Lagrange family the shape functions may be written as the triple product of the Lagrange polynomials in the three-dimensional isoparametric coordinates ξ, η, ζ. Thus, for node I we obtain

$$N_I(\xi, \eta, \zeta) = L_I(\xi)L_I(\eta)L_I(\zeta), \qquad (3.5.2.1)$$

where $L_I(\xi)$, $L_I(\eta)$, and $L_I(\zeta)$ denote the Lagrange polynomials in ξ, η, and ζ, respectively.

Unfortunately, for quadratic and higher-order variations, the Lagrange elements contain many interior nodes (Fig. 3.13a). For this reason, such elements are seldom used. Thus, with quadratic or cubic interpolations, it is advantageous to use the elements in the serendipity family because they contain only exterior nodes. The basis functions for these elements can be derived by the direct method (Section 3.2.3) or by inspection. For the 20-node quadratic and the 32-node cubic elements, one obtains

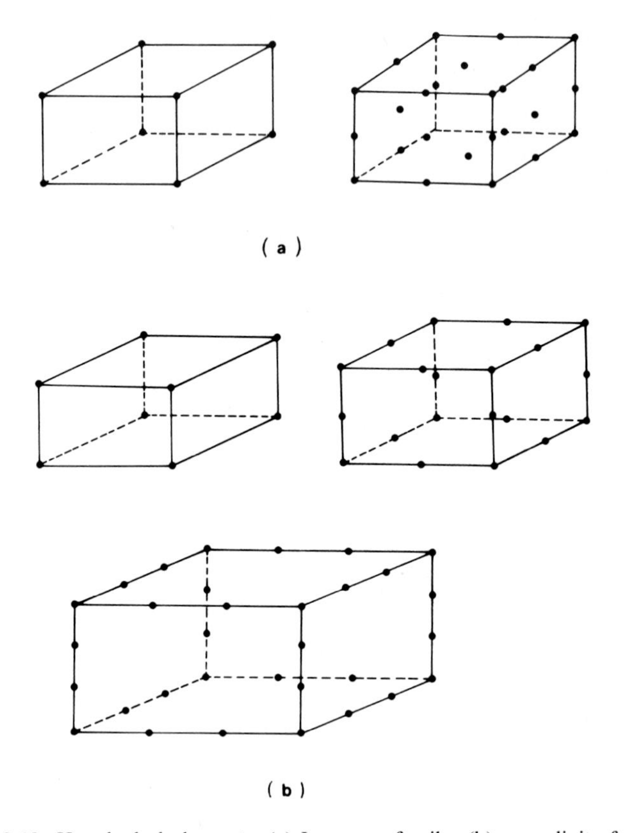

Fig. 3.13. Hexahedral elements: (a) Lagrange family; (b) serendipity family.

(1) *Linear element*

$$N_I = \tfrac{1}{8}(1 + \xi\xi_I)(1 + \eta\eta_I)(1 + \zeta\zeta_I);$$

(2) *Quadratic element*

(a) Corner nodes:

$$N_I = \tfrac{1}{8}(1 + \xi\xi_I)(1 + \eta\eta_I)(1 + \zeta\zeta_I)(\xi\xi_I + \eta\eta_I + \zeta\zeta_I - 2);$$

(b) Midside nodes, located at $\xi_I = 0$, $\eta_I = \zeta_I = \pm 1$:

$$N_I = \tfrac{1}{4}(1 - \xi^2)(1 + \eta\eta_I)(1 + \xi\xi_I);$$

(3) *Cubic element*

(a) Corner nodes:

$$N_I = \tfrac{1}{64}(1 + \xi\xi_I)(1 + \eta\eta_I)(1 + \zeta\zeta_I)[9(\xi^2 + \eta^2 - \zeta^2) - 19];$$

(b) Midface nodes, located at $\xi_I = \pm\frac{1}{3}$, $\eta_I = \pm 1$, $\zeta_I = \pm 1$:

$$N_I = \tfrac{9}{64}(1 - \xi^2)(1 + 9\xi\xi_I)(1 + \eta\eta_I)(1 + \zeta\zeta_I).$$

3.5.3 Triangular Prism Elements

Difficulties can arise in discretizing a complex three-dimensional domain using only hexahedral elements because they may not fit the boundary well. Rather than using an excessive number of hexahedral elements, it is more computationally efficient to grade the mesh using triangular prism elements. The shape function for this type of element can be derived by taking the product of the shape function for triangles with that for quadrilaterals. For the typical linear and quadratic elements shown in Fig. 3.14, the shape functions are

(1) *Linear elements*

$$N_I(L_i, \zeta) = \tfrac{1}{2}L_i(1 + \zeta\zeta_I) \qquad \text{for} \quad I = 1, \ldots, 6;$$

(2) *Quadratic element*

(a) Typical corner node:

$$N_1 = \tfrac{1}{2}L_1(2L_1 - 1)(1 + \zeta) - \tfrac{1}{2}L_1(1 - \zeta^2);$$

(b) Typical midside of quadrilateral:

$$N_7 = L_1(1 - \zeta^2);$$

(c) Typical midside of triangle:

$$N_{10} = 2L_1L_2(1 + \zeta).$$

3.6 Numerical Integration

In forming the element matrix equation, we face the task of evaluating integrals over each element subdomain. Often the integrals contain lengthy complicated expressions and can be very tedious to evaluate. Despite the existence of simple exact analytical formulas for triangles and tetrahedra it is often advantageous to employ numerical integration. Several types of numerical integration formulas are available. We shall discuss only "Gaussian" (or "Gauss–Legendre") quadrature for lines, squares, and cubes, and other quadrature formulas for triangles and tetrahedra.

3.6.1 Gaussian Quadrature Formulas

By the Gaussian quadrature technique, a definite integral of the function f is approximated by the weighted sum of values of f at selected points.

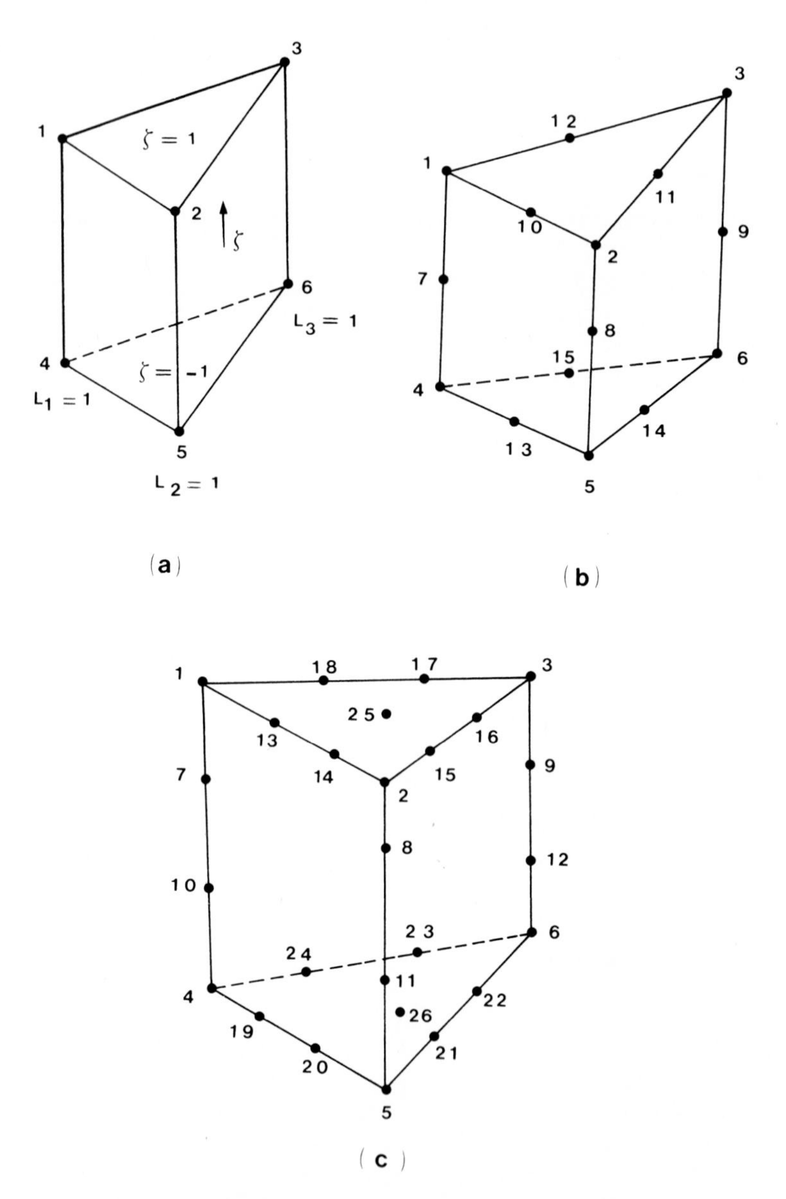

Fig. 3.14. Triangular prism elements: (a) linear; (b) quadratic; (c) cubic.

Consider in particular the definite integral $\int_{-1}^{1} f(\xi)\, d\xi$. The Gaussian quadrature formula for this integral takes the form

$$\int_{-1}^{1} f(\xi)\, d\xi = \sum_{i=1}^{m} W_i f(\xi_i), \tag{3.6.1.1}$$

where $f(\xi_i)$ is the value of the function f at the "Gaussian" point ξ_i, m is the number of Gaussian points, and W_i ($i = 1, 2, \ldots m$) are the weighting coefficients. Table 3.1 shows the locations and the weighting coefficients of the Gaussian points for $m = 2$, 3, and 4. It should be noted that the Gaussian points are not evenly spaced, but are selected in such a way that the weighted sum of m functional values yields the exact value of the integral of a polynomial of degree $2m - 1$ or less. In a case where the function f is a polynomial of degree greater than $2m - 1$, the error in using m Gaussian points is of the order $O(h^{2m})$, where h is the length of the integration interval (Heubner, 1975).

To illustrate the application of Gaussian quadrature, we consider the integral

$$\int_{-1}^{1} f(\xi)\, d\xi = \int_{-1}^{1} (\xi^3 + \xi^2 + \xi)\, d\xi,$$

where $f(\xi) = \xi^3 + \xi^2 + \xi$. To evaluate this integral exactly, we use $m = 2$. From Table 3.1, $\xi_1 = -0.57735$, $\xi_2 = 0.57735$, and $W_1 = W_2 = 1$. Thus, we obtain

$$\int_{-1}^{1} f(\xi)\, d\xi = (1.00)f(-0.57735) + (1.00)f(0.57735) = 0.66666,$$

which is exact to five significant figures.

Table 3.1
One-Dimensional Gauss–Legendre Quadrature Points and Weights.

$\pm\xi_i$	m	W_i
0.577350269189626	2	1.00
0.774596669241483	3	0.555555555555556
0.000000000000000		0.888888888888889
0.861136311594053	4	0.347854845137454
0.339981043584856		0.652145154862546

In two dimensions, the definite integral of the form

$$I = \int_{-1}^{1}\int_{-1}^{1} f(\xi, \eta)\, d\xi\, d\eta \tag{3.6.1.2}$$

can be evaluated by first evaluating the inner integral and then evaluating the outer integral. The inner integral is evaluated by holding η constant, i.e.,

$$\int_{-1}^{1} f(\xi, \eta)\, d\xi = \sum_{j=1}^{m} W_j f(\xi_j, \eta) = \Psi(\eta).$$

The outer integral can now be obtained as

$$\int_{-1}^{1} \Psi(\eta)\, d\eta = \sum_{i=1}^{m} W_i \Psi(\eta_i) = \sum_{i=1}^{m}\sum_{j=1}^{m} W_i W_j f(\xi_j, \eta_i).$$

Thus,

$$\int_{-1}^{1}\int_{-1}^{1} f(\xi, \eta)\, d\xi\, d\eta = \sum_{i=1}^{m}\sum_{j=1}^{m} W_i W_j f(\xi_j, \eta_i). \tag{3.6.1.3}$$

In a similar manner we can extend this procedure to three dimensions. Thus,

$$\int_{-1}^{1}\int_{-1}^{1}\int_{-1}^{1} f(\xi, \eta, \zeta)\, d\xi\, d\eta\, d\zeta = \sum_{i=1}^{m}\sum_{j=1}^{m}\sum_{k=1}^{m} W_i W_j W_k f(\xi_i, \eta_j, \zeta_k). \tag{3.6.1.4}$$

3.6.2 Quadrature Formulas for Triangular and Tetrahedral Regions

For triangles, the integrals written in natural coordinates take the form

$$I = \int_{0}^{1}\int_{0}^{1-L_2} f(L_1, L_2, L_3)\, dL_1\, dL_2. \tag{3.6.2.1}$$

This integral can be approximated by

$$\int_{0}^{1}\int_{0}^{1-L_2} f(L_1, L_2, L_3)\, dL_1\, dL_2 = \tfrac{1}{2}\sum_{i=1}^{m} W_i f_i, \tag{3.6.2.2}$$

where f_i is the value of function f at the integration point i, W_i is the weighting coefficient for integration point i, and m is the number of integration points. Table 3.2 shows the locations of integration points and values of the weighting coefficients for $m = 1$, 3, and 7. For tetrahedra,

Table 3.2
Quadrature Points and Weights for a Triangular Domain.

m	Order	Figure	Error	Points	Triangular Coordinates	Weights W_i
1	Linear	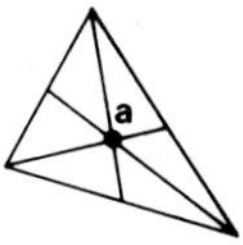	$R = O(h^2)$	a	1/3, 1/3, 1/3	1
3	Quadratic		$R = O(h^3)$	a	1/2, 1/2, 0	1/3
				b	0, 1/2, 1/2	1/3
				c	1/2, 0, 1/2	1/3
3	Quadratic		$R = O(h^3)$	a	2/3, 1/6, 1/6	1/3
				b	1/6, 2/3, 1/6	1/3
				c	1/6, 1/6, 2/3	1/3
7	Quintic		$R = O(h^6)$	a	1/3, 1/3, 1/3	0.225
				b	$\alpha_1, \beta_1, \beta_1$	0.13239415
				c	$\beta_1, \alpha_1, \beta_1$	
				d	$\beta_1, \beta_1, \alpha_1$	
				e	$\alpha_2, \beta_2, \beta_2$	0.12593918
				f	$\beta_2, \alpha_2, \beta_2$	
				g	$\beta_2, \beta_2, \alpha_2$	

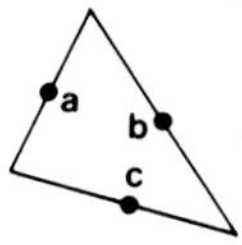

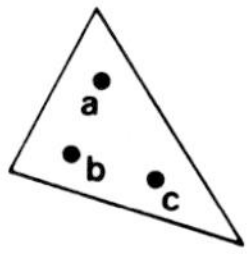

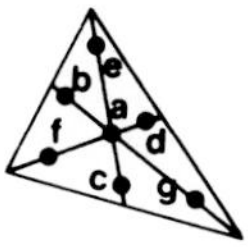

With
$\alpha_1 = 0.05961587$
$\beta_1 = 0.47014206$
$\alpha_2 = 0.79742699$
$\beta_2 = 0.10128651$

the integral takes the form

$$I = \int_0^1 \int_0^{1-L_1} \int_0^{1-L_2-L_1} f(L_1, L_2, L_3, L_4)\, dL_3\, dL_2\, dL_1, \quad (3.6.2.3)$$

which can be approximated by

$$I = \tfrac{1}{6} \sum_{i=1}^{m} W_i f_i. \quad (3.6.2.4)$$

The locations and weighting coefficients of integration points for m = 1, 4, and 5 are shown in Table 3.3.

Table 3.3
Quadrature Points and Weights for a Tetrahedral Domain (after Zienkiewicz, 1977).

m	Order	Fig.	Error	Points	Tetrahedral Coordinates	Weights W_i
1	Linear		$R = O(h^2)$	a	1/4, 1/4, 1/4, 1/4	1
4	Quadratic		$R = O(h^3)$	a	$\alpha, \beta, \beta, \beta$	1/4
				b	$\beta, \alpha, \beta, \beta$	1/4
				c	$\beta, \beta, \alpha, \beta$	1/4
				d	$\beta, \beta, \beta, \alpha$	1/4
					$\alpha = 0.58541020$	
					$\beta = 0.13819660$	
5	Cubic		$R = O(h^4)$	a	1/4, 1/4, 1/4, 1/4	−4/5
				b	1/3, 1/6, 1/6, 1/6	9/20
				c	1/6, 1/3, 1/6, 1/6	9/20
				d	1/6, 1/6, 1/3, 1/6	9/20
				e	1/6, 1/6, 1/6, 1/3	9/20

References

Ergatoudis, I., Irons, B. M., and Zienkiewicz, O. C. (1968). Curved isoparametric quadrilateral elements for finite element analysis. *Int. J. Solids Struct.* **4,** 31–42.

Felippa, C. A. (1966). Refined finite element analysis of linear and non-linear two-dimensional structure. Report 66-22, Dept. of Civil Engineering, University of California, Berkeley.

Gallagher, R. H. (1975). "Finite Element Analysis Fundamentals." Prentice-Hall, Englewood Cliffs, New Jersey.

Hammer, P. C., and Stroud, A. H. (1958). Numerical Evaluation of Multiple Integrals. *Math. Tables Aids Comput.* **12.**

Huebner, K. H. (1975). "Finite Element Method for Engineers." Wiley, New York.

Pinder, G. F., and Gray, W. G. (1977). "Finite Elements in Surface and Subsurface Hydrology." Academic Press, New York.

Van Genuchten, Th. M., Pinder, G. F., and Frind, E. O. (1977). Simulation of two-dimensional contaminant transport with isoparametric Hermitian finite elements. *Water Resour. Res.* **13** (2), 451–458.

Zienkiewicz, O. C. (1977). "The Finite Element Method." 3rd ed. McGraw-Hill, London.

4

Finite Element Simulation of Isothermal Flow in Porous Media

4.1 Introduction

In this chapter, we present finite element approximations for the equations describing various problems involving isothermal flow of homogeneous fluids in porous media [Readers who are not familiar with these equations should refer to Bear (1979) and Corey (1977)]. The first part of the chapter deals with saturated flow of water, oil, or gas in underground reservoirs. The second part deals with water movement in variably saturated soils and multiphase flow of immiscible liquids in petroleum reservoirs. The problems for which solution methods will be presented involve one- and two-dimensional flow in the vertical and horizontal planes and axisymmetric flow around pumped wells. For each class of problem we begin the presentation with a brief description of the governing equations and initial and boundary conditions. We then derive the finite element equations via the Galerkin approach and develop solution algorithms for a variety of linear and nonlinear cases commonly encountered in practice.

The discussion proceeds from the specific to the more general because this also leads us from relatively simple to more complex problems. Although there is a great deal of flexibility in choosing the type of finite elements to be used in the discretization, experience indicates that for most of the subsurface problems treated here, it is sufficient and often advantageous to employ linear triangular or quadrilateral elements. This is particularly true for nonlinear problems, where much of the overall computational effort is spent on the formulation of matrix coefficients. The use of higher-order elements may result in excessive computational cost.

4.2 General Governing Equations for Saturated Flow

4.2.1 Slightly Compressible Flow

Three-Dimensional and Axisymmetric Flow

A mathematical description of fluid flow in a porous medium can be obtained by combining the equation of continuity of mass and Darcy's law. Invoking various simplifying assumptions, one can write the continuity equation for three-dimensional flow in the form [see Hantush (1964), or Walton (1970) for derivation]

$$\partial V_i/\partial x_i = -(\alpha + \phi\beta)\,\partial p/\partial t, \qquad i = 1, 2, 3, \tag{4.2.1.1}$$

where V_i are Darcy velocity components along the Cartesian coordinate axes, p the fluid pressure, and ϕ the porosity of the porous medium. The coefficients α and β defined below are the compressibilities of the rock matrix and the fluid, respectively, where $\alpha = d\phi/dp$ and $\beta = (1/\rho)(d\rho/dp)$.

The mathematical statement of Darcy's law is

$$V_i = -\frac{k_{ij}}{\mu}\left(\frac{\partial p}{\partial x_j} + \rho g \frac{\partial z}{\partial x_j}\right), \tag{4.2.1.2}$$

where k_{ij} are components of the intrinsic permeability tensor, μ is the dynamic viscosity of the fluid, ρ is the fluid density, g is the gravitational acceleration, and z is the elevation above a reference datum.

For the case of homogeneous fluids, Eq. (4.2.1.2) can be reduced to a simpler form by introducing the relation

$$h = \frac{1}{g}\int_{p_o}^{p} \frac{d\xi}{\rho(\xi)} + z. \tag{4.2.1.3}$$

In groundwater hydrology, h is commonly referred to as the "hydraulic head" and is usually approximated by $h = p/(\rho g) + z$. The quantity gh is a potential function commonly referred to as "Hubbert's potential."

We now differentiate Eq. (4.2.1.3) using Leibniz's rule, which may be written as

$$\frac{d}{dx}\int_{a(x)}^{b(x)} f(x, \xi)\, d\xi = \int_{a(x)}^{b(x)} \frac{\partial f}{\partial x}\, d\xi + f(x, b)\frac{db}{dx} - f(x, a)\frac{da}{dx}, \tag{4.2.1.4}$$

where x is an independent variable, $a(x)$ and $b(x)$ are lower and upper limits, respectively, and ξ is the dummy variable. Using Leibniz's rule once again to differentiate Eq. (4.2.1.3) with respect to x_j and t gives

$$\frac{\partial h}{\partial x_j} = \frac{1}{\rho g}\frac{\partial p}{\partial x_j} + \frac{\partial z}{\partial x_j}, \tag{4.2.1.5a}$$

$$\frac{\partial h}{\partial t} = \frac{1}{\rho g}\frac{\partial p}{\partial t}. \tag{4.2.1.5b}$$

Combination of (4.2.1.5b) and the right-hand side of (4.2.1.1) gives

$$-(\alpha + \phi\beta)\frac{\partial p}{\partial t} = -\rho g(\alpha + \phi\beta)\frac{\partial h}{\partial t} = -S_s\frac{\partial h}{\partial t}, \tag{4.2.1.5c}$$

where $S_s = \rho g(\alpha + \phi\beta)$ is called the "specific storage" of the reservoir formation. Substituting Eq. (4.2.1.5a) into (4.2.1.2), we obtain

$$V_i = -\frac{\rho g k_{ij}}{\mu}\frac{\partial h}{\partial x_j}. \tag{4.2.1.6}$$

Combination of Eqs. (4.2.1.1), (4.2.1.5c), and (4.2.1.6) yields

$$\frac{\partial}{\partial x_i}\left(\frac{\rho g k_{ij}}{\mu}\right)\frac{\partial h}{\partial x_j} = S_s\frac{\partial h}{\partial t}. \tag{4.2.1.7}$$

We now define the coefficient of fluid conductivity as

$$K_{ij} = \rho g k_{ij}/\mu. \tag{4.2.1.8}$$

Substitution of Eq. (4.2.1.8) into (4.2.1.7) gives the final form of the governing differential equation

$$\frac{\partial}{\partial x_i}\left(K_{ij}\frac{\partial h}{\partial x_j}\right) = S_s\frac{\partial h}{\partial t}. \tag{4.2.1.9a}$$

In the case of an isotropic porous medium, this expression reduces to

$$\frac{\partial}{\partial x_i}\left(K\frac{\partial h}{\partial x_i}\right) = S_s\frac{\partial h}{\partial t}. \tag{4.2.1.9b}$$

For the case of axisymmetric flow toward a well, the governing equation is written conveniently in cylindrical coordinates, (r, z), viz.,

$$\frac{1}{r}\frac{\partial}{\partial r}\left(K_{rr}r\frac{\partial h}{\partial r}\right) + \frac{\partial}{\partial z}\left(K_{zz}\frac{\partial h}{\partial z}\right) = S_s\frac{\partial h}{\partial t}, \tag{4.2.1.10}$$

where K_{rr} and K_{zz} are principal hydraulic conductivities in the radial and vertical directions, respectively.

Areal Flow

Let us consider the general flow equation written in its three-dimensional form

$$\frac{\partial}{\partial x_\alpha}V_\alpha + S_s\frac{\partial h}{\partial t} = 0, \qquad \alpha = 1, 2, 3, \tag{4.2.1.11a}$$

$$V_\alpha + K_{\alpha\beta}\frac{\partial h}{\partial x_\beta} = 0, \qquad \alpha, \beta = 1, 2, 3. \tag{4.2.1.11b}$$

To reduce the dimensionality of the problem it is often convenient to integrate over the vertical dimension. The nomenclature for the procedure is illustrated in Fig. 4.1. Vertical integration of the continuity equation

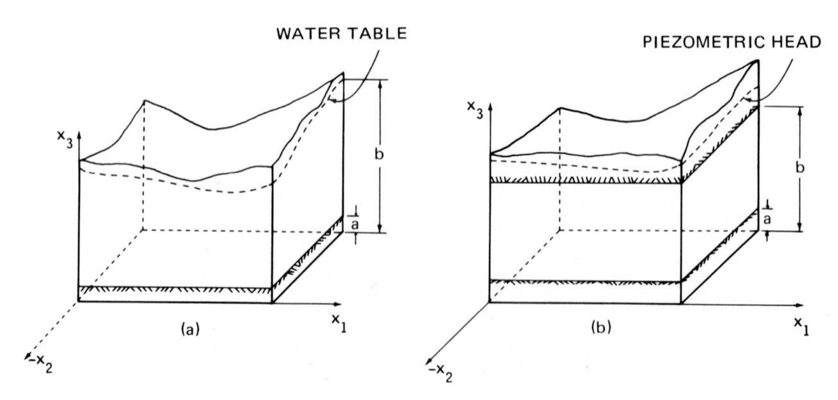

Fig. 4.1. Unconfined and confined groundwater systems: (a) unconfined aquifer; (b) confined aquifer.

yields

$$\int_a^b \left[\frac{\partial}{\partial x_i} V_i + \frac{\partial}{\partial x_3} V_3 + S_s \frac{\partial h}{\partial t}\right] dx_3 = 0, \qquad i = 1, 2. \qquad (4.2.1.12)$$

Assuming that S_s is independent of x_3 and applying, in a reversed sense, the Leibniz rule (4.2.1.4) for differentiation of an integral, one obtains

$$\frac{\partial}{\partial x_i}\int_a^b V_i\, dx_3 - V_i\Big|_b \frac{\partial b}{\partial x_i} + V_i\Big|_a \frac{\partial a}{\partial x_i} + V_3\Big|_b - V_3\Big|_a$$

$$+ S_s\left[\frac{\partial}{\partial t}\int_a^b h\, dx_3 - h\Big|_b \frac{\partial b}{\partial t} + h\Big|_a \frac{\partial a}{\partial t}\right] = 0, \qquad i = 1, 2. \qquad (4.2.1.13)$$

Let us now define a vertically averaged value of a general function ψ as

$$\overline{\psi} = 1/l \int_a^b \psi\, dx_3, \qquad l = b - a. \qquad (4.2.1.14)$$

Substitution of the average defined by (4.2.1.14) into Eq. (4.2.1.13) yields

$$\frac{\partial}{\partial x_i}(\overline{V}_i l) - V_i\Big|_b \frac{\partial b}{\partial x_i} + V_i\Big|_a \frac{\partial a}{\partial x_i} + V_3\Big|_b - V_3\Big|_a$$

$$= S_s\left[\frac{\partial}{\partial t}(\overline{h} l) - h\Big|_b \frac{\partial b}{\partial t} + h\Big|_a \frac{\partial a}{\partial t}\right] = 0, \qquad i = 1, 2. \qquad (4.2.1.15a)$$

Invocation of the assumption that $h|_a \simeq h|_b \simeq \bar{h}$ and the fact that $l = b - a$ enables us to reduce Eq. (4.2.1.15a) to

$$\frac{\partial}{\partial x_i}(\bar{V}_i l) + V_3\bigg|_b - V_i \frac{\partial b}{\partial x_i}\bigg|_b + V_i \frac{\partial a}{\partial x_i}\bigg|_a - V_3\bigg|_a + S_s l \frac{\partial \bar{h}}{\partial t} = 0, \qquad i = 1, 2. \tag{4.2.1.15b}$$

If $K_{\alpha\beta}$ ($\alpha, \beta = 1, 2, 3$) are assumed to be independent of x_3, a similar procedure applied to the momentum balance equation (Darcy's law) (4.2.1.11b) yields the relation

$$l\bar{V}_i + K_{ij}\frac{\partial \bar{h} l}{\partial x_j} - K_{ij}\left[h\bigg|_b \frac{\partial b}{\partial x_j} - h\bigg|_a \frac{\partial a}{\partial x_j}\right] + K_{i3}h\bigg|_b - K_{i3}h\bigg|_a = 0,$$

$$i, j = 1, 2. \tag{4.2.1.16a}$$

Expanding the second term in (4.2.1.16a), invoking the assumption that $h|_a \simeq h|_b \simeq \bar{h}$, and noting that $l = b - a$ allows us to simplify (4.2.1.16a) to

$$l\bar{V}_i + lK_{ij}\,\partial\bar{h}/\partial x_j = 0, \qquad i, j = 1, 2. \tag{4.2.1.16b}$$

Although it is now possible to manipulate (4.2.1.15) and (4.2.1.16b) to obtain several useful expressions, it is advantageous to postpone this step until we introduce the concept of the free surface condition. We define a free surface as the flow boundary between two fluids. A particle initially located on the free surface that separates air and water (i.e., the water table) will continue to reside there in the absence of recharge. This condition is stated mathematically as [see Bear (1979)]

$$\frac{\partial F}{\partial t} + V_\alpha^* \frac{\partial F}{\partial x_\alpha} = 0, \qquad \alpha = 1, 2, 3, \tag{4.2.1.17}$$

where V_α^* is a component of the free surface velocity. Here F is a function (of all three space dimensions and time) that is zero at the free surface, positive above it, and negative below it. Thus, on the free surface,

$$F(x_1, x_2, x_3, t) = x_3 - b(x_1, x_2, t) = 0. \tag{4.2.1.18}$$

Combination of (4.2.1.17) and (4.2.1.18) yields

$$\frac{\partial b}{\partial t} + V_1^* \frac{\partial b}{\partial x_1} + V_2^* \frac{\partial b}{\partial x_2} - V_3^* = 0. \tag{4.2.1.19}$$

Water Table Case

The vertically averaged equations describing the water table problem (Fig. 4.1a) are obtained by combining Eqs. (4.2.1.15b) and (4.2.1.19), after multiplying (4.2.1.19) by the drainable porosity, which we assume here to be the specific yield S_y. The resulting equation is

$$\frac{\partial}{\partial x_i}(\overline{V}_i l) + [V_3 - S_y V_3^*]_b - [V_i - S_y V_i^*]_b \frac{\partial b}{\partial x_i} + V_i \bigg|_a \frac{\partial a}{\partial x_i} - V_3 \bigg|_a + S_s l \frac{\partial \overline{h}}{\partial t} + S_y \frac{\partial b}{\partial t} = 0, \qquad i = 1, 2. \tag{4.2.1.20}$$

Let us now assume

$$\frac{\partial b}{\partial t} = \frac{\partial h}{\partial t}\bigg|_b \simeq \frac{\partial \overline{h}}{\partial t}. \tag{4.2.1.21}$$

The continuity and momentum equations, (4.2.1.20) and (4.2.1.16b), can now be combined along with the assumption of (4.2.1.21) to yield

$$-\frac{\partial}{\partial x_i}\left(l K_{ij} \frac{\partial \overline{h}}{\partial x_j}\right) + [V_3 - S_y V_3^*]_b - [V_i - S_y V_i^*]_b \frac{\partial b}{\partial x_i} + \left[V_i \frac{\partial a}{\partial x_i} - V_3\right]_a + (S_s l + S_y)\frac{\partial \overline{h}}{\partial t} = 0, \qquad i, j = 1, 2. \tag{4.2.1.22}$$

The second and third terms represent the net flow across the water table (the free surface in this instance). It is evident that the net flow is the difference between the fluid velocity and interface velocity. The fourth term represents the net flow across the base of the aquifer. The fifth term represents volume changes with changes in pressure ($S_s l\, \partial\overline{h}/\partial t$) and instantaneous drainage at the free surface $S_y\, \partial\overline{h}/\partial t$.

Confined Aquifer

The confined aquifer case is presented schematically in Fig. 4.1b. The mathematical formulation of this problem is a simplified version of that presented for the water table problem. The difference, of course, is the absence of the free surface and the consequent time invariance of l, which is now the aquifer thickness. Setting V_i^* to zero and recognizing the time invariance of l, we obtain from (4.2.1.22)

$$-\frac{\partial}{\partial x_i}\left(l K_{ij} \frac{\partial \overline{h}}{\partial x_j}\right) + \left[V_3 - V_i \frac{\partial b}{\partial x_i}\right]_b - \left[V_3 - V_i \frac{\partial a}{\partial x_i}\right]_a + S_s l \frac{\partial \overline{h}}{\partial t} = 0, \qquad i, j = 1, 2. \tag{4.2.1.23}$$

The second and third terms now represent flow from the overlying and underlying confining beds, respectively.

It is common to redefine the time invariant coefficients as

$$T_{ij} \equiv lK_{ij} \qquad \text{and} \qquad S_s l \equiv S,$$

where T is called the transmissivity and S the storage coefficient. Equation (4.2.1.23) now becomes

$$-\frac{\partial}{\partial x_i}\left(T_{ij}\frac{\partial \bar{h}}{\partial x_j}\right) + \left[V_3 - V_i\frac{\partial b}{\partial x_i}\right]_b - \left[V_3 - V_i\frac{\partial a}{\partial x_i}\right]_a + S\frac{\partial \bar{h}}{\partial t} = 0, \quad i, j = 1, 2. \tag{4.2.1.24}$$

Coastal Aquifers

Coastal aquifers often contain two fluids, namely, freshwater and saltwater. These two fluids are not completely immiscible, but in many situations they are separated by a relatively thin transition zone that may be approximated by an abrupt change. The freshwater forms a pillow (or lens) that is variable in thickness and underlain by slightly denser saltwater. To describe the areal flow and interaction of these two fluids, we now need to perform the vertical integration for each fluid.

The equation formulation for this problem actually incorporates ideas from the preceding two sections. The nomenclature to be used is presented in Fig. 4.2. Rather than one free surface, such as we observed in the unconfined aquifer problem, we now have two. One is the air–freshwater interface; the other, the freshwater–saltwater interface. The equation governing the fresh water lens can be written immediately. It is simply Eq. (4.2.1.22) with the addition of a subscript f to identify it as freshwater, viz.,

$$-\frac{\partial}{\partial x_i}\left(l_f K_{fij}\frac{\partial \bar{h}_f}{\partial x_j}\right) + [V_{f3} - S_y V_3^*]_c - [V_{fi} - S_y V_i^*]_c \frac{\partial c}{\partial x_i} + \left[V_{fi}\frac{\partial b}{\partial x_i} - V_{f3}\right]_b + (S_{fs} l_f + S_y)\frac{\partial \bar{h}_f}{\partial t} = 0, \quad i, j = 1, 2, \tag{4.2.1.25}$$

where

$$K_{fij} = k_{ij}\rho_f g/\mu_f, \qquad \bar{h}_f = 1/l_f \int_b^c h_f \, dx_3,$$

and

$$l_f = c - b \simeq \bar{h}_f - b.$$

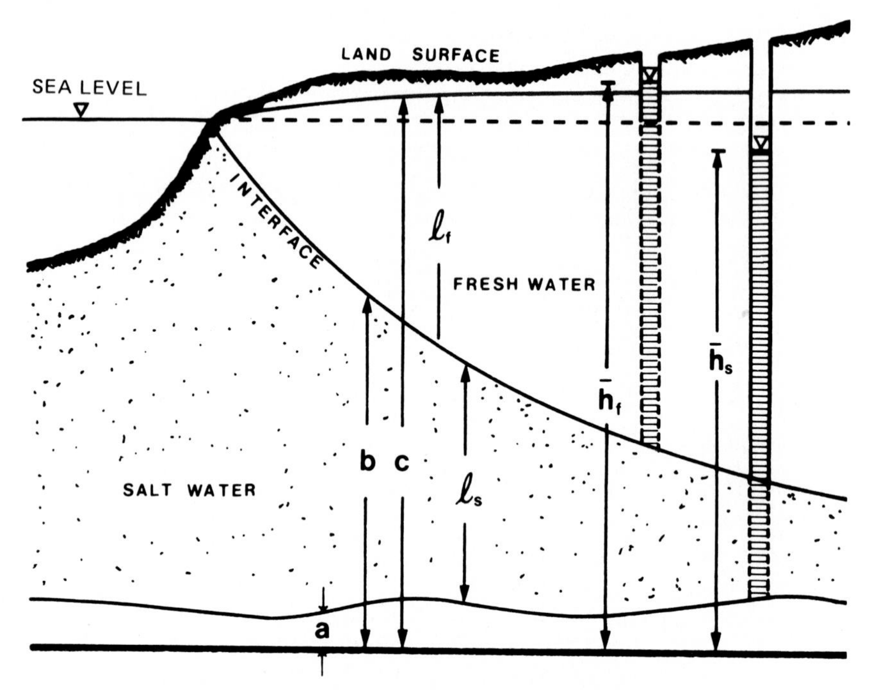

Fig. 4.2. Seawater intrusion in a coastal aquifer system.

The additional free surface condition at the interface between the saltwater and freshwater is written as [c.f. Eq. (4.2.1.17)]

$$\left[\frac{\partial}{\partial t}(x_3 - b) + V_\alpha^0 \frac{\partial}{\partial x_\alpha}(x_3 - b)\right]_{x_3=b}$$

$$= \left[-\frac{\partial b}{\partial t} - V_i^0 \frac{\partial b}{\partial x_i} + V_3^0\right]_{x_3=b} = 0, \qquad \alpha = 1, 2, 3 \text{ and } i = 1, 2, \tag{4.2.1.26}$$

where the superscript 0 indicates interface velocity. Multiplying (4.2.1.26) by porosity ϕ and adding the product to (4.2.1.25) yields

$$-\frac{\partial}{\partial x_i}\left(l_f K_{fij} \frac{\partial \bar{h}_f}{\partial x_j}\right) - q_{fc} + q_{fb} + (S_{fs} l_f + S_y)\frac{\partial \bar{h}_f}{\partial t}$$

$$- \phi \frac{\partial b}{\partial t} = 0, \qquad i, j = 1, 2, \tag{4.2.1.27a}$$

where

$$q_{fc} = [V_{fi} - S_y V_{fi}^*]_c \frac{\partial c}{\partial x_i} - [V_{f3} - S_y V_{f3}^*]_c, \qquad i,j = 1,2, \qquad (4.2.1.27b)$$

and

$$q_{fb} = [V_{fi} - \phi V_i^0]_b \frac{\partial b}{\partial x_i} - [V_{f3} - \phi V_3^0]_b, \qquad i,j = 1,2. \qquad (4.2.1.27c)$$

In the field q_{fc} will represent effective withdrawal from the freshwater aquifer, i.e.,

$$q_{fc} = Q_{fk}\delta(\mathbf{x} - \mathbf{x}_k) - I, \qquad (4.2.1.27d)$$

where Q_{fk} is the well discharge at location $\mathbf{x}_k$, δ is the Dirac delta function, and I is the rate of recharge due to infiltration. Normally, q_{fb} would be zero, because freshwater would not cross the interface.

Flow in the saltwater region is governed by the same basic equation modified to accommodate the impermeable lower unit. The steps required to arrive at the final equation are analogous to those presented previously. Thus, denoting the saltwater variable with the subscript s, we obtain

$$-\frac{\partial}{\partial x_i}\left(l_s K_{sij} \frac{\partial \bar{h}_s}{\partial x_j}\right) - q_{sb} + q_{sa} + S_{ss} l_s \frac{\partial \bar{h}_s}{\partial t} + \phi \frac{\partial b}{\partial t} = 0, \qquad (4.2.1.28a)$$

where q_{sb}, q_{sa}, K_{sij}, $\bar{h}_s$, and l_s are defined as

$$q_{sb} = [V_{si} - \phi V_i^0]_b \frac{\partial b}{\partial x_i} - [V_{s3} - \phi V_3^0]_b, \qquad (4.2.1.28b)$$

$$q_{sa} = \left[V_{si} \frac{\partial a}{\partial x_i} - V_{s3}\right]_a, \qquad (4.2.1.28c)$$

$$K_{sij} = k_{ij} \rho_s g / \mu_s,$$

$$\bar{h}_s = \frac{1}{l_s} \int_a^b h_s \, dx_3,$$

$$l_s = b - a.$$

The q_{sb} term describes flow across the interface and is normally zero except where pumping is encountered. The flow across the aquifer base q_{sa} accounts for vertical leakage from the confining bed.

To close the system of equations requires an additional condition at the interface. The appropriate constraint is pressure continuity across the interface, i.e.,

$$p_s\Big|_b - p_f\Big|_b = 0, \tag{4.2.1.29}$$

where p is the fluid pressure. To express this condition in terms of head we write

$$p_f\Big|_b = \rho_f g\left(h_f\Big|_b - b\right) \simeq \rho_f g(\bar{h}_f - b), \tag{4.2.1.30a}$$

$$p_s\Big|_b = \rho_s g\left(h_s\Big|_b - b\right) \simeq \rho_s g(\bar{h}_s - b). \tag{4.2.1.30b}$$

Combination of Eqs. (4.2.1.30a) and (4.2.1.30b) yields the required condition

$$\rho_f g(\bar{h}_f - b) = \rho_s g(\bar{h}_s - b), \tag{4.2.1.31a}$$

or

$$b = (\rho_s \bar{h}_s - \rho_f \bar{h}_f)/(\rho_s - \rho_f). \tag{4.2.1.31b}$$

Let us define a dimensionless density as

$$\rho_s^* = \rho_s/(\rho_s - \rho_f), \qquad \rho_f^* = \rho_f/(\rho_s - \rho_f). \tag{4.2.1.31c}$$

Substitution of this definition into (4.2.1.31b) and subsequent time differentiation yields

$$\partial b/\partial t = \rho_s^* \,\partial \bar{h}_s/\partial t - \rho_f^* \,\partial \bar{h}_f/\partial t. \tag{4.2.1.32}$$

The final set of governing equations is obtained through combination of (4.2.1.27a), (4.2.1.28a) and (4.2.1.32):

$$\frac{\partial}{\partial x_i}\left(l_f K_{fij} \frac{\partial \bar{h}_f}{\partial x_j}\right) + q_{fc} - (l_f S_{fs} + S_y + \phi\rho_f^*)\frac{\partial \bar{h}_f}{\partial t} + \phi\rho_s^* \frac{\partial \bar{h}_s}{\partial t} = 0, \tag{4.2.1.33a}$$

$$\frac{\partial}{\partial x_i}\left(l_s K_{sij} \frac{\partial \bar{h}_s}{\partial x_j}\right) - q_{sa} + q_{sb} - (l_s S_{ss} + \phi\rho_s^*)\frac{\partial \bar{h}_s}{\partial t} + \phi\rho_f^* \frac{\partial \bar{h}_f}{\partial t} = 0. \tag{4.2.1.33b}$$

The nonlinearities and coupling associated with these equations is best illustrated by rewriting (4.2.1.33) in matrix form, viz.,

$$\begin{bmatrix} \frac{\partial}{\partial x_i}(l_f K_{fij}) + l_f K_{fij}\frac{\partial}{\partial x_i} & 0 \\ 0 & \frac{\partial}{\partial x_i}(l_s K_{sij}) + l_s K_{sij}\frac{\partial}{\partial x_i} \end{bmatrix} \begin{Bmatrix} \frac{\partial \bar{h}_f}{\partial x_j} \\ \frac{\partial \bar{h}_s}{\partial x_j} \end{Bmatrix}$$

$$- \begin{bmatrix} l_f S_{fs} + S_y + \phi \rho_f^* & -\phi \rho_s^* \\ -\phi \rho_f^* & l_s S_{ss} + \phi \rho_s^* \end{bmatrix} \begin{Bmatrix} \frac{\partial \bar{h}_f}{\partial t} \\ \frac{\partial \bar{h}_s}{\partial t} \end{Bmatrix} + \begin{Bmatrix} q_{fc} \\ q_{sb} - q_{sa} \end{Bmatrix} = \begin{Bmatrix} 0 \\ 0 \end{Bmatrix}.$$

The equations are nonlinear because l_f and l_s depend upon $\bar{h}_f$ and $\bar{h}_s$. The full matrix associated with the time derivative demonstrates that the coupling between the two partial differential equations occurs not only through the thickness terms l_f and l_s but also through the time derivatives.

4.2.2 Compressible Flow

In this section, we consider the compressible flow of gas in porous media. The governing equation may be derived by combining the general mass continuity equation with the equation of state and Darcy's law. The continuity equation takes the form

$$-\partial \rho V_i / \partial x_i = \partial \phi \rho / \partial t. \tag{4.2.2.1}$$

Combining Eq. (4.2.2.1) with Darcy's law, Eq. (4.2.1.2), we obtain

$$\frac{\partial}{\partial x_i}\left[\frac{k_{ij}}{\mu}\rho\left(\frac{\partial p}{\partial x_j} + \rho g \frac{\partial z}{\partial x_j}\right)\right] = \frac{\partial}{\partial t}(\phi \rho). \tag{4.2.2.2}$$

Ideal Gas Flow

For ideal gas, the equation of state is given by

$$\rho = Mp/RT, \tag{4.2.2.3a}$$

where M is the molecular weight, R the universal gas constant, and T the absolute temperature. If the flow takes place under isothermal conditions, then the relation between gas density and pressure is given by

$$\frac{\partial \rho}{\partial t} = \frac{M}{RT}\frac{\partial p}{\partial t}. \tag{4.2.2.3b}$$

Substitution of Eqs. (4.2.2.3a) and (4.2.2.3b) into (4.2.2.2) yields

$$\frac{\partial}{\partial x_i}\left(\frac{k_{ij}}{\mu} p \frac{\partial p}{\partial x_j}\right) + \frac{\partial}{\partial x_i}\left(\frac{k_{ij}}{\mu} p\rho g \frac{\partial z}{\partial x_j}\right) = \frac{\partial}{\partial t}(\phi p), \tag{4.2.2.4}$$

which can be written as

$$\frac{\partial}{\partial x_i}\left(\frac{k_{ij}}{\mu} p \frac{\partial p}{\partial x_j}\right) + \frac{\partial}{\partial x_i}\left(\frac{k_{ij}}{\mu} p\rho g e_j\right) = (p\alpha + \phi)\frac{\partial p}{\partial t}, \tag{4.2.2.5}$$

where e_j is the upward positive vertical unit vector and α the compressibility of the rock matrix. When the effect of gravity and rock compressibility are neglected, Eq. (4.2.2.5) reduces to

$$\frac{\partial}{\partial x_i}\left(\frac{k_{ij}}{\mu} p \frac{\partial p}{\partial x_j}\right) = \phi \frac{\partial p}{\partial t}, \tag{4.2.2.6a}$$

which may be written in the form

$$\frac{\partial}{\partial x_i}\left(\frac{k_{ij}}{\mu} \frac{\partial u}{\partial x_j}\right) = \frac{\phi}{p} \frac{\partial u}{\partial t}, \tag{4.2.2.6b}$$

where $u = p^2$.

Real Gas Flow

In the case of real gas, the equation of state is given by

$$\rho = Mp/\sigma RT, \tag{4.2.2.7}$$

where $\sigma = \sigma(p)$ is the gas deviation factor. When the gravity effect and rock compressibility are neglected, the governing differential equation (4.2.2.2) can be reduced to a simple form by using the "Leibnezon" transformation [see Thomas (1982)]:

$$m(p) = \int_{p_o}^{p} 2\xi \, d\xi/\sigma(\xi)\mu(\xi), \tag{4.2.2.8a}$$

where the function m is called the "real gas pseudopressure function."

Applying Leibniz's rule (4.2.1.4) to (4.2.2.8a), we obtain

$$\frac{\partial m}{\partial x_j} = \frac{2p}{\mu\sigma}\frac{\partial p}{\partial x_j}, \tag{4.2.2.8b}$$

$$\frac{\partial m}{\partial t} = \frac{2p}{\mu\sigma}\frac{\partial p}{\partial t}. \tag{4.2.2.8c}$$

From Eq. (4.2.2.7), it follows that

$$\frac{\partial \rho}{\partial t} = \frac{M}{RT}\left[\frac{1}{\sigma}\frac{\partial p}{\partial t} - \frac{p}{\sigma^2}\frac{d\sigma}{dp}\frac{\partial p}{\partial t}\right], \tag{4.2.2.9a}$$

or

$$\frac{\partial \rho}{\partial t} = \frac{M}{\sigma RT}\left[1 - \frac{p}{\sigma}\frac{d\sigma}{dp}\right]\frac{\partial p}{\partial t}. \tag{4.2.2.9b}$$

Substituting Eq. (4.2.2.7) into (4.2.2.9b) and rearranging the terms, we obtain

$$\frac{\partial \rho}{\partial t} = \rho\left[\frac{1}{p} - \frac{1}{\sigma}\frac{d\sigma}{dp}\right]\frac{\partial p}{\partial t}. \tag{4.2.2.10a}$$

Equation (4.2.2.10a) may be written in the form

$$\frac{\partial \rho}{\partial t} = \rho\beta\frac{\partial p}{\partial t}, \tag{4.2.2.10b}$$

where β is the gas compressibility factor defined as

$$\beta = \frac{1}{\rho}\frac{d\rho}{dp} = \left[\frac{1}{p} - \frac{1}{\sigma}\frac{d\sigma}{dp}\right]. \tag{4.2.2.10c}$$

Combination of Eq. (4.2.2.10b), (4.2.2.8c), and (4.2.2.7) yields

$$\frac{\partial \rho}{\partial t} = \frac{M\beta\mu}{2RT}\frac{\partial m}{\partial t}. \tag{4.2.2.10d}$$

From Eq. (4.2.2.8b) and (4.2.2.7), it follows that

$$\frac{\partial p}{\partial x_j} = \frac{\mu M}{2\rho RT}\frac{\partial m}{\partial x_j}. \tag{4.2.2.10e}$$

Now neglecting the gravity effect in Darcy's law and rock compressibility in the continuity equation, we can write the governing differential equation (4.2.2.2) in the form

$$\frac{\partial}{\partial x_i}\left[\rho\frac{k_{ij}}{\mu}\frac{\partial p}{\partial x_j}\right] = \phi\frac{\partial \rho}{\partial t}. \tag{4.2.2.11}$$

Substitution of Eqs. (4.2.2.10d), (4.2.2.10e), and (4.2.2.7) into (4.2.2.11) yields

$$\frac{\partial}{\partial x_i}\left[k_{ij}\frac{\partial m}{\partial x_j}\right] = \phi\beta\mu\frac{\partial m}{\partial t}, \tag{4.2.2.12}$$

which is a more attractive form of Eq. (4.2.2.11) because it has only one unknown parameter, m; but it is nonlinear because the coefficient on the right-hand side depends on the unknown variable.

When the gravity effect must be incorporated, Eq. (4.2.2.2) may be reduced to a simpler form by introducing the piezometric head

$$h = \frac{1}{g}\int_{p_o}^{p} \frac{d\xi}{\rho(\xi)} + z. \tag{4.2.2.13}$$

Differentiating Eq. (4.2.2.13) with respect to x_j and t and substituting the result into (4.2.2.2), we obtain

$$\frac{\partial}{\partial x_i}\left[\rho K_{ij}\frac{\partial h}{\partial x_j}\right] = \frac{\partial}{\partial t}\phi\rho = \phi\beta\rho^2 g\frac{\partial h}{\partial t}, \tag{4.2.2.14}$$

where it is assumed that the rock compressibility is negligible and K_{ij} is, once again, the fluid conductivity tensor, given by $K_{ij} = \rho g k_{ij}/\mu$. Equation (4.2.2.14) is a nonlinear partial differential equation containing the variable coefficients ρ and β, which are related to the fluid pressure by Eqs. (4.2.2.7) and (4.2.2.10c).

4.3 Saturated Groundwater Flow

In this section, we deal with several groundwater flow problems which are described by the equations derived previously. We start by analyzing the local problems of flow toward single pumped wells and then proceed to deal with the larger scale basin-wide problems. Application of the finite element method to these types of problems can also be found in Javandel and Witherspoon (1968) and Pinder and Frind (1972). Boundary conditions pertaining to each type of problem will be identified, and special techniques for handling these boundary conditions will be discussed.

In all of the problems considered in this chapter, Darcy's law is assumed to be valid although, in a few situations of local well flow, non-Darcy effects may occur near the well. Extension of the finite element formulation presented in this section to deal with non-Darcy flow can be found in Huyakorn (1973) and Volker (1969).

4.3.1 Axisymmetric Flow to a Well in a Confined Aquifer

Assuming that the principal axes of the hydraulic conductivity tensor lie in the radial and vertical directions, the governing equation for confined aquifer well flow is Eq. (4.1.2.10) rewritten as

$$\frac{1}{r}\frac{\partial}{\partial r}\left[K_{rr}r\frac{\partial h}{\partial r}\right]+\frac{\partial}{\partial z}\left[K_{zz}\frac{\partial h}{\partial z}\right]=S_{s}\frac{\partial h}{\partial t}. \qquad (4.3.1.1)$$

The initial condition commonly encountered is the known hydraulic head distribution just prior to the start of the pumping. For a situation where the undisturbed initial head is uniform, it is a common practice in well hydraulics to use this value as the datum in referring to the hydraulic head at later times. In such a case, the computed hydraulic head corresponds to the drawdown s, which is defined as $s = H_o - h$, where H_o is the initial head.

To describe the flow problem completely, we also need to specify boundary conditions. Three types of conditions are encountered. Two of these are the prescribed head and the prescribed flux across the boundary. Their finite element treatment is quite standard, and this has been described in Chapter 2. The third type of boundary condition corresponds to the condition of specified total discharge at the well. Because the hydraulic head (or drawdown) at the well is uniform but unknown, this type of boundary condition requires special treatment.

To obtain the Galerkin finite element approximation of Eq. (4.3.1.1), we subdivide the domain into finite elements and construct a trial function of the form

$$\hat{h}(r, z, t) = N_J(r, z)h_J(t), \qquad J = 1, 2, \ldots, n, \qquad (4.3.1.2)$$

where repeated indices indicate nodal summation and n is the total number of nodes in the finite element network. By Galerkin's method, we obtain

$$\int_R N_I\left[\frac{1}{r}\frac{\partial}{\partial r}\left(K_{rr}r\frac{\partial \hat{h}}{\partial r}\right)+\frac{\partial}{\partial z}\left(K_{zz}\frac{\partial \hat{h}}{\partial z}\right)-S_s\frac{\partial \hat{h}}{\partial t}\right]dR = 0,$$
$$I = 1, 2, \ldots, n, \qquad (4.3.1.3)$$

where for an axisymmetric problem $dR = 2\pi r\, dr\, dz$. After applying Green's theorem and substituting for $\hat{h}$, we obtain the discretized equation of the form

$$\sum_{e=1}^{m}\left[\int_{R^e}\left(K_{rr}\frac{\partial N_I}{\partial r}\frac{\partial N_J}{\partial r}+K_{zz}\frac{\partial N_I}{\partial z}\frac{\partial N_J}{\partial z}\right)h_J 2\pi r\, dr\, dz\right.$$
$$+\int_{R^e} S_s N_I N_J \frac{dh_J}{dt} 2\pi r\, dr\, dz$$
$$\left.+\int_{B_2^e} qN_I\, \mathrm{dB}\right] = 0, \qquad I, J = 1, 2, \ldots, n, \qquad (4.3.1.4)$$

where q is the outward normal flux, $q = V_i n_i$, and our sign convention is such that discharge is positive and recharge is negative.

In dealing with well flow in confined aquifers, it is more convenient to formulate the problem in terms of drawdown s. This can be achieved by rewriting Eq. (4.3.1.4) in the form

$$\sum_{e=1}^{m} \left[\int_{R^e} \left(K_{rr} \frac{\partial N_I}{\partial r} \frac{\partial N_J}{\partial r} + K_{zz} \frac{\partial N_I}{\partial z} \frac{\partial N_J}{\partial z} \right) s_J 2\pi r \, dr \, dz \right.$$

$$+ \int_{R^e} S_s N_I N_J \frac{ds_J}{dt} 2\pi r \, dr \, dz$$

$$\left. + \int_{B_2^e} q N_I \, dB \right] = 0, \qquad I, J = 1, 2, \ldots, n, \tag{4.3.1.5}$$

where s_J is the nodal drawdown, taken to be the difference between h_J and the initial head H_o. Using finite differences to perform time integration, we obtain a system of algebraic equations that may be written in matrix form as

$$[E]\{s\}^{k+1} = \{R\}, \tag{4.3.1.6a}$$

where

$$[E] = \sum_{e=1}^{m} [E]^e, \qquad \{R\} = \sum_{e=1}^{m} \{R\}^e. \tag{4.3.1.6b}$$

$\{s\}^{k+1}$ contains the nodal values of drawdown at the current time level, and the typical elements of $[E]^e$ and $\{R\}^e$ are given by

$$E_{IJ}^e = \theta \int_{R^e} \left(K_{rr} \frac{\partial N_I}{\partial r} \frac{\partial N_J}{\partial r} + K_{zz} \frac{\partial N_I}{\partial z} \frac{\partial N_J}{\partial z} \right) 2\pi r \, dr \, dz$$

$$+ \int_{R^e} \frac{S_s}{\Delta t} N_I N_J 2\pi r \, dr \, dz, \tag{4.3.1.7a}$$

$$R_I^e = \left[(\theta - 1) \int_{R^e} \left(\mathrm{K}_{rr} \frac{\partial N_I}{\partial r} \frac{\partial N_J}{\partial r} + K_{zz} \frac{\partial N_I}{\partial z} \frac{\partial N_J}{\partial z} \right) 2\pi r \, dr \, dz \right.$$

$$\left. + \int_{R^e} \frac{S_s}{\Delta t} N_I N_J 2\pi r \, dr \, dz \right] s_J^k - \int_{B_2^e} N_I [\theta q^{k+1} + (1 - \theta) q^k] \, dB, \tag{4.3.1.7b}$$

in which θ is the time-weighting factor. To ensure unconditional stability, the value θ selected should be in the range $0.5 \leq \theta \leq 1.0$. The integrals in (4.3.1.6) are performed over subregions that are obtained by rotating plane areas about the axis of the well. Figure 4.3 illustrates the typical axisymmetric ring element produced by rotating a plane triangle about the z axis. For this particular element, the integrals are easily evaluated if one replaces r by the centroidal radius $\bar{r}$, defined as

$$\bar{r} = \tfrac{1}{3}(r_1 + r_2 + r_3).$$

As an example, consider the integral

$$\int_{R^e} K_{rr} \frac{\partial N_I}{\partial r} \frac{\partial N_J}{\partial r} r\, dr\, dz = \bar{r} \int_{R^e} K_{rr} \frac{\partial N_I}{\partial r} \frac{\partial N_J}{\partial r} dr\, dz, \qquad (4.3.1.8a)$$

in which the right-hand-side integral is easily evaluated using the formulas developed for the plane triangle.

The surface integral in Eq. (4.3.1.7b) exists at the well boundary. This integral corresponds to the fluid flux entering the well. Consider the element at the well boundary shown in Fig. 4.4. Taking into account the radial symmetry, we can write for node I ($I = 1, 2$)

$$\int_{B_2} qN_I\, dB\big|_{r=r_w} = 2\pi \int_{z_2}^{z_1} V_r N_I r_w\, dz = Q_I^e, \qquad (4.3.1.8b)$$

where V_r is the radial velocity. The quantity Q_I^e is referred to as "nodal" flux at node I, contributed by element e. On evaluating the boundary integral, we obtain

$$Q_I^e = 2\pi r_w(z_1 - z_2)(V_r/2). \qquad (4.3.1.8c)$$

The total nodal flux Q_I is obtained by adding the contributions from

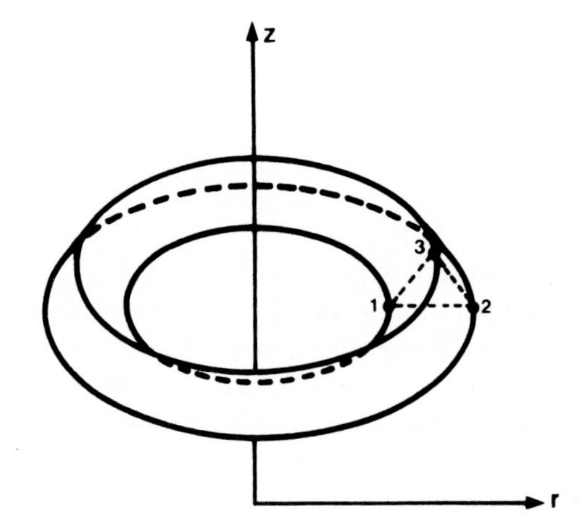

Fig. 4.3. Triangular ring element.

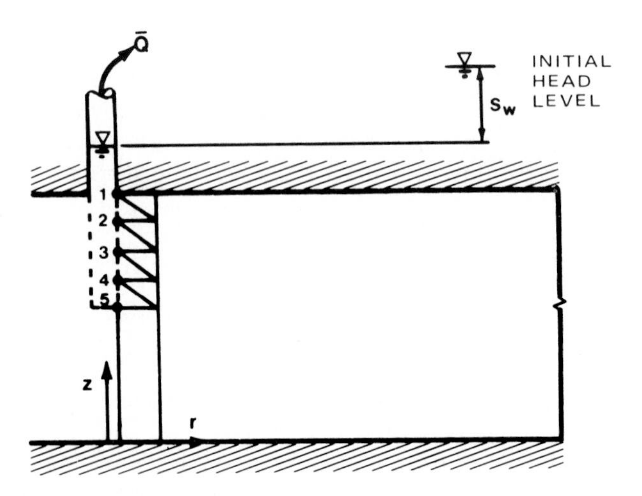

Fig. 4.4. Boundary conditions along a partially penetrating well.

adjoining elements. The sum of all nodal fluxes entering the well is equal to the total calculated discharge Q.

Treatment of Prescribed Well Discharge Condition

When the well is pumped at a prescribed rate $\overline{Q}$, the well boundary condition requires careful consideration. In a case involving a fully penetrating well in a uniform formation, the distribution of velocity can be assumed as uniform along the well bore. Given the prescribed $\overline{Q}$, we can determine V_r from

$$V_r = \overline{Q}/(2\pi r_w l), \tag{4.3.1.8d}$$

where l is the thickness of the formation. Knowing the value of V_r, Eq. (4.3.1.6) can be solved using one of the time-stepping schemes described earlier.

In the more general case of a nonuniform velocity distribution at the well bore, the prescribed discharge condition may be treated using one of the following methods:

METHOD 1 The principle of superposition can be used to correct the computed drawdown values to obtain the prescribed discharge $\overline{Q}$. We first describe the method introduced by Neuman and Witherspoon (1969). Consider the partially penetrating well illustrated in Fig. 4.4. At the well bore, the drawdown distribution may be assumed uniform. Thus, we can write

$$s_1(t) = s_2(t) = \cdots = s_M(t) = s_w(t), \tag{4.3.1.8e}$$

where M is the number of nodes on the well boundary and s_w is the well drawdown. The value of s_w is unknown, but it must be determined such that the calculated discharge Q equals the prescribed discharge $\overline{Q}$. To incorporate the uniform drawdown condition at the well bore, we write the system of simultaneous equations (4.3.1.6a) using indicial notation as

$$E_{IJ}s_J = R_I, \qquad I, J = 1, 2, \ldots, n, \tag{4.3.1.9}$$

where E_{IJ} and R_I are elements of the coefficient matrix and the right-hand-side vector, respectively, and s_J are nodal values of drawdown at the current time level. The preceding set of equations is now decomposed into two subsets given by

$$E_{\alpha J}s_J = R_\alpha = Q_\alpha + F_\alpha, \qquad \alpha = 1, 2, \ldots, M, \tag{4.3.1.10a}$$

$$E_{iJ}s_J = R_i, \qquad i = M + 1, \ldots, n, \tag{4.3.1.10b}$$

where Q_α and F_α are the nodal flux and the remaining terms of R_α, respectively. The first subset consists of the first M equations, and the second subset consists of the remaining $n - M$ equations. If we assume a starting value of well drawdown, then (4.3.1.10b) can be rewritten in the form

$$\begin{aligned} E_{ij}s_j &= R_i - E_{i\alpha}s_\alpha \\ &= R_i^* \qquad (i, j = M + 1, \ldots, n), \end{aligned} \tag{4.3.1.10c}$$

where

$$R_i^* = R_i - E_{i\alpha}s_\alpha = R_i - \left(\sum_{\alpha=1}^{M} E_{i\alpha}\right)s_w. \tag{4.3.1.10d}$$

Equation (4.3.1.10c) represents the reduced set of algebraic equations. We can solve this reduced system to obtain the values of the unknowns $s_{M+1}, \ldots, s_n$ and use these values to calculate the nodal fluxes from the first subset of equations given by (4.3.1.10a). Knowing the nodal fluxes we can calculate the total well discharge from

$$Q = Q_1 + Q_2 + \cdots + Q_M. \tag{4.3.1.11}$$

Because the assumed value of well drawdown, in general, results in an incorrect calculated discharge, we must adjust the computer nodal values s_J for inaccuracy owing to this error.

For the first time level, the drawdown at node J, $\bar{s}_J(t_1)$, can be corrected using the formula

$$\bar{s}_J(t_1) = (\overline{Q}/Q)s_J(t_1), \tag{4.3.1.12a}$$

where $\bar{s}_J(t_1)$ is the adjusted drawdown at node J. Equation (4.3.1.12a) is obtained simply by invoking the principle of superposition as illustrated in Fig. 4.5a. Superposition is permissible because the drawdown for any nodal point J is proportional to the discharge. Thus, the calculated and adjusted drawdowns at t_1 are given by

$$s_J(t_1) = \bar{s}_J(t_0) + Qf_J(t_1 - t_0) \tag{4.3.1.12b}$$

and

$$\bar{s}_J(t_1) = \bar{s}_J(t_0) + \bar{Q}f_J(t_1 - t_0), \tag{4.3.1.12c}$$

where $f_J(t_1 - t_0)$ denotes a well "response" function at node J. Assuming that $\bar{s}_J(t_0) = 0$ and $t_0 = 0$, it follows that

$$s_J(t_1) = Qf_J(t_1) \tag{4.3.1.12d}$$

and

$$\bar{s}_J(t_1) = \bar{Q}f_J(t_1). \tag{4.3.1.12e}$$

Combining Eq. (4.3.1.12e) and (4.3.1.12d) yields the formula in Eq. (4.3.1.12a).

For a subsequent time level k, we solve the reduced system of Eqs. (4.3.1.10c) using an assumed value of $s_w(t_k)$. Next we correct the calculated

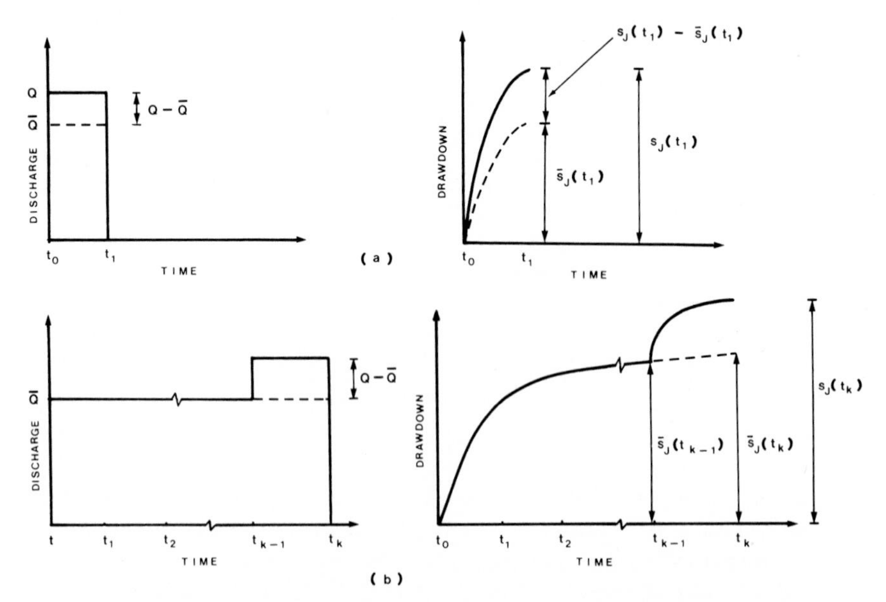

Fig. 4.5. Illustration of superposition technique for treating prescribed well discharge boundary condition: (a) for the first time step; (b) for the kth time step.

drawdowns using the formula

$$\bar{s}_J(t_k) = s_J(t_k) - [(Q - \bar{Q})/\bar{Q}]\bar{s}_J(\Delta t_k), \qquad (4.3.1.13a)$$

where $\Delta t_k = t_k - t_{k-1}$, and $\bar{s}_J(\Delta t_k)$ is known, provided the time step Δt_k is selected such that its value is equal to a previous time value, that is, $\Delta t_k = t_m$, $m < k$. To derive Eq. (4.3.1.13a), we refer to Fig. 4.5b. It follows that

$$s_J(t_k) = \bar{s}_J(t_k) + (Q - \bar{Q})f_J(\Delta t_k). \qquad (4.3.1.13b)$$

Now, if we select $\Delta t_k = t_m$, then

$$f_J(\Delta t_k) = f_J(t_m) = \bar{s}_J(t_m)/\bar{Q}. \qquad (4.3.1.13c)$$

Substituting Eq. (4.3.1.13c) into (4.3.1.13b) and rearranging the terms, we obtain

$$\bar{s}_J(t_k) = s_J(t_k) - [(Q - \bar{Q})/\bar{Q}]\bar{s}_J(t_m), \qquad (4.3.1.13d)$$

which, for $t_m = \Delta t_k$, is identical to Eq. (4.3.1.13a).

It is apparent that we cannot select the time steps arbitrarily. Indeed, this is the main limitation of the preceding procedure. We now present the following procedure, which does not have this limitation.

METHOD 2 This method was introduced by Huang (1973). It is also based on a superposition principle. For a time level k, we start by assuming an arbitrary value of well drawdown s_w, and solving the system of equations written in matrix notation as

$$[E]\{s\} = \{R\}. \qquad (4.3.1.14a)$$

Certainly in solving (4.3.1.14a) we take advantage of the fact that $s_1 = s_2 = \cdots = s_M = s_w$ and employ the matrix reduction technique to determine only the unknowns $s_{M+1}, \ldots, s_n$. Next, we consider a second matrix equation given by

$$[E]\{s^*\} = \{0\}, \qquad (4.3.1.14b)$$

where $\{s^*\}$ contains a second set of drawdown values, the first M of these values being equal to the same assumed value of s_w. Because the two systems of equations (4.3.1.14a) and (4.3.1.14b) contain the same coefficient matrix $[E]$, substantial saving in computational effort can be achieved by decomposing the matrix $[E]$ only once and performing two back substitutions for each time step. Taking a linear combination of Eqs. (4.3.1.14a) and (4.3.1.14b), we obtain

$$[E](\{s\} + \lambda\{s^*\}) = \{R\}, \qquad (4.3.1.14c)$$

or, equivalently,

$$[E]\{\bar{s}\} = \{R\}, \tag{4.3.1.14d}$$

where

$$\{\bar{s}\} = \{s\} + \lambda\{s^*\} \tag{4.3.1.14e}$$

and λ is an undetermined coefficient. If we can determine λ so that the calculated discharge Q equals the prescribed discharge $\bar{Q}$, then $\{\bar{s}\}$ will be the correct solution. To determine λ, we write the first subset of the matrix equation (4.3.1.14a), involving only the discharging nodes at the well bore, in the form

$$E_{\alpha J} s_J = F_\alpha + Q_\alpha, \qquad \alpha = 1, 2, \ldots, M, \tag{4.3.1.15a}$$

where Q_α is the nodal flux and F_α the remaining terms of R_α. The summation convention is implied on subscript J. Equation (4.3.1.15a) represents M equations, which may be summed together to give

$$\sum_{\alpha=1}^{M} E_{\alpha J} s_J = \sum_{\alpha=1}^{M} (F_\alpha + Q_\alpha) = \sum_{\alpha=1}^{M} F_\alpha + Q, \tag{4.3.1.15b}$$

where Q is the calculated discharge, which is equal to $\Sigma_{\alpha=1}^{M} Q_\alpha$.

In a similar manner, we can take the first subset of the matrix equation (4.3.1.14d) and sum the M equations in this subset to obtain

$$\sum_{\alpha=1}^{M} E_{\alpha J} \bar{s}_J = \sum_{\alpha=1}^{M} F_\alpha + \bar{Q}, \tag{4.3.1.15c}$$

where $\bar{Q}$ is the prescribed discharge as we assume that the $\bar{s}_J$ represent the correct drawdown. Substituting Eq. (4.3.1.14e) into (4.3.1.15c) and subtracting it from (4.3.1.15b), we obtain

$$-\lambda\left(\sum_{\alpha=1}^{M} E_{\alpha J} s_J^*\right) = Q - \bar{Q},$$

from which λ is given by

$$\lambda = (\bar{Q} - Q) \Big/ \sum_{\alpha=1}^{M} E_{\alpha J} s_J^* . \tag{4.3.1.16}$$

We can now summarize the procedure of method 2 as follows:

(1) For a time level k, assume an arbitrary value of well drawdown and solve the matrix equations (4.3.1.14a) and (4.3.1.14b) to obtain two sets of drawdown values $\{s\}$ and $\{s^*\}$.

(2) The correct drawdown values are obtained by taking the linear combination $\{\bar{s}\} = \{s\} + \lambda\{s^*\}$, which λ is determined from Eq. (4.3.1.16)

4.3.2 *Axisymmetric Flow to a Well in an Unconfined Aquifer*

The governing equation for unconfined well flow is the same as that for confined well flow. However, in the present case, it is a common practice to neglect the specific storage term due to the fact that the compressibility effect in unconfined flow is usually small compared to the transient effects due to gravity drainage or delayed yield. With such an assumption, the governing equation reduces to

$$\frac{1}{r}\frac{\partial}{\partial r}\left[K_{rr}\, r\frac{\partial h}{\partial r}\right] + \frac{\partial}{\partial z}\left[K_{zz}\frac{\partial h}{\partial z}\right] = 0, \qquad (4.3.2.1)$$

and the transient behavior is introduced through the conditions at the free surface moving boundary, as illustrated in Fig. 4.6. Along the screened section of the well, there exists a "seepage" face where the water exits from the aquifer into the well.

The initial condition required for this type of flow situation is the prevailing position of the free surface at time $t = 0$. This is given by

$$\zeta(r, t = 0) = \zeta_o(r), \qquad (4.3.2.2)$$

where ζ denotes the elevation of the free surface above the aquifer base, taken as the datum. The boundary conditions to be satisfied are as follows (see Fig. 4.6): along B_1, the portion of the well boundary below the seepage face, the hydraulic head is equal to the prescribed elevation of the water level in the well. On the boundary portions B_2 the normal flux is arbitrarily prescribed to be zero for this example. Because we assume no vertical flow across the aquifer base B_3, the flux thereupon is also set to zero. Along the seepage face B_s the head at a point is equal to the elevation of that point. These three boundary conditions can be expressed as

$$h(r_w, z, t) = h_w(t) \qquad \text{on} \quad B_1, \qquad (4.3.2.3a)$$

$$K_{rr}\frac{\partial h}{\partial r}n_r + K_{zz}\frac{\partial h}{\partial z}n_z = 0 \qquad \text{on} \quad B_2 \text{ and } B_3, \qquad (4.3.2.3b)$$

$$h(r_w, z, t) = z \qquad \text{on} \quad B_s, \qquad (4.3.2.3c)$$

where n_r and n_z are the radial and vertical components of the unit normal vector.

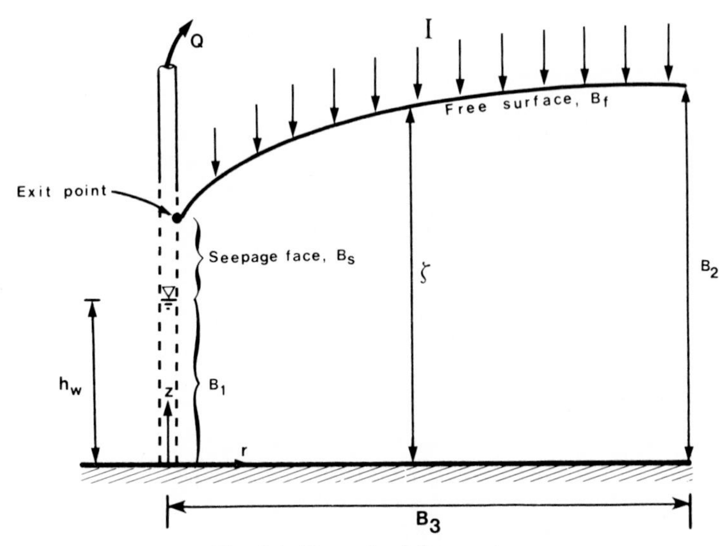

Fig. 4.6. Unconfined flow region.

The remaining boundary conditions exist on the free surface. Because the position of the free surface is unknown a priori, two conditions must be satisfied:

$$h(r, \zeta, t) = \zeta(r, t) \qquad \text{on} \quad B_{\mathrm{f}}, \qquad (4.3.2.3\mathrm{d})$$

$$\left[K_{rr}\frac{\partial h}{\partial r}n_r + K_{zz}\frac{\partial h}{\partial z}n_z\right] = \left(I - S_y\frac{\partial \zeta}{\partial t}\right)n_z \qquad \text{on} \quad B_{\mathrm{f}}, \qquad (4.3.2.3\mathrm{e})$$

where I is the rate of infiltration into the free surface. Equations (4.3.2.3d) and (4.3.2.3e) express the atmospheric pressure condition and the continuity condition on the free surface.

A finite element solution of the transient free surface problem can be obtained using either an implicit or an explicit scheme for calculating the position of the free surface.

Implicit Scheme

One procedure (Neuman and Witherspoon, 1971) solves the problem by iteration at each time step with two distinct stages in each iteration. Each iteration uses a fixed element mesh that reflects the latest estimate of the free surface elevation at the new time level. The first iteration can use the free surface position of the old time level. The first stage of an iteration solves (4.3.2.1) using prescribed head conditions on B_{f} and B_{s}, and the second stage solves again using flux conditions on B_{f} and B_{s}.

STAGE 1 We solve (4.3.2.1) with fluxes on B_2 and B_3 prescribed by Eq. (4.3.2.3b), and the heads on B_1, B_s, and B_f prescribed by (4.3.2.3a), (4.3.2.3c), and (4.3.2.3d). With these boundary conditions, Eq. (4.3.2.1) is a familiar elliptic equation. Applying the Galerkin method to Eq. (4.3.2.1) leads to the following system of finite element equations:

$$E_{IJ} h_J^{k+1} = 0. \tag{4.3.2.4}$$

Here the superscript $k + 1$ refers to the new time level, and

$$E_{IJ} = \sum_{e=1}^{m} \int_{R^e} \left(K_{rr} \frac{\partial N_I}{\partial r} \frac{\partial N_J}{\partial r} + K_{zz} \frac{\partial N_I}{\partial z} \frac{\partial N_J}{\partial z} \right) dR. \tag{4.3.2.5}$$

When we introduce the prescribed head conditions on B_1, B_f, and B_s, we solve only the equations of interior nodes and nodes on B_2 or B_3. Once the h_J^{k+1} are known, we compute the nodal fluxes at the seepage surface B_s by substituting h_J^{k+1} into the subset of the equations (4.3.2.4) that belong to boundary nodes on B_s.

$$Q_\alpha^{k+1} = E_{\alpha J} h_J^{k+1}. \tag{4.3.2.6a}$$

Here α is a prescribed head, seepage face node. Note that these equations are some of the ones not solved earlier in this stage. The Q_α^{k+1} means

$$Q_\alpha^{k+1} = -\sum_e \int_{B^e} N_\alpha q \, dB. \tag{4.3.2.6b}$$

Here q is the outward-directed normal component of the boundary flux. Incidentally, when the need arises to know the total flux into the well, we can sum the Q_α^{k+1} for all nodes α on B_1 as well as B_s.

STAGE 2 We solve (4.3.2.1) once again with a change to flux boundary conditions on B_f and B_s and leave conditions on the other boundaries as they were in the first stage. On B_s we prescribe the fluxes Q_α^{k+1} computed in the first stage. On B_f we use the unknown flux of Eq. (4.3.2.3e). The system of finite element equations becomes

$$E_{IJ} h_J^{k+1} + \sum_e \int_{B_f^e} N_I S_y \frac{\partial \zeta}{\partial t} n_z \, dB = \sum_e \int_{B_f^e} N_I I n_z \, dB - \sum_e \int_{B_s^e} N_I q \, dB. \tag{4.3.2.7a}$$

Here the subscript I refers to nodes other than the remaining prescribed head nodes on B_1. The term containing $\partial\zeta/\partial t$ adds transient behavior to the elliptic equation (4.3.2.1). The free surface nodes move during the current time interval so a time difference of ζ at a node would approximate a total derivative, $d\zeta/dt$. Because $\zeta = \zeta(r, t)$, the chain rule gives

$$\frac{d\zeta}{dt} = \frac{\partial\zeta}{\partial t} + \frac{\partial\zeta}{\partial r}\frac{dr}{dt}, \tag{4.3.2.7b}$$

where dr/dt in this case is the radial component of nodal velocity. For simplicity, we constrain the nodes to move only in the vertical direction, so that $\partial\zeta/\partial t = d\zeta/dt$. The free surface definition (4.3.2.3d) requires that $d\zeta/dt = dh/dt$ for a point on the free surface. The moving nodes on B_f make the basis function N_J depend on t in addition to r and z, but

$$\frac{d\hat{h}}{dt} = N_J\frac{dh_J}{dt} + \frac{dN_J}{dt}h_J.$$

Lynch and Gray (1980) show that, for a point moving with the element velocity (as the free surface does), $dN_J/dt = 0$. The conclusion is that we can represent $\partial\zeta/\partial t$ by

$$\frac{\partial\zeta}{\partial t} \simeq \frac{d\hat{h}}{dt} = N_J\frac{dh_J}{dt}. \tag{4.3.2.7c}$$

Substituting Eq. (4.3.2.7c) into (4.3.2.7a) and using a backward difference time-stepping scheme, one produces the following system of equations:

$$\left(E_{IJ} + \frac{1}{\Delta t}D_{IJ}\right)h_J^{k+1} = \frac{1}{\Delta t}D_{IJ}h_J^k + F_I^{k+1} + Q_{sI}^{k+1}, \tag{4.3.2.8}$$

where

$$D_{IJ} = \sum_e \int_{B_f^e} S_y N_I N_J n_z \, dB, \qquad F_I^{k+1} = \sum_e \int_{B_f^e} N_I I n_z \, dB$$

and

$$Q_{sI}^{k+1} = -\sum_e \int_{B_s^e} N_I q \, dB.$$

The integrals D_{IJ} and F_I^{k+1} actually do not vary with time (except to the extent that S_y and I may vary) even though nodes on B_f move vertically with time. The length of the free surface boundary between nodes exceeds the radial distance between nodes and varies with time, but the factor n_z, the vertical component of the boundary's unit normal vector, exactly cancels that change in the length of a boundary segment with its slope. Therefore

$$n_z \, dB = n_z r \, d\theta \, ds = r \, d\theta \, dr \qquad \text{on} \quad B_f.$$

Here $r\,d\theta$ is a differential increment of the angular coordinate and ds is a differential increment of free surface length, but we only need to integrate in the radial direction and never need to compute n_z.

To approximate the integrals Q_{sI}^{k+1} we use the integrals Q_α^{k+1} defined in (4.3.2.6b) and computed in the first stage from (4.3.2.6a). An error occurs for the node at the intersection between B_s and B_f because the nodal flux Q_I^{k+1} includes a contribution from B_f. Neuman and Witherspoon (1971) suggest correcting this error by positioning the intersection node in line with the two nearest nodes on B_f. We then solve Eq. (4.3.2.8) for the h_J^{k+1}, which include the head values on the free surface and the seepage face.

At the end of Stage 2, a maximum error $\varepsilon_{\max}$ on the free surface elevation is computed from

$$\varepsilon_{\max} = \max_{B_f} \left| h_J^{k+1} - \zeta_J^{k+1} \right|. \tag{4.3.2.9}$$

If this error is greater than a permissible head tolerance, we shift the free surface to a new position by setting $\zeta_J^{k+1} = h_J^{k+1}$. This completes one iteration cycle. For the next iteration, we modify the finite element mesh, reform all the coefficient matrices in Eqs. (4.3.2.4) and (4.3.2.8), and then repeat stages 1 and 2. Iteration is performed until $\varepsilon_{\max}$ is within the prescribed tolerance.

Explicit Scheme

The solution procedure in this scheme (France *et al.*, 1971) can be summarized as follows:

(1) At the beginning of a new time step, the position of the free surface is assumed to correspond to that at the old time level. Equation (4.3.2.1) is then solved subject to the boundary conditions in Eqs. (4.3.2.3a)–(4.3.2.3d).

(2) The Darcy velocity components at the free surface are computed by applying Darcy's law to the heads computed in (1). Dividing the Darcy velocities by the specific yield S_y gives average pore velocities. Assuming no infiltration crosses the free surface ($I = 0$), the velocity component normal to the assumed free surface serves to locate the new surface (i.e., the normal free surface displacement can be obtained from $\Delta\zeta_n = V_n \Delta t / S_y$). One way to derive the outward-directed unit normal vector for this computation is to note that $\partial(x_3 - \zeta)/\partial x_i$ is an outward-directed normal vector that can be scaled to have a unit magnitude. The resulting radial and vertical components of the free surface displacement,

$\Delta\zeta_r$ and $\Delta\zeta_z$, are

$$\Delta\zeta_r = -\Delta\zeta_z \frac{\partial\zeta}{\partial r} \quad \text{and}$$
$$\Delta\zeta_z = \left[\frac{K_{rr}(\partial h/\partial r)(\partial\zeta/\partial r) - K_{zz}(\partial h/\partial z)}{1 + (\partial\zeta/\partial r)^2}\right]\Delta t, \tag{4.3.2.10}$$

where $\partial\zeta/\partial r$ is the tangential slope of the free surface. France *et al.* (1971) recommend computing these displacements at several points on the interface and fitting a polynomial to the new points. In view of the free surface flux condition (4.3.2.3e), we can alternatively compute purely vertical displacements $\Delta\zeta$ at each node and take account of infiltration I:

$$\Delta\zeta = \Delta t \frac{I + K_{rr}(\partial h/\partial r)(\partial\zeta/\partial r) - K_{zz}(\partial h/\partial z)}{S_y}.$$

(3) The free surface is moved to a new position and the finite element mesh modified for the next time level. Stages 1 and 2 can then be repeated.

Although the explicit scheme is simple to implement, it is only conditionally stable. The free surface oscillates and complete numerical instability may result unless the time step is very small. For isotropic aquifers, the critical time step size Δt_{cr} is proportional to $(\Delta r)^2/K$.

Treatment of Prescribed Well Discharge Condition

In the unconfined flow situation, the well discharge–drawdown relation is nonlinear. Consequently, we cannot use the principle of superposition to obtain the correct value of well drawdown. Two general methods are now presented, each of which employs an iterative procedure to obtain the correct value of drawdown or hydraulic head in the well.

METHOD 1 (COOLEY, 1971) Let us assume that the discharge into the well is derived entirely from the unconfined aquifer. Moreover, let us employ the (chord–slope) iterative approach to obtain the value of well drawdown that corresponds to the prescribed discharge $\overline{Q}$. Suppose the well discharge–drawdown relation at time t_{k+1} is as shown in Fig. 4.7. We start with a trial value of well drawdown s_{w1} and use this to obtain the finite element solution leading to the computed discharge Q_1. If $|Q_1 - \overline{Q}| > \varepsilon_Q$, where $\overline{Q}$ is the prescribed discharge and ε_Q is the discharge tolerance, then the drawdown is adjusted according to

$$s_{w2} = (\overline{Q}/Q_1)s_{w1}. \tag{4.3.2.11}$$

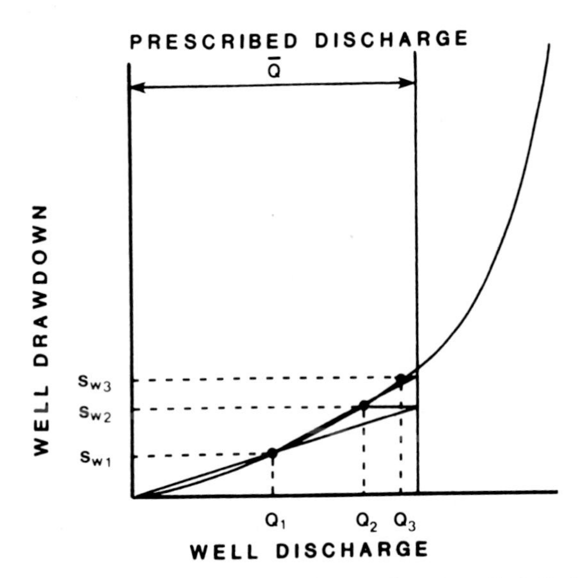

Fig. 4.7. Illustration of the secant method for handling prescribed discharge boundary condition of the unconfined well flow problem. $s_w = H_0 - h_w$; H_0 = initial hydraulic head in aquifer.

Next, we use s_{w2} to repeat the finite element solution which gives Q_2. If $|Q_2 - \overline{Q}| > \varepsilon_Q$, the drawdown for the next iteration is adjusted according to

$$s_{wj} = s_{wj-1} + (s_{wj-1} - s_{wj-2})(\overline{Q} - Q_2)/(Q_2 - Q_1), \quad (4.3.2.12)$$

where j denotes the iteration number. Derivation of this equation follows from the geometry shown in Fig. 4.7. We perform the discharge iteration until the convergence criterion, $|Q_j - \overline{Q}| \leq \varepsilon_Q$, is satisfied.

METHOD 2 (NEUMAN AND WITHERSPOON, 1971) This method takes into account the effect of the well bore storage. The total prescribed discharge is calculated as the sum of the discharge from the aquifer and the discharge from storage in the well. Consider the well shown in Fig. 4.8. Assuming that the discharges vary linearly in the time interval $[t_k, t_{k+1}]$, the continuity equation can be written as

$$\tfrac{1}{2}(\overline{Q}_k + \overline{Q}_{k+1}) = \tfrac{1}{2}(Q_k + Q_{k+1}) + \pi(r_w^2 - r_t^2)\,\Delta L/\Delta t, \quad (4.3.2.13)$$

from which we obtain

$$\Delta L = \frac{\Delta t}{2\pi(r_w^2 - r_t^2)}[\overline{Q}_k + \overline{Q}_{k+1} - Q_k - Q_{k+1}], \quad (4.3.2.14)$$

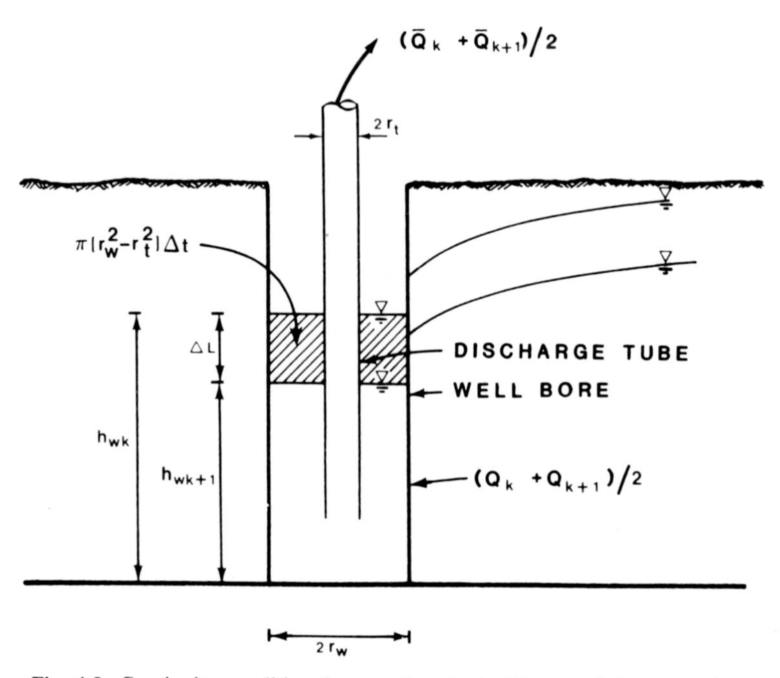

Fig. 4.8. Continuity condition for a well with significant well bore storage.

where $\overline{Q}_{k+1}$ and Q_{k+1} denote the prescribed total discharge from aquifer and well bore storage and the aquifer discharge alone at the new time level, respectively, and ΔL is the change in the water level of the well over time increment Δt.

The iterative procedure simply starts with a trial value of ΔL, which is used to obtain the prescribed value of the hydraulic head in the well and the computed value of aquifer discharge Q_{k+1}. Next we compute the new value of ΔL from

$$\Delta L^{j+1} = \frac{\Delta t}{2\pi(r_{\mathrm{w}}^2 - r_t^2)} [\overline{Q}_k + \overline{Q}_{k+1} - Q_k - Q_{k+1}^j], \quad (4.3.2.15)$$

where once again the superscript j denotes the iteration number. Knowing ΔL^{j+1}, we can obtain $h_{\mathrm{w}k+1}^{j+1}$ from

$$h_{\mathrm{w}k+1}^{j+1} = h_{\mathrm{w}k} - \Delta L^{j+1} \quad (4.3.2.16)$$

and use this to obtain the finite element solution and a newly calculated value of aquifer discharge. The discharge iteration is performed until

$$|\Delta L^{j+1} - \Delta L^j| \leqslant \varepsilon_{\mathrm{h}}, \quad (4.3.2.17)$$

where ε_{h} is the head tolerance.

4.3.3 *Areal Groundwater Flow*

Areal flow is perceived to occur over a much larger region than local well flow. Consequently, we consider areal flow simulation to deal with the regional flow pattern within an extensive aquifer basin and not specifically with localized flow behavior near individual wells. In the finite element mesh, the well are represented by nodal points, and hence no data on well geometry is required [see Pinder and Frind (1972) for an illustration of a typical mesh]. For completeness, we shall now describe a general formulation for transient flow in both confined and unconfined aquifers.

Areal Confined Flow

The governing equation for flow in a confined formation (4.2.1.24) is

$$\frac{\partial}{\partial x_i}\left(T_{ij}\frac{\partial \bar{h}}{\partial x_j}\right) - \left(V_3 - V_i\frac{\partial b}{\partial x_i}\right)_b + \left(V_3 - V_i\frac{\partial a}{\partial x_i}\right)_a - S\frac{\partial \bar{h}}{\partial t} = 0,$$

$$i, j = 1, 2. \qquad (4.3.3.1a)$$

For convenience, we assume that the bottom of the formation is underlain by a completely impermeable stratum. This reduces Eq. (4.3.3.1a) to

$$\frac{\partial}{\partial x_i}\left(T_{ij}\frac{\partial \bar{h}}{\partial x_j}\right) - q' - S\frac{\partial \bar{h}}{\partial t} = 0, \qquad (4.3.3.1b)$$

where

$$q' = \left[V_3 - V_i\frac{\partial b}{\partial x_i}\right]_b$$

is the remaining vertical leakage term.

To derive the finite element equations, we apply the Galerkin procedure to (4.3.3.1b). Let $\hat{h}$ be a trial solution written in the form

$$\hat{h}(x_1, x_2, t) = N_J(x_1, x_2)\bar{h}_J(t), \qquad J = 1, 2, \ldots, n. \qquad (4.3.3.2)$$

We now require that the residual generated when $\hat{h}$ is substituted into (4.3.3.1b) be minimized in the following sense:

$$\int_R \left[\frac{\partial}{\partial x_i}\left(T_{ij}\frac{\partial \hat{h}}{\partial x_j}\right) - q' - S\frac{\partial \hat{h}}{\partial t}\right] N_I \, dR = 0,$$

$$I = 1, 2, \ldots n, \quad i, j = 1, 2. \qquad (4.3.3.3a)$$

Application of Green's theorem and introduction of (4.3.3.2) into the resulting expression yields

$$\int_R T_{ij} \frac{\partial N_I}{\partial x_i} \frac{\partial N_J}{\partial x_j} h_J \, dR + \int_R SN_I N_J \frac{dh_J}{dt} \, dR$$
$$= -\int_{B_2} N_I q \, dB - \int_R N_I q' \, dR, \tag{4.3.3.3b}$$

where q is the outward directed normal flux associated with the Neumann boundary condition, $q = -(T_{ij} \, \partial h/\partial x_j) n_i$, along B_2. Equation (4.3.3.3b) can be written in matrix form as

$$[A]\{H\} + [B]\{dH/dt\} = \{F\}, \tag{4.3.3.3c}$$

where

$$[A] = \sum_{e=1}^{m} [A]^e, \qquad [B] = \sum_{e=1}^{m} [B]^e, \qquad \{F\} = \sum_{e=1}^{m} \{F\}^e$$

and the typical elements of matrices $[A]^e$, $[B]^e$, and $\{F\}^e$ are given by

$$A_{IJ}^e = \int_{R^e} T_{ij} \frac{\partial N_I}{\partial x_i} \frac{\partial N_J}{\partial x_j} \, dR,$$

$$B_{IJ}^e = \int_{R^e} SN_I N_J \, dR,$$

$$F_I^e = -\int_{B_2^e} N_I q \, dB - \int_{R^e} N_I q' \, dR.$$

The time derivative in (4.3.3.3c) can be approximated using finite differences to yield

$$\left(\theta[A] + \frac{1}{\Delta t}[B]\right)\{H\}^{k+1} = \left((\theta - 1)[A] + \frac{1}{\Delta t}[B]\right)\{H\}^k$$
$$+ \theta\{F\}^{k+1} + (1 - \theta)\{F\}^k,$$

where k and $k + 1$ are the old and new time levels, respectively, and θ is the time-weighting factor. Because $\{F\}^{k+1}$ depends on the values of the fluxes q and q' at the current time level, it can be evaluated only if these values are specified. In groundwater hydrology, the leakage flux q' is usually dependent on the unknown hydraulic head. Consider a case shown in Fig. 4.9 in which an aquifer is overlain by a semipervious layer called an aquitard. If the hydraulic conductivity of the aquitard is several orders of magnitude less than that of the aquifer, leakage from the aquitard to the aquifer may be assumed to be vertical. The leakage flux can be

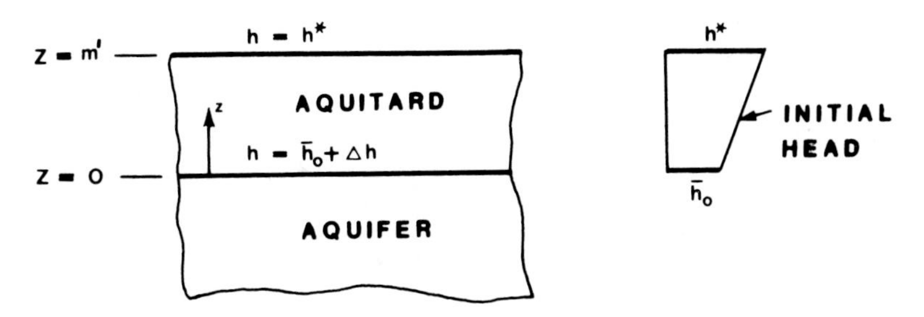

Fig. 4.9. Initial and boundary conditions for one-dimensional flow in aquitard.

determined analytically. If a step head change Δh is introduced at the top of the aquifer and the head at the top of the aquitard is kept constant, the vertical flow in the aquitard is described by (Pinder and Gray, 1977)

$$K' \, \partial^2 h'/\partial z^2 = S_s' \, \partial h'/\partial t \tag{4.3.3.4}$$

subject to the initial and boundary conditions (see Fig. 4.9)

$$h'(z, 0) = (h^* - \bar{h}_o)(z/m') + \bar{h}_o, \tag{4.3.3.5a}$$

$$h'(0, t) = \bar{h}_o + \Delta h, \tag{4.3.3.5b}$$

$$h'(m', t) = h^*, \tag{4.3.3.5c}$$

where h' is the hydraulic head in the aquitard, K' and S_s' are the aquitard hydraulic conductivity and specific storage, respectively, $\bar{h}_o$ is the initial head at the top of the aquifer, and h^* is the hydraulic head at the top of the aquitard. The solution to the preceding system of equations is given by

$$h' = \left[\bar{h}_o + (h^* - \bar{h}_o)\frac{z}{m'}\right] + \Delta h\left(1 - \frac{z}{m'}\right) - \sum_{n=1}^{\infty} \frac{2\Delta h}{n\pi} \exp(-\alpha_n t) \sin\frac{n\pi z}{m'}, \tag{4.3.3.6a}$$

where

$$\alpha_n = n^2\pi^2 K'/S_s' m'^2.$$

For the case where the top boundary of the aquitard is impermeable, the corresponding solution for h' can be found in Carslaw and Jaeger (1959). To obtain the flux at the aquifer–aquitard boundary we must obtain the spatial derivative of h' at $z = 0$ and subsequently calculate $q' = q_z|_{z=0} = -K \, \partial h'/\partial z'|_{z=0}$. Using Eq. (4.3.3.6a) to evaluate q', we obtain

$$q' = -\frac{K'}{m'}(h^* - \bar{h}_o - \Delta h) + \frac{K'\Delta h}{m'}\sum_{n=1}^{\infty} 2\exp(-\alpha_n t). \qquad (4.3.3.6b)$$

In an actual situation, the change in head at $z = 0$, denoted Δh, will be a continuous function of time rather than an instantaneous step change. Therefore, one must employ superposition (or convolution) to evaluate q'. The expression for q' takes the form

$$q' = -\frac{K'}{m'}(h^* - \bar{h}_o) + \frac{K'}{m'}\int_0^t \left.\frac{\partial h}{\partial t}\right|_\tau d\tau + 2\frac{K'}{m'}\sum_{n=1}^{\infty}\int_0^t \left.\frac{\partial h}{\partial t}\right|_\tau \exp[-\alpha_n(t - \tau)]\,d\tau. \qquad (4.3.3.7a)$$

We now present a procedure for approximating the convolution integral. At a current time level $k + 1$, q'^{k+1} is given by

$$q'^{k+1} = -\frac{K'}{m'}(h^* - \bar{h}_o) + \frac{K'}{m'}(\hat{h}^{k+1} - \bar{h}_o) + 2\frac{K'}{m'}\sum_{n=1}^{\infty}\hat{I}_n^{k+1}, \qquad (4.3.3.7b)$$

where $\hat{h}^{k+1}$ is the value of the trial function $\hat{h}$ at time t_{k+1} and $\hat{I}_n^{k+1}$ is a convolution integral defined as

$$\hat{I}_n^{k+1} = \int_0^{t_{k+1}} \frac{\partial \hat{h}}{\partial \tau}\exp[-\alpha_n(t_{k+1} - \tau)]\,d\tau. \qquad (4.3.3.8a)$$

The integral in Eq. (4.3.3.8a) can be approximated by

$$\begin{aligned}\hat{I}_n^{k+1} &= \int_0^{t_{k+1}} \frac{\partial h}{\partial \tau}\exp[-\alpha_n(t_{k+1} - \tau)]\,d\tau \\ &\simeq \left\langle\frac{\Delta\hat{h}}{\Delta t}\right\rangle_1 \int_0^{t_1}\exp[-\alpha_n(t_{k+1} - \tau)]\,d\tau \\ &\quad + \left\langle\frac{\Delta\hat{h}}{\Delta t}\right\rangle_2 \int_{t_1}^{t_2}\exp[-\alpha_n(t_{k+1} - \tau)]\,d\tau + \cdots \\ &\quad + \left\langle\frac{\Delta\hat{h}}{\Delta t}\right\rangle_{k+1} \int_{t_k}^{t_{k+1}}\exp[-\alpha_n(t_{k+1} - \tau)]\,d\tau,\end{aligned} \qquad (4.3.3.8b)$$

where

$$\left\langle\frac{\Delta\hat{h}}{\Delta t}\right\rangle_{k+1} = \frac{\hat{h}^{k+1} - \hat{h}^k}{t_{k+1} - t_k} = \frac{\hat{h}^{k+1} - \hat{h}^k}{\Delta t_{k+1}}. \qquad (4.3.3.8c)$$

Noting that Δt_{k+1} becomes $\Delta t_1 = t_1$ when $k = 0$, a recurrence relation between I_n^{k+1} and I_n^k can be developed by induction as follows: for $k + 1 = 1$,

$$\hat{I}_n^1 = \left\langle \frac{\Delta \hat{h}}{\Delta t} \right\rangle_1 \int_0^{t_1} \exp[-\alpha_n(t_1 - \tau)]\, d\tau$$

$$= \frac{1}{\alpha_n} \left\langle \frac{\Delta \hat{h}}{\Delta t} \right\rangle_1 \{1 - \exp[-\alpha_n(\Delta t)_1]\}; \tag{4.3.3.8d}$$

for $k + 1 = 2$,

$$\hat{I}_n^2 = \left\langle \frac{\Delta \hat{h}}{\Delta t} \right\rangle_1 \int_0^{t_1} \exp[-\alpha_n(t_2 - \tau)]\, d\tau + \left\langle \frac{\Delta \hat{h}}{\Delta t} \right\rangle_2 \int_{t_1}^{t_2} \exp[-\alpha_n(t_2 - \tau)]\, d\tau$$

$$= \frac{1}{\alpha_n} \left\langle \frac{\Delta \hat{h}}{\Delta t} \right\rangle_1 \exp[-\alpha_n(\Delta t)_2]\{1 - \exp[-\alpha_n(\Delta t)_1]\}$$

$$+ \frac{1}{\alpha_n} \left\langle \frac{\Delta \hat{h}}{\Delta t} \right\rangle_2 \{1 - \exp[-\alpha_n(\Delta t)_2]\}$$

$$= \exp[-\alpha_n(\Delta t)_2]\, \hat{I}_n^1 + \frac{1}{\alpha_n} \left\langle \frac{\Delta \hat{h}}{\Delta t} \right\rangle_2 \{1 - \exp[-\alpha_n(\Delta t)_2]\}. \tag{4.3.3.8e}$$

This leads to a recurrence formula of the form

$$\hat{I}_n^{k+1} = \exp[-\alpha_n(\Delta t)_{k+1}]\hat{I}_n^k + \frac{1}{\alpha_n} \left\langle \frac{\Delta \hat{h}}{\Delta t} \right\rangle_{k+1} \{1 - \exp[-\alpha_n(\Delta t)_{k+1}\}. \tag{4.3.3.8f}$$

Substituting Eq. (4.3.3.8f) into (4.3.3.7b), one obtains

$$q'^{k+1} = -\frac{K'}{m'}(h^* - \bar{h}_o) + \frac{K'}{m'}(\hat{h}^{k+1} - \bar{h}_o)$$

$$+ 2\frac{K'}{m'} \sum_{n=1}^{\infty} \exp[-\alpha_n(\Delta t)_{k+1}]\hat{I}_n^k$$

$$+ 2\frac{K'}{m'} \sum_{n=1}^{\infty} \frac{1}{\alpha_n}\{1 - \exp[-\alpha_n(\Delta t)_{k+1}]\} \frac{\hat{h}^{k+1} - \hat{h}^k}{(\Delta t)_{k+1}}, \tag{4.3.3.9a}$$

which can be written in the form

$$q'^{k+1} = \frac{K'}{m'}(\gamma + 1)\hat{h}^{k+1} - \frac{K'}{m'}(h^* + \gamma\hat{h}^k) + \frac{K'}{m'}\hat{\beta}^k, \tag{4.3.3.9b}$$

where

$$\gamma = \frac{2}{(\Delta t)_{k+1}} \sum_{n=1}^{\infty} \frac{1}{\alpha_n}\{1 - \exp[-\alpha_n(\Delta t)_{k+1}]\}, \tag{4.3.3.9c}$$

$$\hat{\beta}^k = 2 \sum_{n=1}^{\infty} \exp[-\alpha_n(\Delta t)_{k+1}]\hat{I}_n^k, \tag{4.3.3.9d}$$

$$\hat{I}_n^k = \exp[-\alpha_n(\Delta t)_k]\hat{I}^{k-1} + \frac{1}{\alpha_n}\left\langle\frac{\Delta h}{\Delta t}\right\rangle_k (1 - \exp[-\alpha_n(\Delta t)_k]). \tag{4.3.3.9e}$$

Because the expression for q' contains the unknown head $\hat{h}^{k+1}$, the coefficient matrix $[A]$ in the finite element matrix equation (4.3.3.3c) will contain additional terms contributed by the unknown part of q'. The new values of the element matrix $[A]^e$ and the right-hand-side vector $\{F\}^e$ are given by

$$A_{IJ}^e = \int_{R^e} \left[T_{ij} \frac{\partial N_I}{\partial x_i} \frac{\partial N_J}{\partial x_j} + \frac{K'}{m'}(\gamma + 1)N_I N_J \right] dR \tag{4.3.3.10a}$$

and

$$F_I^e = -\int_{B_2^e} N_I q \, dB + \int_{R^e} N_I \frac{K'}{m'} [h^* + \gamma N_J h_J^k - N_J \hat{\beta}_J^k] \, dR. \tag{4.3.3.10b}$$

For a case in which the aquifer is pumped by a constant discharge well, a remarkably accurate approximation can be obtained without employing the convolution integrals. This is achieved by considering the leakage problem to correspond roughly to a case where a step change in head is applied at one-half the elapsed time, i.e., at $t/2$. Thus, q' at t is given by

$$q' \simeq -\frac{K'}{m'}(h^* - \bar{h}_o - \Delta h) + \frac{K'}{m'} \Delta h \sum_{n=1}^{\infty} 2 \exp(-\alpha_n t/2) \tag{4.3.3.11a}$$

and q'^{k+1} becomes

$$q'^{k+1} = \frac{K'}{m'}(\lambda + 1)\hat{h}^{k+1} - \frac{K'}{m'}(\lambda \bar{h}_o + h^*), \tag{4.3.3.11b}$$

where

$$\lambda = \sum_{n=1}^{\infty} 2 \exp(-\alpha_n t/2). \tag{4.3.3.11c}$$

Incorporation of Eq. (4.3.3.11b) into the finite element matrix gives

$$A_{IJ}^{e} = \int_{R^e} \left[T_{ij} \frac{\partial N_I}{\partial x_i} \frac{\partial N_J}{\partial x_j} + \frac{K'}{m'} (\lambda + 1) N_I N_J \right] dR \qquad (4.3.3.12a)$$

and

$$F_I^{e} = -\int_{B_2^e} N_I q \, dB + \int_{R^e} \frac{K'}{m'} N_I (h^* + \lambda \bar{h}_o) \, dR. \qquad (4.3.3.12b)$$

Once all the coefficient matrices and right-hand-side vectors have been determined, Eq. (4.3.3.3c) can be solved using either the Crank–Nicolson or the backward difference finite difference scheme. It is apparent that the evaluation of the leakage terms of A_{IJ}^e and F_I^e may involve considerable computational effort unless the infinite exponential series in Eq. (4.3.3.6b) converges rapidly. The rate of convergence of this series depends directly on the dimensionless parameter $t^* = (K't/S_s'm'^2)$. A small value of t^* (a small t or a large value of m') leads to a slow rate of convergence. Experience indicates that the computational efficiency can be improved greatly by replacing the entire thickness of the aquitard m' by a smaller effective thickness m_o', defined as the thickness of the zone of influence of the aquitard. (Within this zone the hydraulic head is affected by the change of head at $z = 0$.) It is sufficiently accurate to determine m_o' such that

$$t^* = K't/(S_s'm_o'^2) = 0.5,$$

or

$$m_o' = [2K't/S_s']^{1/2}.$$

In practice, the numerical solution remains approximately the same whether one uses m' or m_o' in the computation. The advantage in using m_o', however, is the improvement of convergence of the series, particularly at small time values where m_o' is much less than m'. In most cases, it is usually sufficient to consider only the first three or four leading terms of the exponential series.

Areal Unconfined Flow

The governing equation for horizontal flow in an isotropic, unconfined formation (4.2.1.22) is rewritten in the form

$$\frac{\partial}{\partial x_i}\left(T_{ij} \frac{\partial \bar{h}}{\partial x_j} \right) - (S_s l + S_y) \frac{\partial \bar{h}}{\partial t} + I = 0, \qquad (4.3.3.13)$$

where $l = \bar{h} - a$, $T_{ij} = (\bar{h} - a)K_{ij}$ and

$$I = [S_y V_3^* - V_3]_b - [S_y V_i^* - V_i]_b \frac{\partial b}{\partial x_i} + \left[V_3 - V_i \frac{\partial a}{\partial x_i}\right]_a.$$

Note that (4.3.3.13) is nonlinear because T_{ij} depends on $\bar{h}$. The Galerkin formulation of Eq. (4.3.3.13) can be obtained in the same manner in the case of confined flow. If the backward difference time-stepping scheme is used, the resulting matrix equation takes the form

$$\left([A] + \frac{1}{\Delta t}[B]\right)\{H\}^{k+1} = \frac{1}{\Delta t}[B]\{H\}^k + \{F\}^{k+1}, \qquad (4.3.3.14)$$

where

$$[A] = \sum_{e=1}^{m} [A]^e,$$

$$[B] = \sum_{e=1}^{m} [B]^e, \qquad \{F\} = \sum_{e=1}^{m} \{F\}^e.$$

The typical elements of $[A]^e$, $[B]^e$, and $\{F\}^e$ are given by

$$A_{IJ}^e = \int_{R^e} T_{ij} \frac{\partial N_I}{\partial x_i} \frac{\partial N_J}{\partial x_j}\, dR,$$

$$B_{IJ}^e = \int_{R^e} (S_s l + \phi) N_I N_J \, dR,$$

$$F_I^e = \int_{R^e} N_I I \, dR - \int_{B_2^e} N_I q \, dB.$$

Equation (4.3.3.14) represents a system of nonlinear algebraic equations that can be solved via an iterative procedure. To describe this procedure, it is convenient to write this system in the form

$$\left([A] + \frac{1}{\Delta t}[B]\right)_r \{H\}_{r+1}^{k+1} = \frac{1}{\Delta t}[B]\{H\}^k + \{F\}^{k+1}, \qquad (4.3.3.15)$$

where the subscript r indicates the iteration number at the current time level $k + 1$.

At the start of the time level $k + 1$, we calculate the matrix $[A]$ using known values of the hydraulic head at the previous time level k. Then we solve Eq. (4.3.3.10) for the unknown nodal values of the piezometric head and use these values to update matrix $[A]$ and resolve for more accurate values of the head. The iterative procedure is repeated until the following closure criterion is satisfied:

$$\delta h_{\max} \leqslant \varepsilon,$$

where $\delta h_{\max} = \max_J |(h_J)_{r+1} - (h_J)_r|$ and ε is the prescribed head tolerance.

Treatment of Wells

When a number of discharging or recharging wells appear in the modeled aquifer region, the wells are idealized as point sinks or sources, respectively. The finite element mesh is usually set up in such a way that certain nodal coordinates correspond to the well locations. The volumetric flow rate at each well is treated as the Dirac delta function. Thus, the discharging well at a location (x_{1k}, x_{2k}) can be described by $Q_k(t)\delta(x_1 - x_{1k})\delta(x_2 - x_{2k})$, where Q_k is the volumetric flow rate at the well. If we adopt the sign convention that discharge is positive and recharge is negative, then the term $Q_k(t)\delta(x_1 - x_{1k})\delta(x_2 - x_{2k})$ should appear on the right-hand side of the governing equations (4.3.3.1) and (4.3.3.13) for confined and unconfined flow, respectively. In the Galerkin finite element equation for node I, the added term is

$$\int_R N_I Q_k(t)\delta(x_1 - x_{1k})\delta(x_2 - x_{2k})\, dR = Q_k(t).$$

A common practice is to make each well coincide with a node so that its entire discharge pertains to the relevant nodal equation.

4.3.4 *Sharp Interface Seawater Intrusion*

Unlike the groundwater flow problems described in the foregoing sections, the areal simulation of seawater intrusion in coastal aquifers is governed by two coupled nonlinear partial differential equations. Thus, we must introduce in the Galerkin approximation two trial functions, each representing one of the dependent variables. Let this trial function be expressed as

$$\hat{h}_{\rm f} = N_J(x_1, x_2)h_{{\rm f}J}(t), \qquad J = 1, 2, \ldots, n, \tag{4.3.4.1a}$$

and

$$\hat{h}_{\rm s} = N_J(x_1, x_2)h_{{\rm s}J}(t), \qquad J = 1, 2, \ldots, n, \tag{4.3.4.1b}$$

where $h_{{\rm f}J}(t)$ and $h_{{\rm s}J}(t)$ denote the nodal values of the freshwater and saltwater heads, respectively. Substituting Eqs. (4.3.4.1a) and (4.3.4.1b) into the governing equations (4.2.1.33a), and (4.2.1.33b), we obtain the following two residuals:

$$R_{\mathrm{f}} \equiv \frac{\partial}{\partial x_i}\left(\hat{T}_{\mathrm{f}ij}\frac{\partial \hat{h}_{\mathrm{f}}}{\partial x_j}\right) - (S_{\mathrm{y}} + \phi\rho_{\mathrm{f}}^*)\frac{\partial \hat{h}_{\mathrm{f}}}{\partial t} + \phi\rho_{\mathrm{s}}^*\frac{\partial \hat{h}_{\mathrm{s}}}{\partial t} - I, \qquad i,j = 1,2, \tag{4.3.4.2a}$$

$$R_{\mathrm{s}} \equiv \frac{\partial}{\partial x_i}\left(\hat{T}_{\mathrm{s}ij}\frac{\partial \hat{h}_{\mathrm{s}}}{\partial x_j}\right) + \phi\rho_{\mathrm{f}}^*\frac{\partial \hat{h}_{\mathrm{f}}}{\partial t} - \phi\rho_{\mathrm{s}}^*\frac{\partial \hat{h}_{\mathrm{s}}}{\partial t}, \qquad i,j = 1,2, \tag{4.3.4.2b}$$

where we have assumed S_{sf} and $S_{\mathrm{ss}} \ll \phi$, $q_{sa} = q_{sb} \simeq 0$ and have defined

$$T_{\mathrm{f}ij} = l_{\mathrm{f}}K_{\mathrm{f}ij}, \qquad T_{\mathrm{s}ij} = l_{\mathrm{s}}K_{sij}, \qquad \text{and} \qquad I = -q_{\mathrm{f}c}.$$

Application of the Galerkin procedure and Green's theorem results in the approximating equations

$$\begin{aligned}\int_R \hat{T}_{\mathrm{f}ij}\frac{\partial N_I}{\partial x_i}\frac{\partial N_J}{\partial x_j}h_{\mathrm{f}J}\,dR &+ \int_R\left[(S_y + \phi\rho_{\mathrm{f}}^*)N_I N_J\frac{dh_{\mathrm{f}J}}{dt} - \phi\rho_{\mathrm{s}}^* N_I N_J\frac{dh_{\mathrm{s}J}}{dt}\right]dR \\ &+ \int_{B_2} q_{\mathrm{f}}N_I\,dB + \int_R IN_I\,dR = 0\end{aligned} \tag{4.3.4.3a}$$

and

$$\begin{aligned}\int_R \hat{T}_{\mathrm{s}ij}\frac{\partial N_I}{\partial x_i}\frac{\partial N_J}{\partial x_j}h_{\mathrm{s}J}\,dR &+ \int_R\left[(-\phi\rho_f^*)N_I N_J\frac{dh_{\mathrm{f}J}}{dt} + \phi\rho_{\mathrm{s}}^* N_I N_J\frac{dh_{\mathrm{s}J}}{dt}\right]dR \\ &+ \int_{B_2} q_{\mathrm{s}}N_I\,dB = 0,\end{aligned} \tag{4.3.4.3b}$$

where q_{f} and q_{s} are the outward-directed normal boundary fluxes of freshwater and seawater, respectively. Once again, we adopt a sign convention whereby discharge is positive and recharge negative.

In practice, the boundary fluxes q_{f} and q_{s} usually correspond to leakage occurring along those segments of the boundary that intercept streams and the ocean (Fig. 4.10). The leakage conditions are given by

$$q_{\mathrm{f}} = \alpha_{\mathrm{f}}(h_{\mathrm{f}} - H_{\mathrm{f}}), \qquad q_{\mathrm{s}} = \alpha_{\mathrm{s}}(h_{\mathrm{s}} - H_{\mathrm{s}}), \tag{4.3.4.4}$$

where α_{f} and α_{s} are leakage coefficients and H_{f} and H_{s} are the hydraulic heads in the stream bed and the sea, respectively (see Fig. 4.10). Note that at a coastal boundary H_{f} and H_{s} may both be represented by the head of the sea. The boundary integral terms now become

$$\int_{B_2} q_l N_I \, dB = \int_{B_2} N_I \alpha_l N_J (h_{lJ} - H_{lJ}) \, dB, \qquad l = \mathrm{f, s}, \qquad (4.3.4.5)$$

which is formally a third-type boundary condition. Experience from a simulation of a coastal aquifer system at Long Island (Pinder and Page, 1977) indicates that the use of the approximation given by (4.3.4.5) often leads to a solution that exhibits mild spatial numerical oscillations, particularly near the coast. These oscillations can often be damped by lumping (diagonalizing) the terms as

$$\int_{B_2} q_l N_I \, dB = \sum_{J=1}^{n} \int_{B_2} N_I \alpha_l N_J (h_{lI} - H_{lI}) \, dB$$

$$= \int_{B_2} N_I \alpha_l (h_{lI} - H_{lI}) \, dB, \qquad l = \mathrm{f, s}. \qquad (4.3.4.6)$$

Note that the summation convention is not used on the repeated subscript l in Eq. (4.3.4.6). Substituting Eq. (4.3.4.6) into Eqs. (4.3.4.3a) and (4.3.4.3b) and rearranging the terms, we obtain

$$\int_R \hat{T}_{\mathrm{f}ij} \frac{\partial N_I}{\partial x_i} \frac{\partial N_J}{\partial x_j} h_{\mathrm{f}J} \, dR + \int_{B_2} N_I \alpha_{\mathrm{f}} h_{\mathrm{f}I} \, dB$$

$$+ \int_R \left[(S_y + \phi \rho_{\mathrm{f}}^*) N_I N_J \frac{dh_{\mathrm{f}J}}{dt} - \phi \rho_{\mathrm{s}}^* N_I N_J \frac{dh_{\mathrm{s}J}}{dt} \right] dR$$

$$= \int_R I N_I \, dR + \int_{B_2} N_I \alpha_{\mathrm{f}} H_{fI} \, dB \qquad (4.3.4.7\mathrm{a})$$

and

$$\int_R \hat{T}_{\mathrm{s}ij} \frac{\partial N_I}{\partial x_i} \frac{\partial N_J}{\partial x_j} h_{\mathrm{s}J} \, dR + \int_{B_2} N_I \alpha_{\mathrm{s}} h_{\mathrm{s}I} \, dB$$

$$+ \int_R \left[(-\phi \rho_{\mathrm{f}}^*) N_I N_J \frac{dh_{\mathrm{f}J}}{dt} + \phi \rho_{\mathrm{s}}^* N_I N_J \frac{dh_{\mathrm{s}J}}{dt} \right] dR$$

$$= \int_{B_2} N_I \alpha_{\mathrm{s}} H_{\mathrm{s}I} \, dB. \qquad (4.3.4.7\mathrm{b})$$

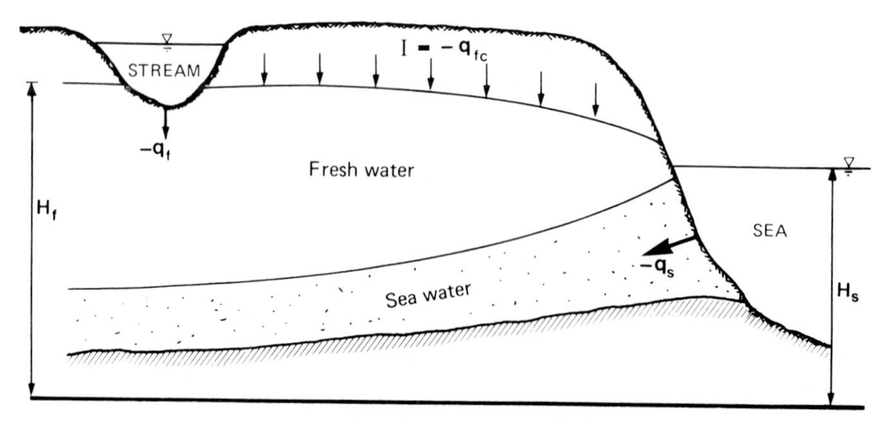

Fig. 4.10. Sharp interface seawater intrusion problem showing various boundary fluxes.

Equations (4.3.4.7a) and (4.3.4.7b) represent a system of ordinary differential equations. To save computational time, it is desirable to make the coefficient matrices symmetric if possible.

For this particular system, it so happens that symmetry can be achieved by scaling Eq. (4.3.4.7b) with the factor $\Lambda = \rho_s^*/\rho_f^*$. The resulting matrix equation may be expressed in the form

$$[E]\{H\} + [B]\{dH/dt\} = \{R\}, \tag{4.3.4.8}$$

where $[E] = [A] + [D]$, $[D]$ is a diagonal matrix, and typical matrix elements are

$$[A_{IJ}] = \sum_{e=1}^{m} \begin{bmatrix} \int_{R^e} \hat{T}_{fij} \frac{\partial N_I}{\partial x_i} \frac{\partial N_J}{\partial x_j}\, dR & 0 \\ 0 & \Lambda \int_{R^e} \hat{T}_{sij} \frac{\partial N_I}{\partial x_i} \frac{\partial N_J}{\partial x_j}\, dR \end{bmatrix},$$

$$[D_{II}] = \sum_{e=1}^{m} \begin{bmatrix} \int_{B_2^e} N_I \alpha_f h_{fI}\, dB & 0 \\ 0 & \Lambda \int_{B_2^e} N_I \alpha_s h_{sI}\, dB \end{bmatrix},$$

$$[B_{IJ}] = \sum_{e=1}^{m} \begin{bmatrix} \int_{R^e} (S_y + \phi\rho_f^*) N_I N_J\, dR & \int_{R^e} -\phi\rho_s^* N_I N_J\, dR \\ \int_{R^e} -\phi\rho_s^* N_I N_J\, dR & \int_{R^e} \Lambda\phi\rho_s^* N_I N_J\, dR \end{bmatrix},$$

$$\{H_J\} = \begin{Bmatrix} h_{fJ} \\ \\ h_{sJ} \end{Bmatrix}, \qquad \left\{\frac{dH_J}{dt}\right\} = \begin{Bmatrix} \dfrac{dh_{fJ}}{dt} \\ \\ \dfrac{dh_{sJ}}{dt} \end{Bmatrix},$$

$$\{R_I\} = \sum_{e=1}^{m} \begin{Bmatrix} \displaystyle\int_{B_2^e} N_I \alpha_f H_f \, dB + \int_{R^e} I N_I \, dR \\ \\ \displaystyle\Lambda \int_{B_2^e} N_I \alpha_s H_{sI} \, dB \end{Bmatrix}.$$

Application of the weighted average finite difference approximation to Eq. (4.3.4.8) yields

$$\left(\theta[E] + \frac{1}{\Delta t}[B]\right)\{H\}^{k+1} = \left((\theta - 1)[E] + \frac{1}{\Delta t}[B]\{H\}^k\right) + \theta\{R\}^{k+1} + (1 - \theta)\{R\}^k. \qquad (4.3.4.9)$$

Equation (4.3.4.9) represents a system of nonlinear algebraic equations. This system can be solved using an iterative solution procedure similar to that described previously for the areal confined flow problem. We recommend the use of the backward difference time-stepping scheme ($\theta = 1$) to ensure a stable solution with a minimum of oscillations in the profiles of the interface. Experience indicates that convergence to the final solution can be accelerated by adopting the following procedure:

(i) Use a linear extrapolation formula to calculate the first estimate of $\{H\}$ at a new time level, that is,

$$\{H\}^{k+1} = \{H\}^k + \sigma^* \frac{\Delta t}{\Delta t'}(\{H\}^k - \{H\}^{k-1}). \qquad (4.3.4.10)$$

Here $\Delta t = t_{k+1} - t_k$, $\Delta t' = t_k - t_{k-1}$, and σ^* is an extrapolation coefficient selected between 0 and 1.

(ii) Use the following relaxation formula to calculate the next iterate

$$\{H\}_{r+1}^{k+1} = \omega\{H\}_{r+1}^{k+1} + (\omega - 1)(\{H\}_{r+1}^{k+1} - \{H\}_r^{k+1}), \qquad (4.3.4.11)$$

where $r + 1$ and r denote the current and previous iterations and ω is the head overrelaxation factor $1 \leqslant \omega < 2$. The use of $\sigma^* = 0.5$ and $\omega = 1.5$ is recommended. More details about the derivation and solution of these equations appear in Pinder and Page (1977) and Page (1979).

4.4 Single-Phase Oil and Gas Reservoir Simulation

4.4.1 Oil Reservoirs

The flow in a reservoir containing only "liquid" oil can be regarded as single-phase slightly compressible flow. Consequently, Eq. (4.2.1.9a), developed in Section 4.2.1, is applicable. In petroleum engineering, it is conventional to write the flow equation in terms of pressure. Thus, by the transformation $h = p/\rho g + z$ and expansion of the parameter K_{ij}, Eq. (4.2.1.9a) can be converted to

$$\frac{\partial}{\partial x_i}\left[\frac{k_{ij}}{\mu}\left(\frac{\partial p}{\partial x_j} + \rho g \frac{\partial z}{\partial x_j}\right)\right] = (\alpha + \phi\beta)\frac{\partial p}{\partial t}, \tag{4.4.1.1}$$

where the fluid pressure is measured relative to atmospheric pressure.

Apart from the fact that now we have to solve for pressure instead of the piezometric head, the finite element solution described for ground-water flow in confined aquifers is applicable. The only caution one needs to exercise is in representation of the production (or injection) rate of the well. Allowance must be made for the fact that there may exist significant changes in the volume of oil upon transition from reservoir to surface conditions. This is particularly true if the reservoir oil contains dissolved gas. When a quantity of oil containing gas is brought to the surface, gas will come out of solution, thus making the volume of oil in the stock tank less than the original volume under reservoir conditions. Such a volume change can be taken into account by the introduction of the "oil formation volume factor" B_o. This factor is defined as the ratio of the volume of oil plus its dissolved gas (measured at reservoir conditions) to the volume of stock tank oil measured at standard surface conditions. Thus, when using the finite element model to simulate single-phase flow in oil reservoirs, one should convert the oil production rate at the surface to the value corresponding to reservoir conditions by the relation

$$Q_o = Q_{os}B_o, \tag{4.4.1.2}$$

where Q_{os} is the rate of production of stock tank oil. The value of Q_o obtained is subsequently employed in the simulation.

4.4.2 Gas Reservoirs

For real gas, the governing equation is Eq. (4.2.2.12), which may be rewritten as

$$\frac{\partial}{\partial x_i}\left[k_{ij}\frac{\partial m}{\partial x_j}\right] = \phi\beta\mu\frac{\partial m}{\partial t} = \lambda\frac{\partial m}{\partial t}, \tag{4.4.2.1}$$

where β is the gas compressibility and m the real gas pseudopressure, defined as

$$m = \int_{p_0}^{p} 2\xi \, d\xi/\sigma\mu, \tag{4.4.2.2}$$

in which μ is the dynamic viscosity of the gas. In addition to initial and boundary conditions, we require several nonlinear constitutive relations to solve (4.4.2.1). These supplementary relations are plotted in Fig. 4.11. The curve of $m(p)$ is derived by numerical integration of $2p/(\mu(p)\sigma(p))$ versus p.

The Galerkin finite element approximation of Eq. (4.4.2.1) can be obtained using the trial functions

$$\hat{m}(x_i, t) = N_J(x_i)m_J(t), \tag{4.4.2.3a}$$

$$\hat{\lambda}(x_i, t) = N_J(x_i)\lambda_J(t). \tag{4.4.2.3b}$$

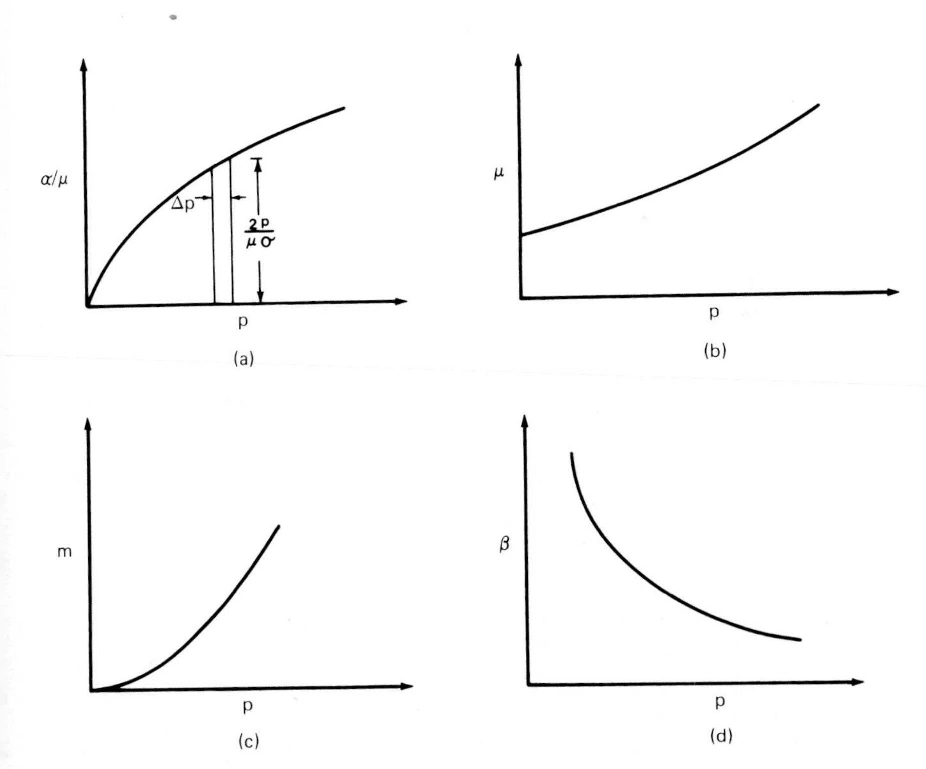

Fig. 4.11. Typical constitutive relations for real gas: (a) α/μ vs. p; (b) μ vs. p; (c) m vs. p; (d) β vs. p.

Application of the Galerkin criterion and Green's theorem to Eq. (4.4.2.1) leads to

$$\int_R k_{ij} \frac{\partial N_I}{\partial x_i} \frac{\partial N_J}{\partial x_j} m_J \, dR + \int_R \lambda_L N_L N_I N_J \frac{dm_J}{dt} \, dR - \int_{B_2} N_I k_{ij} \frac{\partial \hat{m}}{\partial x_j} n_i \, dB = 0. \tag{4.4.2.4}$$

To evaluate the boundary integral in terms of the outward normal flux, we first obtain from Eq. (4.4.2.2)

$$\frac{\partial m}{\partial x_j} = \frac{2p}{\mu\sigma} \frac{\partial p}{\partial x_j}. \tag{4.4.2.5a}$$

Thus,

$$-k_{ij} \frac{\partial m}{\partial x_j} n_i = -\left(\frac{2p}{\sigma}\right)\left(\frac{k_{ij}}{\mu} \frac{\partial p}{\partial x_j} n_i\right) = \frac{2p}{\sigma} V_i n_i = \alpha q, \tag{4.4.2.5b}$$

where q is the outward normal flux and

$$\alpha = 2p/\sigma. \tag{4.4.2.5c}$$

We can express (4.4.2.5b) in terms of an expansion of α, i.e., $\hat{\alpha} = \alpha_L N_L$, and the trial function $\hat{m}$ as

$$-k_{ij} \frac{\partial \hat{m}}{\partial x_j} n_i = \hat{\alpha} q = N_L \alpha_L q. \tag{4.4.2.5d}$$

Substituting Eq. (4.4.2.5d) into (4.4.2.4), we obtain

$$\int_R k_{ij} \frac{\partial N_I}{\partial x_i} \frac{\partial N_J}{\partial x_j} m_J \, dR + \int_R \lambda_L N_L N_I N_J \frac{dm_J}{dt} \, dR = -\int_{B_2} N_I N_L \alpha_L q \, dB. \tag{4.4.2.6}$$

It should be noted that the first integral is linear. The other two integrals are nonlinear because of their dependence on λ_L and α_L, which, in turn, are functions of m_L. Equation (4.4.2.6) can be converted into a system of nonlinear algebraic equations using finite differences. These nonlinear equations can be solved using an iteration and time-stepping procedure similar to that described in Section 4.3.3 for areal unconfined flow. As in the simulation of oil reservoirs, all production or injection rates measured at the surface must be converted, by the gas volume formation factor, to the corresponding reservoir values. An outline of the simulation procedure is illustrated in Fig. 4.12.

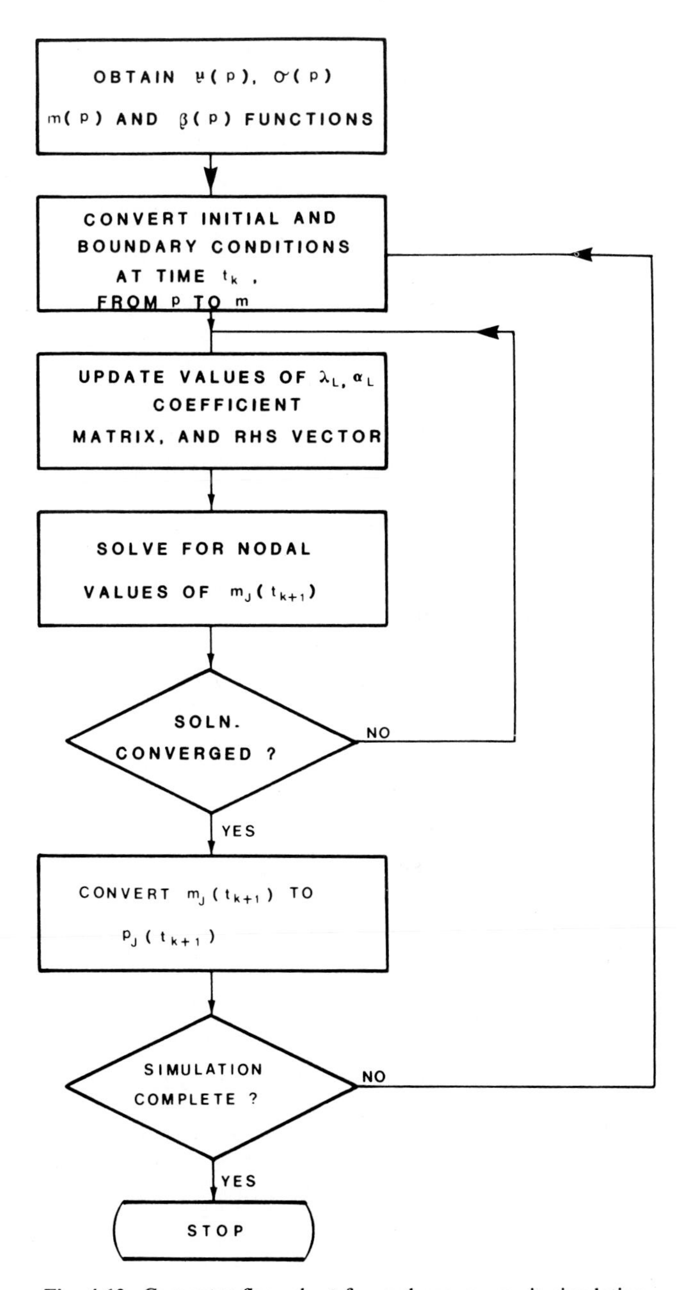

Fig. 4.12. Computer flow chart for real gas reservoir simulation.

4.5 Governing Equations for Variably Saturated Flow

Thus far, only single-phase flow through a porous medium has been considered. Even the situation of seawater intrusion, described in Section 4.3.4, is generally regarded as single-phase flow in two subdomains separated by an interface. In this section, we consider problems in which the porous medium contains two or more fluids flowing together as separate phases. Two important examples of such phenomena are the flow of water through a soil containing continuous channels occupied by air and the flow of oil, gas, and water in underground reservoirs. The first example has been the focus of attention of soil physicists for many years. The problem is commonly treated by a simplified model based on the assumption that the dynamics of the air phase play a minor role in defining water movement through the unsaturated zone. The second example is extremely important to petroleum engineers. This problem is treated by a general model which requires the simultaneous solution of coupled partial differential equations, one written for each fluid phase or component. It will be apparent that this case is similar in many respects to the seawater intrusion formulation.

4.5.1 Approximate Unsaturated Zone Flow Model

The governing equation for water flow in variably saturated soils (containing water and air) can be obtained by combining a special form of Darcy's law and the continuity equation written for the water phase. Darcy's law takes the form

$$V_{iw} = -\frac{k_{ij}k_{rw}}{\mu_w}\left(\frac{\partial p_w}{\partial x_j} + \rho_w g \frac{\partial z}{\partial x_j}\right), \tag{4.5.1.1}$$

where i and j are coordinate subscripts, k_{rw} is relative permeability with respect to the water phase, ρ_w and μ_w are density and dynamic viscosity of water, p_w is water pressure, and the remaining symbols are as defined previously.

The continuity equation is presented in the form

$$-\frac{\partial}{\partial x_i}(\rho_w V_{iw}) = \frac{\partial}{\partial t}(\rho_w S_w \phi), \tag{4.5.1.2}$$

where S_w is saturation of water ($0 \leqslant S_w \leqslant 1$) and ϕ is porosity of the soil medium.

Substituting Eq. (4.5.1.2) into (4.5.1.1), we obtain

$$\frac{\partial}{\partial x_i}\left[\frac{\rho_w k_{ij} k_{rw}}{\mu_w}\left(\frac{\partial p_w}{\partial x_j} + \rho_w g \frac{\partial z}{\partial x_j}\right)\right] = \frac{\partial}{\partial t}(\rho_w S_w \phi). \qquad (4.5.1.3)$$

In groundwater hydrology and soil physics, it is customary to rewrite Eq. (4.5.1.3) in terms of a pressure head ψ, defined as

$$\psi = (p_w - p_a)/\rho_w g, \qquad (4.5.1.4)$$

where p_a is pressure in the air phase. From this definition, it is clear that ψ is positive in the saturated zone and negative in the unsaturated zone. (Some authors use the opposite sign convention.)

Because the movement of air is assumed to be insignificant, p_a may be taken as constant and equal to the atmospheric pressure. Substitution of Eq. (4.5.1.4) into (4.5.1.3) results in

$$\frac{\partial}{\partial x_i}\left[K_{ij} k_{rw} \rho_w \left(\frac{\partial \psi}{\partial x_j} + \frac{\partial z}{\partial x_j}\right)\right] = \rho_w \phi \frac{\partial S_w}{\partial t} + S_w \frac{\partial}{\partial t}(\rho_w \phi), \qquad (4.5.1.5)$$

where K_{ij} is the saturated hydraulic conductivity tensor, defined as

$$K_{ij} = \rho_w g k_{ij}/\mu_w.$$

If it is assumed that the soil medium is only slightly compressible, Eq. (4.5.1.5) reduces to

$$\frac{\partial}{\partial x_i}\left[K_{ij}\, k_{rw}\left(\frac{\partial \psi}{\partial x_j} + \frac{\partial z}{\partial x_j}\right)\right] = \phi \frac{\partial S_w}{\partial t} + S_w S_s \frac{\partial \psi}{\partial t}, \qquad (4.5.1.6)$$

where S_s is the coefficient of specific storage. Equation (4.5.1.6) contains three unknowns: ψ, S_w, and k_{rw}. Thus, to solve it, we require two auxiliary equations. These equations depend on the soil properties and may be written in functional form as follows:

$$k_{rw} = k_{rw}(S_w), \qquad (4.5.1.7a)$$

$$S_w = S_w(\psi). \qquad (4.5.1.7b)$$

If hysteresis effects occurring in soils subjected to wetting and drying are neglected, then k_{rw} and S_w may be considered as single-valued functions. Typical relative permeability and water content curves for fine grain soil are illustrated in Fig. 4.13a. Differentiation of Eq. (4.5.1.7b) with respect to time gives

$$\frac{\partial S_w}{\partial t} = \frac{dS_w}{d\psi}\frac{\partial \psi}{\partial t} = \frac{C}{\phi}\frac{\partial \psi}{\partial t}, \qquad (4.5.1.8a)$$

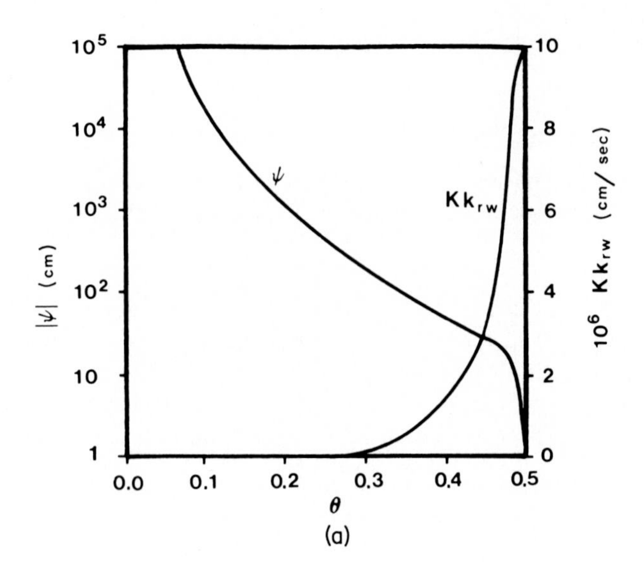

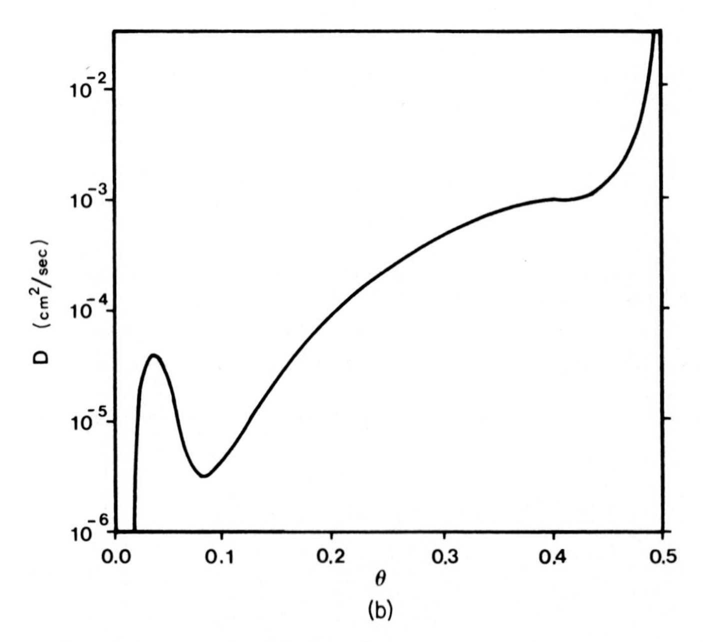

Fig. 4.13. Constitutive relations for Yolo light clay: (a) ψ and Kk_{rw} versus θ; (b) D versus θ. (Adapted from Philip, 1969.)

where C is called the specific moisture capacity, defined as

$$C = \phi(dS_w/d\psi). \tag{4.5.1.8b}$$

Substitution of Eq. (4.5.1.8a) into (4.5.1.6) yields

$$\frac{\partial}{\partial x_i}\left[K_{ij}k_{rw}\left(\frac{\partial \psi}{\partial x_j} + \frac{\partial z}{\partial x_j}\right)\right] = (C + S_w S_s)\frac{\partial \psi}{\partial t}. \tag{4.5.1.9}$$

Equations (4.5.1.9), (4.5.1.7a), and (4.5.1.7b) form the required governing equations and auxiliary equations for variably saturated flow problems. Because of the highly nonlinear nature of the relations between relative permeability, water saturation, and pressure head, the governing equation is highly nonlinear in the unsaturated zone.

For a complete description of a particular flow situation, Eq. (4.5.1.9) must also be supplemented by the appropriate initial and boundary conditions. The initial conditions are simply

$$\psi(x_i, 0) = \psi_o(x_i), \tag{4.5.1.10a}$$

where ψ_o is a prescribed function of x_i.

The boundary conditions take the form of either prescribed head or prescribed normal flux, and these are given by

$$\psi(x_i, t) = \bar{\psi} \quad \text{on} \quad B_1, \tag{4.5.1.10b}$$

$$-k_{rw}K_{ij}\left(\frac{\partial \psi}{\partial x_j} + \frac{\partial z}{\partial x_j}\right)n_i = q \quad \text{on} \quad B_2, \tag{4.5.1.10c}$$

where $\bar{\psi}$ and q are prescribed functions of x_i and t, and n_i is the unit outward normal vector on B_2. If the flow takes place in a homogeneous soil medium that is unsaturated ($S_w < 1$), it is possible to write the unsaturated flow equation in an alternative form. A function called hydraulic diffusivity is defined as

$$D_{ij}(\theta) = K_{ij}k_{rw}\frac{d\psi}{d\theta}, \tag{4.5.1.11}$$

where $\theta \equiv S_w\phi$ is the volumetric water content (the ratio of the water volume to the bulk volume of the porous medium) and $D_{ij}(\theta)$ is assumed to be a unique function of θ. Note that such a uniqueness may be obtained only if k_{rw} and ψ are unique functions of θ, and $d\psi/d\theta$ is bounded. A typical curve of D versus θ is illustrated in Fig. 4.13b. Because the porous medium is unsaturated, the substantial derivative of ρ_w, $\partial\rho_w/\partial t + (V_{iw}/S_w\phi)\,\partial\rho_w/\partial x_i$, plays a relatively insignificant role in Eq. (4.5.1.2).

Thus, the fluid flow equation reduces to

$$\frac{\partial}{\partial x_i}\left[K_{ij}k_{rw}\left(\frac{\partial \psi}{\partial x_j} + \frac{\partial z}{\partial x_j}\right)\right] = \phi \frac{\partial S_w}{\partial t} = \frac{\partial \theta}{\partial t}. \tag{4.5.1.12}$$

Using the chain rule,

$$\frac{\partial \psi}{\partial x_j} = \frac{d\psi}{d\theta}\frac{\partial \theta}{\partial x_j},$$

and substituting Eq. (4.5.1.11) into (4.5.1.12), we obtain

$$\frac{\partial}{\partial x_i}\left(D_{ij}\frac{\partial \theta}{\partial x_j}\right) + \frac{\partial}{\partial x_i}\left(K_{ij}k_{rw}\frac{\partial z}{\partial x_j}\right) = \frac{\partial \theta}{\partial t}. \tag{4.5.1.13}$$

Darcy's law can also be written in terms of D and θ as

$$V_i = -D_{ij}\frac{\partial \theta}{\partial x_j} - K_{ij}k_{rw}\frac{\partial z}{\partial x_j}. \tag{4.5.1.14}$$

For a case of horizontal flow (adsorption case), Eq. (4.5.1.14) becomes equivalent to Fick's first law of diffusion and Eq. (4.5.1.13) is just a nonlinear diffusion equation with a concentration dependent diffusivity. The coefficient D_{ij} tends to infinity as θ approaches the saturated value. Thus, to avoid numerical difficulties, we should make sure that the calculated values of θ lie within a prescribed range, $\theta_{min} \leqslant \theta \leqslant \theta_{max}$ ($\theta_{max} < \theta_{sat}$), in which the value of D_{ij} is well defined.

4.5.2 Generalized Multiphase Flow Models

A completely general formulation of the equations for multiphase flow in oil reservoirs requires a detailed consideration of spatial and temporal variations of each material component in the hydrocarbon system. Numerical models that are based on such a formulation are usually referred to as "compositional simulators." These models are typically employed in studying the behavior of condensate and volatile oil reservoirs. Our approach to the multiphase flow problems will be based on a simplified formulation which neglects phase transfers due to vaporization or condensation. In this formulation, we consider three fluid phases: gas, oil, and water. It is assumed that a one-way phase transfer occurs in the form of dissolution of gas in oil. In the hydrocarbon (oil–gas) system, only two components are considered. The "oil" component refers to the residual liquid at atmospheric pressure left after a differential vaporization, and the "gas" component is the remaining fluid. Models that develop solutions of the flow equations resulting from such a formulation are called "black oil simulators." To develop a system of governing equations, we first write Darcy's law in the form

$$V_{if} = -\frac{k_{ij}k_{rf}}{\mu_f}\left(\frac{\partial p_f}{\partial x_j} + \rho_f g \frac{\partial z}{\partial x_j}\right), \qquad f = w, o, g. \tag{4.5.2.1}$$

Next, the continuity equations are obtained for individual components. In deriving these continuity equations, it is important to recognize the fact that a given mass of each fluid occupies different volumes under reservoir and surface conditions. This is particularly true of gas and oil. Gas can change its volume dramatically because of its high compressibility. On the other hand, oil at high prevailing reservoir pressure and temperature contains substantial amounts of dissolved gas which is released under the much lower pressure value at the surface. Figure 4.14 illustrates the changes that take place in the production from such a hydrocarbon system. To account for volume changes due to different pressures at the reservoir and the surface, it is convenient to introduce a "formation volume factor" for each phase B_f and the dissolved gas–oil ratio R_s. The formation volume factor B_f is defined for "fluid phase f" as the ratio of its volume measured at reservoir conditions to its volume $\mathbf{V}$ measured at standard surface conditions. Thus,

$$B_f = \mathbf{V}_f/\mathbf{V}_{fs}, \qquad f = o, w, g, \tag{4.5.2.2a}$$

where the subscript s denotes the surface conditions.

The dissolved gas–oil ratio R_s is defined as the volume of gas, measured at standard conditions, dissolved at reservoir conditions in a unit volume of stock tank oil. Thus,

$$R_s = \mathbf{V}_{dgs}/\mathbf{V}_{os}, \tag{4.5.2.2b}$$

where the subscript dg denotes dissolved gas. Typical relations of B_o, B_g, and R_s versus pressure are plotted in Fig. 4.15. As illustrated, at pressure values below the bubble point pressure p_b, the value of B_o and R_s increase steadily with increasing pressure because more and more gas is dissolved into the oil volume, causing that volume to swell. At pressure values greater than p_b, the value of B_o decreases very slightly with increasing pressure because the effect of gas dissolution ceases and compressibility of oil starts to play a role. On the other hand, the value of B_g decreases drastically with increasing pressure, which reflects the effect of high compressibility. Performing volumetric balances with respect to the surface conditions, we obtain the following continuity equations for the individual components:

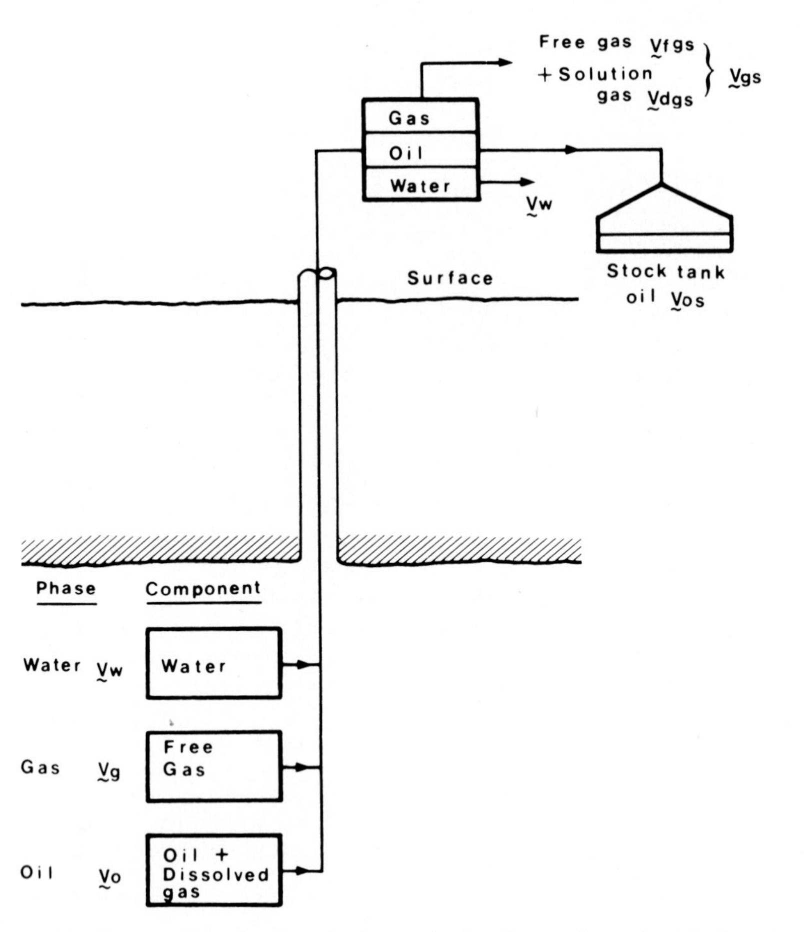

Fig. 4.14. Changes that take place in the production from a three-phase hydrocarbon system. ($\underset{\sim}{V}$ denotes volume.)

(1) Water:

$$-\frac{\partial}{\partial x_i}\left(\frac{V_{iw}}{B_w}\right) = \frac{\partial}{\partial t}\left(\frac{\phi S_w}{B_w}\right) + Q_{ws}; \tag{4.5.2.3a}$$

(2) Oil:

$$-\frac{\partial}{\partial x_i}\left(\frac{V_{io}}{B_o}\right) = \frac{\partial}{\partial t}\left(\frac{\phi S_o}{B_o}\right) + Q_{os}; \tag{4.5.2.3b}$$

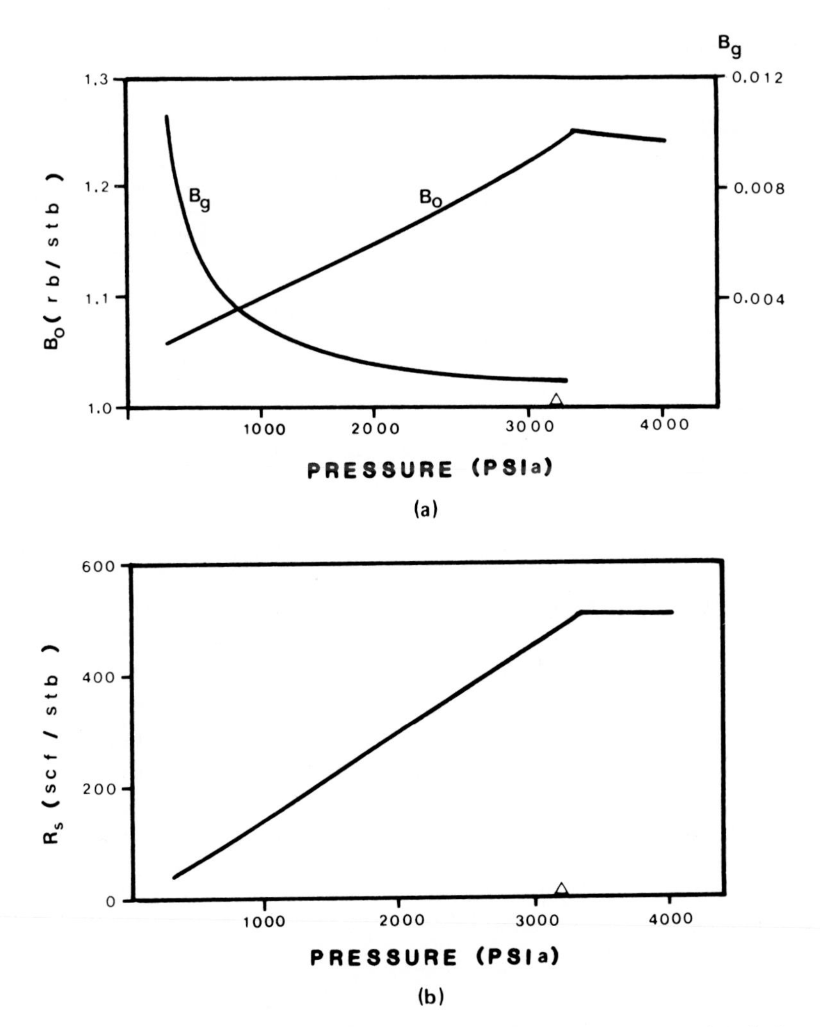

Fig. 4.15. Relations of B_o, B_g, and R_s vs. p. The point where $P_b = 3300$ psia is indicated by a triangle.

(3) Gas:

$$-\frac{\partial}{\partial x_i}\left(\frac{V_{ig}}{B_g}\right) - \frac{\partial}{\partial x_i}\left(\frac{R_s V_{io}}{B_o}\right) = \frac{\partial}{\partial t}\left[\phi\left(\frac{S_g}{B_g} + \frac{R_s S_o}{B_o}\right)\right] + Q_{gs}; \quad (4.5.2.3c)$$

where Q_{ws}, Q_{gs} and Q_{os} are the production rates of water, gas, and oil per unit volume of the porous medium.

Combination of Eqs. (4.5.2.3) and (4.5.2.1) yields the following equations for the three components:

(1) Water:

$$\frac{\partial}{\partial x_i}\left[\frac{k_{ij}\lambda_w}{B_w}\left(\frac{\partial p_w}{\partial x_j} + \rho_w g \frac{\partial z}{\partial x_j}\right)\right] = \frac{\partial}{\partial t}\left(\frac{\phi S_w}{B_w}\right) + Q_{ws}; \qquad (4.5.2.4a)$$

(2) Oil:

$$\frac{\partial}{\partial x_i}\left[\frac{k_{ij}\lambda_o}{B_o}\left(\frac{\partial p_o}{\partial x_j} + \rho_o g \frac{\partial z}{\partial x_j}\right)\right] = \frac{\partial}{\partial t}\left(\frac{\phi S_o}{B_o}\right) + Q_{os}; \qquad (4.5.2.4b)$$

(3) Gas:

$$\frac{\partial}{\partial x_i}\left[\frac{k_{ij}\lambda_g}{B_g}\left(\frac{\partial p_g}{\partial x_j} + \rho_g g \frac{\partial z}{\partial x_j}\right)\right] + \frac{\partial}{\partial x_i}\left[\frac{R_s k_{ij}\lambda_o}{B_o}\left(\frac{\partial p_o}{\partial x_j} + \rho_o g \frac{\partial z}{\partial x_j}\right)\right]$$
$$= \frac{\partial}{\partial t}\left(\frac{\phi S_g}{B_g} + \frac{R_s \phi S_o}{B_o}\right) + Q_{gs}; \qquad (4.5.2.4c)$$

where λ_w, λ_o, and λ_g are called "mobility factors" for water, oil, and gas, respectively. For fluid phase f, λ_f is defined as $\lambda_f = k_{rf}/\mu_f$.

Equations (4.5.2.4a)–(4.5.2.4c) form the required system of partial differential equations for three-phase flow in the "black oil" reservoir system. Examination of these equations reveals that there are nine unknowns, namely, p_w, p_o, p_g, k_{rw}, k_{ro}, k_{rg}, S_w, S_o, and S_g. Thus, we require six auxiliary equations to obtain a general solution. These auxiliary equations are merely two sets of constitutive relations. The first set consists of the following relations:

$$k_{rw} = k_{rw}(S_o, S_w), \qquad (4.5.2.5a)$$

$$k_{ro} = k_{ro}(S_o, S_w), \qquad (4.5.2.5b)$$

$$k_{rg} = k_{rg}(S_o, S_w), \qquad (4.5.2.5c)$$

$$S_w + S_o + S_g = 1. \qquad (4.5.2.5d)$$

The second set consists of the relations between capillary pressures and saturations. Capillary pressures exist because of the interfacial tensions between the fluids and the contact angles between the rock and the fluid phases. In general, a capillary pressure associated with two fluid phases is defined as the difference between the pressures of the nonwetting and wetting phases. Because we have three fluids (water, oil, and gas) flowing together, it is necessary to introduce two capillary pressures, $p_{o/w}$ between oil and water and $p_{g/o}$ between gas and oil. These are defined as

$$p_{o/w} = p_o - p_w, \qquad p_{g/o} = p_g - p_o.$$

The relations between these capillary pressures and saturations may be written in the form

$$p_{o/w} = p_{o/w}(S_o, S_w), \tag{4.5.2.5e}$$

$$p_{g/o} = p_{g/o}(S_o, S_g). \tag{4.5.2.5f}$$

The capillary pressure is generally a function of the rock and fluid properties and is determined by laboratory measurements. When laboratory data are unavailable, the capillary pressure is often correlated by the Leverett J function. For example, $p_{o/w}$ is given by

$$J(S_w) = \frac{p_{o/w}}{\sigma_{o/w}} \sqrt{\frac{k}{\phi}},$$

where $\sigma_{o/w}$ is fluid interfacial tension.

The capillary pressure may be estimated from the height of the transition zone in a reservoir as determined by well logging or other measurements. That is,

$$p_{o/w} = (\rho_w - \rho_o)gH,$$

where H is the height above the water–oil contact. The reader is referred to Dake (1978, pp. 344–352) for more detailed discussion. Apart from the six auxiliary constitutive relations, the governing equations must also be supplemented by initial and boundary conditions for a complete description of a particular flow situation. The initial conditions take the form

$$p_f(x_i, 0) = p_f^0(x_i), \qquad f = o, w, g, \tag{4.5.2.6a}$$

where p_f^0 is the initial pressure of fluid f.

The boundary conditions appear in the form of prescribed pressures on B_1 and prescribed fluid flux on B_2. Thus, for f = o, w, g, we obtain

$$p_f(x_i, t) = \bar{p}_f \qquad \text{on} \quad B_1, \tag{4.5.2.6b}$$

$$V_{if}n_i = q_f \qquad \text{on} \quad B_2, \tag{4.5.2.6c}$$

where $\bar{p}_f$ and q_f are prescribed fluid pressure and outward normal flux of fluid f. In a simple case where the reservoir contains hardly any dissolved or free gas, one may be able to neglect density variations with pressure. Then the set of governing equations reduces to a smaller set consisting of only the oil and water equations, which now take the form

$$\frac{\partial}{\partial x_i}\left[k_{ij}\lambda_w\left(\frac{\partial p_w}{\partial x_j}+\rho_w g\frac{\partial z}{\partial x_j}\right)\right]=\frac{\partial}{\partial t}(\phi S_w)+Q_w \tag{4.5.2.7a}$$

and

$$\frac{\partial}{\partial x_i}\left[k_{ij}\lambda_o\left(\frac{\partial p_o}{\partial x_j}+\rho_o g\frac{\partial z}{\partial x_j}\right)\right]=\frac{\partial}{\partial t}(\phi S_o)+Q_o, \tag{4.5.2.7b}$$

in which $Q_f = Q_{fs}/B_f$, f = w, o, and it is assumed that B_w and B_o are constant. Equations (4.5.2.7a) and (4.5.2.7b) thus represent the governing equations for two-phase flow with a single component in each phase. Their supplementary equations can be obtained by reducing Eqs. (4.5.2.5a)–(4.5.2.5f) to the following set:

$$k_{rw} = k_{rw}(S_w), \tag{4.5.2.8a}$$

$$k_{ro} = k_{ro}(S_o), \tag{4.5.2.8b}$$

$$S_o + S_w = 1, \tag{4.5.2.8c}$$

$$p_{o/w} = p_o - p_w. \tag{4.5.2.8d}$$

The initial and boundary conditions are as given by Eqs. (4.5.2.6a)–(4.5.2.6c) with f = w, o.

4.6 Iterative Methods for Solving Nonlinear Equations

Owing to nonlinearities in the partial differential equations describing variably saturated flow, the system of algebraic equations derived from the numerical approximation is also nonlinear. Consequently, it is necessary to employ iterative methods to obtain a solution. In this section, we present a general description of three iterative methods that are commonly applied to highly nonlinear problems. All methods call for an initial estimate of the solution to start, and each uses a different algorithm to produce a new and, we hope, closer estimate.

4.6.1 Picard Method

We have used the Picard method without generalization in the solution of the nonlinear finite element equations obtained for the foregoing problems of areal unconfined groundwater flow, seawater intrusion, and gas flow.

The general algorithm for Picard iteration can be described as follows. Consider a set of nonlinear equations written in the form

$$f_I(x_1, x_2, \ldots, x_m) = 0, \qquad I = 1, 2, \ldots, n, \tag{4.6.1.1}$$

where $x_1, x_2, \ldots, x_n$ are the unknowns. We first construct a set of auxiliary functions $g_I(x_1, x_2, \ldots, x_n)$ such that [see Carnahan *et al.* (1969)]

$$E_{IJ}x_J = g_I, \tag{4.6.1.2}$$

where the repeated subscripts indicate summation and each of the E_{IJ} and g_I is a function of $x_1, x_2, \ldots, x_n$. We start the iteration by assuming an initial solution $(x_1^1, x_2^1, \ldots, x_n^1)$ and use this to evaluate the left-hand coefficients and right-hand side of Eq. (4.6.1.2). Thus, this equation becomes a set of linear equations, and we can solve for the next set of x_J values. The solution can be expressed in the form

$$x_J^{r+1} = (E_{JI}^r)^{-1} g_I^r, \tag{4.6.1.3}$$

where r is an iteration counter with the initial value one and the $(E_{JI}^r)^{-1}$ denotes elements of the inverse matrix $[\mathrm{E}^r]^{-1}$. Equation (4.6.1.3) thus provides the means for obtaining successive solutions of x_J. At each iterative cycle we must update the left-hand coefficients and the right-hand side. The iteration is performed until satisfactory convergence is achieved. A typical criterion for checking convergence is given by

$$(\max_J|x_J^{r+1} - x_J^r|)/(\max_J|x_J^{r+1}|) \leq \varepsilon, \tag{4.6.1.4}$$

where ε is a prescribed tolerance for the x values. For the simplest case of one unknown, the iteration sequence is graphically illustrated in Fig. 4.16. The iterative procedure just described is applicable directly to the system of nonlinear algebraic equations obtained from the finite element approximation of any nonlinear problem. In this case, the system of equations is already in the form given by (4.6.1.2). Thus the auxiliary functions g_I need not be constructed.

Having described the algorithm of the Picard method, it is of practical interest to determine how the errors decrease at each step of the iteration. First, we define the error at the $(r+1)$th iteration, e_J^{r+1}, as follows:

$$e_J^{r+1} = x_J - x_J^{r+1}, \tag{4.6.1.5}$$

where x_J denotes the exact solution of Eq. (4.6.1.1). Equation (4.6.1.3) can now be written in the form

$$x_J - x_J^{r+1} = e_J^{r+1} = x_J - (E_{JI}^r)^{-1}g_I^r,$$

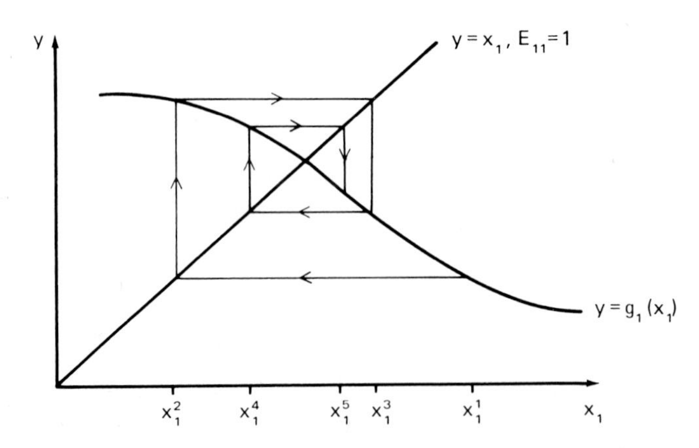

Fig. 4.16. Graphical illustration of Picard iterative method in one variable.

or

$$e_J^{r+1} = x_J - [E_{JI}(x_1 - e_1^r, \ldots, x_n - e_n^r)]^{-1} \times g_I(x_1 - e_1^r, \ldots, x_n - e_n^r). \tag{4.6.1.6}$$

Expanding the second term on the right-hand side of Eq. (4.6.1.6) in a Taylor's series and neglecting higher-order terms, we obtain as $e^r \rightarrow 0$,

$$e_J^{r+1} \simeq x_J - \left[\left(E_{JI}\right)^{-1} - e_L^r \frac{\partial}{\partial x_L}\left(E_{JI}\right)^{-1}\right]\left[g_I - e_k^r \frac{\partial g_I}{\partial x_k}\right], \tag{4.6.1.7}$$

where E^{-1}, g, and their derivatives are evaluated at the exact solution, i.e., at $x_1, x_2, \ldots, x_n$.

Since $x_J = (E_{JI})^{-1} g_I$, Eq. (4.6.1.7) becomes

$$e_J^{r+1} \simeq e_k^r \left(E_{JI}\right)^{-1} \frac{\partial g_I}{\partial x_k} + e_L^r \frac{\partial}{\partial x_L}\left(E_{JI}\right)^{-1} g_I + e_k^r e_L^r \frac{\partial}{\partial x_L}\left(E_{JI}\right)^{-1} \frac{\partial g_I}{\partial x_k}. \tag{4.6.1.8a}$$

From Eq. (4.6.1.8a), it can be deduced that, as $e^r \rightarrow 0$,

$$\|e^{r+1}\| \simeq \gamma \|e^r\|, \tag{4.6.1.8b}$$

where $\|e\|$ is the norm of the error vector and γ is a positive number. The value of γ must be less than unity to ensure convergence.

Thus, it is evident that the error decreases linearly with the error at the previous iteration. Such a rate of convergence is called "linear" or "first-order" convergence.

4.6.2 Newton–Raphson Method

For some highly nonlinear problems, the first-order convergence rate obtained from the Picard method may be too inefficient. In such a case, it is desirable to employ the Newton–Raphson method, which normally converges much more rapidly. To describe this method, we consider once again the system of nonlinear equations represented by Eq. (4.6.1.1). Assuming that the functions f_I are continuous, we can make a Taylor series expansion about the starting point $(x_1^r, x_2^r, \ldots, x_n^r)$. Truncating the second- and higher-order terms, we obtain

$$f_I^{r+1} = f_I(x_1^{r+1}, x_2^{r+1}, \ldots, x_n^{r+1}) = f_I(x_1^r + \Delta x_1^{r+1}, \ldots, x_n^r + \Delta x_n^{r+1})$$

$$\simeq f_I^r + \left(\frac{\partial f_I}{\partial x_J}\right)^r \Delta x_J^{r+1}, \qquad (4.6.2.1)$$

where Δx_J^{r+1} is a displacement vector, defined as

$$\Delta x_J^{r+1} = x_J^{r+1} - x_J^r. \qquad (4.6.2.2)$$

In Eq. (4.6.2.1), we employ the summation convention on subscript J. The coefficients $(\partial f_I/\partial x_J)$ in this equation represent tangential slopes (gradients) of f_I with respect to each element of the vector x_J. These coefficients form the "Jacobian matrix" of the system in (4.6.1.1), i.e.,

$$J_{IJ}^r = \left(\frac{\partial f_I}{\partial x_J}\right)^r.$$

Although we do not know the value of f_I^{r+1}, we realize that under optimal conditions of good fortune we would like it to vanish. This, according to (4.6.1.1), is our ultimate goal. As a first approximation we assume $f_I^{r+1} = 0$ and proceed to solve for Δx_J^{r+1} using the remaining elements of (4.6.2.1), i.e.,

$$\frac{\partial f_I}{\partial x_J} \Delta x_J^{r+1} = -f_I^r, \qquad I = 1, 2, \ldots, n. \qquad (4.6.2.3)$$

Because the Jacobian matrix is known, Eq. (4.6.2.3) represents a set of linear equations whose solution exists, provided that the Jacobian matrix is nonsingular, i.e., its determinant is nonzero. Thus, we can solve for Δx_J^{r+1} from

$$\Delta x_J^{r+1} = -(J_{JI}^r)^{-1} f_I^r, \qquad (4.6.2.4)$$

where $(J_{JI}^r)^{-1}$ denotes the elements of the inverse of the Jacobian matrix.

Knowing the displacement vector, we can compute the next iterate from Eq. (4.6.2.2). Thus,

$$x_J^{r+1} = x_J^r + \Delta x_J^{r+1}. \tag{4.6.2.5}$$

Once the initial values x_J^1 are provided, the successive solutions of x_J at the $(r+1)$th iteration can be obtained using Eqs. (4.6.2.4) and (4.6.2.5). In the conventional Newton–Raphson method, the Jacobian is updated at every iterative cycle. However, at the expense of slower convergence, the initial Jacobian J_{IJ}^1 may be kept and used in all subsequent iterations. Such a procedure is often referred to as the "initial slope" Newton–Raphson method. For the simple case of one unknown, the operations of both versions of the Newton–Raphson method can be graphically represented as shown in Fig. 4.17.

To check convergence of the iteration, we can use the same criterion given in (4.6.1.4). It should be noted that each iterative cycle of the Newton–Raphson method usually requires more computational effort than that of the Picard method [not only because the Jacobian matrix is always asymmetric, but also because its coefficients usually contain more terms than those of matrix $[E]$ of Eq. (4.6.1.2). The matrix $[E]$ can be symmetric or asymmetric depending on the type of problem]. Nonetheless, because of its fast convergence rate, the use of the Newton–Raphson method is still preferable for some highly nonlinear problems. To evaluate the rate of convergence of the Newton–Raphson procedures we let e_J^r and e_J^{r+1} be error vectors at the rth and $(r+1)$th steps. Thus, if x_J represents

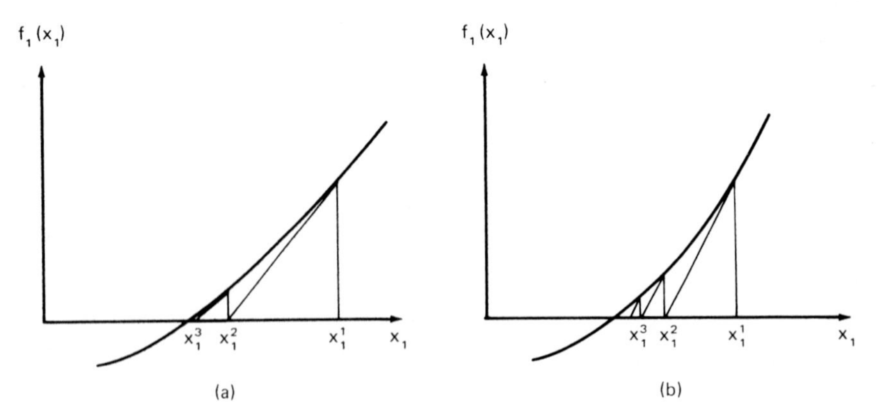

Fig. 4.17. Graphical illustration of the Newton–Raphson method in one variable: (a) variable slope; (b) constant (initial) slope.

the true solution, we obtain

$$e_J^r = x_J - x_J^r, \qquad (4.6.2.6a)$$

$$e_J^{r+1} = x_J - x_J^{r+1}. \qquad (4.6.2.6b)$$

Equation (4.6.2.4) is now written in the form

$$x_J^{r+1} = x_J^r - (J_{JI}^r)^{-1} f_I^r,$$

or

$$x_J - x_J^{r+1} = x_J - x_J^r + (J_{JI}^r)^{-1} f_I^r. \qquad (4.6.2.7)$$

Substitution of Eqs. (4.6.2.6a) and (4.6.2.6b) into (4.6.2.7), gives, as $e^r \to 0$,

$$e_J^{r+1} = e_J^r + [J_{JI}(x_1 - e_1^r, \ldots, x_n - e_n^r)]^{-1} \times f_I(x_1 - e_1^r, \ldots, x_n - e_n^r). \qquad (4.6.2.8)$$

Taking a Taylor series expansion and neglecting higher-order terms, we obtain

$$e_J^{r+1} = e_J^r + \left[(J_{JI})^{-1} - e_L^r \frac{\partial}{\partial x_L} (J_{JI})^{-1} \right] \left[f_I - e_K^r \frac{\partial f_I}{\partial x_K} + \frac{1}{2} e_K^r e_M^r \frac{\partial^2 f_I}{\partial x_K \, \partial x_M} \right], \qquad (4.6.2.9)$$

where $(J_{JI})^{-1}$, f_I, and their derivatives are evaluated at the exact solution, i.e., at $x_1, x_2, \ldots, x_n$. Because $f_I = 0$ and $\partial f_I / \partial x_K = J_{IK}$, Eq. (4.6.2.9) reduces to

$$e_J^{r+1} \simeq \frac{1}{2} e_K^r e_M^r (J_{JI})^{-1} \frac{\partial J_{IK}}{\partial x_M} + e_K^r e_L^r \frac{\partial}{\partial x_L} (J_{JI})^{-1} J_{IK} + \cdots$$

$$= -\frac{1}{2} e_K^r e_M^r (J_{JI})^{-1} \frac{\partial J_{IK}}{\partial x_M}. \qquad (4.6.2.10)$$

Using the standard order notation Eq. (4.6.2.10) can be represented by

$$e^{r+1} = O\{(e^r)^2\}. \qquad (4.6.2.11)$$

This indicates that the $(r+1)$th error is approximately proportional to the square of the rth error. In other words, the rate of convergence is second order. Thus, when the Newton–Raphson method converges, it usually converges much faster than the Picard method if the error is small.

To illustrate the Newton–Raphson procedure, we solve the following set of equations: $f_1(x_1, x_2) = x_1^2 + x_2^2 - 5 = 0$, $f_2(x_1, x_2) = x_1 + x_2 - 1 = 0$. It is readily seen that

$$\frac{\partial f_1}{\partial x_1} = 2x_1, \quad \frac{\partial f_1}{\partial x_2} = 2x_2,$$

$$\frac{\partial f_2}{\partial x_1} = 1, \quad \frac{\partial f_2}{\partial x_2} = 1.$$

The displacements Δx_1 and Δx_2 are determined directly from (4.6.2.1) as

$$\frac{\partial f_1}{\partial x_1}\Delta x_1 + \frac{\partial f_1}{\partial x_2}\Delta x_2 = -f_1,$$

$$\frac{\partial f_2}{\partial x_1}\Delta x_1 \frac{\partial f_2}{\partial x_2}\Delta x_2 = -f_2.$$

Thus, Δx_1 and Δx_2 can be obtained explicitly as

$$\Delta x_1 = \frac{f_2 \partial f_1/\partial x_2 - f_1 \partial f_2/\partial x_2}{D}, \qquad \Delta x_2 = \frac{f_1 \partial f_2/\partial x_1 - f_2\, \partial f_1/\partial x_1}{D},$$

where

$$D = \frac{\partial f_1}{\partial x_1}\frac{\partial f_2}{\partial x_2} - \frac{\partial f_1}{\partial x_2}\frac{\partial f_2}{\partial x_1}.$$

The result of the calculation, given in Carnahan *et al.* (1969), is presented in the accompanying tabulation.

r	x_1	x_2	f_1	f_2	$\frac{\partial f_1}{\partial x_1}$	$\frac{\partial f_1}{\partial x_2}$	$\frac{\partial f_2}{\partial x_1}$	$\frac{\partial f_2}{\partial x_2}$	D	Δx_1	Δx_2
1	1	0	−4	0	2	0	1	1	2	2	−2
2	3	−2	8	0	6	−4	1	1	10	−0.8	0.8
3	2.2	−1.2	1.28	0	4.4	−2.4	1	1	6.8	−0.2	0.2
4	2	−1	0	0							

4.6.2 Chord Slope Method

This method is commonly applied to transient nonlinear problems. It is similar to Newton–Raphson with the exception that the Taylor series expansion of the nonlinear functions is made about the values at the previous time level and the gradients of the nonlinear functions are computed using chord slopes. To describe the chord slope method, we consider the following set of nonlinear equations written at the current time level $k+1$:

$$f_I(x_1, x_2, \ldots, x_n) = 0, \qquad I = 1, 2, \ldots, n. \tag{4.6.3.1}$$

Let $(x_1^{r+1}, x_2^{r+1}, \ldots, x_n^{r+1})$ be the values at a new iteration $r+1$ for time level $k+1$. In addition, let

$$x_J^{r+1} = x_J^k + \Delta x_J^{r+1}, \tag{4.6.3.2}$$

where x_J^k represents the values of x at the previous time level and Δx_J^{r+1} represents the increment vector, i.e., the change of the nodal values over Δt as estimated by $r+1$ iterations. Assuming that the functions f_I, $I = 1, 2, \ldots, n$, are continuous, we can write

$$f_I^{r+1} = f_I(x_1^{r+1}, x_2^{r+1}, \ldots, x_n^{r+1}) \simeq f_I^k + \Delta f_I^{r+1}, \tag{4.6.3.3}$$

where

$$f_I^k = f_I(x_1^k, x_2^k, \ldots, x_n^k), \tag{4.6.3.4a}$$

$$\Delta f_I^{r+1} = \left(\frac{\partial f_I}{\partial x_J}\right)_r \Delta x_J^{r+1}, \tag{4.6.3.4b}$$

and repeated subscript J indicates summation. As distinct from the Newton–Raphson method, we now approximate $(\partial f_I/\partial x_J)_r$ using the chord slopes, which can be expressed by

$$(\partial f_I/\partial x_J)_r = (f_I^r - f_I^k)/(x_J^r - x_J^k). \tag{4.6.3.4c}$$

Having calculated $(\partial f_I/\partial x_J)_r$, we next determine Δx_J^{r+1} by setting f_I^{r+1}, the left-hand term in (4.6.3.3), to zero and solving the following set of linear equations:

$$(\partial f_I/\partial x_J)_r \, \Delta x_J^{r+1} = -f_I^k. \tag{4.6.3.5}$$

Here f_I^k is the residual vector, which is evaluated using the values of x at the previous time level. It is apparent that the right-hand side of (4.6.3.5) needs to be calculated only once at the beginning of the new iteration. After solving Δx_J^{r+1}, we calculate x_J^{r+1} using Eq. (4.6.3.2) and then repeat the iterative process until the following convergence criterion is satisfied:

$$\frac{\max_J |x_J^{r+1} - x_J^r|}{\max_J |x_J^{r+1}|} \leq \varepsilon, \tag{4.6.3.6}$$

where ε is a prescribed tolerance for the x values.

4.7 Fluid Flow in Variably Saturated Media

4.7.1 Moisture Movement in Desiccated Soils

Consider a situation of two-dimensional unsaturated flow in a homogeneous soil medium that is relatively dry. For this type of situation there usually exist extremely large variations in the magnitude of ψ. It is sometimes more advantageous to use the θ-based governing Eq. (4.5.1.13) rather than the ψ-based equation (4.5.1.6). Equation (4.5.1.13) can be rewritten in the form

$$\frac{\partial}{\partial x_i}\left(D_{ij}\frac{\partial \theta}{\partial x_j}\right) + \frac{\partial}{\partial x_i}\left(K_{ij}k_{rw}\right)e_j = \frac{\partial \theta}{\partial t}, \qquad i = 1, 2, \tag{4.7.1.1}$$

where e_j is the unit vector in the upward vertical direction, $e_1 = 0$, and $e_2 = 1$. To apply the Galerkin method, we represent θ, D_{ij}, and k_{rw} by trial functions of the form

$$\hat{\theta}(x_i, t) = N_J(x_i)\theta_J(t), \tag{4.7.1.2a}$$

$$\hat{D}_{ij}(x_i, t) = N_J(x_i)D_{ijJ}(t), \tag{4.7.1.2b}$$

$$\hat{k}_{rw}(x_i, t) = N_J(x_i)k_{rwJ}(t). \tag{4.7.1.2c}$$

Application of the Galerkin criterion and Green's theorem leads to

$$\int_R \left[\hat{D}_{ij}\frac{\partial N_I}{\partial x_i}\frac{\partial N_J}{\partial x_j}\theta_J + K_{ij}\frac{\partial N_I}{\partial x_i}\hat{k}_{rw}\, e_j\right] dR + \int_R N_I N_J \frac{d\theta_J}{dt}\, dR$$

$$- \int_{B_2} N_I \left(\hat{D}_{ij}\frac{\partial \hat{\theta}}{\partial x_j} + K_{ij}\hat{k}_{rw}e_j\right) n_i \, dB = 0. \tag{4.7.1.3}$$

Using the flux boundary condition on B_2, we obtain

$$\int_R \hat{D}_{ij} \frac{\partial N_I}{\partial x_i} \frac{\partial N_J}{\partial x_j} \theta_J \, dR + \int_R N_I N_J \frac{d\theta_J}{dt} \, dR$$

$$+ \int_R K_{ij} \frac{\partial N_I}{\partial x_i} \hat{k}_{rw} e_j \, dR + \int_{B_2} N_I q \, dB = 0, \qquad (4.7.1.4)$$

where q is the outward normal flux. In matrix form, Eq. (4.7.1.4) becomes

$$[A]\{\theta\} + [B]\{d\theta/dt\} = \{F\}, \qquad (4.7.1.5)$$

where

$$A_{IJ} = \sum_e \int_{R^e} \hat{D}_{ij} \frac{\partial N_I}{\partial x_i} \frac{\partial N_J}{\partial x_j} \, dR,$$

$$B_{IJ} = \sum_e \int_{R^e} N_I N_J \, dR,$$

$$F_I = \sum_e \left(- \int_{R^e} K_{ij} k_{rw} \frac{\partial N_I}{\partial x_i} e_j \, dR - \int_{B_2^e} N_I q dB \right).$$

Equation (4.7.1.5) represents a system of ordinary differential equations that are highly nonlinear for dry soils. To obtain stable numerical results of acceptable accuracy, the time-stepping and iterative solution must be carefully controlled. In order to obtain second-order accuracy, the time integration should be performed using the Crank–Nicolson scheme

$$\frac{1}{2} [A]^{k+1/2} (\{\theta\}^{k+1} + \{\theta\}^k) + \frac{[B]^{k+1}}{\Delta t} (\{\theta\}^{k+1} - \{\theta\}^k) = \{F\}^{k+1/2},$$

which can be arranged in the form

$$\left([A]^{k+1} + \frac{2}{\Delta t} [B]^{k+1} \right) \{\theta\}^{k+1} = 2\{F\}^{k+1/2} - \left([A]^k + \frac{2}{\Delta t} [B]^k \right) \{\theta\}^k, \quad (4.7.1.6a)$$

where $k+1/2$ represents the time level midway between the current $(k+1)$ and old (k) time levels.

Equation (4.7.1.6a) now represents a system of nonlinear equations with symmetric left-hand-side matrix coefficients. In view of the fact that the dispersion coefficient D may not be a smooth function of the moisture content θ when the soil is relatively dry, we recommend the use of the Picard method, which, unlike the Newton–Raphson method, does not require continuity of the gradient $dD/d\theta$ to converge. Let us now outline the solution procedure.

At the beginning of each time step, the values of $\theta_J^{k+1/2}$ to be used in the Picard iteration are calculated using the linear extrapolation

$$\theta_J^{k+1/2} = \theta_J^k + \frac{\Delta t^k}{2\,\Delta t^{k-1}}(\theta_J^k - \theta_J^{k-1}). \tag{4.7.1.6b}$$

These values are now used in the evaluation of the coefficient matrices. The solution of the resulting set of linear algebraic equations is achieved by a direct Gaussian elimination algorithm. For the next iteration, an improved estimate of $\theta_J^{k+1/2}$ is derived from

$$\theta_J^{k+1/2} = (\theta_J^r + \theta_J^k)/2. \tag{4.7.1.6c}$$

where θ_J^r denotes the rth iterate of θ_J^{k+1}. The iterative procedure is repeated until the following convergence criterion is satisfied:

$$\max_J \left|\Delta\theta_J^{k+1}\right| \Big/ \max_J \left|\theta_J^{k+1}\right| < \varepsilon,$$

where $\Delta\theta_J^{k+1}$ denotes the difference between two successive iterations and ε is a prescribed tolerance.

Owing to several nonlinearities encountered in problems of flow in dry soils, the iterative procedure may not converge unless Δt is unusually small. It is a good practice to incorporate an automatic time-stepping control in the computer program. One simple way to control the time-step size is to select a maximum number of iterations allowed per time step. This can be used as a criterion for reducing or increasing the time-step value. If it takes more than the prescribed number of iterations (say five iterations) to obtain convergence, the time step would be reduced to half and the solution reinitiated at the old time level. The time-step reduction is continued until satisfactory convergence is achieved. On the other hand, if convergence is reached within a smaller number of iterations (e.g., two or three), then we have the option of either setting the next time step to the same value as the current time step or increasing by some percentage. Typical examples of unsaturated flow simulations performed using the θ-based equation and the Galerkin solution approach can be found in Bruch (1977).

4.7.2 Flow in Saturated–Unsaturated Soils

In a situation involving a layered porous medium or a flow domain consisting of saturated and unsaturated zones, the θ-based equation is not applicable. Thus, it is necessary to employ the ψ-based equation (4.5.1.6), which, assuming that the specific storage term is negligible, may be written as

$$\frac{\partial}{\partial x_i}\left[K_{ij}k_{rw}\left(\frac{\partial \psi}{\partial x_j}+\frac{\partial z}{\partial x_j}\right)\right]=\phi\frac{\partial S_w}{\partial t}, \qquad i,j=1,2. \tag{4.7.2.1}$$

To apply the Galerkin method, the pressure head ψ is approximated by the trial function of the form

$$\hat{\psi}=N_J(x_i)\Psi_J(t), \qquad J=1,2,\ldots,n. \tag{4.7.2.2a}$$

In addition, we approximate the nonlinear functions k_{rw} and S_w by $\hat{k}_{rw}$ and $\hat{S}_w$, which take the form

$$\hat{k}_{rw}(x_i,t)=N_J(x_i)k_{rwJ}(t), \tag{4.7.2.2b}$$

$$\hat{S}_w(x_i,t)=N_J(x_i)S_{wJ}(t). \tag{4.7.2.2c}$$

Application of the Galerkin criterion and Green's theorem leads to

$$\int_R \frac{\partial N_I}{\partial x_i}\left[K_{ij}\hat{k}_{rw}\left(\frac{\partial \hat{\psi}}{\partial x_j}+\frac{\partial z}{\partial x_j}\right)\right]dR+\int_R \phi N_I\frac{\partial \hat{S}_w}{\partial t}\,dR$$

$$-\int_{B_2} N_I K_{ij}\hat{k}_{rw}\left(\frac{\partial \hat{\psi}}{\partial x_j}+\frac{\partial z}{\partial x_j}\right)n_i\,dB=0, \qquad I,J=1,2,\ldots,n. \tag{4.7.2.3}$$

After substituting for $\hat{\Psi}$ and incorporating the flux boundary condition, we obtain

$$\int_R\left(K_{ij}\hat{k}_{rw}\frac{\partial N_I}{\partial x_i}\frac{\partial N_J}{\partial x_j}\,dR\right)\psi_J+\int_R N_I\phi\frac{\partial \hat{S}_w}{\partial t}\,dR$$

$$=-\int_R K_{ij}\hat{k}_{rw}\frac{\partial N_I}{\partial x_i}e_j\,dR-\int_{B_2}N_I q\,dB, \qquad I,J=1,2,\ldots,n, \tag{4.7.2.4}$$

where $e_j = \partial z/\partial x_j$ is once again the unit vector in the upward vertical direction. The resulting system of nonlinear equations can be treated by one of the three methods described in Section 4.6.2. Here, we choose to present the application of the Picard and the chord slope methods. For this type of flow situation, however, the Picard method is the one most widely used.

Application of Picard Method

Before applying the Picard method, we first write Eq. (4.7.2.4) in the form

$$\int_R \left(K_{ij}\hat{k}_{rw} \frac{\partial N_I}{\partial x_i} \frac{\partial N_J}{\partial x_j} \, dR \right) \psi_J + \int_R N_I C \frac{\partial \hat{\psi}}{\partial t} \, dR = -\int_R K_{ij}\hat{k}_{rw} \frac{\partial N_I}{\partial x_i} e_j \, dR - \int_{B_2} N_I q \, dB, \tag{4.7.2.5}$$

where C is the specific moisture capacity defined by (4.5.1.8b). Next, we replace C by $\hat{C}$, where $\hat{C} = N_I(x_i)C_I(t)$, and use Eq. (4.7.2.2a) to substitute for ψ. The result is given by

$$\int_R \left(K_{ij}\hat{k}_{rw} \frac{\partial N_I}{\partial x_i} \frac{\partial N_J}{\partial x_j} \, dR \right) \psi_J + \int_R (\hat{C} N_I N_J \, dR) \frac{d\psi_J}{dt} = -\int_R K_{ij}\hat{k}_{rw} \frac{\partial N_I}{\partial x_i} e_j \, dR - \int_{B_2} N_I q \, dB. \tag{4.7.2.6}$$

Equation (4.7.2.6) now represents a system of nonlinear ordinary differential equations which can be written in matrix form as

$$[A]\{\psi\} + [B]\{d\psi/dt\} = \{F\}, \tag{4.7.2.7}$$

in which typical matrix elements are given by

$$A_{IJ} = \sum_e \int_{R^e} K_{ij}\hat{k}_{rw} \frac{\partial N_I}{\partial x_i} \frac{\partial N_J}{\partial x_j} \, dR,$$

$$B_{IJ} = \sum_e \int_{R^e} \hat{C} N_I N_J \, dR, \tag{4.7.2.8a}$$

$$F_I = -\sum_e \left[\int_{R^e} K_{ij}\hat{k}_{rw} \frac{\partial N_I}{\partial x_i} e_j \, dR + \int_{B_2^e} N_I q \, dB \right]. \tag{4.7.2.8b}$$

At this stage, it is worthwhile to mention that there exist two alternative schemes for forming the mass matrix $[B]$ when linear triangular or quadrilateral elements are used. In the first scheme, the matrix $[B]$ is formed according to Eq. (4.7.2.8a) and is referred to as a "consistent" mass matrix. In the second scheme, $[B]$ is diagonalized by a procedure known as "lumping." According to this procedure, we calculate the matrix elements as

$$B_{II} = \sum_e \int_{R^e} \hat{C} N_I \, dR, \qquad B_{IJ} = 0, \quad I \neq J.$$

Experience indicates that a more stable but usually less accurate solution is obtained with the lumped mass matrix. A lumping procedure similar to that described was successfully applied by Neuman (1973) to a number of practical problems.

We now return to Eq. (4.7.2.7) and perform the time integration using finite differences as follows:

$$[A]^{k+\varepsilon}(\varepsilon\{\psi\}^{k+1} + (1-\varepsilon)\{\Psi\}^{k}) + \frac{[B]^{k+\varepsilon}}{\Delta t}(\{\psi\}^{k+1} - \{\psi\}^{k}) = \{F\}^{k+\varepsilon}, \tag{4.7.2.9a}$$

where, as before, $k+1$ and k refer to the current and previous time levels and ε is the time-weighting factor, which we use rather than θ to avoid confusion. Equation (4.7.2.9a) may be arranged in the standard form

$$[E]^{k+\varepsilon}\{\psi\}^{k+1} = \{G\}^{k+\varepsilon}, \tag{4.7.2.9b}$$

where

$$[E]^{k+\varepsilon} = \varepsilon[A]^{k+\varepsilon} + \frac{[B]^{k+\varepsilon}}{\Delta t}$$

and

$$\{G\}^{k+\varepsilon} = \{F\}^{k+\varepsilon} + \left[(\varepsilon - 1)[A]^{k+\varepsilon} + \frac{[B]^{k+\varepsilon}}{\Delta t}\right]\{\psi\}^{k}.$$

In applying the Picard iteration method to Eq. (4.7.2.9b), it is convenient to once again introduce superscripts r and $r+1$ to denote the old and new iterations, respectively, at the current time level $k+1$. Thus, Eq. (4.7.2.9b) is now written as

$$[E]^{k+\varepsilon}\{\psi\}^{r+1} = \{G\}^{k+\varepsilon}. \tag{4.7.2.10a}$$

As mentioned previously, there may be a large variation in the magnitude of ψ for the situation of flow in dry soils. To reduce the resulting computer round-off errors, it is preferable to solve the equation in terms of $\Delta\psi$. Thus,

$$[E]^{k+\varepsilon}\{\Delta\psi\}^{r+1} = \{R\}^{k+\varepsilon}, \tag{4.7.2.10b}$$

where

$$\{\Delta\psi\}^{r+1} = \{\psi\}^{r+1} - \{\psi\}^{r}$$

and $\{R\}^{k+\varepsilon}$ is the residual vector given by

$$\{R\}^{k+\varepsilon} = \{G\}^{k+\varepsilon} - [E]^{k+\varepsilon}\{\psi\}^{r}.$$

To solve for $\{\Delta\psi\}^{r+1}$, we linearize (4.7.2.10) by evaluating $[E]^{k+\varepsilon}$ and $\{R\}^{k+\varepsilon}$ using the head values obtained by linear interpolation over the time step

$$\{\psi\}^{k+\varepsilon} = \varepsilon\{\psi\}^{k} + (1 - \varepsilon)\{\psi\}^{k+1}. \tag{4.7.2.11}$$

Note, in solving for $\{\Delta\psi\}^{r+1}$, that all prescribed boundary values of ψ must be converted to the corresponding prescribed values of $\Delta\psi$. The latter are always equal to zero because the prescribed values of ψ do not vary with the iteration number. After $\{\Delta\psi\}^{r+1}$ has been prescribed, the current iterate $\{\psi\}^{r+1}$ is calculated from

$$\{\psi\}^{r+1} = \{\psi\}^{r} + \{\Delta\psi\}^{r+1}. \tag{4.7.2.12}$$

Next, we use $\{\psi\}^{r+1}$ to update the coefficient matrix and the residual vector and repeat the iterative process until convergence is achieved. We employ two convergence criteria, the head closure and the residual or mass balance check. These criteria are

$$\frac{\max_J|\Delta\psi_J^{r+1}|}{\max_J|\psi_J^r|} \leqslant \delta \qquad \text{and} \qquad \frac{\sum_{J=1}^{n}|R_J^r|}{\sum_{J=1}^{n}|Q_J^r|} \leqslant \delta', \tag{4.7.2.13}$$

where δ and δ' are prescribed tolerances and Q_J are nodal discharges or recharges. Experience indicates that this numerical procedure gives satisfactory stability and convergence when the backward difference time scheme, $\varepsilon = 1$, is used. Nonetheless, improved accuracy may be achieved by an alternative scheme that combines central difference with backward difference time stepping. In this particular scheme, the nodal points of the flow region are classified into three categories (Neuman, 1973);

(1) nodal points that remain unsaturated during the time interval Δt,
(2) nodal points that remain saturated,
(3) nodal points that undergo a change from a state of saturation to a state of unsaturation during the time interval Δt.

For nodal points in category 1, we employ central time stepping, $\varepsilon = \frac{1}{2}$, and thus calculate the coefficient matrix $[E]^{k+1/2}$ and vector $\{R\}^{k+1/2}$ using the head values given by

$$\{\psi\}^{k+1/2} = \tfrac{1}{2}\left(\{\psi\}^{r} + \{\psi\}^{k}\right).$$

For nodal points in category 2, it is necessary to employ backward difference time stepping, $\varepsilon = 1$. This is because the governing equation becomes elliptic because $(\partial S_w/\partial t = 0)$ at these saturated points. Under this condition, the most appropriate numerical analog can only be obtained with the backward difference time-stepping procedure that eliminates the spatial term $(1 - \varepsilon)[A]\{\psi\}^k$ in the original matrix equation (4.7.2.9a).

For nodal points in category 3, we also employ central stepping but make provision for the fact that the change in moisture content $(\partial\theta/\partial t) = C(\partial\psi/\partial t)$ occurs only within the negative range of ψ values (because C is zero when ψ is positive). To allow for this, we replace ψ_I^k by zero, in forming the residual vector $\{R\}$, whenever node I falls into the third category as the iteration progresses.

Application of the Chord Slope Method

Another efficient procedure commonly used by petroleum engineers to solve the highly nonlinear, multiphase, variably saturated flow problem is the chord slope method. To apply this method to our current problem we first use the backward difference time-stepping scheme to obtain the following equation for a current time level $k+1$:

$$\int_R K_{ij} \frac{\partial N_I}{\partial x_i} \frac{\partial N_J}{\partial x_j} N_L k_{rwL}^{k+1} \psi_J^{k+1}\, dR + \int_R \frac{\phi}{\Delta t} N_I N_J (S_{wJ}^{k+1} - S_{wJ}^k)\, dR$$

$$+ \int_R K_{ij} \frac{\partial N_I}{\partial x_i} e_j N_L k_{rwL}^{k+1}\, dR + \int_{B_2} N_I q\, dB = 0. \qquad (4.7.2.14)$$

For convenience, let G_I^{k+1} represent the left side of Eq. (4.7.2.14). We now obtain

$$G_I^{k+1} = G_I^k + (\partial G_I/\partial\psi_J)^r\, \Delta\psi_J^{r+1} = 0, \qquad (4.7.2.15)$$

where, as earlier, r and $r+1$ denote the old and new iterations of time $k+1$ and

$$\Delta\psi_J^{r+1} = \psi_J^{r+1} - \psi_J^k. \qquad (4.7.2.16)$$

Equation (4.7.2.15) can be written in matrix form as

$$[E]^r \{\Delta\psi\}^{r+1} = \{R\}^k, \qquad (4.7.2.17)$$

where typical elements of the Jacobian matrix and right-hand-side vector are given by

$$E_{IJ}^r = (\partial G_I/\partial \psi_J)^r = A_{IJ}^r + D_{IJ}^r,$$

$$A_{IJ}^r = \int_R K_{ij} \frac{\partial N_I}{\partial x_i} \frac{\partial N_J}{\partial x_j} N_L k_{rwL}^k \, dR$$

$$+ \int_R K_{ij} \frac{\partial N_I}{\partial x_i} \frac{\partial N_L}{\partial x_j} \psi_L^k \left(N \frac{dk_{rw}}{d\psi} \right)_J^r + \int_R K_{ij} \frac{\partial N_I}{\partial x_i} e_j \left(N \frac{dk_{rw}}{d\psi} \right)_J^r dR,$$

$$D_{IJ}^r = \int_R \frac{\phi}{\Delta t} N_I \left(N \frac{dS_w}{d\psi} \right)_J^r dR, \tag{4.7.2.18}$$

$$R_I^k = -\int_R K_{ij} \frac{\partial N_I}{\partial x_i} \frac{\partial N_J}{\partial x_j} N_L k_{rwL}^k \psi_J^k \, dR - \int_R K_{ij} \frac{\partial N_I}{\partial x_i} e_j N_L k_{rwL}^k \, dR$$

$$- \int_{B_2} N_I q \, dB.$$

The slopes of k_{rw} and S_w at node J are chord slopes given by

$$(N\, dk_{rw}/d\psi)_J^r = N_J (k_{rwJ}^r - k_{kwJ}^r)/(\psi_J^r - \psi_J^k),$$

$$(N\, dS_w/d\psi)_J^r = N_J (S_{wJ}^r - S_{wJ}^k)/(\psi_J^r - \psi_J^k).$$

Note that the Jacobian matrix is asymmetric, $E_{IJ} \neq E_{JI}$. Equation (4.7.2.18) represents a system of linear equations which can be solved for the head increment $\{\Delta\psi\}^{r+1}$. After $\{\Delta\psi\}^{r+1}$ has been determined, $\{\psi\}^{r+1}$ is calculated from

$$\{\psi\}^{r+1} = \{\psi\}^k + \{\Delta\psi\}^{r+1}. \tag{4.7.2.19}$$

Next, we use $\{\psi\}^{r+1}$ to update the coefficient matrix $[E]^r$ and iterate until satisfactory convergence is achieved. We recommend the use of the following convergence criterion:

$$\max_J \left| \frac{\Delta\psi_J^{r+1} - \Delta\psi_J^r}{\psi_J^{r+1}} \right| \leq \delta, \tag{4.7.2.20}$$

where δ is a prescribed tolerance.

4.7.3 *Immiscible Displacements in Oil Reservoirs*

For simplicity, we consider a situation involving a reservoir containing two fluids, oil and water. Thus, there exists two-phase flow with small compressibility. If compressibility effects are assumed to be negligible, then the system of governing equations (4.5.2.4a)–(4.5.2.4c) reduces to

$$\frac{\partial}{\partial x_i}\left[k_{ij}\lambda_w\left(\frac{\partial p_w}{\partial x_j} + \rho_w g\frac{\partial z}{\partial x_j}\right)\right] = \phi\frac{\partial S_w}{\partial t} \tag{4.7.3.1a}$$

$$\frac{\partial}{\partial x_i}\left[k_{ij}\lambda_o\left(\frac{\partial p_o}{\partial x_j} + \rho_o g\frac{\partial z}{\partial x_j}\right)\right] = \phi\frac{\partial S_o}{\partial t} = -\phi\frac{\partial S_w}{\partial t}, \tag{4.7.3.1b}$$

where, without loss of generality, the source terms Q_{ws} and Q_{os} may be conveniently dropped. These source terms can be treated separately in the finite element formulation.

Next, we introduce two functions defined as follows:

$$\Phi_w = p_w + \rho_w g z, \tag{4.7.3.2a}$$

$$\Phi_o = p_o + \rho_o g z. \tag{4.7.3.2b}$$

These two functions are related by

$$\Phi_o - \Phi_w = (p_o - p_w) + gz(\rho_o - \rho_w) = p_{o/w} + zg(\rho_o - \rho_w), \tag{4.7.3.2c}$$

where $p_{o/w}$ is the capillary pressure of the oil and water phases and $p_{o/w} = p_{o/w}(S_w)$. Substitution of Eqs. (4.7.3.2a) and (4.7.3.2b) into Eqs. (4.7.3.1a) and (4.7.3.1b) yields

$$\frac{\partial}{\partial x_i}\left(k_{ij}\lambda_w\frac{\partial \Phi_w}{\partial x_j}\right) = \Phi\frac{\partial S_w}{\partial t}, \tag{4.7.3.3a}$$

$$\frac{\partial}{\partial x_i}\left(k_{ij}\lambda_o\frac{\partial \Phi_o}{\partial x_j}\right) = -\Phi\frac{\partial S_w}{\partial t}. \tag{4.7.3.3b}$$

We now solve these equations by the finite element method.

Experience indicates that the numerical approximation obtained by the standard Galerkin criterion often yields results that exhibit numerical difficulties for situations involving negligible capillary effects. The numerical difficulties manifest themselves as oscillations in the neighborhood of sharp fronts and convergence in an iterative sense to the wrong solution. These difficulties can be overcome by adopting a numerical scheme that produces upstream weighting of the mobilities. The term "upstream weighting" is used to denote the concept in which the mobilities between two adjacent nodes are evaluated by placing greater weight on the node that is upstream. Several upstream weighting techniques have been used in finite difference simulations of multiphase flow [see, for example, Settari and Aziz (1975), Todd *et al.* (1972) and Yasonik and McCraken (1976)].

In this section, we present two alternative techniques for achieving upstream weighting in the finite element formulation. These techniques have been demonstrated to produce good results with the use of linear triangular and rectangular elements.

Upstream Galerkin Technique

In this particular technique (Dalen, 1979) we first apply the Galerkin criterion to derive the element equations, group the terms in each equation so that a discrete difference analog is obtained, and then evaluate the mobilities of each connecting pair of points using an "upstream weighting formula."

To elaborate on the procedure, let us first express the trial functions of the potentials and saturation in the form

$$\hat{\Phi}_{\mathrm{m}}(x_i, t) = N_J(x_i)\Phi_{\mathrm{m}J}(t), \qquad \mathrm{m} = \mathrm{w}, \mathrm{o}, \tag{4.7.3.4a}$$

$$\hat{S}_{\mathrm{w}}(x_i, t) = N_J(x_i)S_{\mathrm{w}J}(t). \tag{4.7.3.4b}$$

Application of the Galerkin criterion leads to element equations of the form

$$T^e_{\mathrm{w}IJ}\Phi_{\mathrm{w}J} + M^e_{IJ}\, dS_{\mathrm{w}J}/dt = F^e_{\mathrm{w}I}, \tag{4.7.3.5a}$$

$$T^e_{\mathrm{o}IJ}\Phi_{\mathrm{o}J} - M^e_{IJ}\, dS_{\mathrm{w}J}/dt = F^e_{\mathrm{o}I}, \tag{4.7.3.5b}$$

where the superscript e is used to denote the fact that the equations are written at the element level and

$$\begin{aligned} T^e_{\mathrm{m}IJ} &= \int_{R^e} k_{ij}\lambda_{\mathrm{m}} \frac{\partial N_I}{\partial x_i}\frac{\partial N_J}{\partial x_j}\, dR \\ &\simeq \lambda_{\mathrm{m}} k_{ij} \int_{R^e} \frac{\partial N_I}{\partial x_i}\frac{\partial N_J}{\partial x_j}\, dR = \lambda_{\mathrm{m}} A^e_{IJ}, \qquad \mathrm{m} = \mathrm{w}, \mathrm{o}, \\ M^e_{IJ} &= \int_{R^e} \phi N_I N_J\, dR, \\ F^e_{\mathrm{m}I} &= -\int_{B^e_2} N_I q_{\mathrm{m}}\, dB, \qquad \mathrm{m} = \mathrm{w}, \mathrm{o}. \end{aligned}$$

If it is assumed that k_{ij} is isotropic, then for linear triangular and quadrilateral elements Eqs. (4.7.3.5a) and (4.7.3.5b) can be simply integrated and expressed in the form

$$\sum_{J=1}^{n^e} [\lambda_{\mathrm{w}} A^e_{IJ}(\Phi_{\mathrm{w}I} - \Phi_{\mathrm{w}J}) + M^e_{IJ}\, dS_{\mathrm{w}J}/dt] = F_{\mathrm{w}I}, \tag{4.7.3.6a}$$

$$\sum_{J=1}^{n^e} [\lambda_{\mathrm{o}} A^e_{IJ}(\Phi_{\mathrm{o}I} - \Phi_{\mathrm{o}J}) - M^e_{IJ}\, dS_{\mathrm{w}J}/dt] = F_{\mathrm{o}I}, \tag{4.7.3.6b}$$

where n^e is the number of nodes in element e.

To obtain upstream weighting, the values of the mobilities in the direction IJ are calculated from

$$\lambda_m = \tfrac{1}{2}[(1 + \alpha)\lambda_{mI} + (1 - \alpha)\lambda_{mJ}], \qquad m = w, o, \qquad (4.7.3.7)$$

where α is called an "upstream weighting" factor and λ_{mI} and λ_{mJ} are values of mobilities at nodes I and J, respectively. The magnitude of α lies between 0 and 1. Its sign depends on the flow direction. According to Eq. (4.7.3.7), α is positive if the flow is from I to J. To achieve full upstream weighting $|\alpha|$ should be set equal to 1.

The system of element equations (4.7.3.6a) and (4.7.3.6b) contains two variables, Φ_w and Φ_o. The oil potential Φ_o can be expressed in terms of the water potential Φ_w using Eq. (4.7.3.2c). Equations (4.7.3.6a) and (4.7.3.6b) can thus be written as

$$\sum_{J=1}^{n^e} \left\{ [\tfrac{1}{2}(1 + \alpha)\lambda_{wI} + \tfrac{1}{2}(1 - \alpha)\lambda_{wJ}]A^e_{IJ}(\Phi_{wI} - \Phi_{wJ}) + M^e_{IJ}\frac{dS_{wJ}}{dt} \right\} - F_{wI} = 0, \qquad (4.7.3.8a)$$

$$\sum_{J=1}^{n^e} \left\{ [\tfrac{1}{2}(1 + \alpha)\lambda_{oI} + \tfrac{1}{2}(1 - \alpha)\lambda_{oJ}]A^e_{IJ}(\Phi_{wI} - \Phi_{wJ} + p_{o/wI} - p_{o/wJ}) + (\rho_o - \rho_w)g(z_I - z_J)] - M^e_{IJ}\frac{dS_{wJ}}{dt} \right\} - F_{oI} = 0, \qquad (4.7.3.8b)$$

where λ_{wI}, λ_{oI}, and $p_{o/w}$ are nonlinear functions by S_{wI}. This nonlinear system of ordinary differential equations can be treated using either the chord slope or the Newton–Raphson method. Experience indicates that the Picard method is not efficient for this type of problem. It is more advantageous to use either the chord slope or the Newton–Raphson method. If the chord slope method is used, the resulting global matrix equation is given by

$$[E]^r \{\Delta x\}^{r+1} = -\{R\}^k, \qquad (4.7.3.9)$$

where k denotes the old time level and r and $r+1$ denote the previous and current iterations of the new time level $k+1$, then

$$[E]^r = \begin{bmatrix} \dfrac{\partial G_I}{\partial \Phi_{wJ}} & \dfrac{\partial G_I}{\partial S_{wJ}} \\ \dfrac{\partial H_I}{\partial \Phi_{wJ}} & \dfrac{\partial H_I}{\partial S_{wJ}} \end{bmatrix}^r, \qquad \{\Delta x\}^{r+1} = \begin{Bmatrix} \Delta\Phi_{wJ} \\ \Delta S_{wJ} \end{Bmatrix}^{r+1}, \qquad \{R\}^k = \begin{Bmatrix} G_I \\ H_I \end{Bmatrix}^k,$$

in which G_I and H_I represent the left sides of Eqs. (4.7.3.8a) and (4.7.3.8b), respectively. Once the Jacobian matrix $[E]^r$ and the right-hand-side vector $\{R\}^k$ are formed, the solution for $\{\Delta x\}^{r+1}$ can be obtained using a direct solver that handles an asymmetric coefficient matrix. Iteration can then be performed until satisfactory convergence is achieved. The convergence criterion recommended is given by Eq. (4.7.2.20), except that ψ is replaced by Φ_w and S_w.

Upstream Weighted Residual Technique

In this technique, we apply the general weighted residual criterion to Eqs. (4.7.3.1a) and (4.7.3.1b). The application is made in such a manner that the spatial derivative terms in both equations are weighted using asymmetric weighting functions and the time derivative terms are weighted using the standard basis functions (Huyakorn and Pinder, 1977). Typical asymmetric weighting functions for linear one-dimensional and rectangular elements are illustrated in Fig. 4.18. These functions are given in Table 4.1. They may be obtained by adding quadratic functions to the linear basis functions. The expressions for the quadratic functions contain "up-

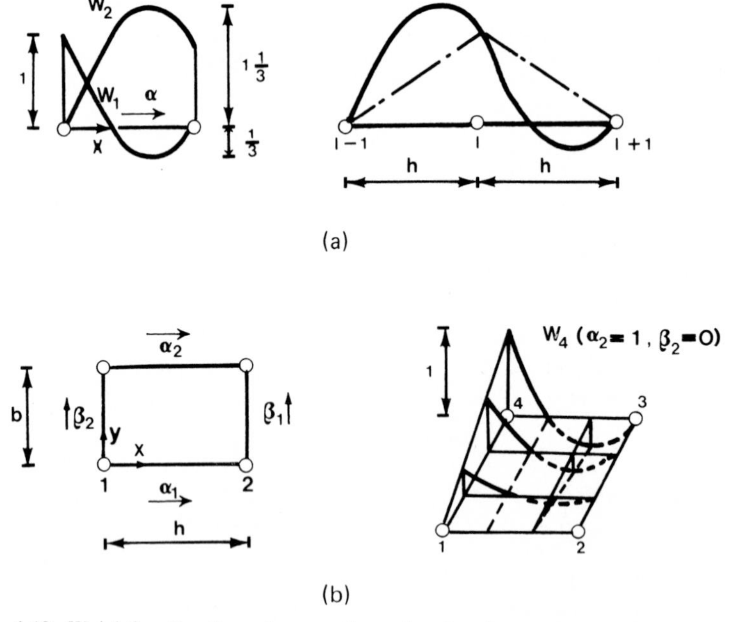

Fig. 4.18. Weighting functions for one-dimensional and rectangular elements: (a) one-dimensional element, —— indicates $W_I(\alpha = 1)$; — – — indicates N_I; (b) rectangular element.

Table 4.1
Expressions of Weighting Functions for Linear One-Dimensional and Rectangular Element in Fig. 4.18[a]

Element	*Function*
One-dimensional $0 \leq x \leq h$	$W_1 = f_1\left(\frac{x}{h}, \alpha\right)$
	$W_2 = f_2\left(\frac{x}{h}, \alpha\right)$
	where
	$f_1(\xi, \alpha) = 1 - \xi + 3\alpha(\xi^2 - \xi)$
	$f_2(\xi, \alpha) = \xi - 3\alpha(\xi^2 - \xi)$
Rectangular $0 \leq x \leq h, \quad 0 \leq y \leq b$	$W_1 = f_1\left(\frac{x}{h}, \alpha_1\right)f_1\left(\frac{y}{b}, \beta_2\right)$
	$W_2 = f_2\left(\frac{x}{h}, \alpha_1\right)f_1\left(\frac{y}{b}, \beta_1\right)$
	$W_3 = f_2\left(\frac{x}{h}, \alpha_2\right)f_2\left(\frac{y}{b}, \beta_1\right)$
	$W_4 = f_1\left(\frac{x}{h}, \alpha_2\right)f_2\left(\frac{y}{b}, \beta_2\right)$

[a] Note that experience indicates that, to obtain satisfactory numerical solutions, the derivatives of the two-dimensional weighting functions must be evaluated so that $\partial W_I(\alpha_i, \beta_j)/\partial x = \partial W_I(\alpha_i, 0)/\partial x$ and $\partial W_I(\alpha_i, \beta_j)/\partial y = \partial W_I(0, \beta_j)/\partial y$.

stream'' parameters α_i and β_i, which are dependent on the flow directions parallel to the element sides:

$$\hat{\Phi}_{\mathrm{m}}(x_i, t) = N_J(x_i)\Phi_{\mathrm{m}J}(t), \qquad \mathrm{m} = \mathrm{w}, \mathrm{o}, \tag{4.7.3.10a}$$

$$\hat{S}_{\mathrm{w}}(x_i, t) = N_J(x_i)S_{\mathrm{w}J}(t), \tag{4.7.3.10b}$$

$$\hat{\lambda}_{\mathrm{m}}(x_i, t) = N_J(x_i)\lambda_{\mathrm{m}J}(t), \qquad \mathrm{m} = \mathrm{w}, \mathrm{o}. \tag{4.7.3.10c}$$

We now construct the weighted residual equations as

$$\int_R W_I\left[\frac{\partial}{\partial x_i}\left(k_{ij}\hat{\lambda}_{\mathrm{w}}\frac{\partial \hat{\Phi}_{\mathrm{w}}}{\partial x_j}\right)\right] dR - \int_R N_I \phi \frac{\partial \hat{S}_{\mathrm{w}}}{\partial t}\, dR = 0, \tag{4.7.3.11a}$$

$$\int_R W_I\left[\frac{\partial}{\partial x_i}\left(k_{ij}\hat{\lambda}_{\mathrm{o}}\frac{\partial \hat{\Phi}_{\mathrm{o}}}{\partial x_j}\right)\right] dR + \int_R N_I \phi \frac{\partial \hat{S}_{\mathrm{w}}}{\partial t}\, dR = 0, \tag{4.7.3.11b}$$

where W_I denotes the weighting function for node I.

Application of Green's theorem and a backward difference time-stepping scheme to (4.7.3.3a) and (4.7.3.3b) leads to

$$\int_R k_{ij}\hat{\lambda}_w \frac{\partial W_I}{\partial x_i}\frac{\partial N_J}{\partial x_j}\Phi_{wJ}^{k+1}\,dR - \int_R \frac{\phi}{\Delta t} N_I N_J (S_{wJ}^{k+1} - S_{wJ}^k)\,dR + \int_{B_2} W_I q_w^{k+1}\,dB = 0, \quad (4.7.3.12a)$$

$$\int_R k_{ij}\hat{\lambda}_o \frac{\partial W_I}{\partial x_i}\frac{\partial N_J}{\partial x_j}\Phi_{oJ}^{k+1}\,dR + \int_R \frac{\phi}{\Delta t} N_I N_J (S_{wJ}^{k+1} - S_{wJ}^k)\,dR + \int_{B_2} W_I q_o^{k+1}\,dB = 0, \quad (4.7.3.12b)$$

where B_2 denotes the flow boundary, q_w^{k+1} and q_o^{k+1} denote the outward-directed fluxes of water and oil at time level $k+1$.

We now apply the Newton–Raphson technique to the above system of nonlinear equations. For convenience, let G_I and H_I denote the left sides of Eqs. (4.7.3.12a) and (4.7.3.12b), respectively. By the Newton–Raphson procedure, we obtain

$$\frac{\partial G_I^r}{\partial \Phi_{wJ}}\Delta\Phi_{wJ}^{r+1} + \frac{\partial G_I^r}{\partial S_{wJ}}\Delta S_{wJ}^{r+1} = -G_I^r, \quad (4.7.3.13a)$$

$$\frac{\partial H_I^r}{\partial \Phi_{wJ}}\Delta\Phi_{wJ}^{r+1} + \frac{\partial H_I^r}{\partial S_{wJ}}\Delta S_{wJ}^{r+1} = -H_I^r, \quad (4.7.3.13b)$$

where r and $r+1$ denote the old and new iterations of time level $k+1$. Equations (4.7.3.13a) and (4.7.3.13b) can be expressed in the form

$$[E]^r\{\Delta x\}^{r+1} = -\{R\}^r, \quad (4.7.3.14)$$

where

$$\{\Delta x\}^{r+1} = \begin{Bmatrix}\Delta\Phi_{wJ}\\ \Delta S_{wJ}\end{Bmatrix}^{r+1}, \qquad \{R\}^r = \begin{Bmatrix}-G_I\\ -H_I\end{Bmatrix}^r,$$

$$[E] = \begin{bmatrix}\dfrac{\partial G_I}{\partial \Phi_{wJ}} & \dfrac{\partial G_I}{\partial S_{wJ}}\\[2ex] \dfrac{\partial H_I}{\partial \Phi_{wJ}} & \dfrac{\partial H_I}{S_{wJ}}\end{bmatrix}^r$$

The gradients of G_I and H_I form the elements of the Jacobian matrix

$[E]$. These gradients may be obtained by differentiating the left sides of Eqs. (4.7.3.12a) and (4.7.3.12b) with respect to the nodal values Φ_{wJ} and S_{wJ}. Once the Jacobian matrix $[E]^r$ and the right-hand vector $\{R\}^r$ are formed, one can solve for $\{\Delta x\}^{r+1}$ and iterate until satisfactory convergence is achieved. The convergence criterion recommended is given by Eq. (4.7.2.13) with the exception that ψ_J is replaced by Φ_{wJ} and S_{wJ} and one may wish to add a constraint on the residual.

4.8 Concluding Remarks

In this chapter, we have covered a lot of material. The fundamental equations describing many subsurface flow phenomena have been examined. Techniques for their solution based on finite element theory have been presented. When nonlinear equations were encountered we described appropriate methodology to accommodate them. Although the detailed expressions and techniques presented for each problem type are unique, it should be evident to the reader that there is a common thread linking them all together.

References

Bear, J. (1979). "Hydraulics of Groundwater." McGraw-Hill, New York.

Bruch, J. C., Jr. (1977). Two-dimensional unsteady water flow in unsaturated porous media. *In* "Proceedings of the First International Conference on Finite Elements in Water Resources," 3.3–3.19, Pentech Press, Plymouth, England.

Carnahan, B., Luther, H. A., and Wilkes, J. O. (1969). "Applied Numerical Methods." Wiley, New York.

Carslaw, H. S., and Jaeger, J. C. (1959). "Conduction of Heat in Solids." Oxford University Press, Oxford.

Cooley, R. L. (1971). A finite difference method for unsteady flow in variably saturated porous media: Application to a single pumping well. *Water Resour. Res.* **7** (6), 1607–1625.

Corey, A. T. (1977). "Mechanics of Heterogeneous Fluids in Porous Media". Water Resources Publications, Fort Collins, Colorado.

Dake, L. P. (1978). "Fundamentals of Reservoir Engineering." Elsevier, Amsterdam.

Dalen, V. (1979). Simplified finite-element models for reservoir flow problems. *Soc. Pet. Eng. J.* **19** (5), 333–343.

France, P. W., Parekh, C. J., Peters, J. C., and Taylor, C. (1971). Numerical analysis of free surface seepage problems. *ASCE J. Irrig. Drain. Div.* **97** (IR1), 165–179.

Hantush, M. S. (1964). Hydraulics of wells. *In* "Advances in Hydroscience," Vol. 1, 281–442, V. T. Chow, ed. Academic Press, New York.

Henry, H. R. (1964). Interfaces between salt water and fresh water in coastal aquifers. Water-supply paper 1613-3, U.S. Geological Survey.

Huang, Y. H. (1973). Unsteady flow toward an artesian well. *Water Resour. Res.* **9** (2), 426–433.

Huyakorn, P. S. (1973). Finite element solution of two-regime flow toward wells. Report No. 137, Water Research Lab., University of New South Wales, Australia.

Huyakorn, P. S., and Pinder, G. F. (1977). Solution of two-phase flow using a new finite element technique. *In* "Proceedings of the International Conference on Numerical Modeling," 375–390. University of Southampton, U.K..

Javandel, I., and Witherspoon, P. A. (1968). Application of the finite element method to transient flow in porous media. *Soc. Pet. Eng. J.* **8** (3), 241–252.

Mercer, J. W., and Faust, C. R. (1977). The application of the finite element method to immiscible flow in porous media. *In* "Proceedings of the First International Conference on Finite Elements in Water Resources," 1.21–1.57. Pentech Press, Plymouth, England.

Lynch, D. R., and Gray, W. G. (1980). Finite element simulation of flow in deforming regions. *J. Comput. Phys.* **36,** 135–153.

Neuman, S. P. (1973). Saturated–Unsaturated seepage by finite elements. *ASCE J. Hydraul. Div.* **99**(HY2), 2233–2251.

Neuman, S. P., and Witherspoon, P. A. (1969). Transient flow of ground water to wells in multiple-aquifer systems. Report 69-1, Geotechnical Engineering, University of California, Berkeley.

Neuman, S. P., and Witherspoon, P. A. (1971). Analysis of nonsteady flow with a free surface using the finite element method. *Water Resour. Res.* **7** (3), 611–623.

Page, R. H. (1979). An areal model for sea-water intrusion in a coastal aquifer: Program documentation. Report 79-WR-11, Princeton University Water Resources Program.

Philip, J. R. (1969). Theory of infiltration. *In* "Advances in Hydroscience," Vol. 5, V. T. Chow, ed., 216–291. Academic Press, New York.

Pinder, G. F., and Frind, E. O. (1972). Application of Galerkin's procedure to aquifer analysis. *Water Resour. Res.* **8** (1), 108–120.

Pinder, G. F., and Gray, W. G. (1977). "Finite element simulation in surface and subsurface hydrology." Academic Press, New York.

Pinder, G. F., and Page, R. H. (1977). Finite element simulation of salt water intrusion on the south fork of Long Island. *In* "Proceedings of the First International Conference on Finite Elements in Water Resources, 2.51–2.69. Pentech Press, Plymouth, England.

Rumer, R. R., and Shiau, J. C. (1968). Salt water interface in a layered coastal aquifer. *Water Resour. Res.* **4** (6), 1235–1247.

Reeves, M., and Duguid, J. O. (1975). Water Movement through Saturated–Unsaturated Porous Media: A Finite Element Galerkin Model. Report 4927, Oak Ridge Nat. Lab.

Settari, A., and Aziz, K. (1975). Treatment of nonlinear terms in the numerical solution of partial differential equations for multiphase flow in porous media. *Int. J. Multiphase Flow* **1,** 817–844.

Thomas, G. W. (1982). Principles of Hydrocarbon Reservoir Simulation, International Human Resources Development Corporation, 207 pp.

Todd, M. R., O'Dell, P. M., and Hirasaki, G. J. (1972). Methods for increasing accuracy in numerical simulation. *Soc. Pet. Eng. J.,* **12,** 515–530.

Volker, R. E. (1969). Nonlinear flow in porous media by finite elements. *ASCE J. Hydraul. Div.* **95** (HY6), 2093–2112.

Walton, W. C. (1970). "Groundwater Resource Evaluation." McGraw-Hill, New York.

Yasonik, J. L., and McCracken, T. A. (1976). A nine-point finite difference reservoir simulator for realistic prediction of unfavorable mobility ratio displacements. Paper no. SPE 5734, Fourth Soc. Pet. Eng. Symposium on Numerical Simulation of Reservoir Performance, Los Angeles, California, 209–230.

5

Finite Element Simulation of Solute and Energy Transport in Porous Media

5.1 Introduction

In this section we consider the important problem of mass and energy transport in porous media. To simulate transport one must, of course, have an appropriate representation of the velocity field. We assume at the outset that the fluid potential or pressure field has been made available through procedures described in the preceding section and that macroscopic fluid velocities are, therefore, available. In the later sections of the chapter, we relax this assumption and also consider the flow equations.

5.2 Mass Transport in Single-Phase Flow

5.2.1 Governing Equations

The equations describing species transport in porous media are obtained by formal integration over the porous media of the species transport equations for each porous medium phase α. In this context each immiscible fluid is a phase and the solid is another phase. Typically, we are concerned with at most three phases, which we shall denote as liquid (l), gas (g), and solid (s). Additional fluid phases may also be encountered, particularly in oil reservoir simulation.

The macroscopic equations for reacting solutes are presented in Domenico and Palciauskas (1979) and Nguyen *et al.* (1982). Derivation of the governing transport equations for radioactive solutes are presented in Harada *et al.* (1980). For species k in phase α we can write the transport equation as

$$\frac{\partial}{\partial x_i}\left(n^\alpha \rho^\alpha D_{ij}^\alpha \frac{\partial}{\partial x_j}\,\omega_k^\alpha\right) - \frac{\partial}{\partial x_i}(\rho^\alpha \omega_k^\alpha V_i^\alpha) + n^\alpha \rho^\alpha r_k^\alpha$$

$$+ n^\alpha \rho^\alpha e^\alpha(\rho\omega_k) + n^\alpha \rho^\alpha I_k^\alpha - \frac{\partial}{\partial t}(n^\alpha \rho^\alpha \omega_k^\alpha) = 0. \qquad (5.2.1.1)$$

Note that repetition of the phase and species indices does not imply summation. In Eq. (5.2.1.1), ρ^α is the phase density, D_{ij}^α the hydrodynamic dispersion coefficient with respect to phase α, ω_k^α the mass fraction of species k, V_i^α the macroscopic Darcy velocity, and r_k^α the rate of production of species k by homogeneous reaction, written in terms of mass per unit volume of solution per unit time. The superscript α denotes the phase associated with each variable. The volume fraction $n^\alpha \equiv \mathbf{V}^\alpha/\mathbf{V}$, where $\mathbf{V} = \Sigma_{\alpha=1}^N \mathbf{V}^\alpha$ and N is the number of phases. The fourth term in (5.2.1.1), $n^\alpha\rho^\alpha e^\alpha(\rho\omega_k)$, describes the transfer of species k from other phases to phase α via phase change. The transfer of species k to phase α from other phases by diffusion across a phase boundary is described by the fifth term $n^\alpha\rho^\alpha I_k^\alpha$.

Define the heterogeneous rate of reaction term for species k as

$$\overline{R}_k \equiv \sum_{\alpha=1}^{N} R_k^\alpha = \sum_{\alpha=1}^{N} n^\alpha\rho^\alpha[e^\alpha(\rho\omega_k) + I_k^\alpha]. \tag{5.2.1.2a}$$

The conservation of reacting mass requires

$$\sum_{k=1}^{M} (n^\alpha\rho^\alpha r_k^\alpha + \overline{R}_k) = 0, \tag{5.2.1.2b}$$

where M is the number of species.

Because the sum of mass fluxes of all species into phase α must be equal to the total change of mass in phase α, we can write

$$\sum_{\alpha=1}^{N} n^\alpha\rho_k^\alpha \sum_{k=1}^{M} [e^\alpha(\rho\omega_k) + I_k^\alpha] = 0. \tag{5.2.1.2c}$$

Let us now consider the case of a single-phase fluid l with adsorption of a solute on the porous matrix solid phase s. This rather simple case reduces to the following set of equations, which are derived directly from (5.2.1.1):

$$\frac{\partial}{\partial x_i}\left(\phi\rho^l D_{ij}\frac{\partial}{\partial x_j}\omega_k^l\right) - \frac{\partial}{\partial x_i}\rho^l\omega_k^l V_i^l + \phi\rho^l r_k^l + \phi\rho^l I_k^l$$
$$+ \phi\rho^l e^l(\rho\omega_k) - \frac{\partial}{\partial t}\phi\rho^l\omega_k^l = 0, \tag{5.2.1.3a}$$

$$(1-\phi)\rho^s r_k^s + (1-\phi)\rho^s I_k^s + (1-\phi)\rho^s e^s(\rho\omega_k)$$
$$- \frac{\partial}{\partial t}(1-\phi)\rho^s\omega_k^s = 0, \tag{5.2.1.3b}$$

where

$$\phi = \mathbf{V}^l/\mathbf{V} \equiv n^l, \qquad 1 - \phi = \mathbf{V}^s/\mathbf{V} \equiv n^s.$$

We assume in writing (5.2.1.3b) that the velocity of the solid grains V_i^s, is negligible.

We now sum (5.2.1.3a) and (5.2.1.3b) to give a conservation equation for species k over both phases; that is,

$$\frac{\partial}{\partial x_i}\left(\phi\rho^l D_{ij}\frac{\partial}{\partial x_j}\omega_k^l\right) - \frac{\partial}{\partial x_i}\rho^l\omega_k^l V_i^l + [\phi\rho^l r_k^l + (1-\phi)\rho^s r_k^s]$$
$$+ [\phi\rho^l I_k^l + (1-\phi)\rho^s I_k^s] + [\phi\rho^l e^l(\rho\omega_k) + (1-\phi)\rho^s e^s(\rho\omega_k)]$$
$$- \frac{\partial}{\partial t}[\phi\rho^l\omega_k^l + (1-\phi)\rho^s\omega_k^s] = 0. \qquad (5.2.1.4)$$

From the definition of the heterogeneous rate of reaction (5.2.1.2a) and assuming no homogeneous reactions, we can rewrite (5.2.1.4) as

$$\frac{\partial}{\partial x_i}\left(\phi\rho^l D_{ij}\frac{\partial}{\partial x_j}\omega_k^l\right) - \frac{\partial}{\partial x_i}(\rho^l\omega_k^l V_i^l) + \overline{R}_k$$
$$- \frac{\partial}{\partial t}[(1-\phi)\rho^s\omega_k^s + \phi\rho^l\omega_k^l] = 0. \qquad (5.2.1.5)$$

The two unknown mass fractions for phase k, ω_k^l and ω_k^s, still require two independent equations. The overall conservation equations (5.2.1.5) can serve as an alternative to one of the individual phase equations (5.2.1.3a) or (5.2.1.3b).

5.2.2 Adsorption

Let us now consider the classical case of transport with physical adsorption. A commonly used empirical relation for the adsorption of a species k is the Freundlich equilibrium isotherm (Bear, 1979, p. 240)

$$\omega_k^s = \beta_k(\omega_k^l)^{1/m}, \qquad (5.2.2.1)$$

where ω_k^s is the mass fraction of species k on the solid phase, ω_k^l the mass fraction concentration of species k in the solution, and β_k and m empirically derived constants. The coefficient β_k is usually expressed as $\beta_k = \rho_l K_{Dk}$, where K_{Dk} denotes the "distribution coefficient" for species K and ρ_l is the fluid density.

Recall that the heterogeneous reaction term $\overline{R}_k$ is made up of N terms, which in our special case we can write as

$$\overline{R}_k \equiv \sum_{\alpha = s,\, l} R_k^\alpha. \tag{5.2.2.2}$$

This term vanishes if there are no homogeneous reactions and the pore fluid is in chemical equilibrium with the solid particles. In general, however, this is not likely to be the case, owing to the transport conditions within the pore. A more general constitutive description is presented by Nguyen *et al.* (1982), namely,

$$\overline{R}_k = \overline{R}_k\,(n^l\rho^l,\, V_i^s - V_i^l,\, \partial\omega_k^l/\partial x_i,\, \omega_k^l), \qquad i = 1, 2, 3. \tag{5.2.2.3}$$

Owing to a lack of information regarding the functional form of (5.2.2.3), we assume in this development that $\overline{R}_k = 0$.

Combination of (5.2.1.5) and (5.2.2.1) with the assumption that $m = 1$ and introduction of a new variable $c_k = \rho^l\omega_k^l$ yields

$$\frac{\partial}{\partial x_i}\left(\phi D_{ij}\frac{\partial c_k}{\partial x_j}\right) - \frac{\partial}{\partial x_i}V_i c_k = \frac{\partial}{\partial t}\left[\frac{(1-\phi)\rho^s\beta_k}{\rho^l}c_k + \phi c_k\right], \tag{5.2.2.4}$$

where, without loss of generality, superscript l is dropped. Note that in Eq. (5.2.2.4) the variable c_k denotes mass concentration of species k per unit volume of the fluid phase l. In addition, repeated indices k do not indicate summation. Equation (5.2.2.4) can be written in a more compact form as

$$\frac{\partial}{\partial x_i}\left(\phi D_{ij}\frac{\partial c_k}{\partial x_j}\right) - \frac{\partial}{\partial x_i}V_i c_k = \frac{\partial}{\partial t}\phi\kappa_k c_k, \tag{5.2.2.5}$$

where κ_k is commonly referred to as the "retardation coefficient of species k." The coefficient κ_k is defined as

$$\kappa_k = \left(\frac{1-\phi}{\phi}\right)\frac{\rho^s\beta_k}{\rho^l} + 1 = \left(\frac{1-\phi}{\phi}\right)\rho^s K_{Dk} + 1. \tag{5.2.2.6}$$

5.2.3 Transport of Radioactive Nuclides

Transport of radioactive nuclides involves species decay and generation owing to parent-to-daughter transformation of components. It is thus necessary to modify the original governing equation (5.2.2.5) to account for these effects. The modified equation can be written in the form

$$\frac{\partial}{\partial x_i}\left(\phi D_{ij}\frac{\partial c_k}{\partial x_j}\right) - \frac{\partial}{\partial x_i}(V_i c_k) - \frac{\partial}{\partial t}\phi\kappa_k c_k - \phi\lambda_k\kappa_k c_k$$

$$+ \sum_{m=1}^{M} \phi\xi_{km}\lambda_m\kappa_m c_m = 0, \qquad (5.2.3.1)$$

where λ_k is the decay constant of radionuclide species k, ξ_{km} the fraction of (parent) component m transforming into (daughter) component k, and here M is the number of parent components transforming into component k. The decay coefficient λ_k is defined in terms of the half-life, $t_{1/2}$ of species k as

$$\lambda_k = \ln 2/(t_{1/2})_k. \qquad (5.2.3.2)$$

As an example, consider a simple decay chain of the following type:

$$1 \rightarrow 2 \rightarrow 3$$

where 1 , 2 and 3 are component numbers. For this simple case, Eq. (5.2.3.1) represents a system of three equations. These equations take the form

$$\frac{\partial}{\partial x_i}\left(\phi D_{ij}\frac{\partial c_1}{\partial x_j}\right) - \frac{\partial}{\partial x_i}V_i c_1$$

$$= \frac{\partial}{\partial t}(\phi\kappa_1 c_1) + \phi\lambda_1\kappa_1 c_1, \qquad (5.2.3.3a)$$

$$\frac{\partial}{\partial x_i}\left(\phi D_{ij}\frac{\partial c_2}{\partial x_j}\right) - \frac{\partial}{\partial x_i}V_i c_2$$

$$= \frac{\partial}{\partial t}\phi\kappa_2 c_2 + \phi\lambda_2\kappa_2 c_2 - \phi\lambda_1\kappa_1 c_1, \qquad (5.2.3.3b)$$

$$\frac{\partial}{\partial x_i}\left(\phi D_{ij}\frac{\partial c_3}{\partial x_j}\right) - \frac{\partial}{\partial x_i}V_i c_3$$

$$= \frac{\partial}{\partial t}\phi\kappa_3 c_3 + \phi\lambda_3\kappa_3 c_3 - \phi\lambda_2\kappa_2 c_2. \qquad (5.2.3.3c)$$

Another form of the solute transport equation is obtainable by combining Eq. (5.2.3.1) with the following continuity equation for incompressible fluid flow in a compressible medium:

$$\frac{\partial V_i}{\partial x_i} + \frac{\partial \phi}{\partial t} = 0. \qquad (5.2.3.4)$$

Thus, Eq. (5.2.3.1) reduces to

$$\frac{\partial}{\partial x_i}\left(\phi D_{ij} \frac{\partial c_k}{\partial x_j}\right) - V_i \frac{\partial c_k}{\partial x_i} - \phi \kappa_k \frac{\partial c_k}{\partial t} - \phi \lambda_k \kappa_k c_k + \sum_{m=1}^{M} \phi \xi_{km} \lambda_m \kappa_m c_m = 0. \qquad (5.2.3.5)$$

5.2.4 Transport without Adsorption or Radioactive Decay

The transport of nonradioactive solutes in the absence of significant adsorption can be described by a reduced form of (5.2.3.1). In the absence of radioactive decay and adsorption in a solution devoid of chemical reactions, the terms containing λ_k vanish, and $\kappa_k = 1$. Equation (5.2.3.1) thus becomes

$$\frac{\partial}{\partial x_i}\left(\phi D_{ij} \frac{\partial c}{\partial x_j}\right) - \frac{\partial}{\partial x_i} V_i c = \frac{\partial}{\partial t} \phi c. \qquad (5.2.4.1)$$

where, without loss of generality, the subscript k has been dropped and thus once again c denotes mass concentration of the solute per unit volume of the fluid. Note that Eq. (5.2.4.1) could have been derived without the assistance of the equation describing the transport of mass of the solids.

5.2.5 Dispersion Coefficient

In fluid transport through porous media the flow paths at the microscopic level are tortuous and complex. Because this detailed flow information is important in transport but is lost in passing to the macroscopic level of observation, this physical phenomenon must be captured through a constitutive relation. The classic work by Scheidegger (1961) in the derivation of the functional form of the dispersion relation remains the primary reference on this subject, although several formulations based on constitutive theory have recently appeared [see, for example, Shapiro (1981) and Nguyen *et al.* (1982)]. The functional form of the components of the dispersion tensor that is in general use is that for a porous medium, isotropic with respect to dispersivity. Thus, D_{ij} is expressed as (Bear, 1972)

$$\phi D_{ij} = D_{\mathrm{T}}|V|\delta_{ij} + (D_{\mathrm{L}} - D_{\mathrm{T}})\frac{V_i V_j}{|V|} + \phi D_{\mathrm{d}}\tau\delta_{ij}, \qquad (5.2.5.1)$$

where D_d is molecular diffusion, τ the tortuosity, and V_i is the ith component of Darcy velocity. The D_{11} term, for example, would read

$$\begin{aligned}\phi D_{11} &= D_{\mathrm{T}}|V| + (D_{\mathrm{L}} - D_{\mathrm{T}})\frac{V_1 V_1}{|V|} + \phi D_{\mathrm{d}}\tau \\ &= D_{\mathrm{T}}\frac{V_1^2 + V_2^2 + V_3^2}{|V|} + (D_{\mathrm{L}} - D_{\mathrm{T}})\frac{V_1^2}{|V|} + \phi D_{\mathrm{d}}\tau \\ &= D_{\mathrm{L}}\frac{V_1^2}{|V|} + D_{\mathrm{T}}\frac{V_2^2}{|V|} + D_{\mathrm{T}}\frac{V_3^2}{|V|} + \phi D_{\mathrm{d}}\tau, \end{aligned} \qquad (5.2.5.2)$$

where D_{L} and D_{T} denote the coefficients of longitudinal and transverse dispersivity, respectively, and $|V|$ is the absolute Darcy velocity.

5.2.6 *Initial and Boundary Conditions*

In obtaining the solution of the solute transport equation, one must specify as an initial condition the concentration distribution at some initial time $t = 0$ at all points in the flow domain. In addition, boundary conditions must also be specified at all times. The types of boundary conditions of practical interest include the conditions of prescribed concentration and prescribed material flux. These can be expressed for the solute species k as

$$c_k(\mathbf{x}, t) = \tilde{c}_k \qquad \text{on boundary portion } B_1 \qquad (5.2.6.1)$$

and

$$-\phi D_{ij}\frac{\partial c_k}{\partial x_j}n_i + qc_k = q\tilde{c}_k \qquad \text{on boundary portion } B_2, \qquad (5.2.6.2)$$

where n_i is the outward unit normal vector and $q = V_i n_i$ the outward fluid flux on B_2.

Equations (5.2.6.1) and (5.2.6.2) are usually referred to as first-type and third-type boundary conditions, respectively. In the special case where B_2 is impermeable, q becomes zero, and Eq. (5.2.6.2) becomes the Neumann (second-type) boundary condition.

These boundary conditions may be either steady state or transient, depending on the physical nature of the transport problem considered. As a general example, if one is dealing with transport of radioactive nuclides, then $\tilde{c}_k$ is time dependent and is described by the following set of mass-balance equations:

$$d\tilde{c}_k/dt = -\lambda_k\tilde{c}_k + \sum_{m=1}^{M} \xi_{km}\lambda_m\tilde{c}_m. \tag{5.2.6.3}$$

For a straight chain of radionuclides, Eq. (5.2.6.3) becomes

$$d\tilde{c}_k/dt = -\lambda_k\tilde{c}_k + \lambda_{k-1}\tilde{c}_{k-1}. \tag{5.2.6.4}$$

The analytical solution of Eq. (5.2.6.4) is known as Bateman's equation (Bateman, 1910). This solution takes the form

$$\tilde{c}_k = \tilde{c}_k^0 e^{(-\lambda_k t)} + \lambda_{k-1}\tilde{c}_{k-1}^0 \sum_{m=k-1}^{k} e^{(-\lambda_m t)} \Bigg/ \left[\prod_{\substack{l=k-1 \\ l\neq m}}^{k} (\lambda_l - \lambda_m)\right]$$

$$+ \cdots + \lambda_{k-1}\lambda_{k-2}\cdots\lambda_1\tilde{c}_1^0 \sum_{m=1}^{k} e^{(-\lambda_m t)} \Bigg/ \left[\prod_{\substack{l=1 \\ l\neq m}}^{k} (\lambda_l - \lambda_m)\right] \tag{5.2.6.5}$$

More detailed discussion of the boundary conditions pertaining to radionuclide transport can be found in Harada *et al.* (1980).

5.3 Mass Transport in Multiphase Flow

The formulation of the mass balance equations for multiphase flow is a direct extension of the single-phase flow equations. In fact, (5.2.1.1) and (5.2.1.2) describe the transport of a species k in phase α where α ranges over N phases. The challenging aspect of multiphase transport is the development of the constitutive relations for complicated systems. Perhaps the most commonly encountered systems of this kind are associated with oil reservoirs. In this work, we restrict our attention to the relatively simple case of mass transport in single-phase flow. The flow problem was discussed in Chapter 4 and the associated species transport equation is solved in the manner outlined in Section 5.6.

5.4 Energy Transport in Single-Phase Flow

The macroscopic energy balance is derived from its microscopic counterpart through the vehicle of mass and volume averaging. For details on volume averaging, the reader is referred to Slattery (1972) or Whitaker (1969, 1973). The derivation procedure is the same as that used to derive the mass balance equation (5.2.1.1). In the case of energy, however, the resulting expressions are very complicated and considerably more difficult to interpret.

5.4.1 General Energy Balance Equations

The macroscopic energy balance for an arbitrary phase α is written (Hassanizadeh and Gray, 1979)

$$n^\alpha \rho^\alpha \frac{D^\alpha E^\alpha}{Dt} - \sigma_{ij}^\alpha \frac{\partial}{\partial x_j} V_i^\alpha + \frac{\partial}{\partial x_i} q_i^\alpha - n^\alpha \rho^\alpha Q^\alpha - n^\alpha \rho^\alpha \hat{Q}^\alpha$$

$$- n^\alpha \rho^\alpha [e^\alpha(\rho \hat{E}) - e^\alpha(\rho) E^\alpha] = 0, \qquad (5.4.1.1)$$

where

$$\frac{D(\ )}{Dt} \equiv \frac{\partial(\ )}{\partial t} + \frac{V_i^\alpha}{n^\alpha} \frac{\partial(\ )}{\partial x_i},$$

$e^\alpha(\rho)$ is the mass exchange term between phases, $e^\alpha(\rho\hat{E})$ the transfer of energy owing to phase change, E^α the internal energy, σ_{ij} the stress tensor, q_i the heat flux vector, Q the total macroscopic rate of heat supply from the external source, $\hat{Q}$ the production of energy owing to mechanical interactions at the phase boundary, and the superscript α once again indicates the phase under consideration. The last term in (5.4.1.1) is the exchange of internal energy due to phase change. Let us consider first the special case of a water-saturated soil. The more general steam–water system will be considered in the next section.

The energy equation for the liquid water and solid matrix is obtained by substitution of l and s, respectively, for the phase index α in (5.4.1.1). An equation for the combined system is obtained by summation of the resulting two equations:

$$\phi \rho^l \frac{D^l E^l}{Dt} + (1 - \phi)\rho^s \frac{D^s E^s}{Dt} - \sigma_{ij}^l \frac{\partial}{\partial x_j} V_i^l - \sigma_{ij}^s \frac{\partial}{\partial x_j} (V_i^s)$$

$$+ \frac{\partial}{\partial x_i} q_i^l + \frac{\partial}{\partial x_i} q_i^s - \phi \rho^l Q^l - (1 - \phi)\rho^s Q^s = 0. \qquad (5.4.1.2)$$

Because total energy must be conserved at phase boundaries, it can be shown (Pinder and Shapiro, 1982) that the last two terms arising from

(5.4.1.1) must vanish when summed over all phases. By the arguments presented by Garg and Pritchett (1977) we neglect the macroscopic viscous dissipation effects, i.e., the term containing σ_{ij}^{l}. Moreover, because we consider only the liquid water phase, the change in internal energy may be written as a function of temperature and pressure (Slattery, 1972, p. 292):

$$\rho^{\alpha} \frac{D^{\alpha}E^{\alpha}}{Dt} = \rho^{\alpha}\hat{c}_{\mathrm{V}}^{\alpha} \frac{D^{\alpha}T^{\alpha}}{Dt} + \left[T^{\alpha}\left(\frac{\partial p}{\partial T}\right)_{\mathrm{V}}^{\alpha} - p^{\alpha}\right] \frac{\partial}{\partial x_i} V_i^{\alpha}, \quad (5.4.1.3)$$

where $\hat{c}_{\mathrm{V}}$ is the specific heat at constant volume, T is absolute temperature, and $(\partial p/\partial T)_{\mathrm{V}}$ is the partial derivative at constant volume.

Except at singularities the divergence of the velocity in porous medium flow can be neglected in the energy equation. The preceding arguments reduce (5.4.1.2) to the form

$$\phi\rho^{l}\hat{c}_{\mathrm{V}}^{l} \frac{D^{l}T^{l}}{Dt} + (1 - \phi)\rho^{\mathrm{s}}\hat{c}_{\mathrm{V}}^{\mathrm{s}} \frac{D^{\mathrm{s}}T^{\mathrm{s}}}{Dt} + \frac{\partial}{\partial x_i}(q_i^{l} + q_i^{\mathrm{s}})$$

$$- [\phi\rho^{l}Q^{l} + (1 - \phi)\rho^{\mathrm{s}}Q^{\mathrm{s}}] = 0. \quad (5.4.1.4)$$

In most groundwater flow problems it is reasonable to neglect grain velocity and assume thermal equilibrium between phases. Complete constitutive relations for the energy flux q_i^{α} are not generally available, and a simple Fourier-type expression is commonly employed:

$$q_i^{l} + q_i^{\mathrm{s}} = -K_{ij}\,\partial T/\partial x_j. \quad (5.4.1.5)$$

where, as assumed previously, $T = T^{l} = T^{\mathrm{s}}$, and K_{ij} is the hydrodynamic thermal dispersion tensor for the combined fluid–solid medium. Thus, K_{ij} is given by

$$K_{ij} = \phi K_{ij}^{l} + (1 - \phi)K_{ij}^{\mathrm{s}}, \quad (5.4.1.6)$$

where K_{ij}^{s} corresponds to the thermal conductivity of solid and K_{ij}^{l} corresponds to the hydrodynamic thermal dispersion of fluid. In an analogous manner to the treatment of the hydrodynamic dispersion tensor for solute transport, K_{ij}^{l} is composed of portions contributed by mechanical dispersion and molecular diffusion. Thus, it can be written as [see also Eq. (5.2.5.1)]

$$K_{ij}^{l} = \rho^{l}\hat{c}_{\mathrm{V}}^{l}\left[D_{\mathrm{T}}|V|\delta_{ij} + (D_{\mathrm{L}} - D_{\mathrm{T}})\frac{V_iV_j}{|V|}\right] + \phi\tau\lambda_{ij}^{l}, \quad (5.4.1.7)$$

where λ_{ij}^{l} is the thermal conductivity of the fluid and, as before, D_{L} and D_{T} are the longitudinal and transverse dispersivities, respectively.

Employing Eqs. (5.4.1.5)–(5.4.1.7) to evaluate the energy flux and introducing the assumptions indicated previously, the energy equation (5.4.1.4) becomes

$$[\phi\rho^l\hat{c}_V^l + (1-\phi)\rho^s\hat{c}_V^s]\frac{\partial T}{\partial t} + \rho^l\hat{c}_V^l V_i^l \frac{\partial T}{\partial x_i} - \frac{\partial}{\partial x_i}\left(K_{ij}\frac{\partial T}{\partial x_j}\right)$$

$$- [\phi\rho^l Q^l + (1-\phi)\rho^s Q^s] = 0. \qquad (5.4.1.8)$$

5.5 Energy Transport in Multiphase Flow

Development of the energy transport equation for multiple fluid phases begins with the general energy balance equation (5.4.1.1). Although there are several approaches to obtaining a useful field equation, those employed in geothermal simulation can be catalogued in a general sense into "two-equation" and "three-equation" formulations. The two-equation approach [see Witherspoon *et al.* (1975)] represents the thermodynamic system in terms of enthalpy and pressure or internal energy and density.The three-equation scheme uses the dependent variables saturation, temperature, and pressure. Three thermodynamic variables are required when temperature is treated as a dependent variable; two others are required to describe uniquely the system in the two-phase steam–water region. We consider only the two-equation formulation; the three-equation development can be found in Thomas and Pierson (1976) or Coats (1977).

5.5.1 Two-Equation Formulation

The energy transport equations are readily obtained by substitution of s, l, and g for α in (5.4.1.1) to represent the solid, liquid, and gas phases, respectively. Summation of the three equations yields (neglecting the viscous dissipation term)

$$n^s\rho^s\frac{D^sE^s}{Dt} + n^l\rho^l\frac{D^lE^l}{Dt} + n^g\rho^g\frac{D^gE^g}{Dt} + \frac{\partial}{\partial x_i}(q_i^s + q_i^l + q_i^g)$$

$$- [n^s\rho^s Q^s + n^l\rho^l Q^l + n^g\rho^g Q^g] = 0. \qquad (5.5.1.1)$$

We now define a phase-averaged internal energy for the liquid and steam (gas)

$$E = \frac{1}{\rho}[S^l\rho^l E^l + S^g\rho^g E^g], \qquad (5.5.1.2a)$$

where

$$\rho = S^l\rho^l + S^g\rho^g, \tag{5.5.1.2b}$$

$$n^l = S^l\phi, \tag{5.5.1.2c}$$

$$n^g = S^g\phi, \tag{5.5.1.2d}$$

$$S^l + S^g = 1;$$

here S^α is the saturation of the α phase. The rate of change of the internal energy of the solid phase is expanded as [see Eq. (5.4.1.3)]

$$\rho^s \frac{D^sE^s}{Dt} \simeq \rho^s\hat{c}_V^s \frac{D^sT}{Dt} \simeq \rho^s\hat{c}_V^s \frac{\partial T}{\partial t} = \rho^s\hat{c}_V^s \left[\frac{\partial T}{\partial E}\frac{\partial E}{\partial t} + \frac{\partial T}{\partial \rho}\frac{\partial \rho}{\partial t}\right], \tag{5.5.1.3}$$

where we assume $T = T(E, \rho)$.

The energy flux q_i^α is represented using a relation similar to that employed for the single-phase fluid. Assuming that the hydrodynamic thermal dispersion tensor is isotropic, we obtain

$$q_i^s + q_i^l + q_i^g \equiv -K\,\partial T/\partial x_i, \tag{5.5.1.4}$$

where K is the coefficient of hydrodynamic dispersion for the combined liquid–gas–solid medium, that is, $K = n^lK^l + n^gK^g + n^sK^s$. A chain rule expansion of (5.5.1.4) yields

$$q_i^s + q_i^l + q_i^g = -K\left(\frac{\partial T}{\partial E}\frac{\partial E}{\partial x_i} + \frac{\partial T}{\partial \rho}\frac{\partial \rho}{\partial x_i}\right). \tag{5.5.1.5}$$

Combination of (5.5.1.1)–(5.5.1.5) provides the following form of the energy equation:

$$(1 - \phi)\rho^s\hat{c}_V^s\left(\frac{\partial T}{\partial E}\frac{\partial E}{\partial t} + \frac{\partial T}{\partial \rho}\frac{\partial \rho}{\partial t}\right) + S^l\phi\rho^l\frac{D^lE^l}{Dt} + (1 - S^l)\phi\rho^g\frac{D^gE^g}{Dt}$$
$$- \frac{\partial}{\partial x_i}\left[K\left(\frac{\partial T}{\partial E}\frac{\partial E}{\partial x_i} + \frac{\partial T}{\partial \rho}\frac{\partial \rho}{\partial x_i}\right)\right] = 0, \tag{5.5.1.6a}$$

where it has been assumed for simplicity that the terms associated with heat supplied from an external source are zero. The mass continuity equations for the liquid and gas phases are now written as

$$\frac{\partial}{\partial t}\phi S^l\rho^l + \frac{\partial}{\partial x_i}\rho^lV_i^l = 0, \tag{5.5.1.6b}$$

$$\frac{\partial}{\partial t}\phi S^g\rho^g + \frac{\partial}{\partial x_i}\rho^gV_i^g = 0. \tag{5.5.1.6c}$$

Using the definition of substantial derivatives given in Section 5.4.1 and combining Eq. (5.5.1.6a) with (5.5.1.6b) and (5.5.1.6c), one obtains

$$(1-\phi)\rho^{s}\hat{c}_{V}^{s}\left(\frac{\partial T}{\partial E}\frac{\partial E}{\partial t}+\frac{\partial T}{\partial \rho}\frac{\partial \rho}{\partial t}\right)+\frac{\partial}{\partial t}\phi\rho E+\frac{\partial}{\partial x_i}V_i^l\rho^l E^l$$

$$+\frac{\partial}{\partial x_i}V_i^g\rho^g E^g-\frac{\partial}{\partial x_i}K\left(\frac{\partial T}{\partial E}\frac{\partial E}{\partial x_i}+\frac{\partial T}{\partial \rho}\frac{\partial \rho}{\partial x_i}\right)=0. \qquad (5.5.1.6d)$$

which is the required form of the first governing equation in the two-equation formulation.

The second governing equation is a flow equation obtained by combining Darcy's law and the continuity equations, which are now represented by

$$\frac{\partial}{\partial t}(\phi\rho)+\frac{\partial}{\partial x_i}(\rho V_i)=0. \qquad (5.5.1.7)$$

where ρ is the phase averaged density given by (5.5.1.2b) and V_i is the phase averaged velocity given by

$$V_i=\frac{1}{\rho}(\rho^g V_i^g+\rho^l V_i^l) \qquad (5.5.1.8)$$

and

$$V_i^\alpha=\frac{-k_{ij}k_r^\alpha}{\mu^\alpha}\left(\frac{\partial p}{\partial x_j}+\rho^\alpha g_j\right),\qquad \alpha=l,g, \qquad (5.5.1.9)$$

where $g_j \equiv g(\partial z/\partial x_j)$, with g and z being the gravitation constant and the height above a given datum, respectively.

In the pressure gradient term, capillary pressure is neglected, so that $p^l=p^g=p$. Assuming $p=p(\rho, E)$, one again employs the chain rule to rewrite (5.5.1.9) for isotropic porous media as

$$V_i^\alpha=\frac{-kk_r^\alpha}{\mu^\alpha}\left(\frac{\partial p}{\partial E}\frac{\partial E}{\partial x_i}+\frac{\partial p}{\partial \rho}\frac{\partial \rho}{\partial x_i}+\rho^\alpha g_i\right),\qquad \alpha=l,g. \qquad (5.5.1.10)$$

In addition to Eqs. (5.5.1.6)–(5.5.1.10), other constitutive and thermodynamic relations are required to close the system of equations. The thermodynamic information is normally derived from the steam tables (Meyer *et al.*, 1968). Empirical equations as well as FORTRAN computer subroutines for describing thermodynamic properties of steam can be found in McClintock and Silvestri (1968). Figure 5.1 provides this thermodynamic information in graphical form. Additional constitutive information will be introduced in our discussion of solution methodology.

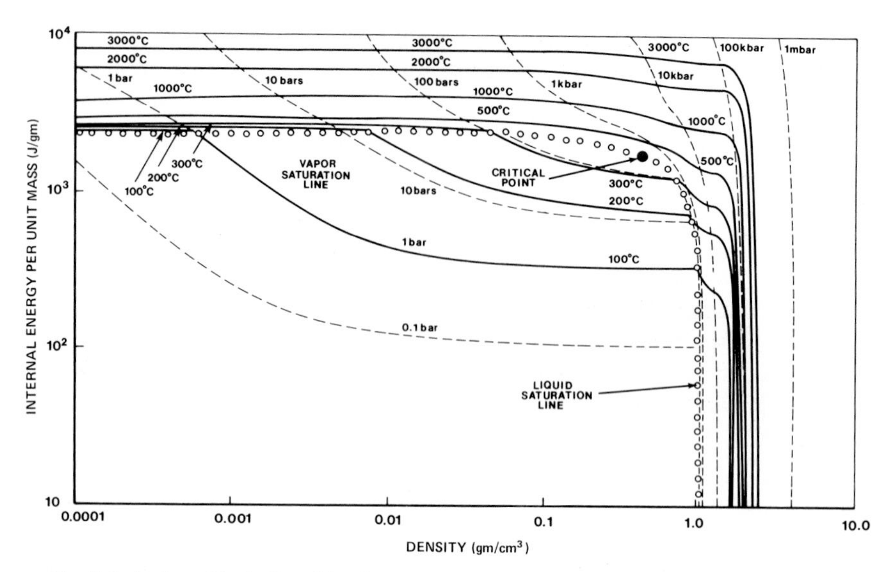

Fig. 5.1. Isobars (dashed) and isotherms (solid) in the density–energy plane [after Pritchett *et al.* (1975)].

An alternative form of the two-equation formulation uses enthalpy and pressure as the two dependent variables. The principal reason for this preference is that these variables are more often measured in the field and consequently auxiliary conditions are more readily specified. To transform the foregoing equations into an enthalpy–pressure form we note that

$$h^\alpha = E^\alpha + (p^\alpha/\rho^\alpha), \qquad \alpha = l, g, \tag{5.5.1.11a}$$

$$h^s = E^s. \tag{5.5.1.11b}$$

Substitution of (5.5.1.11) into (5.5.1.1) yields

$$n^s\rho^s \frac{D^s h^s}{Dt} + n^l\rho^l \frac{D^l}{Dt}\left(h^l - \frac{p^l}{\rho^l}\right) + n^g\rho^g \frac{D^g}{Dt}\left(h^g - \frac{p^g}{\rho^g}\right) + \frac{\partial}{\partial x_i}(q_i^s + q_i^l + q_i^g) = 0, \tag{5.5.1.12}$$

where, once again, it has been assumed that the external heat terms are negligible.

This expression can be written in a more convenient form when the mass-continuity relations, (5.5.1.6b) and (5.5.1.6c), are introduced. This substitution yields

$$n^s \rho^s \frac{D^s h^s}{Dt} + \frac{\partial}{\partial t} n^l \rho^l h^l + \frac{\partial}{\partial t} n^g \rho^g h^g$$

$$+ \frac{\partial}{\partial x_i} V_i^l \rho^l h^l + \frac{\partial}{\partial x_i} V_i^g \rho^g h^g + \frac{\partial}{\partial x_i} (q_i^s + q_i^l + q_i^g)$$

$$- n^l \rho^l \frac{D^l}{Dt} \frac{p^l}{\rho^l} - n^g \rho^g \frac{D^g}{Dt} \frac{p^g}{\rho^g} = 0. \qquad (5.5.1.13)$$

The last two terms of (5.5.1.13) describe compressible work and are generally neglected in field applications.

To obtain a more useful form of (5.5.1.13) we now average over the fluid phases using the same procedure employed for internal energy, that is,

$$h = \frac{1}{\rho} [S^l \rho^l h^l + S^g \rho^g h^g]. \qquad (5.5.1.14)$$

Substitution of (5.5.1.14) and (5.5.1.4) into (5.5.1.13) and the use of commonly encountered thermodynamic relations yields

$$(1 - \phi) \rho^s \frac{\partial h^s}{\partial t} + \frac{\partial}{\partial t} \phi \rho h$$

$$+ \frac{\partial}{\partial x_i} V_i^l \rho^l h^l + \frac{\partial}{\partial x_i} V_i^g \rho^g h^g$$

$$- \frac{\partial}{\partial x_i} \left[K \left(\frac{\partial T}{\partial h} \frac{\partial h}{\partial x_i} + \frac{\partial T}{\partial p} \frac{\partial p}{\partial x_i} \right) \right] = 0, \qquad (5.5.1.15)$$

where we have neglected the compressible work term in (5.5.1.13).

The second governing equation in enthalpy and pressure is the flow equation that is obtained by combining (5.5.1.7) with (5.5.1.8) and expanding the time derivative as

$$\frac{\partial (\phi \rho)}{\partial t} = \phi \frac{\partial \rho}{\partial t} + \rho \frac{\partial \phi}{\partial t}$$

$$= \phi \left(\frac{\partial \rho}{\partial h} \frac{\partial h}{\partial t} + \frac{\partial \rho}{\partial p} \frac{\partial p}{\partial t} \right) + \rho \frac{d\phi}{dp} \frac{\partial p}{\partial t}, \qquad (5.5.1.16)$$

with V_i^α given by (5.5.1.9). The standard thermodynamic information for this formulation is given in Fig. 5.2.

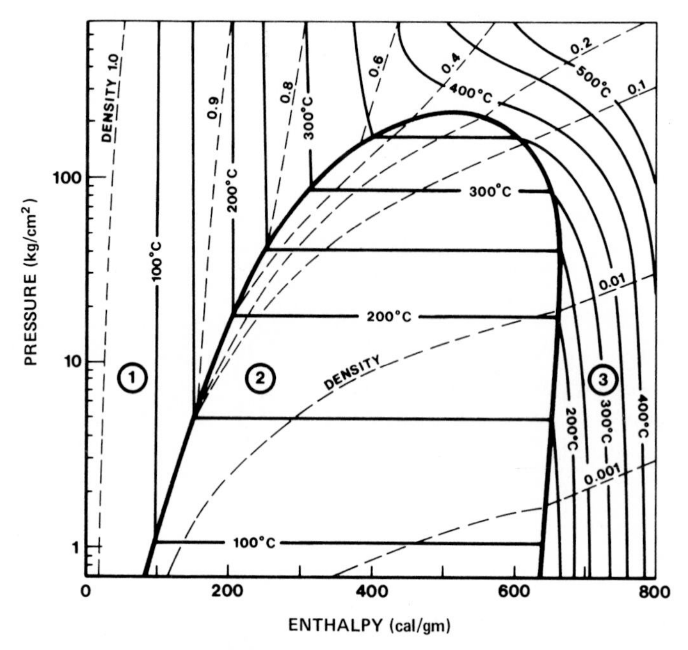

Fig. 5.2. Pressure–enthalpy diagram for water and steam with thermodynamic regions: (1) compressed water; (2) two-phase steam and water; and (3) superheated steam [adapted from Mercer and Faust (1975), and White *et al.* (1971)].

5.6 Finite Element Simulation of Mass and Energy Transport in Single-Phase Flow

5.6.1 Introduction

The equations describing mass and energy transport in single-phase flow [(i.e., Eqs. (5.2.3.5) and (5.4.1.8)] are of the same fundamental form. We can write them as

$$A\frac{\partial u}{\partial t} + B_i\frac{\partial u}{\partial x_i} + \frac{\partial}{\partial x_i}\left(E_{ij}\frac{\partial u}{\partial x_j}\right) + Fu + G = 0. \qquad (5.6.1.1)$$

where for the mass transport equation (5.2.3.5)

$$A \equiv -\phi\kappa_k, \qquad (5.6.1.2a)$$

$$B_i \equiv -V_i, \qquad (5.6.1.2b)$$

$$E_{ij} \equiv \phi D_{ij}, \qquad (5.6.1.2c)$$

$$F \equiv -\phi\lambda_k\kappa_k, \tag{5.6.1.2d}$$

$$G \equiv \sum_{m=1}^{M} \phi\xi_{km}\lambda_m\kappa_m c_m, \tag{5.6.1.2e}$$

$$u \equiv c_k, \tag{5.6.1.2f}$$

where c_k is the concentration of the kth species in the liquid phase. For the energy transport equation (5.4.1.8)

$$A \equiv \phi\rho^l\hat{c}_V^l + (1 - \phi)\rho^s\hat{c}_V^s, \tag{5.6.1.3a}$$

$$B_i \equiv \rho^l\hat{c}_V^l V_i^l, \tag{5.6.1.3b}$$

$$E_{ij} \equiv -K_{ij}, \tag{5.6.1.3c}$$

$$F \equiv 0, \tag{5.6.1.3d}$$

$$G \equiv -[\phi\rho^l Q^l + (1 - \phi)\rho^s Q^s], \tag{5.6.1.3e}$$

$$u \equiv T \qquad \text{(the medium temperature)}, \tag{5.6.1.3f}$$

with the Darcy velocity V_i^l defined as

$$V_i^l = \frac{-k_{ij}}{\mu}\left(\frac{\partial p}{\partial x_j} + \rho^l g_j\right). \tag{5.6.1.3g}$$

In addition to the transport equation, we require a flow equation. The flow equation for the density-dependent case can be written (see Section 4.2.1)

$$H\frac{\partial p}{\partial t} + \frac{\partial}{\partial x_i}\left[M_{ij}\left(\frac{\partial p}{\partial x_j} + \rho^l g_j\right)\right] = 0, \tag{5.6.1.4}$$

where $H = -\rho^l(\alpha + \phi\beta)$ and $M_{ij} = \rho^l k_{ij}/\mu$. Both the fluid flow and the solute or energy transport equations, (5.6.1.1) and (5.6.1.4), may be either linear or weakly nonlinear, depending upon the assumptions introduced in the constitutive relations required to close the system of equations.

5.6.2 Density-Dependent Mass and Energy Transport

The transport of solutes in a flow field dependent upon concentration or temperature is described by (5.6.1.1) and (5.6.1.4). For brevity, we summarize them as $L_u(u) = 0$ and $L_p(p) = 0$, where

$$L_u(u) \equiv A\frac{\partial u}{\partial t} + B_i\frac{\partial u}{\partial x_i} + \frac{\partial}{\partial x_i}\left(E_{ij}\frac{\partial u}{\partial x_j}\right) + Fu + G, \tag{5.6.2.1a}$$

$$L_p(p) \equiv H\frac{\partial p}{\partial t} + \frac{\partial}{\partial x_i}\left[M_{ij}\left(\frac{\partial p}{\partial x_j} + \rho^l g_j\right)\right]. \tag{5.6.2.1b}$$

Density variations with temperature or concentration are assumed to have the form

$$\rho^l = \rho_0^l + \frac{d\rho^l}{du}\,du. \tag{5.6.2.1c}$$

For the case of saltwater intrusion, u is the mass fraction of chlorides and $d\rho^l/du \simeq \rho_0^l(1 - \varepsilon)$ with $\varepsilon = 0.03$ for concentrations up to that of sea water. When modeling thermal transport, $d\rho^l/du$ is the thermal volume expansion. In addition, the dynamic viscosity of the fluid is also a function of temperature. Typical empirical expressions can be obtained from Ramey *et al.* (1974) and Mercer *et al.* (1975).

Sequential Solution Procedure

There are several schemes available for solving the set of equations (5.6.1.1) and (5.6.1.4). The most popular is the sequential solution procedure wherein (5.6.1.4) is solved using an estimate of u and then (5.6.1.1) is solved using the current value of V_i. This procedure is repeated for the same time increment until u and p stabilize. To describe the approach we employ the Galerkin formulation.

Let the two dependent variables u and p be expressed in a finite series of the form

$$u(\mathbf{x}, t) \simeq \hat{u}(\mathbf{x}, t) = u_J(t)N_J(\mathbf{x}), \qquad J = 1, 2, \ldots, n, \tag{5.6.2.2a}$$

$$p(\mathbf{x}, t) \simeq \hat{p}(\mathbf{x}, t) = p_J(t)N_J(\mathbf{x}), \qquad J = 1, 2, \cdots, n, \tag{5.6.2.2b}$$

where the summation notation convention once again applies to the repeated indices and N_J are the standard shape functions defined over an appropriate finite element (see Chapter 3). Employing the operators L_p and L_u defined in (5.6.2.1), the approximating equations arising from Galerkin's method can be written as

$$\sum_e \int_{R^e} L_u(\hat{u})N_I(\mathbf{x})\,dR = 0, \qquad I = 1, 2, \ldots, n, \tag{5.6.2.3a}$$

$$\sum_e \int_{R^e} L_p(\hat{p})N_J(\mathbf{x})\,dR = 0, \qquad J = 1, 2, \ldots, n, \tag{5.6.2.3b}$$

where

$$\sum_e \int_{R^e} dR = \int_R dR \qquad \text{and} \qquad \sum_e \int_{B^e} dB = \int_B dB;$$

L_u $(\hat{u})$ and L_p $(\hat{p})$ are residuals that can be attributed to the use of $\hat{u}$ and $\hat{p}$ rather than u and p in L_u and L_p, respectively, and R^e is the subregion of the space domain identified with element e. Equations (5.6.2.3) require that these residuals, when weighted by the shape functions, be minimized in an integral sense over the volume $R = \Sigma_e R^e$. We now substitute (5.6.1.1), (5.6.1.4), and (5.6.2.2) into (5.6.2.3):

$$\sum_e \int_{R^e} \left[A \frac{\partial \hat{u}}{\partial t} + B_i \frac{\partial \hat{u}}{\partial x_i} + \frac{\partial}{\partial x_i} \left(E_{ij} \frac{\partial \hat{u}}{\partial x_j} \right) + F\hat{u} + G \right] N_I \, dR = 0,$$

$$I = 1, 2, \ldots, n; \tag{5.6.2.4a}$$

$$\sum_e \int_{R^e} \left\{ H \frac{\partial \hat{p}}{\partial t} + \frac{\partial}{\partial x_i} \left[M_{ij} \left(\frac{\partial \hat{p}}{\partial x_j} + \rho^l g_j \right) \right] \right\} N_I \, dR = 0,$$

$$I = 1, 2, \ldots, n. \tag{5.6.2.4b}$$

Application of Green's theorem to the second-derivative term of (5.6.2.4a) yields

$$\sum_e \int_{R^e} \left[\left(A \frac{\partial \hat{u}}{\partial t} + B_i \frac{\partial \hat{u}}{\partial x_i} + F\hat{u} + G \right) N_I - E_{ij} \frac{\partial \hat{u}}{\partial x_j} \frac{\partial N_I}{\partial x_i} \right] dR$$

$$+ \sum_e \int_{B_2^e} E_{ij} \frac{\partial \hat{u}}{\partial x_j} n_i N_I \, dB = 0, \qquad I = 1, 2, \ldots, n, \tag{5.6.2.5}$$

where B_2 is the segment of the boundary of R^e at which a Neumann or a third-type boundary condition is specified. Note that one might also consider applying Green's theorem to the convective term. Although this would provide a convenient way to accommodate third-type boundary conditions, the method is sensitive to the manner in which the velocity is determined and is not recommended.

Green's theorem can also be used in the flow equation (5.6.2.4b) to give

$$\sum_e \int_{R^e} \left[H \frac{\partial \hat{p}}{\partial t} N_I - M_{ij} \left(\frac{\partial \hat{p}}{\partial x_j} + \rho^l g_j \right) \frac{\partial N_I}{\partial x_i} \right] dR$$

$$+ \sum_e \int_{B_2^e} M_{ij} \left(\frac{\partial \hat{p}}{\partial x_j} + \rho^l g_j \right) n_i N_I \, dB = 0, \qquad I = 1, 2, \ldots, n. \tag{5.6.2.6}$$

The last term in (5.6.2.6) incorporates the specified flux boundary condition.

Let us now substitute the appropriate series representation of $\hat{u}$ and $\hat{p}$ from (5.6.2.2) into (5.6.2.5) and (5.6.2.6):

$$\sum_e \int_{R^e} \left[A \frac{du_J}{dt} N_J N_I + u_J \left(B_i \frac{\partial N_J}{\partial x_i} N_I - E_{ij} \frac{\partial N_J}{\partial x_j} \frac{\partial N_I}{\partial x_i} + F N_J N_I \right) + G N_I \right] dR + \sum_e \int_{B_2^e} E_{ij} \frac{\partial \hat{u}}{\partial x_j} n_i N_I \, dB = 0, \qquad I = 1, 2, \ldots, n \tag{5.6.2.7a}$$

$$\sum_e \int_{R^e} \left\{ H \frac{dp_J}{dt} N_J N_I - p_J M_{ij} \frac{\partial N_J}{\partial x_j} \frac{\partial N_I}{\partial x_i} - M_{ij} \rho^l g_j \frac{\partial N_I}{\partial x_i} \right\} dR + \sum_e \int_{B_2^e} M_{ij} \left(\frac{\partial \hat{p}}{\partial x_j} + \rho^l g_j \right) n_i N_I \, dB = 0, \qquad I = 1, 2, \ldots, n. \tag{5.6.2.7b}$$

In matrix form (5.6.2.7a) becomes

$$[Q]\{du/dt\} + [S]\{u\} + \{f\} = 0, \tag{5.6.2.8a}$$

where

$$Q_{IJ} = \sum_e \int_{R^e} A N_I N_J \, dR, \tag{5.6.2.8b}$$

$$S_{IJ} = \sum_e \int_{R^e} \left\{ B_i \frac{\partial N_J}{\partial x_i} N_I - E_{ij} \frac{\partial N_J}{\partial x_j} \frac{\partial N_I}{\partial x_i} + F N_J N_I \right\} dR, \tag{5.6.2.8c}$$

$$f_I = \sum_e \int_{R^e} G N_I \, dR + \sum_e \int_{B_2^e} E_{ij} \frac{\partial \hat{u}}{\partial x_j} n_i N_I \, dB, \tag{5.6.2.8d}$$

and, similarly, (5.6.2.7b) becomes

$$[T]\{dp/dt\} + [V]\{p\} + \{g\} = 0, \tag{5.6.2.9a}$$

where

$$T_{IJ} = \sum_e \int_{R^e} H N_I N_J \, dR, \tag{5.6.2.9b}$$

$$V_{IJ} = - \sum_e \int_{R_e} M_{ij} \frac{\partial N_I}{\partial x_i} \frac{\partial N_J}{\partial x_j} \, dR, \tag{5.6.2.9c}$$

$$g_I = - \sum_e \int_{R^e} M_{ij} \rho^l g_j \frac{\partial N_I}{\partial x_i} \, dR + \sum_e \int_{B_2^e} M_{ij} \left(\frac{\partial \hat{p}}{\partial x_j} + \rho^l g_j \right) n_i N_I \, dB. \tag{5.6.2.9d}$$

Equations (5.6.2.8a) and (5.6.2.9a) are both sets of n ordinary differential equations. Before proceeding to present the solution algorithm, it is appropriate to describe briefly the procedure for handling boundary conditions and source and sink terms associated with the solute transport equation. For the flow equation, such treatment has been presented in Chapter 4 and hence is not repeated here.

For the transport equation, only the treatment of third-type boundary conditions needs elaboration. The first- (Dirichlet) and second- (Neumann) type boundary conditions are treated in the same manner as described previously for the flow equation.

To handle the third-type inlet boundary condition, we simply return to Eq. (5.2.6.2) and rewrite it as

$$E_{ij}\frac{\partial u}{\partial x_j}n_i = q(u - \tilde{u}),$$

where $\tilde{u}$ denotes concentration of the fluid entering the boundary. This equation is used to evaluate the boundary integral in Eq. (5.6.2.8d).

In a case where there is a point source such as an injection well located at node I, the transport equation (5.6.2.1a) contains an additional flux term of the form

$$q(u - \tilde{u}) \equiv Q_{\mathrm{w}}\delta(\mathbf{x} - \mathbf{x}_I)(u - \tilde{u}),$$

where Q_{w} is the volumetric flow rate of the well and δ is the Dirac delta function. In such a case, it can be shown that the resulting finite element equation (5.6.2.7a) also contains an additional term given by

$$\int_R N_I q(\hat{u} - \tilde{u})\,dR \equiv \int_B N_I Q_{\mathrm{w}}\delta(\mathbf{x} - \mathbf{x}_I)\,(\hat{u} - \tilde{u})\,dR = Q_{\mathrm{w}}(\hat{u}_I - \tilde{u}_I).$$

This term vanishes if the well becomes a pumped well because $\tilde{u}_I$, the concentration of the discharging fluid, is obviously the same as $\hat{u}_I$, the concentration in the porous medium reservoir.

Next, we return once again to Eqs. (5.6.2.8a) and (5.6.2.9a), and proceed to solve the set of ordinary differential equations by finite difference time stepping. Employing first-order correct representations of the time derivatives, one obtains two sets of algebraic equations as follows:

$$\frac{1}{\Delta t}[Q]^{k+\theta}(\{u\}^{k+1} - \{u\}^k) + \theta[S]^{k+1}\{u\}^{k+1} + (1 - \theta)[S]^k\{u\}^k$$
$$+ \theta\{f\}^{k+1} + (1 - \theta)\{f\}^k = 0, \tag{5.6.2.10a}$$

$$\frac{1}{\Delta t}[T]^{k+\theta}(\{p\}^{k+1} - \{p\}^k) + \theta[V]^{k+1}\{p\}^{k+1} + (1 - \theta)[V]^k\{p\}^k$$
$$+ \theta\{g\}^{k+1} + (1 - \theta)\{g\}^k = 0, \tag{5.6.2.10b}$$

where k and $k+1$ denote the previous and current time levels, Δt the current time step, θ a time-weighting factor, and the coefficient matrices $[Q]^{k+\theta}$ and $[T]^{k+\theta}$ are evaluated using the weighted average of the current and previous nodal values of $\hat{u}$ and $\hat{p}$. Because the flow equation is solved before the transport equation in the sequential solution algorithm, it is appropriate at this stage to refer to (5.6.2.10b) before (5.6.2.10a). Thus they are now rewritten as

$$[F_1]\{p\}^{k+1} + \{f_1\} = 0, \tag{5.6.2.11}$$

$$[F_2]\{u\}^{k+1} + \{f_2\} = 0, \tag{5.6.2.12}$$

where

$$[F_2] = \frac{1}{\Delta t}[Q]^{k+\theta} + \theta[S]^{k+1},$$

$$\{f_2\} = -\frac{1}{\Delta t}[Q]^{k+\theta}\{u\}^k + (1-\theta)[S]^k\{u\}^k + \theta\{f\}^{k+1} + (1-\theta)\{f\}^k,$$

$$[F_1] = \frac{1}{\Delta t}[T]^{k+\theta} + \theta[V]^{k+1},$$

$$\{f_1\} = -\frac{1}{\Delta t}[T]^{k+\theta}\{p\}^k + (1-\theta)[V]^k\{p\}^k + \theta\{g\}^{k+1} + (1-\theta)\{g\}^k.$$

The solution procedure for each time step can be described as follows. First, the system of n equations represented by (5.6.2.11) can be solved efficiently using a solver that takes account of both the banded and symmetric features of the coefficient matrix. Nodal values of pressure $\{p\}^{k+1}$ obtained are then used with Darcy's law to determine velocities. Gauss point values rather than nodal point values of velocity components should be computed. [Gauss points have been shown to be, in a certain sense, optimal sampling points; see Zienkiewicz (1977, pp. 280–284)]. The velocity values obtained are supplied as input to the transport equation. Thus, (5.6.2.12) is then solved for $\{u\}^{k+1}$ using a general solver for an asymmetric banded matrix. In the case where transport is independent of density change, one may proceed directly to the next time step. However, for density dependent transport, (5.6.2.11) and (5.6.2.12) are nonlinear, and it is generally necessary to iterate within the same time level t_{k+1}. Between each iteration the fluid properties, velocities, and all quantities depending upon these parameters must be updated. The flow and the transport equations must be resolved sequentially. Iterations must be performed until successive changes in nodal values of $\hat{p}$ and $\hat{u}$ are within a prescribed tolerance. After convergence is achieved one can then proceed to the next time level.

We must now revisit the integrals appearing in (5.6.2.8) and (5.6.2.9) to establish methodology for their evaluation. There are two basic approaches. The simplest approach is to assume the coefficients in the integrand are uniform within each element. As an example, one could write

$$T_{IJ} = \sum_e \int_{R^e} HN_J N_I \, dR \simeq \sum_e H^e \int_{R^e} N_J N_I \, dR. \qquad (5.6.2.13)$$

When the assumption of element-wise constant coefficients is physically reasonable, this approach is the methodology of choice. An alternative approach is to represent H as a polynomial in the same way that we treat the dependent variables:

$$H(\mathbf{x}) \simeq \hat{H}(\mathbf{x}) = H_K N_K(\mathbf{x}), \qquad (5.6.2.14)$$

where the H_K are nodal values of the parameter H.

Substitution of this expression into the integrand of (5.6.2.9b) yields

$$T_{IJ} = \sum_e \int_{R^e} HN_I N_J \, dR \simeq \sum_e \int_{R^e} H_K N_K N_I N_J \; dR$$

$$= H_K \sum_e \int_{R^e} N_K N_I N_J \, dR. \qquad (5.6.2.15)$$

One now evaluates the integral either directly or using a Gaussian quadrature scheme.

Although this approach seems reasonable, it is usually less attractive than the following minor variant on this concept. Consider once again the term in (5.6.2.6) that gives rise to T_{IJ}, that is,

$$\sum_e \int_{R^e} H \frac{\partial \hat{p}}{\partial t} N_I \; dR. \qquad (5.6.2.16a)$$

Now consider the interpolation relation

$$H(\mathbf{x}) p(\mathbf{x}, t) \simeq (Hp)_J N_J(\mathbf{x}), \qquad (5.6.2.16b)$$

where $(Hp)_J$ depends on t.

Equation (5.6.2.16b), when substituted into (5.6.2.16a), gives

$$\sum_e \int_{R^e} H \frac{\partial \hat{p}}{\partial t} N_I \, dR = \sum_e \left(H \frac{dp}{dt} \right)_J \int_{R^e} N_J N_I \, dR. \qquad (5.6.2.17)$$

The important point here is that the interpolating polynomial in (5.6.2.17) is of lower degree than in (5.6.2.15), given the same choice of shape function in each. Empirical evidence indicates that a smoother, more consistent solution results using the formulation of (5.6.2.17), particularly for nonlinear problems.

Coupled Solution Procedure

Consider now the possibility of writing (5.6.2.3) as a single matrix equation. To do this we introduce the constitutive relation for fluid density as a linear function of concentration (5.6.2.1c) in the gravity term of the flow equation. The resulting expression is

$$H\frac{\partial p}{\partial t} + \frac{\partial}{\partial x_i}\left[M_{ij}\left(\frac{\partial p}{\partial x_j} + \left\{\rho_0^l + \rho_0^l\left(1 - \varepsilon\right)\hat{u}\right\}g_j\right)\right] = 0. \quad (5.6.2.18)$$

If we proceed along the same road that led us to Eqs. (5.6.2.7a) and (5.6.2.7b) for the two-equation formulation, we obtain

$$\sum_e \int_{R^e}\left\{A\frac{du_J}{dt}N_JN_I + u_J\left[B_i\frac{\partial N_J}{\partial x_i}N_I - E_{ij}\frac{\partial N_J}{\partial x_j}\frac{\partial N_I}{\partial x_i} + FN_JN_I\right] + GN_I\right\}dR + \sum_e \int_{B_2^e} E_{ij}\frac{\partial \hat{u}}{\partial x_j}n_iN_I\,dB = 0,$$

$$I = 1, 2, \ldots, n, \quad (5.6.2.19a)$$

$$\sum_e \int_{R^e}\left\{H\frac{dp_J}{dt}N_JN_I - p_JM_{ij}\frac{\partial N_J}{\partial x_j}\frac{\partial N_I}{\partial x_i} - M_{ij}\left[\rho_0^l + \rho_0^l\left(1 - \varepsilon\right)u_JN_J\right]g_j\frac{\partial N_I}{\partial x_i}\right\}dR + \sum_e \int_{B_2^e} M_{ij}\left(\frac{\partial \hat{p}}{\partial x_j} + \rho^l g_j\right)n_iN_I\,dB = 0,$$

$$I = 1, 2, \ldots, n. \quad (5.6.2.19b)$$

Equations (5.6.2.19) can be written in matrix form as

$$[Q^*]\{du^*/dt\} + [S^*]\{u^*\} + \{f^*\} = 0, \quad (5.6.2.20)$$

$$Q_{IJ}^* = \begin{bmatrix} \sum_e \int_{R^e}\{AN_JN_I\}\,dR & 0 \\ 0 & \sum_e \int_{R^e}\{HN_JN_I\}\,dR \end{bmatrix}, \quad (5.6.2.21a)$$

$$\left\{\frac{du^*}{dt}\right\}_I = \left\{\begin{matrix} du_I/dt \\ dp_I/dt \end{matrix}\right\}, \tag{5.6.2.21b}$$

$$S^*_{IJ} = \tag{5.6.2.21c}$$

$$\begin{bmatrix} \sum_e \int_{R_e} \left[B_i \frac{\partial N_J}{\partial x_i} N_I - E_{ij} \frac{\partial N_J}{\partial x_j}\frac{\partial N_I}{\partial x_i} + FN_J N_I \right] dR & 0 \\ -\sum_e \int_{R^e} M_{ij}\, \rho_0^l (1-\varepsilon) N_J g_j \frac{\partial N_I}{\partial x_i}\, dR & -\sum_e \int_{R^e} M_{ij} \frac{\partial N_J}{\partial x_j}\frac{\partial N_I}{\partial x_i}\, dR \end{bmatrix},$$

$$f^*_I = \left\{\begin{matrix} \sum_e \int_{R^e} GN_I\, dR + \sum_e \int_{B_2^e} E_{ij} \frac{\partial \hat{u}}{\partial x_j} n_i N_I\, dB \\ -\sum_e \int_{R^e} M_{ij}\rho_0^l g_j \frac{\partial N_I}{\partial x_i}\, dR + \sum_e \int_{B_2^e} M_{ij}\left(\frac{\partial \hat{p}}{\partial x_j} + \rho^l g_j\right) n_i N_I\, dB \end{matrix}\right\}. \tag{5.6.2.21d}$$

The coupling between Eqs. (5.6.2.19a) and (5.6.2.19b) is evident in the elements of the matrices of (5.6.2.20). Because the off-diagonal elements of the time derivatives coefficient matrix vanish, the equations are not explicitly coupled through the time derivative. This was not the case in the simulation of multiphase flow or saltwater intrusion using a sharp interface assumption. The coefficient matrix $[S^*]$ (5.6.2.21c) appears semicoupled. This is indicated by the off-diagonal zero element in the first row. Careful inspection of this equation, however, reveals that this statement of the problem is somewhat arbitrary. The coefficient B_i, defined through (5.6.1.2b) as

$$B_i = -V_i, \tag{5.6.2.22}$$

can be written using Darcy's law as

$$B_i = \frac{k_{ij}}{\mu}\left(\frac{\partial p}{\partial x_j} + \rho^l g_j\right). \tag{5.6.2.23}$$

Thus, there is a nonzero coefficient identified with the pressure. However, B_i multiplies the concentration, and therefore the term is nonlinear. We have, albeit somewhat arbitrarily, elected to lag the velocity dependent coefficient B_i; alternatively we could have produced a full matrix in (5.6.2.21c) by lagging the concentration in this term. The solution of the weakly nonlinear equations described by (5.6.2.20) is readily achieved using Picard iteration. Dealing with a complicated flow field, such as that encountered in coastal aquifer simulation where the velocity may change direction radically within a short distance, calls for a fine element network.

5.6.3 *General Remarks on Numerical Solution of the Transport Equation*

It is well known that the advective–dispersive transport equation is more difficult to solve numerically than the flow equation. The problems are particularly severe when advection dominates over dispersion. In this situation, the Galerkin finite element solution usually exhibits numerical spatial oscillations (overshoot and undershoot) near the concentration front. Overshoot describes the erroneously high values of concentration encountered upstream of the moving front. The analogous behavior on the downstream side is called undershoot. These numerical oscillations tend to be more severe as the front becomes sharper (i.e., as the advection becomes more dominant). A theoretical explanation of such behavior can be found in Pinder and Gray (1977, pp. 150–169). Using Fourier series representations of one-dimensional analytical and numerical solutions, it is shown that the numerical difficulties are caused by the inability of the numerical approximation to propagate accurately short wavelength harmonics of the Fourier series. When the value of the physical dispersion coefficient is decreased, these short wavelength harmonics become more important to the correct description of the front. Theoretical investigation (Price *et al.*, 1966) and our experience indicate that in a case where the dispersion coefficient D is greater than zero, numerical oscillations in the Galerkin finite element solution using linear basis functions can be virtually eliminated if the element size is selected so that its local Peclet number does not exceed 2. This local element Peclet number is defined as $Pe = V\,\Delta l/D$, where Δl is the characteristic length of the finite element. For a one-dimensional problem with linear elements, Δl corresponds to Δx. For a two-dimensional problem Δl can be selected as $\max(\Delta x, \Delta y)$. In most cases involving nonuniform flow, acceptable numerical solutions with very mild oscillations are achieved even when the local Peclet number is as high as 10. For example, in Pinder and Gray (1977, p. 163), a Galerkin finite element solution obtained using $Pe = 0.369/0.069 = 5.35$ is shown to be oscillation free and in excellent agreement with the exact analytical solution.

In situations where advection-dominated problems must be solved and computational costs are not of prime importance, a simple guideline for selecting the finite element mesh and time step size can be offered. The element size should be selected so that $\Delta l \leqslant 10D/V$, and the time-step size Δt should be selected so that the local Courant number, defined as $Cr = V\Delta t/\Delta l$, is less than or equal to 1, i.e., the Δt selected should be such that $\Delta t \leqslant \Delta l/V$. To obtain a solution that is second-order accurate in time, the Crank–Nicolson time stepping, $\theta = 0.5$, should be used.

Although one can eliminate numerical oscillations by refining the finite element grid, there are also many cases where the computer cost and storage resulting from using very refined grids become economically and computationally excessive. For these cases, the use of an upstream-weighted finite element technique, described in the next section, is recommended. It should be noted, however, that this upstream weighting, like all upstream-weighted finite difference techniques, curbs numerical oscillations by introducing numerical dispersion and thus leads to smearing of the concentration front. Because advection-dominated transport problems are difficult to solve numerically, the user of computer models needs to test the adequacy of selected mesh and time-step discretizations to avoid obtaining misleading or meaningless numerical solutions. For this reason, we recommend the use of available analytical solutions for calibration. Some useful analytical solutions can be found in the references cited next. For solute transport in saturated media, a comprehensive list of simple one-dimensional and two-dimensional analytical solutions is presented in Bear (1979, pp. 263–276). Other supplementary solutions are also available in Fried (1975, pp. 130–133). Specialized solutions for adsorptive porous media are presented in Cameron and Klute (1977), Cleary and Adrian (1973), Coats and Smith (1964), and van Genuchten and Wierenga (1976). Several of these solutions have been programmed for the computer by Yeh (1981). A one-dimensional solution for solute transport in fractured media is also available in Rasmuson and Neretneiks (1981) and Tang *et al.* (1981).

For solute transport in unsaturated media, a number of analytical solutions can be found in van Genuchten (1982) and Warrick *et al.* (1971). For transport of radionuclide chains, one-dimensional solutions can be found in Lester *et al.* (1975). More complex two- and three-dimensional solutions can be found in Harada *et al.* (1980) and Pigford *et al.* (1980).

For energy transport in single-phase flow, one-dimensional and radial flow solutions are presented in Avdonin (1964) and Carslaw and Jaeger (1959, pp. 385–391). More complex solutions involving energy transport via a well doublet system and transport in fractured media can be found in Gringarten *et al.* (1975) and Gringarten and Sauty (1975).

5.6.4 Upstream Weighted Finite Element Formulation

For completeness, we now present in brief the upstream weighted finite element formulation mentioned earlier. Once again, consider the general form of the solute transport equation. Equation (5.6.1.1) is now rewritten as

$$\frac{\partial}{\partial x_i}\left(E_{ij}\frac{\partial u}{\partial x_j}\right) + B_i\frac{\partial u}{\partial x_i} + A\frac{\partial u}{\partial t} + Fu + G = 0, \qquad (5.6.4.1)$$

where $u \equiv c_k$ and the coefficients A, B_i, E_{ij}, F, and G are defined by Eqs. (5.6.1.2a)–(5.6.1.2e).

In the upstream weighted finite element formulation employed by Huyakorn and Nilkuha (1979), a weighted residual approximation of Eq. (5.6.4.1) is obtained as follows:

$$\sum_e \int_{R^e} W_I \left[\frac{\partial}{\partial x_i}\left(E_{ij} \frac{\partial \hat{u}}{\partial x_j} \right) + B_i \frac{\partial \hat{u}}{\partial x_i} \right] dR$$

$$+ \sum_e \int_{R^e} N_I \left[A \frac{\partial \hat{u}}{\partial t} + F\hat{u} + G \right] dR = 0, \qquad I = 1, 2, \ldots, n, \tag{5.6.4.2}$$

where the spatial derivative terms are weighted using an asymmetric upstream weighting function W_I and the remaining terms are weighted using the standard basis function N_I. For linear one-dimensional and two-dimensional rectangular elements, the weighting functions W_I are described in Section 4.7.3. Next, Green's theorem is applied in the usual manner to the first integral of Eq. (5.6.4.2). This and the definition (5.6.2.2a) for $\hat{u}$ lead to

$$\sum_e \int_{R^e} \left[-E_{ij} \frac{\partial W_I}{\partial x_i} \frac{\partial N_J}{\partial x_j} + B_i W_I \frac{\partial N_J}{\partial x_i} \right] u_J \, dR$$

$$+ \sum_e \int_{R^e} N_I \left[A N_J \frac{du_J}{dt} + F N_J u_J + G \right] dR$$

$$+ \sum_e \int_{B_2^e} \left(E_{ij} \frac{\partial \hat{u}}{\partial x_j} n_i \right) W_I \, dB = 0, \qquad I = 1, 2, \ldots, n. \tag{5.6.4.3}$$

The remaining steps in the finite element solution are performed in the same manner as described in Section 5.6.2.

Experience in applying the upstream weighting functions using rectangular elements indicates that to obtain a satisfactory solution it is necessary to evaluate the two-dimensional weighting functions and their derivatives as follows:

$$\frac{\partial W_I}{\partial x}(\alpha_i, 0) \quad \text{replaces} \quad \frac{\partial W_I}{\partial x}(\alpha_i, \beta_j), \tag{5.6.4.4a}$$

$$\frac{\partial W_I}{\partial y}(0, \beta_j) \quad \text{replaces} \quad \frac{\partial W_I}{\partial y}(\alpha_i, \beta_j), \tag{5.6.4.4b}$$

$$V_x W_I(\alpha_i, 0) \quad \text{replaces} \quad V_x W_I(\alpha_i, \beta_j), \tag{5.6.4.4c}$$

$$V_y W_I(0, \beta_j) \quad \text{replaces} \quad V_y W_I(\alpha_i, \beta_j), \tag{5.6.4.4d}$$

where x and y are the coordinates of the element and α_i and β_j are upstream weighting factors in the x and y directions, respectively.

In other words, according to Eqs. (5.6.4.4a) and (5.6.4.4b), the derivatives of W_I are evaluated such that when differentiation is taken with respect to one particular coordinate, the values of the upstream parameters along the other coordinate are set to zero. Similarly, according to Eqs. (5.6.4.4c) and (5.6.4.4d), W_I is evaluated so that when it is associated with the velocity component in one particular coordinate, the values of the upstream parameters along the other coordinate are set to zero. Examples showing the behavior of the numerical solutions obtained using W_I can be found in Huyakorn and Nilkuha (1979).

5.6.5 Material Balance Calculation

Because of the aforementioned difficulties in the numerical solution of the convection-dominated transport equation, a calculation of mass balance over the entire solution region should be provided as an option in the computer program. It is essential that the resulting mass balance errors be sufficiently small for the numerical solutions of both the flow and transport equations. A procedure for computing mass balance error is described next for the solute transport equation. This same procedure can easily be adapted to the flow and energy transport equations. Consider as an example the transport of a nonconservative chemical species in a two-dimensional region. For clarity, the governing equation (5.2.3.1) is written in the form

$$\frac{\partial}{\partial x_i}\left[\phi D_{ij}\frac{\partial c}{\partial x_j}\right] - \frac{\partial}{\partial x_i}(V_i c) - \frac{\partial}{\partial t}(\phi\kappa c) - \phi\lambda\kappa c - M = 0, \qquad (5.6.5.1)$$

where M is included to take into account the contribution of point sinks (production wells) that may exist in the solution domain. In general, M can be expressed as

$$M = \sum_{I'=1}^{n_w} Q_{I'}\delta(\mathbf{x} - \mathbf{x}_{I'})\bar{c}_{I'}, \qquad (5.6.5.2)$$

where $Q_{I'}$ and $\bar{c}_{I'}$ are the flow rate and concentration at well I', n_w is the number of wells, and δ is the Dirac delta function.

The mass balance over the whole region R is obtained by integrating (5.6.5.1) and applying Green's (divergence) theorem to both the dispersive and advective terms. Thus, one obtains

$$-\int_B \left(V_i c - \phi D_{ij}\frac{\partial c}{\partial x_j}\right) n_i \, dB - \int_R \frac{\partial}{\partial t}(\phi\kappa c)\, dR$$

$$-\int_R \phi\lambda\kappa \, dR - \sum_{I'=1}^{n_w} (Q_{I'}\cdot\bar{c}_{I'}) = 0. \qquad (5.6.5.3)$$

It should be noted that the first integral in (5.6.5.3) represents the net material flux across the whole boundary. The second and third integrals represent, respectively, the rate of mass storage and the rate of mass decay in region R. The last term represents the net rate of mass production owing to well injection and pumping.

If the exact solution of the transport equation (5.6.5.1) is substituted into (5.6.5.3), that equation will be satisfied exactly. However, if an approximate finite element solution $\hat{c} = c_J(t)N_J(\mathbf{x})$ is substituted, the left-hand side of (5.6.5.3) will sum to a nonzero result and will correspond to the rate of material loss $\dot{\varepsilon}_M$:

$$\dot{\varepsilon}_{\mathrm{M}} \equiv \int_B -\left(V_i\hat{c} - \phi D_{ij}\frac{\partial \hat{c}}{\partial x_j}\right)n_i\, dB - \int_R \phi\kappa N_J\left(\frac{dc_J}{dt} + \lambda c_J\right) dR - \sum_{I'=1}^{n_{\mathrm{W}}} Q_{I'}\cdot c_{I'}. \tag{5.6.5.4}$$

Next, we make use of the following properties of the basis functions:

$$\sum_{I^*=1}^{n_{\mathrm{B}}} N_{I^*}(\mathbf{x}) = 1 \qquad \text{for all } \mathbf{x} \text{ on } B, \tag{5.6.5.5}$$

$$\sum_{I=1}^{n} N_I(\mathbf{x}) = 1 \qquad \text{for all } \mathbf{x} \text{ in } R, \tag{5.6.5.6}$$

where n_B is the number of nodes on the whole boundary and n is the total number of nodes in the whole solution region. Combination of Eqs. (5.6.5.4)–(5.6.5.6) yields

$$\dot{\varepsilon}_{\mathrm{M}} = -\sum_{I^*=1}^{n_{\mathrm{B}}} \int_B N_{I^*}\left(V_i\hat{c} - \phi D_{ij}\frac{\partial \hat{c}}{\partial x_j}\right)n_i\, dB - \sum_{I=1}^{n} \int_R \phi\kappa N_I N_J\left(\frac{dc_J}{dt} + \lambda c_J\right)dR - \sum_{I'=1}^{n_{\mathrm{W}}} Q_{I'}c_{I'}. \tag{5.6.5.7}$$

Equation (5.6.5.7) can be expressed in a simple form as

$$\dot{\varepsilon}_{\mathrm{M}} = -\sum_{I^*=1}^{n_{\mathrm{B}}} F_{I^*}^{\mathrm{B}} - \sum_{I=1}^{n} F_I^{\mathrm{S}} - \sum_{I'=1}^{n_{\mathrm{W}}} F_{I'}^{\mathrm{W}}, \tag{5.6.5.8}$$

where

$$F_{I^*}^{\mathrm{B}} = \int_B N_{I^*}\left(V_i\hat{c} - \phi D_{ij}\frac{\partial \hat{c}}{\partial x_j}\right)n_i\, dB, \tag{5.6.5.9a}$$

$$F_I^s = \int_R \phi \kappa N_I N_J \left(\frac{dc_J}{dt} + \lambda c_J \right) dR, \qquad (5.6.5.9b)$$

$$F_{I'}^w = Q_{I'} \tilde{c}_{I'}. \qquad (5.6.5.9c)$$

It is apparent that the evaluation of F_I^s and $F_{I'}^w$ is straightforward. Note in particular that considerable savings in computational time can be achieved by lumping the mass matrix in Eq. (5.6.5.9b).

The evaluation of the boundary material flux is more involved and requires further elaboration. In a case involving a third-type boundary condition, the total material flux distribution is prescribed as

$$\left(V_i \hat{c} - \phi D_{ij} \frac{\partial \hat{c}}{\partial x_j} \right) n_i = q_c^{\mathrm{T}},$$

and thus $F_{I^*}^B$ is given explicitly by

$$F_{I^*}^{\mathrm{B}} = \int_B N_{I^*} q_c^{\mathrm{T}} \, dB. \qquad (5.6.5.10a)$$

In a case involving a second-type boundary condition, the outward dispersive flux distribution is prescribed as

$$-\phi D_{ij} \frac{\partial \hat{c}}{\partial x_j} n_i = q_c^{\mathrm{D}} \qquad (5.6.5.10b)$$

and the outward advective flux is obtainable from

$$V_i n_i \hat{c} = q N_J c_J, \qquad (5.6.5.10c)$$

where q is the outward normal fluid flux distribution. Equations (5.6.5.10b) and (5.6.5.10c) are combined with Eq. (5.6.5.9a) to yield

$$F_{I^*}^{\mathrm{B}} = \int_B N_{I^*} (q N_J c_J + q_c^{\mathrm{D}}) \, dB$$

$$\simeq c_{I^*} \int_B N_{I^*} q \, dB + \int_B N_{I^*} q_c^{\mathrm{D}} \, dB, \qquad (5.6.5.10d)$$

where the summation convention is not applied to subscript I^* and the boundary integral corresponds to the nodal fluid flux at boundary node I^*. Such a nodal flux should be evaluated a priori, in the finite element solution of the governing equation for fluid flow. Its evaluation should be performed by back substitution of the computed hydraulic head (or pressure) into the Galerkin finite element equation written for node I^*. This procedure avoids the problem of discontinuities in the velocity field and hence discontinuities of q at the boundary nodes. Such discontinuities arise if one employs Darcy's law to compute velocities at the boundary nodes of individual elements and uses these velocities to obtain q. In a case involving a first-type boundary condition, the dispersive boundary nodal flux is not known explicitly and should be computed by back substitution of $\hat{c}$ into the nonboundary integral terms of the original finite element approximation of the transport equation at node I^* [that is, Eq. (5.6.2.7a) with I replaced by I^*]. The net result is equal to the dispersive boundary integral. Once the dispersive nodal flux has been computed, the total boundary nodal flux is obtainable from

$$F_{I^*}^{\mathrm{B}} \simeq c_{I^*} \int_B N_{I^*} q \, dB + Q_{cI^*}^{\mathrm{D}}. \tag{5.6.5.10e}$$

Finally, when all the various material flux terms have been computed, the rate of mass loss $\dot{\varepsilon}_{\mathrm{M}}$ can be determined from Eq. (5.6.5.8). In practice, the cumulative mass balance error at a current time level, $\varepsilon_{\mathrm{M}}^{k+1}$, should also be computed. It is given by

$$\varepsilon_{\mathrm{M}}^{k+1} = \sum_{l=1}^{k} (\dot{\varepsilon}_{\mathrm{M}} \Delta t)^l, \tag{5.6.5.11}$$

where l is the time level.

Either $\varepsilon_{\mathrm{M}}^{k+1}$ or its normalized form $\tilde{\varepsilon}_{\mathrm{M}}^{k+1}$ can be used as an indicator of the global accuracy of the numerical solution of the transport equation. For convenience, we define $\tilde{\varepsilon}_{\mathrm{M}}^{k+1}$ as

$$\tilde{\varepsilon}_{\mathrm{M}}^{k+1} = \sum_{l=1}^{k} \left\{ (\dot{\varepsilon}_{\mathrm{M}} \Delta t)^l \Big/ \left[\Delta t \left(\sum_{I=1}^{n} |F_I^{\mathrm{S}}| + \sum_{I'=1}^{n_{\mathrm{W}}} |F_{I'}^{\mathrm{W}}| + \sum_{I^*=1}^{n_{\mathrm{B}}} |F_{I^*}^{\mathrm{B}}| \right) \right] \right\}.$$

5.7 Finite Element Simulation of Mass and Energy Transport in Multiphase Flow

5.7.1 Isothermal Multiphase Species Transport

The simulation of isothermal species transport in multiphase flow is a relatively straightforward extension of the more restrictive single-phase flow case. The appropriate equation is (5.2.1.1) augmented with suitable constitutive relations. The velocity field, assumed known when writing (5.2.1.1), is obtained from the multiphase flow equations. The important concept that should be recognized in the simulation of multiphase transport is that the multiphase medium is a continuum and each species has a single equation that should be used for the entire flow domain. One should not think of the domain as one consisting of seperate zones, each containing one fluid and separated by sharp interfaces.

Because we considered the cases of variably saturated and multiphase flow in Sections 4.5–4.7 and discussed species transport in Section 5.6, we shall not detail the procedure. Rather, we now turn to the more challenging and difficult problem of non-isothermal flow and transport.

5.7.2 Nonisothermal Multiphase Flow: The Geothermal Simulation Problem

In this section we focus on the specific problem of geothermal simulation. First, we consider (5.5.1.7)–(5.5.1.9). Combined in various ways these equations yield for flow

$$\frac{\partial \phi\rho}{\partial t} + \frac{\partial}{\partial x_i}(\rho^g V_i^g + \rho^l V_i^l) = 0. \tag{5.7.2.1a}$$

Substitution of Darcy's law yields

$$\frac{\partial}{\partial t}\phi\rho - \frac{\partial}{\partial x_i}\left[\tau_{ij}^g\left\{\frac{\partial p}{\partial x_j} + \rho^g g_j\right\} + \tau_{ij}^l\left\{\frac{\partial p}{\partial x_j} + \rho^l g_j\right\}\right] = 0. \tag{5.7.2.1b}$$

where

$$\tau_{ij}^\alpha \equiv \rho^\alpha k_{ij} k_r^\alpha / \mu^\alpha.$$

The energy equation, written in terms of pressure and enthalpy (5.5.1.15), is

$$(1-\phi)\rho^s \frac{\partial h^s}{\partial t} + \frac{\partial}{\partial t}\phi\rho h + \frac{\partial}{\partial x_i} V_i^l \rho^l h^l + \frac{\partial}{\partial x_i} V_i^g \rho^g h^g$$

$$- \frac{\partial}{\partial x_i}\left[K\left(\frac{\partial T}{\partial h}\frac{\partial h}{\partial x_i} + \frac{\partial T}{\partial p}\frac{\partial p}{\partial x_i}\right)\right] = 0. \tag{5.7.2.2a}$$

Substitution of Darcy's law in the convection terms (the third and fourth) yields

$$(1-\phi)\rho^s \frac{\partial h^s}{\partial t} + \frac{\partial}{\partial t}\phi\rho h - \frac{\partial}{\partial x_i}\left[\tau_{ij}^l\left(\frac{\partial p}{\partial x_j} + \rho^l g_j\right)h^l\right]$$

$$-\frac{\partial}{\partial x_i}\left[\tau_{ij}^g\left(\frac{\partial p}{\partial x_j} + \rho^g g_j\right)h^g\right] - \frac{\partial}{\partial x_i}\left[K\left(\frac{\partial T}{\partial h}\frac{\partial h}{\partial x_i} + \frac{\partial T}{\partial p}\frac{\partial p}{\partial x_i}\right)\right]$$

$$= 0. \tag{5.7.2.2b}$$

We now adopt the notation of Huyakorn and Pinder (1977) and rewrite

(5.7.2.1b) and (5.7.2.2b) as

$$\frac{\partial}{\partial x_i}\left(\tau_{ij}\frac{\partial p}{\partial x_j}\right) + \frac{\partial G_i}{\partial x_i} - \frac{\partial F}{\partial t} = 0, \qquad (5.7.2.3a)$$

$$\frac{\partial}{\partial x_i}\left(\lambda_{ij}\frac{\partial p}{\partial x_j}\right) + \frac{\partial}{\partial x_i}\left(\beta\frac{\partial h}{\partial x_i}\right) + \frac{\partial P_i}{\partial x_i} - \frac{\partial H}{\partial t} = 0, \qquad (5.7.2.3b)$$

where

$$\tau_{ij} = \tau_{ij}^l + \tau_{ij}^g,$$

$$F = \phi\rho,$$

$$H = \phi\rho h + (1 - \phi)\rho^s h^s,$$

$$\lambda_{ij} = K\frac{\partial T}{\partial p}\delta_{ij} + \tau_{ij}^l h^l + \tau_{ij}^g h^g \qquad (h^l \leqslant h \leqslant h^g) \text{ (two-phase region)},$$

$$\lambda_{ij} = K\frac{\partial T}{\partial p}\delta_{ij} + \tau_{ij}^l h^l + \tau_{ij}^g h^g \qquad (h^l \leqslant h \leqslant h^g) \text{ (two-phase region)},$$

$$\beta = K\frac{\partial T}{\partial h},$$

$$G_i = \tau_{ij}^l\rho^l g_j + \tau_{ij}^g\rho^g g_j,$$

$$P_i = \tau_{ij}^g h^g\rho^g g_j + \tau_{ij}^l h^l\rho^l g_j.$$

With (5.7.2.3) as our governing equations we are now in a position to develop our finite element model. We begin by representing the dependent variables using the finite series

$$p(\mathbf{x}, t) \simeq \hat{p}(\mathbf{x}, t) = p_J(t)N_J(\mathbf{x}), \qquad (5.7.2.4a)$$

$$h(\mathbf{x}, t) \simeq \hat{h}(\mathbf{x}, t) = h_J(t)N_J(\mathbf{x}), \qquad (5.7.2.4b)$$

where $N_J(\mathbf{x})$ are suitable basis or shape functions and repeated indices imply summation. Moreover, the following variables are also interpolated elementwise over the domain using the same shape functions:

$$\rho(\mathbf{x}, t) \simeq \hat{\rho}(\mathbf{x}, t) = \rho_J(t)N_J(\mathbf{x}), \qquad (5.7.2.4c)$$

$$F(\mathbf{x}, t) \simeq \hat{F}(\mathbf{x}, t) = F_J(t)N_J(\mathbf{x}), \qquad (5.7.2.4d)$$

$$\lambda(\mathbf{x}, t) \simeq \hat{\lambda}(\mathbf{x}, t) = \lambda_J(t)N_J(\mathbf{x}), \qquad (5.7.2.4e)$$

$$\beta(\mathbf{x}, t) \simeq \hat{\beta}(\mathbf{x}, t) = \beta_J(t)N_J(\mathbf{x}), \tag{5.7.2.4f}$$

$$H(\mathbf{x}, t) \simeq \hat{H}(\mathbf{x}, t) = H_J(t)N_J(\mathbf{x}), \tag{5.7.2.4g}$$

$$G(\mathbf{x}, t) \simeq \hat{G}(\mathbf{x}, t) = G_J(t)N_J(\mathbf{x}), \tag{5.7.2.4h}$$

$$P(\mathbf{x}, t) \simeq \hat{P}(\mathbf{x}, t) = P_J(t)N_J(\mathbf{x}). \tag{5.7.2.4i}$$

Let us now write (5.7.2.3) combined with (5.7.2.4) using a generalized weighted residual formulation. Specifically, we use asymmetric weighting functions for the spatial derivatives and symmetric weighting functions for the time derivatives. We denote the asymmetric (upstream) weighting function $W_J(\mathbf{x})$ (see Section 4.7.3), and the shape functions $N_J(\mathbf{x})$ are used for the symmetric weighting. The functional forms of $W_J(\mathbf{x})$ are given in Table 4.1. The proposed weighted residual scheme is

$$\sum_e \int_{R^e} \left[W_I \frac{\partial}{\partial x_i}\left(\tau_{ij} \frac{\partial \hat{p}}{\partial x_j}\right) + N_I \left(\frac{\partial \hat{G}_i}{\partial x_i} - \frac{\partial \hat{F}}{\partial t}\right)\right] dR = 0, \qquad I = 1, 2, \ldots, n, \tag{5.7.2.5a}$$

$$\sum_e \int_{R^e} \left\{ W_I \left[\frac{\partial}{\partial x_i}\left(\lambda_{ij} \frac{\partial \hat{p}}{\partial x_j}\right) + \frac{\partial}{\partial x_i}\left(\beta \frac{\partial \hat{h}}{\partial x_i}\right)\right] + N_I \left(\frac{\partial \hat{P}_i}{\partial x_i} - \frac{\partial \hat{H}}{\partial t}\right)\right\} dR = 0, \qquad I = 1, 2, \ldots, n. \tag{5.7.2.5b}$$

These equations are now transformed by applying Green's theorem to the second-derivative terms:

$$\sum_e \int_{R^e} \left[\frac{\partial W_I}{\partial x_i} \tau_{ij} \frac{\partial \hat{p}}{\partial x_j} - N_I\left(\frac{\partial \hat{G}_i}{\partial x_i} - \frac{\partial \hat{F}}{\partial t}\right)\right] dR - \int_{B_2^e} W_I \tau_{ij} \frac{\partial \hat{p}}{\partial x_j} n_i \, dB = 0, \qquad I = 1, 2, \ldots, n, \tag{5.7.2.6a}$$

$$\sum_e \int_{R^e} \left[\frac{\partial W_I}{\partial x_i} \left(\lambda_{ij} \frac{\partial \hat{p}}{\partial x_j} + \beta \frac{\partial \hat{h}}{\partial x_i}\right) - N_I\left(\frac{\partial \hat{P}_i}{\partial x_i} - \frac{\partial \hat{H}}{\partial t}\right)\right] dR - \int_{B_2^e} W_I\left(\lambda_{ij} \frac{\partial \hat{p}}{\partial x_j} + \beta \frac{\partial \hat{h}}{\partial x_i}\right) n_i \, dB = 0, \qquad I = 1, 2, \ldots, n. \tag{5.7.2.6b}$$

The boundary integration (5.7.2.6a) is the fluid flux flowing across the region perimeter B_2^e. To identify the meaning of the boundary integral of (5.7.2.6b), we rewrite it as

$$\sum_e \int_{B_2^e} W_I\left(\lambda_{ij}\frac{\partial \hat{p}}{\partial x_j} + \beta\frac{\partial \hat{h}}{\partial x_i}\right)n_i\,dB = \sum_e \int_{B_2^e} W_I K\frac{\partial \hat{T}}{\partial x_i}n_i\,dB$$

$$+\sum_e \int_{B_2^e} W_I(\tau_{ij}^l h^l + \tau_{ij}^g h^g)\frac{\partial \hat{p}}{\partial x_j}n_i\,dB, \qquad I = 1, 2, \ldots, n. \qquad (5.7.2.7)$$

The first and second terms on the right-hand side of (5.7.2.7) describe the conductive and convective energy fluxes, respectively, across B_2^e. Care must be exercised in the evaluation of these terms to avoid numerical instabilities. Notice that in applying Green's theorem to the term

$$\sum_e \int_{R^e} W_I\frac{\partial}{\partial x_i}\left(\lambda_{ij}\frac{\partial \hat{p}}{\partial x_j}\right)dR$$

we have transformed a convective term. In our discussion of species transport we noted that this transformation may lead to numerical difficulties. A more stable formulation may arise when the convective term is not included in the Green's theorem transformation. This approach would result from the use of (5.7.2.2a) written for enthalpy and pressure before it is transformed using the conservation of fluid mass equation.

The final set of nonlinear algebraic equations is obtained through substitution of (5.7.2.4) into (5.7.2.6) and finite differencing of the time derivative:

$$\sum_e \int_{R^e}\left[\frac{\partial W_I}{\partial x_i}\tau_{ij}p_J\frac{\partial N_J}{\partial x_j} - N_I\frac{\partial \hat{G}_i}{\partial x_i} + N_I(\hat{F}^{k+1} - \hat{F}^k)\Big/\Delta t\right]dR$$

$$-\sum_e \int_{B_2^e} W_I\tau_{ij}\frac{\partial \hat{p}}{\partial x_j}n_i\,dB = 0, \qquad I = 1, 2, \ldots, n, \qquad (5.7.2.8a)$$

$$\sum_e \int_{R^e}\left[\left(\frac{\partial W_I}{\partial x_i}\left(\lambda_{ij}p_J\frac{\partial N_J}{\partial x_j} + \beta h_J\frac{\partial N_J}{\partial x_i}\right)\right.\right.$$

$$\left.-N_I\frac{\partial \hat{P}_i}{\partial x_i} + N_I\left(\hat{H}^{k+1} - \hat{H}^k\right)\Big/\Delta t\right]dR$$

$$-\sum_e \int_{B_2^e} W_I\left(K\frac{\partial \hat{T}}{\partial x_i} + \left[\tau_{ij}^l h^l + \tau_{ij}^g h^g\right]\frac{\partial \hat{p}}{\partial x_j}\right)n_i\,dB, = 0, \quad I = 1, 2, \ldots, n.$$

$$(5.7.2.8b)$$

Let us now define two residuals of the form

$$R_I \equiv \sum_e \int_{R^e} \left[\frac{\partial W_I}{\partial x_i} \tau_{ij} p_J \frac{\partial N_J}{\partial x_j} - N_I \frac{\partial \hat{G}_i}{\partial x_i} + N_I \left(\hat{F}^{k+1} - \hat{F}^k \right) \Big/ \Delta t \right] dR, \qquad I = 1, 2, \ldots, n, \tag{5.7.2.9a}$$

$$G_I \equiv \sum_e \int_{R^e} \left[\frac{\partial W_I}{\partial x_i} \left(\lambda_{ij} p_J \frac{\partial N_J}{\partial x_j} + \beta h_J \frac{\partial N_J}{\partial x_i} \right) - N_I \frac{\partial \hat{P}_i}{\partial x_i} + N_I \left(\hat{H}^{k+1} - \hat{H}^k \right) \Big/ \Delta t \right] dR, \qquad I = 1, 2, \ldots, n. \tag{5.7.2.9b}$$

The boundary terms omitted from (5.7.2.9) will be considered separately.

Recall that in Section 4.6.2 we discussed in considerable detail a methodology for the solution of nonlinear equations called the Newton–Raphson method. The scheme is widely used to solve equations of the form of (5.7.2.8). The reader with a good memory may recall that the objective of the scheme is to reduce the residuals R_I and G_I to zero using a Taylor series expansion about a starting point, R_I^r, say, where r is an iteration index. In our particular problem the expansion takes the form

$$R_I^{r+1} = R_I^r + \left. \frac{\partial R_I}{\partial p_J} \right|_r (p_J^{r+1} - p_J^r)^{k+1} + \left. \frac{\partial R_I}{\partial h_J} \right|_r (h_J^{r+1} - h_J^r)^{k+1} = 0, \tag{5.7.2.10a}$$

$$G_I^{r+1} = G_I^r + \left. \frac{\partial G_I}{\partial p_J} \right|_r (p_J^{r+1} - p_J^r)^{k+1} + \left. \frac{\partial G_I}{\partial h_J} \right|_r (h_J^{r+1} - h_J^r)^{k+1} = 0. \tag{5.7.2.10b}$$

To solve (5.7.2.10) for the iterative increments appearing in the parentheses, one must evaluate the Jacobian matrix that consists of elements of the form $\partial G_I / \partial p_J$ and so on. For the equations of concern these derivatives are

$$\frac{\partial R_I}{\partial p_J} = \sum_e \int_{R^e} \left\{ \frac{\partial W_I}{\partial x_i} \left[\tau_{ij} \frac{\partial N_J}{\partial x_j} + \frac{\partial \tau_{ij}}{\partial p_J} \frac{\partial \hat{p}^{k+1}}{\partial x_j} \right] - N_I \left(\frac{\partial G_i}{\partial p} \right)_J \frac{\partial N_J}{\partial x_i} + \frac{N_I}{\Delta t} \frac{\partial \hat{F}^{k+1}}{\partial p_J} \right\} dR, \tag{5.7.2.11a}$$

$$\frac{\partial R_I}{\partial h_J} = \sum_e \int_{R^e} \left[\frac{\partial W_I}{\partial x_i} \frac{\partial \tau_{ij}}{\partial h_J} \frac{\partial \hat{p}^{k+1}}{\partial x_j} - N_I \left(\frac{\partial G_i}{\partial h} \right)_J \frac{\partial N_J}{\partial x_i} + \frac{N_I}{\Delta t} \frac{\partial \hat{F}^{k+1}}{\partial h_J} \right] dR, \tag{5.7.2.11b}$$

$$\frac{\partial G_I}{\partial p_J} = \sum_e \int_{R^e} \left[\frac{\partial W_I}{\partial x_i} \left\{ \lambda_{ij} \frac{\partial N_J}{\partial x_j} + \frac{\partial \lambda_{ij}}{\partial p_J} \frac{\partial \hat{p}^{k+1}}{\partial x_j} + \frac{\partial \beta}{\partial p_J} \frac{\partial \hat{h}^{k+1}}{\partial x_i} \right\} - N_I \left(\frac{\partial P_i}{\partial p} \right)_J \frac{\partial N_J}{\partial x_i} + \frac{N_I}{\Delta t} \frac{\partial \hat{H}^{k+1}}{\partial p_J} \right] dR, \tag{5.7.2.11c}$$

$$\frac{\partial G_I}{\partial h_J} = \sum_e \int_{R^e} \left[\frac{\partial W_I}{\partial x_i} \left\{ \frac{\partial \lambda_{ij}}{\partial h_J} \frac{\partial \hat{p}^{k+1}}{\partial x_j} + \beta \frac{\partial N_J}{\partial x_i} + \frac{\partial \beta}{\partial h_J} \frac{\partial \hat{h}^{k+1}}{\partial x_i} \right\} - N_I \left(\frac{\partial P_i}{\partial h} \right)_J \frac{\partial N_J}{\partial x_i} + \frac{N_I}{\Delta t} \frac{\partial \hat{H}^{k+1}}{\partial h_J} \right] dR. \tag{5.7.2.11d}$$

The final set of linearized algebraic equations are obtained by combining Eqs. (5.7.2.9)–(5.7.2.11). If we define

$$\Delta p_J \equiv (p_J^{r+1} - p_J^r)^{k+1} \tag{5.7.2.12a}$$

$$\Delta h_J \equiv (h_J^{r+1} - h_J^r)^{k+1}, \tag{5.7.2.12b}$$

the resulting equations are

$$\sum_e \int_{R^e} \left[\frac{\partial W_I}{\partial x_i} \tau_{ij} \frac{\partial N_J}{\partial x_j} p^{k+1,r} - N_I \frac{\partial N_J}{\partial x_i} G_{iJ}^{k+1,r} + \frac{N_I}{\Delta t} (\hat{F}^{k+1,r} - \hat{F}^k) \right] dR$$

$$+ \sum_e \int_{R^e} \left[\frac{\partial W_I}{\partial x_i} \left\{ \tau_{ij} \frac{\partial N_J}{\partial x_j} + \frac{\partial \tau_{ij}}{\partial p_J} \frac{\partial \hat{p}^{k+1,r}}{\partial x_j} \right\} - N_I \left(\frac{\partial G_i}{\partial p} \right)_J \frac{\partial N_J}{\partial x_i} + \frac{N_I}{\Delta t} \frac{\partial \hat{F}^{k+1,r}}{\partial p_J} \right] dR \, \Delta p_J$$

$$+\sum_e \int_{R^e} \left[\frac{\partial W_I}{\partial x_i} \frac{\partial \tau_{ij}}{\partial h_J} \frac{\partial \hat{p}^{k+1,\, r}}{\partial x_j} - N_I \left(\frac{\partial G_i}{\partial h} \right)_J \frac{\partial N_J}{\partial x_i} \right.$$

$$\left. + \frac{N_I}{\Delta t} \frac{\partial \hat{F}^{k+1,\, r}}{\partial h_J} \right] dR\, \Delta h_J = 0, \tag{5.7.2.13a}$$

$$\sum_e \int_{R^e} \left[\frac{\partial W_I}{\partial x_i} \left(\lambda_{ij} p_J^{k+1,\, r} \frac{\partial N_J}{\partial x_j} + \beta h_J^r \frac{\partial N_J}{\partial x_i} \right) - N_I \frac{\partial N_J}{\partial x_i} P_{iJ}^{k+1,\, r} \right.$$

$$\left. + \frac{N_I}{\Delta t} (\hat{H}^{k+1,\, r} - \hat{H}^k) \right] dR$$

$$+\sum_e \int_{R^e} \left[\frac{\partial W_I}{\partial x_i} \left\{ \lambda_{ij} \frac{\partial N_J}{\partial x_j} + \frac{\partial \lambda_{ij}}{\partial p_J} \frac{\partial \hat{p}^{k+1,\, r}}{\partial x_j} + \frac{\partial \beta}{\partial p_J} \frac{\partial \hat{h}^{k+1,\, r}}{\partial x_i} \right\} \right.$$

$$\left. - N_I \left(\frac{\partial P_i}{\partial p} \right)_J \frac{\partial N_J}{\partial x_i} + \frac{N_I}{\Delta t} \frac{\partial \hat{H}^{k+1,\, r}}{\partial p_J} \right] dR\, \Delta p_J$$

$$+\sum_e \int_{R^e} \left[\frac{\partial W_I}{\partial x_i} \left\{ \frac{\partial \lambda_{ij}}{\partial h_J} \frac{\partial \hat{p}^{k+1,\, r}}{\partial x_j} + \beta \frac{\partial N_J}{\partial x_i} + \frac{\partial \beta}{\partial h_J} \frac{\partial \hat{h}^{k+1,\, r}}{\partial x_i} \right\} \right.$$

$$\left. - N_I \left(\frac{\partial P_i}{\partial h} \right)_J \frac{\partial N_J}{\partial x_i} + \frac{N_I}{\Delta t} \frac{\partial \hat{H}^{k+1,\, r}}{\partial h_J} \right] dR\, \Delta h_J = 0. \tag{5.7.2.13b}$$

One now assembles (5.7.2.13) into a matrix equation in the two unknown variables Δp_I and Δh_I. The equations are solved using a direct solution procedure and the solution used to determine $p_I^{k+1,\, r+1}$ and $h_I^{k+1,\, r+1}$ through (5.7.2.12). These estimates are now used to update the nonlinear functions in (5.7.2.13), and the entire process is repeated. This iterative procedure continues until the values of $p_J^{k+1,\, r+1}$ and $h_J^{k+1,\, r+1}$ are sufficiently close to $p_J^{k+1,\, r}$ and $h_J^{k+1,\, r}$ as defined using a convenient norm. One then proceeds to the next time step.

When the boundary integrals appearing in (5.7.2.6) are nonzero we must modify our formulation. Following the development by Huyakorn and Pinder (1977), we modify our residual operators to read

$$R_I^* = R_I + \hat{q}_I, \tag{5.7.2.14a}$$

$$G_I^* = G_I + \hat{Q}_I + \sum_e \int_{B_2^e} W_I (\tau_{ij}^l h^l + \tau_{ij}^g h^g) \frac{\partial \hat{p}}{\partial x_j} n_i\, dB, \tag{5.7.2.14b}$$

where

$$\hat{q}_I = -\sum_e \int_{B_2^e} W_I \tau_{ij} \frac{\partial \hat{p}}{\partial x_j} n_i\, dB,$$

$$\hat{Q}_I = -\sum_e \int_{B_2^e} W_I K \frac{\partial \hat{T}}{\partial x_i} n_i\, dB.$$

Although certain of these terms will be specified in the problem formulation, others may not be. The Newton–Raphson procedure must be applied to the unspecified terms just as it was for the earlier equations (5.7.2.9). The coefficient matrix and right-hand-side vector are subsequently modified to accommodate the new information from the boundary integrals.

The thermodynamic information required to close the system of equations can be obtained from tables or regression equations. A third alternative involves the use of catastrophe theory. The regression approach of Huyakorn and Pinder (1977) is summarized later.

Let us assume pressure is given in dyne/cm^2 $\times$ 10^{-7} (i.e., joule/cm^3) and enthalpy is expressed in joule/gm. The known pressure is used to calculate the saturated enthalpy of steam using the regression formula [see Mercer and Faust (1975)]:

$$h_*^g = 2822.82 - 39.952/p + 2.5434/p^2 - 0.93888p^2. \quad (5.7.2.15)$$

The saturated enthalpy of water, h_*^l, is obtained using linearly interpolated values extracted from the steam table. Knowing h_*^g and h_*^l, one can establish the fluid phase existing at the point in space and time of interest. If the computed values of pressure and enthalpy indicate the existence of a compressed water zone, then the density of water (gm/cm^3) is determined from (Mercer and Faust, 1975)

$$\rho^l = \begin{cases} 0.98988 + 4.0089 \times 10^{-4}p - 4.0049 \times 10^{-5}h + 2.6661/h \\ \quad + 5.4628 \times 10^{-7}ph - 1.2996 \times 10^{-7}h^2 \\ \quad \text{for} \quad h > 200, & (5.7.2.16\text{a}) \\ 1 \quad \text{for} \quad h \leqslant 200. & (5.7.2.16\text{b}) \end{cases}$$

When superheated steam is indicated one obtains

$$\rho^g = -2.2616 \times 10^{-5} + 0.043844p - 1.7909 \times 10^{-5}ph$$
$$+3.6928 \times 10^{-8}p^4 + 5.1764 \times 10^{-13}ph^3. \quad (5.7.2.17)$$

If a two-phase region is indicated, then the saturated enthalpy h_*^l is used for h in (5.7.2.16a) and h_*^g replaces h in (5.7.2.17).

Having calculated the saturated enthalpies and the densities one can now compute the water saturation from the relation

$$S^l = \begin{cases} 1 & \text{for} \quad h \leqslant h_*^l & (5.7.2.18\text{a}) \\ 0 & \text{for} \quad h \geqslant h_*^g & (5.7.2.18\text{b}) \\ \dfrac{\rho^g(h^g - h)}{h(\rho^l - \rho^g) - (h^l\rho^l - h^g\rho^g)} & \text{for} \quad h_*^l < g < h_*^g). & (5.7.2.18\text{c}) \end{cases}$$

The bulk density is now obtained from the equation

$$\rho = \rho^g(1 - S^l) + \rho^l S^l. \tag{5.7.2.19}$$

The fluid mixture temperature can now be calculated as follows (Mercer and Faust, 1975):

(1) two-phase region (Ramey *et al.*, 1974)

$$T = \psi_0(p) = \frac{4667.075}{12.599 - \ln(10p)} - 273.15; \tag{5.7.2.20a}$$

(2) compressed water zone (Mercer and Faust, 1975)

$$T = \psi_1(p, h) + [\psi_0(p) - \psi_1(p, h_*^l)],$$

$$\psi_1(p, h) = -28.152 - 0.13746p + 0.30112h + 3536.4/h$$
$$-4.3192 \times 10^{-5}h^2; \tag{5.7.2.20b}$$

(3) Superheated steam region:

$$T = \psi_2(p, h) + [\psi_0(p) - \psi_2(p, h_*^g)],$$

$$\psi_2(p, h) = -374.67 + 47.992p - 0.63361p^2 + 7.3939 \times 10^{-5}h^2$$
$$-3.3372 \times 10^6/(p^2h^2) + 0.035715/p^3 - 1.1725 \times 10^{-9}h^3p$$
$$-2.2686 \times 10^{15}/h^4. \tag{5.7.2.20c}$$

Having obtained the temperature, one can determine the fluid viscosity and rock enthalpy from the equations (Mercer and Faust, 1975, or Ramey *et al.*, 1974)

$$\mu^l = 10^{-6} \times 239.4 \times 10^{[248.37/(T + 133)]}, \tag{5.7.2.21a}$$

$$\mu^g = (0.407T + 80.4) \times 10^{-6}, \tag{5.7.2.21b}$$

$$h^s = (2041. + 8.79T) \times 10^6, \tag{5.7.2.21c}$$

where μ^α is given in grams per centimeter · second and h^s in Joules per gram.

There are little data available on the relative permeabilities for water and steam. A typical set of constitutive relations for these parameters is

$$k_r^l = (S^l)^2, \tag{5.7.2.22a}$$

$$k_r^g = (1 - S^l)^2. \tag{5.7.2.22b}$$

Another set of experimentally determined constitutive relations can be found in Corey *et al.* (1956). These relationships were used by Faust and Mercer (1979) in their geothermal simulation. The system of equations is now complete and simulation is conceptually possible. However, when one attempts to solve these equations, even with the Newton–Raphson scheme, difficulties arise. At the boundary between the liquid and two-phase region of the pressure–enthalpy diagram (Fig. 5.2) oscillatory behavior is encountered. Several schemes have been developed to overcome this problem [see Huyakorn and Pinder (1977) and Voss (1978)]. The approach of Huyakorn and Pinder (1977) follows:

> The derivatives with respect to the pressure variable are evaluated by taking a negative or backward increment if the point (p_J^r, h_J^r) lies in the two-phase zone and by taking a positive or forward increment if the point (p_J^r, h_J^r) lies in the single-phase zone. On the other hand, the derivatives with respect to the enthalpy variable are evaluated by taking a positive increment if (p_J^r, h_J^r) lies in the two-phase zone and by taking a negative increment if it lies in the single-phase zone.

There is an attractive alternative approach to overcoming the phase transition problem. This concept, recently introduced by Nguyen and Pinder (1980), is based on a branch of mathematical topology called

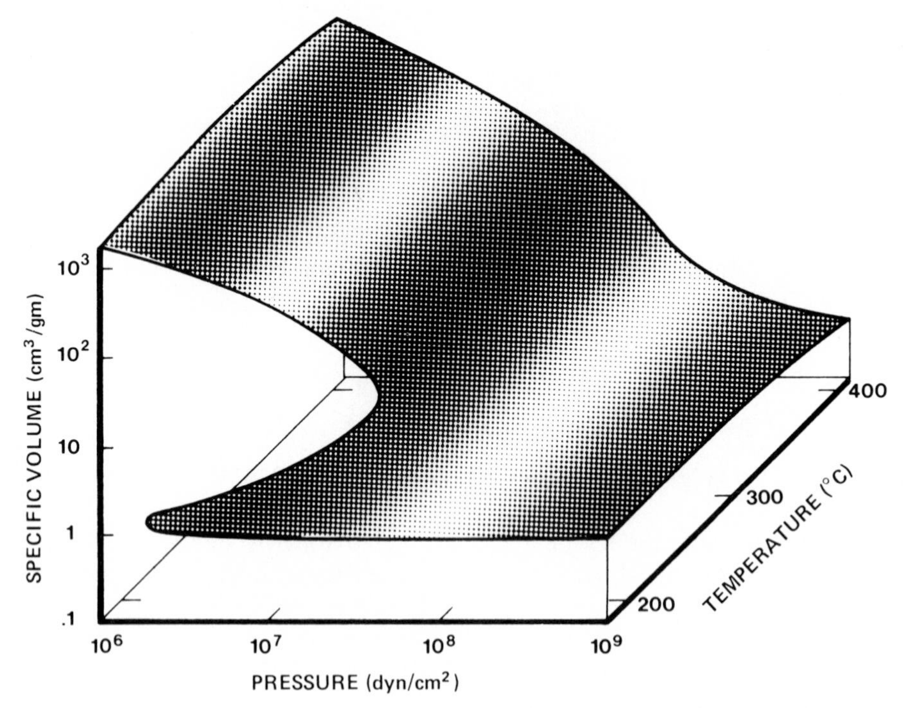

Fig. 5.3. Geometry of the van der Waals equation of state [from Nguyen and Pinder (1980)].

"catastrophe theory." Although the theoretical underpinnings of the method are rather abstract, the practical application is quite straightforward. In this approach, the steam–water system is described thermodynamically by a cubic polynomial called the cusp catastrophe surface. This turns out to be equivalent to the van der Waals equation of state and is presented in Fig. 5.3. Although the thermodynamic evolution of the steam–water system is described by this surface, it is conceptually more appealing to look at a cross section through the pressure-specific volume plane. Such a section is presented as Fig. 5.4.

The most widely accepted theory for describing the dynamics of this system would describe the thermodynamic evolution of a point moving along the 180°C isotherm by the path *A–B–D–F–G*. This is indeed the situation presented previously in the description of the geothermal simulator

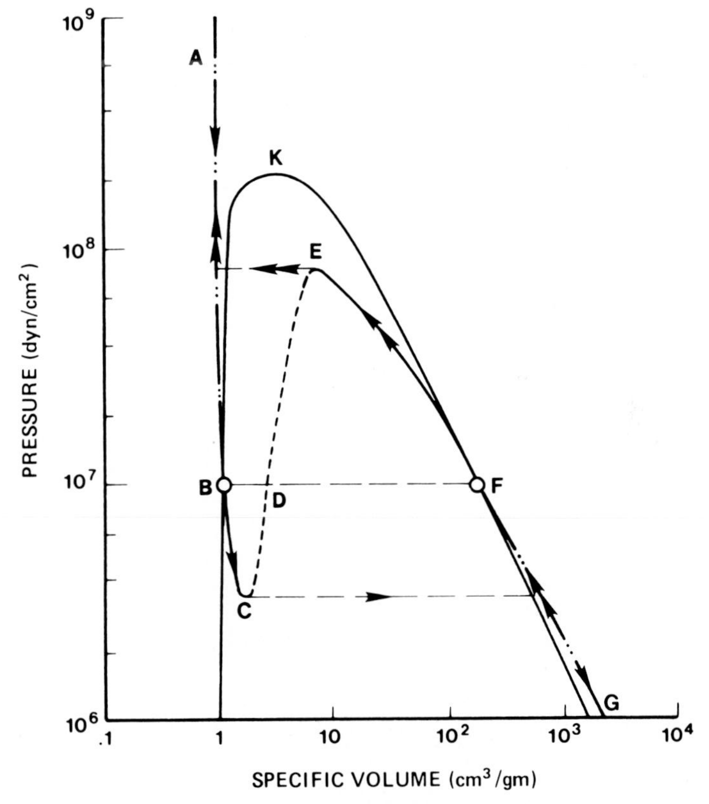

Fig. 5.4. Thermodynamics of a steam–water system employing equilibrium theory and the van der Waals equation of state –··– indicates 180°C isotherm [after Nguyen and Pinder (1980)].

(see Fig. 5.2). The kinks in this line at B and F give rise to discontinuous derivatives and, thereby, to the difficulties encountered in moving across the two-phase boundary.

The approach presented by Nguyen and Pinder circumvents this problem through a redefinition of the thermodynamic behavior in the two-phase region. In this scheme the evolution of the system follows the curve A–B–C as it enters the two-phase region (follow single arrows). At point B, steam begins to appear; it has the properties indicated by point F. Liquid water and steam coexist until point C is reached. At this point only steam can exist, and the system proceeds along the line toward G. This formulation implies the existence of two pressures, one for steam and one for liquid water. Our pooled pressure cannot accommodate this dual pressure, and we neglect the pressure in the bubble. Thus, from point C we must jump horizontally to the line F–G and then proceed along this line toward G. In moving from the steam to the two-phase region a similar procedure is followed (follow double arrows); this time, however, we progress along F–E to E, then horizontally across to A–B, and finally along B–A toward A. Now let us consider how this formulation enhances our ability to move numerically across the two-phase boundary.

Examination of Fig. 5.4 reveals no discontinuity in the derivative $\partial \rho/\partial h$ as we move across the boundary B–K. Similarly, one can deduce from Fig. 5.3 that the derivatives $\partial T/\partial \text{p}$ and $\partial T/\partial \rho$ are also continuous as we enter the two-phase region. To obtain the required thermodynamic information, we use the algebraic form of the cusp surface (van der Waals equation). See Kac *et al.* (1963) for a discussion of this equation,

$$\left(p + \frac{a}{\mathbf{V}^2}\right)(\mathbf{V} - b) = RT, \qquad (5.7.2.23)$$

where $\mathbf{V}$ is the specific volume, R the universal constant, and a and b constants fitted to data. In this instance

$$a = 170.8 \times 10^3 \quad \text{N cm}^4/\text{gm}^2,$$
$$b = 1.694 \quad \text{cm}^3/\text{gm}.$$

The Joule–Kelvin formula is derived from (5.7.2.23) by requiring the system to be thermodynamically irreversible. From this expression we obtain $\partial T/\partial p$:

$$dT = \left[\frac{2a}{RT}\left(\frac{\mathbf{V} - b}{\mathbf{V}}\right)^2 - b\right]\left[c_\text{p}\left\{1 - \left(\frac{\mathbf{V} - b}{\mathbf{V}}\right)^2 \frac{2a}{RT\mathbf{V}}\right\}\right]^{-1} dp, \qquad (5.7.2.24)$$

where $c_p \equiv (\partial h/\partial T)_p$ is the specific heat at constant pressure. Because the saturation of liquid water is much higher than the steam saturation as the compressed water state enters the two-phase zone, and the mixture goes to superheated steam when $S^g = 0.2$ in most cases, the value of $\partial T/\partial h$ obtained from the compressed water zone is used in the two-phase region, i.e., until point C is encountered.

The two densities ρ^l and ρ^g are obtained as the proper roots of the following form of the van der Waals equation (Nguyen and Pinder, 1981):

$$\rho_*^3 + \frac{\rho_c^2}{3}\left(8\frac{T}{T_c} + \frac{p}{p_c} - 9\right)\rho_* + \rho_c^3\left(\frac{8}{3}\frac{T}{T_c} - \frac{2}{3}\frac{p}{p_c} - 2\right) = 0, \qquad (5.7.2.25)$$

where

$$p_c = 2212 \quad \mathrm{N/cm^2}, \qquad T_c = 646\ \mathrm{K}, \qquad \mathbf{V}_c = 3.17 \quad \mathrm{cm^3/g}$$

are the critical triple characteristics of water and

$$\rho_* = \rho - \rho_c = \frac{1}{\mathbf{V}} - \frac{1}{\mathbf{V}_c}.$$

This formulation generates a set of smooth nonlinear coefficients and consequently a more stable solution.

There are now only two temperature relations. For a liquid-to-steam transition we have

$$T = \psi_1(p, h) + [\psi_0(p) - \psi_1(p, h^l)], \qquad (5.7.2.26a)$$

where ψ_1 and ψ_0 are defined as in (5.7.2.20). The steam-to-liquid transition employs the relation

$$T = \psi_2(p, h) + [\psi_0(p) - \psi_2(p, h^g)], \qquad (5.7.2.26b)$$

and $\psi_2(p, h)$ is given by (5.7.2.20c). Other elements of the simulation procedure are the same as outlined earlier for the classical thermodynamic formulation.

In closing this discussion, it is important to point out that there are no meaningful steam or water saturation measurements in the middle region of the two-phase dome. This state is never observed and, according to the catastrophe theory model, is thermodynamically unstable. An example problem using both thermodynamic approaches is presented in Nguyen and Pinder (1980).

5.8 Conclusions

In this chapter, we have examined the methodology employed in the finite element solution of the solute and energy transport equations. These equations are particularly difficult to solve for the convection-dominated case. In this instance, the asymmetric upstream weighting approximation is recommended. The multiphase energy transport problem is highly nonlinear in nature and two of several possible schemes have been presented. Simulation using the solute and energy transport equations generally is not recommended for the inexperienced analyst.

References

Avdonin, N. A. (1964). Some formulas for calculating the temperature field of a formation during thermal injection. *Izv. Vysshikh uchebn. Zavedenii, Neft Gaz* **7** (3), 37–41.

Bateman, H. (1910). The solution of a system of differential equations occurring in the theory of radioactive transformation. *Proc. Cambridge Philos. Soc.* **15.**

Bear, J. (1972). "Dynamics of Fluids in Porous Media," Elsevier, New York.

Bear, J. (1979). "Hydraulics of Groundwater," McGraw-Hill, New York.

Cameron, D. R., and Klute, A. (1977). Convective-dispersion solute transport with combined equilibrium and kinetic adsorption model. *Water Resour. Res.* **13** (1), 183–188.

Carslaw, H. S., and Jaeger, J. C. (1959). "Conduction of Heat in Solids," 2nd ed. Oxford Univ. Press, Oxford.

Cleary, R. W., and Adrian, D. D. (1973). Analytical solution of the convective–dispersive equation for cation adsorption in soils. *Soil Sci. Soc. Am.* **37,** 197–199.

Coats, K. H. (1977). Geothermal Reservoir Modelling. Paper SPE6892, presented at 52nd Annual Fall Technical Conference and Exhibition of the Soc. Pet. Eng. AIME, Denver, Colorado.

Coats, K. H., and Smith, B. D. (1964). Dead-end pore volume and dispersion in porous media. *Soc. Pet. Eng. J.* **4** (4), 73–83.

Corey, A. T., Rathjens, C. H., Henderson, J. H., and Wyllie, M. R. J. (1956). Three-phase relative permeability, Trans. AIME, 207, 349 pp.

Domenico, P. A., and Palciauskas, V. V. (1979). The volume-averaged mass-transport equation for chemical diagenetic models. *In* "Developments in Water Science No. 12," W. Back and D. A. Stephenson, eds., 427–438. Elsevier, Amsterdam.

Faust, C. R., and Mercer, J. W. (1977). Finite-difference models of two-dimensional single- and two-phase heat transport in a porous medium, U.S. Geol. Surv., Open File Report 77–234, 84 pp.

Faust, C. R. and Mercer, J. W. (1979). Geothermal reservoir simulation: 2. Numerical solution techniques for liquid- and vapor-dominated hydrothermal systems. *Water Resour. Res.* **15** (1), 31–46.

Fried, J. (1975). "Groundwater Pollution." Elsevier, New York.

Garg, S. K., and Pritchett, J. W. (1977). On pressure–work viscous dissipation and the energy balance relation for geothermal reservoirs. *Adv. Water Resour.* **1,** 41–47.

Gringarten, A. C., Witherspoon, P. A., and Ohnishi, Y. (1975). Theory of heat extraction from fractured hot dry rock. *J. Geophys. Res.* **80** (8), 1120–1124.

Gringarten, A. C., and Sauty, J. P. (1975). A theoretical study of heat extraction from aquifers with uniform regional flow. *J. Geophys. Res.* **80** (35), 4956–4962.

Harada, M., Chambre, P. L., Foglia, M., Higashi, K., Iwamoto, F., Leung, D., Pigford, T. H. and Ting, D. (1980). Migration of radionuclides through sorbing media: Analytical solutions —I, Battelle Office of Nuclear Waste Isolation, Technical Report ONWI-359.

Hassanizadeh, M., and Gray, W. G. (1979). General conservation equations for multiphase systems 2: Mass momentum energy and entropy equations. *Adv. Water Res.* **2,** 191–203.

Huyakorn, P. S., and Nilkuha, K. (1979). Solution of transient transport equation using an upstream finite element scheme. *Appl. Math. Modelling* **3,** 7–17.

Huyakorn, P. S., and Pinder, G. F. (1977). A pressure–enthalpy finite element model for simulating hydrothermal reservoirs. *In* "Advances in Computational Methods for Partial Differential Equations II: Proceedings of the 2nd IMACS (AICA) Symposium, Rutgers University.

Kac, M., Uhlenbeck, G. E., and Hemmer, P. C. (1963). On the van der Waals theory of the vapor–liquid equilibrium. *J. Math. Phys.* **4** (2), 216–228.

Lapidus, L., and Amundson, N. R. (1952). Mathematics of adsorption in Beds: VI. The effect of longitudinal diffusion and ion exchange in chromatographic columns. *J. Phys. Chem.* **56** (3), 984–988.

Lester, D. H., Jansen, G., and Burkholder, H. C. (1975). Migration of radionuclide chains through an adsorbing medium. *AI ChE Symp. Ser.* 71, **152,** 202–213.

McClintock, R. B., and Silvestri, G. J. (1968). "Formulations and Iterative Procedures for the Calculation of Properties of Steam." ASME, New York.

Mercer, J. M., and Faust, C. R. (1975). Simulation of water- and vapour-dominated hypothermal reservoirs. Paper SPE 5520, presented at the 50th Annual Fall Meeting of the Soc. Pet. Eng. *AIME*, Dallas, Texas.

Mercer, J. W., Pinder, G. F., and Donaldson, I. G. (1975). A Galerkin-finite element analysis of the hydrothermal system at Wairakei, New Zealand. *J. Geophys. Res.* **80** (17), p. 2608–2621.

Meyer, C. A. et al. (1968). "ASME Steam Tables," 2nd ed. ASME, New York.

Nguyen, V. V., and Pinder, G. F. (1980). Is Geothermal Simulation a Catastrophe? *In* "Proceedings of the VI Annual Workshop on Geothermal Reservoir Engineering," 213–217. Stanford University, Stanford, California.

Nguyen, V. V. and Pinder, G. F. (1981). Geothermal reservoir simulation using nonequilibrium thermodynamics. Princeton Water Resources Program Report 81-WR-2.

Nguyen, V. V., Gray, W. G., Pinder, G. F., Botha, J., and Crerar, D. A. (1982). "An analytical investigation of the transport of chemicals in reactive porous media. Submitted for publication in *Water Resources Research.*

Pigford, T. *et al.* (1980). Migration of radionuclides through sorbing media, analytical solution I and II. Technical Report No. LBL-11616, Vol. I and II, Lawrence Berkeley Lab.

Pinder, G. F., and Gray, W. G. (1977). "Finite Element Simulation in Surface and Subsurface Hydrology." Academic Press, New York.

Pinder, G. F., and Shapiro, A. (1982). Physics of flow in geotherml systems. Special Paper 189, *In* "Proceeding of Symposium on Recent Trends in Hydrogeology," 25–30. Geological Society of America, Boulder, Colorado.

Price, H. S., Varga, R. S., and Warren, J. E. (1966). "Application of oscillation matrices to diffusion-convection equations", *J. Math Phys.* **45,** 301–311.

Pritchett, J. W., Garg, S. K., Brownell Jr., D. H., and Levine, H. B. (1975). Geohydrological environmental effects of geothermal power production: Phase 1. Systems Science and Software, Inc., Report SSS-R-75-2733.

Ramey, H. J. Jr., Brigham, W. E., Chen, H. K., Aitkinson, P. G., and Arihara, N. (1974). Thermodynamic properties of hydrothermal systems. Proceedings of NSF Conference on the Utilization of Volcano Energy, Hilo, Hawaii.

Rasmuson, A., and Neretneiks, I. (1981). Migration of radionuclides in fissured rock: The influence of micropore diffusion and longitudinal dispersion. *J. Geophys. Res.* **86** (B5), 3749–3758.

Scheidegger, A. E. (1961). General theory of dispersion in porous media. *J. Geophys. Res.* **66,** 3273–3278.

Shapiro, A. M. (1981). An alternative formulation for hydrodynamic dispersion in porous media. Paper presented at EURO Mech. 143, Flow and Transport in Porous Media, Delft, Netherlands.

Slattery, J. C. (1972). "Momentum, Energy, and Mass Transfer in Continua". McGraw-Hill, New York.

Tang, D. H., Frind, E. O., and Sudicky, E. A. (1981). Contaminant transport in fractured media: Analytical solution for a single fracture. *Water Resour. Res.* **17** (3), 555–564.

Thomas, L. K., and Pierson, R. (1976). Three dimensional geothermal reservoir simulation. Paper SPE6104, presented at the 51st Annual Fall Technical Conference and Exhibition of the Soc. Pet. Eng. AIME, New Orleans.

van Genuchten, M. T., and Wierenga, P. J. (1976). Mass transfer studies in sorbing porous media: I. Analytical solutions. *Soil Sci. Soc. Am. J.* **40** (4), 473–480.

van Genuchten, M. T. (1981). Non-equilibrium transport parameters from miscible displacement experiments. Research report no. 119, U.S. Dept. of Agriculture, Salinity Lab., Riverside, California.

Voss, C. I. (1978). Finite Element Simulation of Multiphase Geothermal Reservoirs. Ph.D. dissertation, Dept. of Civil Engineering, Princeton University, Princeton, New Jersey.

Warrick, A. W., Biggar, J. W., and Nielson, D. R. (1971). Simultaneous solute and water transfer for an unsaturated soil. *Water Resour. Res.* **7** (5), 1216–1225.

Whitaker, S. (1969). Advances in theory of fluid motion in porous media. *Ind. Eng. Chem.* **12,** 14–28.

Whitaker, S. (1973). The transport equations for multiphase systems. *Chem. Eng. Sci.* **28,** 139–147.

White, D. E., Muffler, L. J. P., and Trusdell, A. H. (1971). Vapor-dominated hydrothermal systems compared with hot-water systems. *Econ. Geol.* **66,** p. 75–97.

Witherspoon, P. A., Neuman, S. P., Sorey, M. L., and Lipmann, M. J. (1975). Modeling Geothermal Systems. Technical Report No. LBL-3263, Lawrence Berkeley Lab.

Yeh, G. T. (1981). AT123D: Analytical transient one-, two-, and three-dimensional simulation of waste transport in the aquifer system. Report ORNL-5602, Oak Ridge National Laboratory.

Zienkiewicz, O. C. (1977). "The Finite Element Method," 3rd ed. McGraw-Hill, London.

6

Finite Element Simulation of Fluid Flow and Deformation in Unfractured and Fractured Porous Media

6.1 Introduction

In this chapter we treat various problems of coupled fluid flow and deformation in porous media. The first part of the chapter focuses attention on Biot's mathematical description of three-dimensional consolidation in granular media (soils). We start by presenting a derivation of the governing equations for fluid flow and solid displacement based on the conventional assumption of linear elastic deformation. From the three-dimensional equations we obtain two sets of equations for plane strain and axisymmetric cases and approximate these via the Galerkin finite element technique. We then extend the finite element formulation to deal with elastic–plastic material behavior frequently encountered in practice. For completeness, we also describe the treatment of areal subsidence problems. In this type of situation, the flow equations may be uncoupled from the displacement equations by the introduction of certain simplifying assumptions.

The remaining part of the chapter deals with flow and deformation in fractured porous media. Because this topic is still in active research, we elect to present two alternative conceptual models for fractured porous medium systems. These models are distinct in the sense that one relies on the use of a discrete representation of fractures, whereas the other relies on the use of two continua to represent the fractures and the porous matrix. For a brief review of the theoretical development and practical aspects of fracture flow, the reader is referred to Bear and Braester (1972), Snow (1965), and Streltsova-Adams (1978).

6.2 Fluid Flow and Deformation in an Unfractured Porous Medium

6.2.1 Governing Equations

The phenomena of fluid flow and solid deformation in a porous medium can be described by a set of coupled partial differential equations.

These equations describe fluid or pore pressure distribution and displacement in the porous medium. The following formulation is based on the theory of three-dimensional consolidation developed by Biot in 1941. To grasp the material in this section, the reader should be familiar with the concept of stress versus strain and the basic equilibrium and compatibility equations of continuum mechanics.

Displacement Equations

Consider a volume of an elastic porous medium filled with a homogeneous fluid. The distribution of total stress within this volume can be shown to satisfy the equilibrium equation

$$\frac{\partial \tau_{ij}}{\partial x_j} + F_i = 0, \qquad i = 1, 2, 3, \tag{6.2.1.1}$$

where τ_{ij} is the total stress on the medium and F_i is the body force per unit volume of the medium. The nine stress components are depicted in Fig. 6.1. According to the usual sign convention in the theory of elasticity, the stress components acting on the planes with positive outward unit normals are considered positive when pointing in the positive coordinate directions and negative otherwise. The total stress τ_{ij} can be expressed as

$$\tau_{ij} = \sigma_{ij} - \alpha p \delta_{ij}, \tag{6.2.1.2}$$

where σ_{ij} is referred to as the "effective" stress acting on the solid skeleton, p is the fluid pressure, and α is a physical constant. According

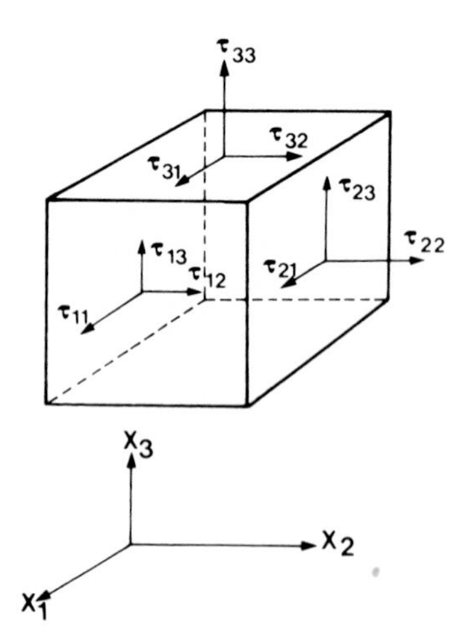

Fig. 6.1. Elementary parallelepiped with total stress components. The stress components are denoted by τ_{ij}, the first index denoting the direction of the outward normal to the plane and the second index denoting the direction of the stress component.

to Biot (1941) the coefficient α may be interpreted as the "ratio of the fluid volume squeezed out to the volume change of the porous medium if the latter is compressed while allowing the fluid to escape." For a saturated porous medium, α approaches unity.

The effective stress after an elapsed time t can be expressed as

$$\sigma_{ij} = \sigma_{ij}^0 + \tilde{\sigma}_{ij}, \tag{6.2.1.3}$$

where σ_{ij}^0 is the initial effective stress and $\tilde{\sigma}_{ij}$ is the effective stress increment.

If it is assumed that the porous medium is isotropic, then the linear elastic stress–strain relation takes the form

$$\tilde{\sigma}_{ij} = 2G\tilde{\varepsilon}_{ij} + \lambda\tilde{\varepsilon}_{kk}\delta_{ij}, \tag{6.2.1.4}$$

where $\tilde{\varepsilon}_{ij}$ is the incremental strain of the solid skeleton and G and λ are called Lame's constants. The parameter G is identified as the shear modulus $G = E/[2(1+\nu)]$ and λ is identified as $E\nu/[(1+\nu)(1-2\nu)]$, where E is Young's modulus and ν is Poisson's ratio for the solid skeleton. In general, $\tilde{\varepsilon}_{ij} = \varepsilon_{ij} - \varepsilon_{ij}^0$, where ε_{ij}^0 is the initial strain, which may be caused by such factors as temperature changes, shrinkage, etc. However, we assume here that these initial strains are negligible and hence that

$$\tilde{\varepsilon}_{ij} = \varepsilon_{ij}. \tag{6.2.1.5}$$

Combining Eqs. (6.2.1.1)–(6.2.1.5) and using $\alpha = 1$, i.e., considering the saturated case,

$$\frac{\partial\sigma_{ij}^0}{\partial x_j} + F_i + 2G\frac{\partial\varepsilon_{ij}}{\partial x_j} + \lambda\frac{\partial\varepsilon_{jj}}{\partial x_i} - \frac{\partial p}{\partial x_i} = 0. \tag{6.2.1.6}$$

For the case of small deformations, the strain components are related to displacements by the linearized relations

$$\varepsilon_{ij} = \frac{1}{2}\left(\frac{\partial u_i}{\partial x_j} + \frac{\partial u_j}{\partial x_i}\right), \tag{6.2.1.7}$$

where u_i denotes the incremental displacement vector of the solid skeleton. Substitution of (6.2.1.7) into (6.2.1.6) yields

$$\frac{\partial\sigma_{ij}^0}{\partial x_j} + F_i + G\frac{\partial^2 u_i}{\partial x_j\,\partial x_j} + (\lambda + G)\frac{\partial^2 u_j}{\partial x_i \partial x_j} - \frac{\partial p}{\partial x_i} = 0. \tag{6.2.1.8}$$

Equation (6.2.1.8) forms the required governing equations for the solid matrix displacement. There are two types of boundary conditions that should be considered. The first type corresponds to the condition of prescribed displacement on the boundary portion B_u, and the second type corresponds to the Cauchy condition of prescribed surface traction or stress on boundary portion B_s. These boundary conditions can be expressed as

$$u_i = U_i \qquad \text{on } B_u$$

and

$$\tau_{ij} n_j = S_i \qquad \text{on } B_s,$$

where S_i denotes the components of surface traction. We now set aside temporarily the equation for solid matrix displacement while we examine the associated fluid flow equation.

Flow Equation

A mathematical description of fluid flow in a compacting porous medium can be obtained by combining Darcy's law with the mass conservation equations of the fluid and solid. We begin the development by defining the relative bulk velocity of the fluid as

$$V_i = \phi \left(\frac{\partial u_{i\mathrm{f}}}{\partial t} - \frac{\partial u_{i\mathrm{s}}}{\partial t} \right) = \phi \, (V_{i\mathrm{f}} - V_{i\mathrm{s}}), \tag{6.2.1.9}$$

where ϕ is porosity, $u_{i\mathrm{f}}$ and $u_{i\mathrm{s}}$ are displacements of the fluid and solid ($u_{i\mathrm{s}} \equiv u_i$), and $V_{i\mathrm{f}}$ and $V_{i\mathrm{s}}$ are velocities of the fluid and solid, respectively. Equation (6.2.1.9) can be rearranged as

$$\phi V_{i\mathrm{f}} = V_i + \phi V_{i\mathrm{s}}. \tag{6.2.1.10}$$

Note that the velocities identified with the fluid and solid phases are intrinsic phase averages, whereas V_i is the relative Darcy velocity.

The relative bulk velocity of the fluid may now be related to the incremental pressure through Darcy's law, which for an isotropic material takes the form

$$V_i = -\frac{k}{\mu} \left(\frac{\partial p}{\partial x_i} + \rho_{\mathrm{f}} g_i \right), \tag{6.2.1.11}$$

where k is the intrinsic permeability of the porous medium, μ is the dynamic viscosity of fluid, g_i is the ith component of gravitational acceleration, and ρ_{f} is the density of fluid.

The continuity equation for solid mass is

$$-\frac{\partial}{\partial x_i}(1 - \phi)\rho_{\mathrm{s}} V_{i\mathrm{s}} = \frac{\partial}{\partial t}(1 - \phi)\rho_{\mathrm{s}}. \tag{6.2.1.12}$$

Equation (6.2.1.12) can be expanded into the form

$$-\left\{(1-\phi)\rho_s \frac{\partial V_{is}}{\partial x_i} + (1-\phi)V_{is}\frac{\partial \rho_s}{\partial x_i} - \rho_s V_{is}\frac{\partial \phi}{\partial x_i}\right\}$$

$$= (1-\phi)\frac{\partial \rho_s}{\partial t} - \rho_s \frac{\partial \phi}{\partial t},$$

or, rearranging,

$$\rho_s\left(\frac{\partial \phi}{\partial t} + V_{is}\frac{\partial \phi}{\partial x_i}\right) = (1-\phi)\left[\frac{\partial \rho_s}{\partial t} + V_{is}\frac{\partial \rho_s}{\partial x_i}\right] + (1-\phi)\rho_s\frac{\partial V_{is}}{\partial x_i}. \tag{6.2.1.13}$$

Introducing the following form of the substantial time derivatives of porosity and solid density:

$$\frac{D\phi}{Dt} = \frac{\partial \phi}{\partial t} + V_{is}\frac{\partial \phi}{\partial x_i}, \qquad \frac{D\rho_s}{Dt} = \frac{\partial \rho_s}{\partial t} + V_{is}\frac{\partial \rho_s}{\partial x_i},$$

we can reduce Eq. (6.2.1.13) to

$$\frac{D\phi}{Dt} = \frac{(1-\phi)}{\rho_s}\frac{D\rho_s}{Dt} + (1-\phi)\frac{\partial V_{is}}{\partial x_i}. \tag{6.2.1.14}$$

The continuity equation of fluid mass may be written as

$$-\frac{\partial}{\partial x_i}\rho_f \phi V_{if} = \frac{\partial}{\partial t}\phi\rho_f. \tag{6.2.1.15}$$

Substitution of (6.2.1.10) into (6.2.1.15) yields

$$-\frac{\partial}{\partial x_i}\rho_f V_i - \frac{\partial}{\partial x_i}\rho_f \phi V_{is} = \frac{\partial}{\partial t}\phi\rho_f, \tag{6.2.1.16}$$

which may be rearranged to give

$$\begin{aligned} -\frac{\partial}{\partial x_i}\rho_f V_i - \rho_f\phi\frac{\partial V_{is}}{\partial x_i} &= \frac{\partial}{\partial t}\phi\rho_f + V_{is}\frac{\partial}{\partial x_i}\phi\rho_f \\ &= \frac{D}{Dt}\phi\rho_f = \phi\frac{D\rho_f}{Dt} + \rho_f\frac{D\phi}{Dt}. \end{aligned} \tag{6.2.1.17}$$

Substitution of (6.2.1.14) into (6.2.1.17) leads to

$$-\frac{\partial}{\partial x_i}\rho_f V_i = \phi\frac{D\rho_f}{Dt} + \rho_f\frac{1-\phi}{\rho_s}\frac{D\rho_s}{Dt} + \rho_f\frac{\partial V_{is}}{\partial x_i}. \tag{6.2.1.18}$$

Next, we introduce the following constitutive relations for the fluid and solid matrix:

$$\frac{1}{\rho_{\mathrm{f}}}\frac{D\rho_{\mathrm{f}}}{Dt} = \beta\frac{Dp}{Dt}, \qquad \frac{1}{\rho_{\mathrm{s}}}\frac{D\rho_{\mathrm{s}}}{Dt} = C_{\mathrm{s}}\frac{Dp}{Dt}, \tag{6.2.1.19}$$

where β and C_{s} are the compressibilities of fluid and solid grains, respectively. Substituting Eq. (6.2.1.19) and Darcy's equation into (6.2.1.18), we obtain

$$\frac{1}{\rho_{\mathrm{f}}}\left\{\frac{\partial}{\partial x_i}\left[\rho_{\mathrm{f}}\frac{\mathrm{k}}{\mu}\left(\frac{\partial p}{\partial x_i} + \rho_{\mathrm{f}}g_i\right)\right]\right\} = \phi\beta\frac{Dp}{Dt} + \frac{\partial V_{i\mathrm{s}}}{\partial x_i} + (1-\phi)C_{\mathrm{s}}\frac{Dp}{Dt}. \tag{6.2.1.20}$$

Finally, we introduce the following assumptions:

$$Dp/Dt = \partial p/\partial t, \qquad C_{\mathrm{s}} \simeq 0,$$

and

$$\frac{1}{\rho_{\mathrm{f}}}\left\{\frac{\partial}{\partial x_i}\left[\rho_{\mathrm{f}}\frac{k}{\mu}\left(\frac{\partial p}{\partial x_i} + \rho_{\mathrm{f}}g_i\right)\right]\right\} \simeq \frac{\partial}{\partial x_i}\left[\frac{k}{\mu}\left(\frac{\partial p}{\partial x_i} + \rho_{\mathrm{f}}g_i\right)\right].$$

Note that the second assumption $C_{\mathrm{s}} \simeq 0$ implies that the solid grains are incompressible, and the third assumption implies that the fluid is slightly compressible. By these assumptions, Eq. (6.2.1.20) reduces to

$$\frac{\partial}{\partial x_i}\left[\frac{k}{\mu}\left(\frac{\partial p}{\partial x_i} + \rho_{\mathrm{f}}g_i\right)\right] = \phi\beta\frac{\partial p}{\partial t} + \frac{\partial V_{i\mathrm{s}}}{\partial x_i}, \tag{6.2.1.21}$$

or, using the definition $V_{i\mathrm{s}} = \partial u_i/\partial t$,

$$\frac{\partial}{\partial x_i}\left[\frac{k}{\mu}\left(\frac{\partial p}{\partial x_i} + \rho_{\mathrm{f}}g_i\right)\right] = \phi\beta\frac{\partial p}{\partial t} + \frac{\partial}{\partial t}\left(\frac{\partial u_i}{\partial x_i}\right). \tag{6.2.1.22}$$

Equation (6.2.1.22) is the governing equation for fluid flow.

6.2.2 Finite Element Formulation of Consolidation Problems

Plane Strain Case

Equations (6.2.1.8) and (6.2.1.22) form the required system of coupled partial differential equations. We now consider a two-dimensional plane strain situation. For convenience, we assume that the material is isotropic and the coefficients λ, G, ϕ, k, μ, and ρ_{f} are constant. This enables us to reduce the governing equations to the form [see Gibson *et al.* (1970) and

Schiffman *et al.* (1969)]

$$G \frac{\partial^2 u_i}{\partial x_j \partial x_j} + (\lambda + G) \frac{\partial^2 u_j}{\partial x_i \partial x_j} - \frac{\partial p}{\partial x_i} = -F_i - \frac{\partial \sigma_{ij}^0}{\partial x_j}, \quad (6.2.2.1a)$$

$$\frac{\partial}{\partial x_i}\left(\frac{k}{\mu}\frac{\partial p}{\partial x_i}\right) - \phi\beta \frac{\partial p}{\partial t} - \frac{\partial}{\partial t}\left(\frac{\partial u_i}{\partial x_i}\right) = 0, \quad (6.2.2.1b)$$

where i and j range from 1 to 2.

We can now approximate Eqs. (6.2.2.1a) and (6.2.2.1b) using the Galerkin finite element procedure. Let trial solutions be expressed in the form

$$\hat{u}_i(x_1, x_2, t) = N_J(x_1, x_2)u_{iJ}(t), \qquad J = 1, 2, \ldots, n, \quad (6.2.2.2a)$$

$$\hat{p}(x_1, x_2, t) = N_J(x_1, x_2)p_J(t), \qquad J = 1, 2, \ldots, n. \quad (6.2.2.2b)$$

Application of Galerkin's criterion to Eq. (6.2.2.1a) leads to

$$\int_R N_I \left[G \frac{\partial^2 \hat{u}_i}{\partial x_j \partial x_j} + (\lambda + G) \frac{\partial^2 \hat{u}_j}{\partial x_i \partial x_j} - \frac{\partial \hat{p}}{\partial x_i} \right] dR$$
$$= -\int_R N_I F_i \, dR - \int_R N_I \frac{\partial \sigma_{ij}^0}{\partial x_j} dR, \qquad I = 1, 2, \ldots, n. \quad (6.2.2.3)$$

Next, we apply Green's theorem to all derivative terms in Eq. (6.2.2.3). The result is given by

$$-\int_R G \frac{\partial N_I}{\partial x_j}\frac{\partial \hat{u}_i}{\partial x_j} dR + \int_B G N_I \frac{\partial \hat{u}_i}{\partial x_j} n_j \, dB - \int_R \lambda \frac{\partial N_I}{\partial x_i}\frac{\partial \hat{u}_j}{\partial x_j} dR$$
$$+ \int_B \lambda N_I \frac{\partial \hat{u}_j}{\partial x_j} n_i \, dB - \int_R G \frac{\partial N_I}{\partial x_j}\frac{\partial \hat{u}_j}{\partial x_i} dR + \int_B G N_I \frac{\partial \hat{u}_j}{\partial x_i} n_j \, dB$$
$$+ \int_R \frac{\partial N_I}{\partial x_i} \hat{p} \, dR - \int_B N_I \hat{p} n_i \, dB$$
$$= -\int_R N_I F_i \, dR + \int_R \frac{\partial N_I}{\partial x_j} \sigma_{ij}^0 \, dR - \int_B N_I \sigma_{ij}^0 n_j \, dB, \qquad I = 1, 2, \ldots, n,$$
$$(6.2.2.4)$$

which may be rearranged to give

$$\int_R \left[G \frac{\partial N_I}{\partial x_j}\left(\frac{\partial \hat{u}_i}{\partial x_j} + \frac{\partial \hat{u}_j}{\partial x_i}\right) + \lambda \frac{\partial N_I}{\partial x_i}\frac{\partial \hat{u}_j}{\partial x_j} - \frac{\partial N_I}{\partial x_i}\hat{p} \right] dR$$
$$- \int_B N_I \left[G\left(\frac{\partial \hat{u}_i}{\partial x_j} + \frac{\partial \hat{u}_j}{\partial x_i}\right) + \lambda \frac{\partial \hat{u}_k}{\partial x_k}\delta_{ij} - \hat{p}\delta_{ij} + \sigma_{ij}^0 \right] n_j \, dB$$
$$= \int_R N_I F_i \, dR - \int_R \frac{\partial N_I}{\partial x_j} \sigma_{ij}^0 \, dR, \qquad I = 1, 2, \ldots, n. \quad (6.2.2.5)$$

From Eqs. (6.2.1.2)–(6.2.1.4) and (6.2.1.7) and from the assumption that $\alpha = 1$, it follows that

$$\tau_{ij} = \sigma_{ij}^0 + G\left(\frac{\partial \hat{u}_i}{\partial x_j} + \frac{\partial \hat{u}_j}{\partial x_i}\right) + \lambda \frac{\partial \hat{u}_k}{\partial x_k}\delta_{ij} - \hat{p}\delta_{ij}. \tag{6.2.2.6}$$

Substituting Eqs. (6.2.2.2a), (6.2.2.2b), and (6.2.2.6) into (6.2.2.5), we obtain

$$\int_R \left[G\frac{\partial N_I}{\partial x_j}\left(\frac{\partial N_J}{\partial x_j}u_{iJ} + \frac{\partial N_J}{\partial x_i}u_{jJ}\right) + \lambda\frac{\partial N_I}{\partial x_i}\frac{\partial N_J}{\partial x_j}u_{jJ} - \frac{\partial N_I}{\partial x_i}N_J p_j\right] dR$$
$$- \int_{B_s} N_I \tau_{ij} n_j \, dB$$
$$= \int_R N_I F_i \, dR - \int_R \frac{\partial N_I}{\partial x_j}\sigma_{ij}^0 \, dR, \qquad I, J = 1, 2, \ldots, n. \tag{6.2.2.7}$$

Using the Cauchy surface traction boundary condition $S_i = \tau_{ij}n_j$ on B_s, and subdividing the region R into m finite elements, Eq. (6.2.2.7) may be rewritten for $I, J = 1, 2, \ldots, n$ as

$$\sum_{e=1}^{m}\left\{\int_{R^e}\left[G\frac{\partial N_I}{\partial x_j}\left(\frac{\partial N_J}{\partial x_j}u_{iJ} + \frac{\partial N_J}{\partial x_i}u_{jJ}\right) + \lambda\frac{\partial N_I}{\partial x_i}\frac{\partial N_J}{\partial x_j}u_{jJ} - \frac{\partial N_I}{\partial x_i}N_J p_J\right] dR\right.$$
$$\left. - \int_{B_s^e} N_I S_i \, dB\right\}$$
$$= \sum_{e=1}^{m}\left\{\int_{R^e} N_I F_i \, dR - \int_{R^e}\frac{\partial N_I}{\partial x_j}\sigma_{ij}^0 \, dR\right\}. \tag{6.2.2.8}$$

Expanding subscripts i and j of Eq. (6.2.2.8), we obtain

(1) for $i = 1$,

$$\sum_{e=1}^{m}\left\{\int_{R^e}\left[(\lambda + 2G)\frac{\partial N_I}{\partial x_1}\frac{\partial N_J}{\partial x_1} + G\frac{\partial N_I}{\partial x_2}\frac{\partial N_J}{\partial x_2}\right]u_{1J}\, dR\right.$$
$$\left. + \int_{R^e}\left(\lambda\frac{\partial N_I}{\partial x_1}\frac{\partial N_J}{\partial x_2} + G\frac{\partial N_I}{\partial x_2}\frac{\partial N_J}{\partial x_1}\right)u_{2J}\, dR - \int_{R^e}\frac{\partial N_I}{\partial x_1}N_J p_J \, dR - \int_{B_s^e} N_I S_1 \, dB\right\}$$
$$= \sum_{e=1}^{m}\left\{\int_{R^e} N_I F_1 \, dR - \int_{R^e}\frac{\partial N_I}{\partial x_j}\sigma_{2j}^0 \, dR\right\}; \tag{6.2.2.9b}$$

(2) for $i = 2$,

$$\sum_{e=1}^{m}\left\{\int_{R^e}\left(\lambda\frac{\partial N_I}{\partial x_2}\frac{\partial N_J}{\partial x_1} + G\frac{\partial N_I}{\partial x_1}\frac{\partial N_J}{\partial x_2}\right)u_{1J}\, dR\right.$$

$$+ \int_{R^e} \left(G \frac{\partial N_I}{\partial x_1} \frac{\partial N_J}{\partial x_1} + (\lambda + 2G) \frac{\partial N_I}{\partial x_2} \frac{\partial N_J}{\partial x_2} \right) u_{2J} \, dR$$

$$- \int_{R^e} \frac{\partial N_I}{\partial x_2} N_J p_J \, dR - \int_{B_s^e} N_I S_2 \, dB \Bigg\}$$

$$= \sum_{e=1}^{m} \left\{ \int_{R^e} N_I F_2 \, dR - \int_{R^e} \frac{\partial N_I}{\partial x_j} \sigma_{2j}^0 \, dR \right\}. \tag{6.2.2.9b}$$

Equations (6.2.2.9a) and (6.2.2.9b) are the required results of the Galerkin approximation of the displacement equations.

In a similar manner, we can approximate the flow equation (6.2.2.1b) using the Galerkin procedure. After transformation of the space derivative using Green's theorem, this approximation becomes

$$\sum_{e=1}^{m} \left[- \int_{R^e} \frac{k}{\mu} \frac{\partial N_I}{\partial x_i} \frac{\partial N_J}{\partial x_i} p_J \, dR - \int_{R^e} \phi \beta N_I N_J \frac{dp_J}{dt} \, dR - \int_{R^e} N_I \frac{\partial N_J}{\partial x_i} \frac{du_{iJ}}{dt} \, dR \right]$$

$$= \sum_{e=1}^{m} \left[\int_{B_2^e} N_I q \, dB \right],$$

where $q = -(k/\mu)(\partial p/\partial x_i)n_i$ is the outward fluid flux normal to boundary surface B_2^e. Upon application of a generally weighted finite difference time-stepping scheme, this equation becomes

$$\sum_{e=1}^{m} \int_{R^e} \left\{ \left(N_I \frac{\partial N_J}{\partial x_i} \right) u_{iJ}^{t+\Delta t} + \phi \beta N_I N_J p_J^{t+\Delta t} + \left(\theta \, \Delta t \frac{k}{\mu} \frac{\partial N_I}{\partial x_i} \frac{\partial N_J}{\partial x_i} \right) p_J^{t+\Delta t} \right\} dR$$

$$= \sum_{e=1}^{m} \left[\int_{R^e} \left\{ N_I \frac{\partial N_J}{\partial x_i} u_{iJ}^t + \phi \beta N_I N_J p_J^t - (1 - \theta) \, \Delta t \frac{k}{\mu} \frac{\partial N_I}{\partial x_i} \frac{\partial N_J}{\partial x_i} p_J^t \right\} dR \right.$$

$$\left. - \int_{B_2^e} N_I (\Delta t) q \, dB \right], \tag{6.2.2.9c}$$

where q is the fluid flux, u_{iJ}^t and p_J^t denote the nodal displacement and pressure values at the old time level, and θ is the time-weighting factor. (The use of $\theta = \frac{1}{2}$ is recommended to obtain stable solutions with second-order accuracy in time.)

Equations (6.2.2.9a)–(6.2.2.9c) can be written in matrix form as

$$\sum_{e=1}^{m} [H]^e \{\chi\}_{t+\Delta t}^e = \sum_{e=1}^{m} \{R\}_t^e, \tag{6.2.2.10}$$

where typical matrix elements are given by

$$\{\chi_I\}^e = \begin{Bmatrix} u_{1I} \\ u_{2I} \\ p_I \end{Bmatrix}^e, \qquad \Big\{R_I\Big\}^e = \begin{Bmatrix} R_{1I} \\ R_{2I} \\ R_{3I} \end{Bmatrix}^e,$$

$$[H_{IJ}]^e = \begin{bmatrix} a_{IJ} & b_{IJ} & -c_{IJ} \\ (b_{IJ})^{\mathrm{T}} & d_{IJ} & -e_{IJ} \\ (c_{IJ})^{\mathrm{T}} & (e_{IJ})^{\mathrm{T}} & f_{IJ} \end{bmatrix}^e,$$

$$a_{IJ} = \int_{R^e} \left[(\lambda + 2G) \frac{\partial N_I}{\partial x_1} \frac{\partial N_J}{\partial x_1} + G \frac{\partial N_I}{\partial x_2} \frac{\partial N_J}{\partial x_2} \right] dR,$$

$$b_{IJ} = \int_{R^e} \left(\lambda \frac{\partial N_I}{\partial x_1} \frac{\partial N_J}{\partial x_2} + G \frac{\partial N_I}{\partial x_2} \frac{\partial N_J}{\partial x_1} \right) dR,$$

$$c_{IJ} = \int_{R^e} \frac{\partial N_I}{\partial x_1} N_J \, dR,$$

$$d_{IJ} = \int_{R^e} \left(G \frac{\partial N_I}{\partial x_1} \frac{\partial N_J}{\partial x_1} + (\lambda + 2G) \frac{\partial N_I}{\partial x_2} \frac{\partial N_J}{\partial x_2} \right) dR,$$

$$e_{IJ} = \int_{R^e} \frac{\partial N_I}{\partial x_2} N_J \, dR,$$

$$f_{IJ} = \int_{R^e} \left(\theta \, \Delta t \frac{k}{\mu} \frac{\partial N_I}{\partial x_i} \frac{\partial N_J}{\partial x_i} + \phi \beta N_I N_J \right) dR;$$

where the superscript T denotes matrix transpose. It is apparent that the matrix $[H]^e$ is not symmetric unless the sign of the third column of $[H_{IJ}]^e$ is reversed. This can be achieved by reversing the sign convention normally adopted for the variable p. Thus, for the coupled problem of flow and deformation, the sign of p is negative for compressive pressure and positive for tensile pressure or suction. The structure of $[H_{IJ}]^e$ now becomes

$$[H_{IJ}]^e = \begin{bmatrix} a_{IJ} & b_{IJ} & c_{IJ} \\ (b_{IJ})^{\mathrm{T}} & d_{IJ} & e_{IJ} \\ (c_{IJ})^{\mathrm{T}} & (e_{IJ})^{\mathrm{T}} & -f_{IJ} \end{bmatrix}^e.$$

The corresponding elements of $\{R_I\}^e$ for this definition of p are

$$R_{1I} = \int_{B_s^e} N_I S_1 \, dB + \int_{R^e} N_I F_1 \, dR - \int_{R^e} \frac{\partial N_I}{\partial x_j} \sigma_{1j}^0 \, dR,$$

$$R_{2I} = \int_{B_s^e} N_I S_2 \, dB + \int_{R^e} N_I F_2 \, dR - \int_{R^e} \frac{\partial N_I}{\partial x_j} \sigma_{2j}^0 \, dR,$$

$$R_{3I} = \int_{R^e} \left[N_I \frac{\partial N_J}{\partial x_i} u_{iJ}^t - \phi\beta N_I N_J p_J^t + (1-\theta)\,\Delta t \frac{k}{\mu} \frac{\partial N_I}{\partial x_i} \frac{\partial N_J}{\partial x_i} p_J^t \right] dR$$

$$+ \int_{B_2^e} N_I\, \Delta t q \, dB.$$

The matrix $[H_{IJ}]^e$ can be partitioned as

$$[H_{IJ}]^e = \left[\begin{array}{cc|c} a_{IJ} & b_{IJ} & c_{IJ} \\ (b_{IJ})^{\mathrm{T}} & d_{IJ} & e_{IJ} \\ \hline (c_{IJ})^{\mathrm{T}} & (e_{IJ})^{\mathrm{T}} & -f_{IJ} \end{array}\right]^e = \left[\begin{array}{c|c} [K_{IJ}] & [A_{IJ}] \\ \hline [A_{IJ}]^{\mathrm{T}} & [-f_{IJ}] \end{array}\right]^e. \qquad (6.2.2.11a)$$

The first submatrix $[K_{IJ}]^e$ corresponds to the conventional stiffness matrix in structural analysis. It can be readily shown that $[K_{IJ}]^e$ can be expressed in the form

$$[K_{IJ}]^e = \int_{R^e} [B_I]^{\mathrm{T}} [D][B_J]\, dR, \qquad (6.2.2.11b)$$

where

$$[B_J] = \begin{bmatrix} \dfrac{\partial N_J}{\partial x_1} & 0 \\ 0 & \dfrac{\partial N_J}{\partial x_2} \\ \dfrac{\partial N_J}{\partial x_2} & \dfrac{\partial N_J}{\partial x_1} \end{bmatrix}, \qquad [B_I]^{\mathrm{T}} = \begin{bmatrix} \dfrac{\partial N_I}{\partial x_1} & 0 & \dfrac{\partial N_I}{\partial x_2} \\ 0 & \dfrac{\partial N_I}{\partial x_2} & \dfrac{\partial N_I}{\partial x_1} \end{bmatrix}, \qquad (6.2.2.11c)$$

and $[D]$ is given by

$$[D] = \frac{\lambda(1-\nu)}{\nu} \begin{bmatrix} 1 & \dfrac{\nu}{1-\nu} & 0 \\ \dfrac{\nu}{1-\nu} & 1 & 0 \\ 0 & 0 & \dfrac{1-2\nu}{2(1-\nu)} \end{bmatrix}. \qquad (6.2.2.11d)$$

The matrix $[B]$ is simply a matrix relating strain and nodal displacements, whereas $[D]$ is an elasticity matrix relating stress and strain for the plane strain situation. Thus it follows that

$$\{\varepsilon\} = [B_J]\{u_J\}$$

and

$$\{\sigma\} = [D]\{\varepsilon\} + \{\sigma^0\},$$

where

$$\{\varepsilon\}^{\mathrm{T}} = \{\varepsilon_{11}, \varepsilon_{22}, 2\varepsilon_{12}\},$$

$$\{\sigma\}^{\mathrm{T}} = \{\sigma_{11}, \sigma_{22}, \sigma_{12}\},$$

$$\{\sigma^0\}^{\mathrm{T}} = \{\sigma_{11}^0, \sigma_{22}^0, \sigma_{12}^0\}.$$

Note the use of the "engineering" strain $\gamma_{12} \equiv 2\varepsilon_{12}$, rather than ε_{12} itself. Equation (6.2.2.11b) provides us with a simple way to construct the stiffness matrix. Its usefulness will be evident in the treatment of nonlinear material properties, described in the next section.

Axisymmetric Case

The preceding formulation can be readily extended to the axisymmetric case using cylindrical coordinates instead of Cartesian coordinates. If we let x_1 correspond to the radial coordinate r, and x_2 correspond to the vertical coordinate z, the governing equations for the radial and vertical displacements and pressure may be shown to take the form [cf. Verruijt (1969, p. 346)]

$$\frac{G}{x_1}\frac{\partial}{\partial x_j}\left(x_1 \frac{\partial u_i}{\partial x_j}\right) + (\lambda + G)\frac{\partial \varepsilon^*}{\partial x_i} - \frac{G}{x_1^2}\delta_{i1} u_1 - \frac{\partial p}{\partial x_i} = -F_i, \qquad (6.2.2.12a)$$

$$\frac{1}{x_1}\frac{\partial}{\partial x_i}\left(\frac{k}{\mu} x_1 \frac{\partial p}{\partial x_i}\right) - \phi\beta \frac{\partial p}{\partial t} - \frac{\partial \varepsilon^*}{\partial t} = 0, \qquad (6.2.2.12b)$$

where it is assumed in Eq. (6.2.2.12a) that the initial stress is zero, δ_{i1} is the Kronecker delta, and ε^* is a volumetric strain (dilation), defined as

$$\varepsilon^* = \frac{\partial u_j}{\partial x_j} + \frac{u_1}{x_1}.$$

After the application of the Galerkin procedure, the global matrix equation can be put in the form of Eq. (6.2.2.10), where the matrix elements are given by

$$\left\{\chi_I\right\}^e = \begin{Bmatrix} u_{1I} \\ u_{2I} \\ p_I \end{Bmatrix}^e, \qquad \left\{R_I\right\}^e = \begin{Bmatrix} R_{1I} \\ R_{2I} \\ R_{3I}^* \end{Bmatrix}^e,$$

$$\left[H_{IJ}\right]^e = \left[\begin{array}{cc|c} a_{IJ}^* & b_{IJ}^* & c_{IJ}^* \\ (b_{IJ}^*)^{\mathrm{T}} & d_{IJ} & e_{IJ} \\ \hline (c_{IJ}^*)^{\mathrm{T}} & (e_{IJ})^{\mathrm{T}} & -f_{IJ} \end{array}\right]^e = \left[\begin{array}{c|c} [K_{IJ}^*] & [A_{IJ}^*] \\ \hline [A_{IJ}^*]^{\mathrm{T}} & [-f_{IJ}] \end{array}\right]^e.$$

If it is understood that x_1 and x_2 are replaced by r and z, respectively, then only the matrix elements with the asterisk symbol are different from the corresponding matrix elements in the plane strain case. We now list the expressions of the matrix elements with the asterisk symbols:

$$R_{3I}^* = R_{3I} + \int_{R^e} \frac{N_I}{r} N_J u_{1J}^t \, dR, \tag{6.2.2.12c}$$

$$c_{IJ}^* = c_{IJ} + \int_{R^e} \frac{N_I}{r} N_J \, dR, \qquad dR = 2\pi r \, dr \, dz. \tag{6.2.2.12d}$$

As in the plane strain case, the remaining matrix elements belong to the stiffness submatrix $[K_{IJ}^*]$, which can be obtained from the formula given by Eq. (6.2.2.11b). For the axisymmetric case, the expressions are

$$[B_I]^{\mathrm{T}} = \begin{bmatrix} 0 & \frac{\partial N_I}{\partial r} & \frac{N_I}{r} & \frac{\partial N_I}{\partial z} \\ \frac{\partial N_I}{\partial z} & 0 & 0 & \frac{\partial N_I}{\partial r} \end{bmatrix}, \qquad [B_J] = \begin{bmatrix} 0 & \frac{\partial N_J}{\partial z} \\ \frac{\partial N_J}{\partial r} & 0 \\ \frac{N_J}{r} & 0 \\ \frac{\partial N_J}{\partial z} & \frac{\partial N_J}{\partial r} \end{bmatrix}, \tag{6.2.2.12e}$$

$$[D] = \frac{\lambda(1-\nu)}{\nu} \begin{bmatrix} 1 & \frac{\nu}{1-\nu} & \frac{\nu}{1-\nu} & 0 \\ & 1 & \frac{\nu}{1-\nu} & 0 \\ & & 1 & 0 \\ \text{sym} & & & \frac{1-2\nu}{2(1-\nu)} \end{bmatrix}, \tag{6.2.2.12f}$$

where sym means the matrix is symmetric. Performing the matrix multiplication, we obtain

$$[K_{IJ}^*] = \begin{bmatrix} a_{IJ}^* & b_{IJ}^* \\ (b_{IJ}^*)^{\mathrm{T}} & d_{IJ} \end{bmatrix},$$

where

$$a_{IJ}^* = a_{IJ} + \int_{R^e} \left[(\lambda + 2G) \frac{N_I N_J}{r^2} + \lambda \left(\frac{N_I}{r} \frac{\partial N_J}{\partial r} + \frac{\partial N_I}{\partial r} \frac{N_J}{r} \right) \right] dR,$$

$$b_{IJ}^* = b_{IJ} + \int_{R^e} \lambda \frac{N_I}{r} \frac{\partial N_J}{\partial z} \, dR,$$

$$c_{IJ}^* = c_{IJ} + \int_{R^e} \frac{N_I}{r} N_J \, dR.$$

Solution Procedure

Once the coefficient matrix and the right-hand-side vector of Eq. (6.2.2.10) are formed, the equation can be solved for the nodal values of the displacements and pore pressure at any time value $t + \Delta t$ using the nodal values at the previous time t. It is important to recognize that the response at time $t = 0^+$ of the system is purely elastic because no dissipation of the fluid has taken place. To capture this initial response, the use of a very small Δt is recommended for the first time step. For the subsequent time steps larger values can be used. It is often advantageous to work in terms of the dimensionless time factor $\bar{t}$, defined as $\bar{t} = 2Gk/(\rho_f g L^2)$, where g is the gravitational constant and L a characteristic length of the flow system. Experience indicates that in consolidation processes the variation of the state variables is logarithmic in time. Thus, it is possible to obtain computational saving by increasing the size of the time steps logarithmically until a new load step is applied.

Caution must be exercised in the choice of Δt for problems where the permeability is low and/or the solid skeleton is stiff. In these cases, there will be several orders of magnitude difference in the values of the elements in the stiffness matrix $[K_{IJ}]$ and the flow matrix $[f_{IJ}]$. As pointed out by Ghaboussi and Wilson (1973), this may result in the loss of significant digits in solving the system of algebraic equations. The use of enhanced precision is highly recommended.

Material Nonlinearity and Plasticity

It is known from experiments that soils generally have nonlinear and partly irreversible stress–strain relations. Several models are available for describing this type of material nonlinearity. The one adopted here is a model for isotropic elastic–plastic solids. Before proceeding to modify the preceding finite element formulation to take account of material nonlinearity, it is appropriate to introduce the fundamental concepts in the theory of plasticity [see also Hill (1950)].

A material is classified as elastic–plastic if, after loading beyond a certain limit, a permanent (plastic) deformation remains after unloading. This situation is illustrated in Fig. 6.2, where the effective stress $\bar{\sigma}$ and the effective strain $\bar{\varepsilon}$ are invariant quantities expressed in terms of stress and strain components as follows (Suh and Turner, 1975):

$$\bar{\sigma} = \{\tfrac{1}{2}[(\sigma_{11} - \sigma_{22})^2 + (\sigma_{22} - \sigma_{33})^2 + (\sigma_{33} - \sigma_{11})^2] + 3(\sigma_{12}^2 + \sigma_{13}^2 + \sigma_{23}^2)\}^{1/2},$$
$$\bar{\varepsilon} = \tfrac{2}{3}\{\tfrac{1}{2}[(\varepsilon_{11} - \varepsilon_{22})^2 + (\varepsilon_{22} - \varepsilon_{33})^2 + (\varepsilon_{33} - \varepsilon_{11})^2] + 3(\varepsilon_{12}^2 + \varepsilon_{13}^2 + \varepsilon_{23}^2)\}^{1/2}.$$

The relation between $\bar{\sigma}$ and $\bar{\varepsilon}$ may be established by a uniaxial test conducted in a laboratory.

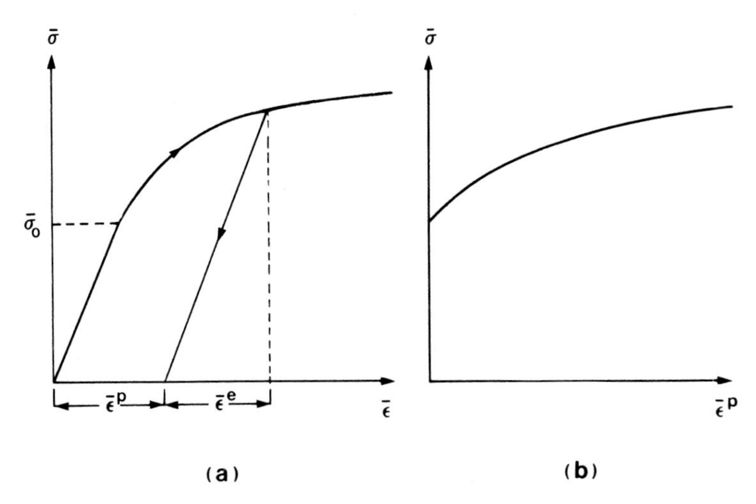

Fig. 6.2. Effective stress–strain relations: (a) $\bar{\sigma}$ versus $\bar{\varepsilon}$: $E = \bar{\sigma}/\bar{\varepsilon}^e$, where $\bar{\varepsilon}^e$ is the elastic strain and $\bar{\varepsilon}^p$ the plastic strain. (b) $\bar{\sigma}$ versus $\bar{\varepsilon}^p$: $S = d\bar{\sigma}/d\bar{\varepsilon}^p$; $\bar{\varepsilon}^p = \varepsilon - \bar{\varepsilon}^e$.

In general, plastic behavior of a solid may be described by three basic concepts in the theory of plasticity: the yield condition, the flow rule, and the material-hardening rule (Prager, 1959).

(a) YIELD CONDITION This is a condition concerning the limit of elasticity under any possible combination of stresses. It is used to indicate the onset of plastic flow. In general, the yield criterion may be expressed as

$$\bar{F}(\{\sigma\}) = \bar{\sigma}_0, \tag{6.2.2.13a}$$

where $\bar{F}$ is a scalar function of the stress components collected in the column vector $\{\sigma\}$ and $\bar{\sigma}_0$ (see Fig. 6.2) is the initial yield stress.

For a given value of $\bar{\sigma}_0$, the function $\bar{F}$ forms a surface in the principal stress space. Several yield criteria are available. Some of the useful criteria for soils and rocks are described by Boonlualohr (1977), Callahan and Fossum (1980), Nayak (1971), and Zienkiewicz and Pande (1977). We elect to present the criterion proposed by Drucker and Prager (1952) and the widely accepted linear Mohr–Coulomb criterion (other well-known criteria include the Tresca and the von Mises criteria). According to the Drucker–Prager or modified von Mises criterion, Eq. (6.2.2.13a) takes the form

$$\alpha J_1 + (J_2')^{1/2} = \bar{\sigma}_0, \tag{6.2.2.13b}$$

where α and $\bar{\sigma}_0$ are constants depending on the cohesion c and the internal angle of friction ϕ of the material (do not confuse this with the porosity of the porous medium, which is also denoted by the symbol ϕ), J_1 is the first invariant of the stress tensor, and J_2' is the second invariant of the deviatoric stress tensor. The two stress invariants are defined as

$$J_1 = \sigma_{11} + \sigma_{22} + \sigma_{33},$$

$$J_2' = \tfrac{1}{6}[(\sigma_{11} - \sigma_{22})^2 + (\sigma_{22} - \sigma_{33})^2 + (\sigma_{33} - \sigma_{11})^2] + \sigma_{12}^2 + \sigma_{23}^2 + \sigma_{31}^2.$$

In terms of the principal stresses σ_1, σ_2, and σ_3, J_1 and J_2' are given by

$$J_1 = \sigma_1 + \sigma_2 + \sigma_3,$$

$$J_2' = \tfrac{1}{6}[(\sigma_1 - \sigma_2)^2 + (\sigma_2 - \sigma_3)^2 + (\sigma_3 - \sigma_1)^2].$$

Note that, given σ_{ij}, we can determine the principal stresses by solving for the roots of the cubic equation (Fung, 1965, p. 73):

$$\sigma^3 - J_1\sigma^2 + J_2\sigma - J_3 = 0,$$

where J_2 and J_3 are the second and third invariants of the stress tensor:

$$J_2 = \begin{vmatrix} \sigma_{11} & \sigma_{12} \\ \sigma_{21} & \sigma_{22} \end{vmatrix} + \begin{vmatrix} \sigma_{22} & \sigma_{23} \\ \sigma_{32} & \sigma_{33} \end{vmatrix} + \begin{vmatrix} \sigma_{11} & \sigma_{13} \\ \sigma_{31} & \sigma_{33} \end{vmatrix}$$

and

$$J_3 = \begin{vmatrix} \sigma_{11} & \sigma_{12} & \sigma_{13} \\ \sigma_{21} & \sigma_{22} & \sigma_{23} \\ \sigma_{31} & \sigma_{32} & \sigma_{33} \end{vmatrix}.$$

Returning now to Eq. (6.2.2.13b), several expressions for the constants α and $\bar{\sigma}_0$ exist in the literature. We elect to use the following representations (Owen and Hinton, 1980, p. 220):

$$\alpha = \frac{2 \sin \phi}{\sqrt{3}\,(3 - \sin \phi)}, \qquad \bar{\sigma}_0 = \frac{6c\,(\cos \phi)}{\sqrt{3}\,(3 - \sin \phi)}.$$

Equation (6.2.2.13b) can now be written in the form

$$\alpha(\sigma_1 + \sigma_2 + \sigma_3) + 1/\sqrt{6}\,[(\sigma_1 - \sigma_2)^2 + (\sigma_2 - \sigma_3)^2 + (\sigma_3 - \sigma_1)^2]^{1/2} = \bar{\sigma}_0, \qquad (6.2.2.13c)$$

which, when plotted in the principal stress space, appears as the yield surface shown in Fig. 6.3a. This surface takes the shape of a cone with the axis pointing in the direction where the three principal stresses are equal.

As an alternative to the Drucker–Prager criterion, we also present the

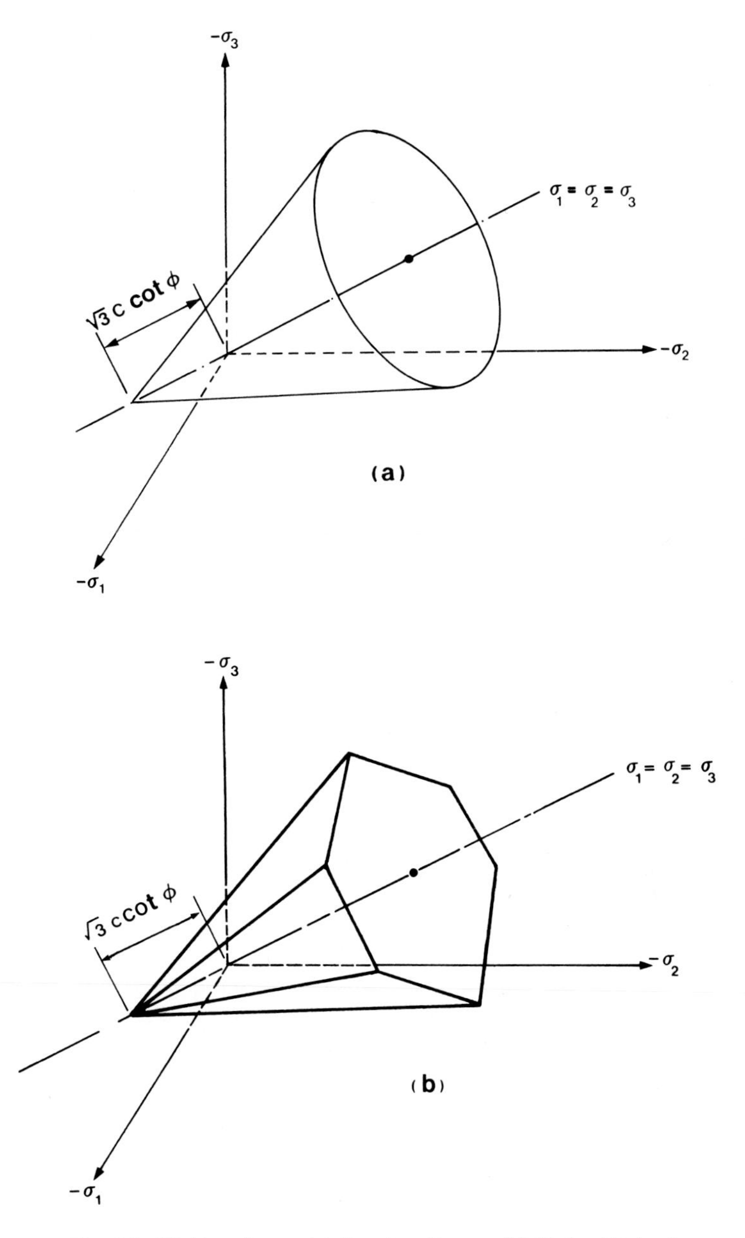

Fig. 6.3. Yield surfaces: (a) Drucker–Prager; (b) Mohr–Coulomb.

linearized Mohr–Coulomb criterion. According to the latter, the yield equation (6.2.2.13a) takes the form (Owen and Hinton, 1980, p. 230)

$$\tfrac{1}{3}(\sin\phi)J_1 + (\cos\theta_0 - (1/\sqrt{3})\sin\theta_0\sin\phi)(J_2')^{1/2} = c(\cos\phi), \tag{6.2.2.13d}$$

where θ_0 is commonly referred to as Lode's angle, and is given by

$$-\pi/6 \leqslant \theta_0 = \tfrac{1}{3}\sin^{-1}[(-3\sqrt{3}/2)(J_3'/J_2')^{3/2}] \leqslant \pi/6, \tag{6.2.2.13e}$$

where J_3' is given by

$$J_3' = \tfrac{1}{3}\sigma_{ij}'\sigma_{jk}'\sigma_{ki}' = \sigma_1'\sigma_2'\sigma_3'$$

and

$$\sigma_i' = \sigma_i - \tfrac{1}{3}(\sigma_1 + \sigma_2 + \sigma_3), \qquad i = 1, 2, 3.$$

The yield surface described by Eq. (6.2.2.13d) is the hexagonal cone shown in Fig. 6.3b.

(b) FLOW RULE This rule enables us to determine the plastic strain increments $\{d\varepsilon^p\}$. According to the flow rule, the strain increments can be derived from a function called the "plastic potential" and represented by $\overline{G}(\{\sigma\})$. First, it is assumed that such a function exists and second, the plastic strain increments are proportional to the gradient of $\overline{G}$. Thus,

$$\{d\varepsilon^{\mathrm{p}}\} = d\lambda\{\partial\overline{G}/\partial\sigma\}, \tag{6.2.2.14}$$

where, for a three-dimensional situation,

$$\left\{\frac{\partial\overline{G}}{\partial\sigma}\right\}^T = \left\{\frac{\partial\overline{G}}{\partial\sigma_{11}}, \frac{\partial\overline{G}}{\partial\sigma_{22}}, \frac{\partial\overline{G}}{\partial\sigma_{33}}, \frac{\partial\overline{G}}{\partial\sigma_{12}}, \frac{\partial\overline{G}}{\partial\sigma_{23}}, \frac{\partial\overline{G}}{\partial\sigma_{31}}\right\}$$

and $d\lambda$, the plasticity multiplier, is a proportionality constant to be determined. A flow rule is said to be "associative" or "associated" when $\overline{G}$ is chosen to be identical to $\overline{F}$; otherwise, it is said to be "nonassociative." For simplicity, we adopt the associative flow rule (cf. Cook, 1974),

$$\{d\varepsilon^{\mathrm{p}}\} = d\bar{\varepsilon}^{\mathrm{p}}\{\partial\overline{F}/\partial\sigma\}, \tag{6.2.2.15a}$$

where $d\bar{\varepsilon}^{\mathrm{p}}$ is the effective plastic strain increment.

One objection to the associative flow rule is the fact that this rule often leads to large extensional volumetric strains that are not observed experimentally, even for dense soils for which some dilation is observed. For this reason, several researchers (Davies, 1968; Hagmann, 1971) have suggested the use of nonassociative flow rules.

(c) MATERIAL HARDENING RULE This rule specifies the subsequent modification of the initial yield surface during plastic deformation. Again, several rules are available. We elect to use the simplest one, called the isotropic strain hardening rule. According to this rule, the subsequent yield surfaces may be represented by

$$\overline{F}(\{\sigma\}) = \overline{\sigma}(\kappa^*), \tag{6.2.2.15b}$$

where κ^*, called a hardening parameter, is assumed to be a function of the effective plastic strain $\overline{\varepsilon}^{\mathrm{p}}$. Because κ^* is a function of $\overline{\varepsilon}^{\mathrm{p}}$, Eq. (6.2.2.15b) can be written in the form

$$\overline{F}(\{\sigma\}) = \overline{\sigma}(\overline{\varepsilon}^{\mathrm{p}}). \tag{6.2.2.15c}$$

Incremental Stress–Strain Relation

The three basic concepts in the theory of plasticity can now be used to develop an incremental stress–strain relation analogous to Hooke's law but valid beyond the elastic limit. First we recognize that the total strain increment consists of the elastic and plastic parts and write

$$\{d\varepsilon\} = \{d\varepsilon^{e}\} + \{d\varepsilon^{\mathrm{p}}\}. \tag{6.2.2.16a}$$

Because the elastic strain and stress increments are related by the generalized Hooke's law, Eq. (6.2.2.16a), becomes

$$\{d\varepsilon\} = [D]^{-1}\{d\sigma\} + \{d\varepsilon^{\mathrm{p}}\}, \tag{6.2.2.16b}$$

where $[D]$ is the elasticity matrix given earlier for the plane strain and axisymmetric cases.

From Eq. (6.2.2.16b) it follows that

$$\{d\sigma\} = [D](\{d\varepsilon\} - \{d\varepsilon^{\mathrm{p}}\}). \tag{6.2.2.17}$$

Combination of Eqs. (6.2.2.15a) and (6.2.2.17) gives

$$\{d\sigma\} = [D](\{d\varepsilon\} - d\overline{\varepsilon}^{\mathrm{p}}\{\partial\overline{F}/\partial\sigma\}). \tag{6.2.2.18}$$

Premultiplying Eq. (6.2.2.18) by $\{\partial\overline{F}/\partial\sigma\}^{\mathrm{T}}$, we obtain

$$\{\partial\overline{F}/\partial\sigma\}^{\mathrm{T}}\{d\sigma\} = \{\partial\overline{F}/\partial\sigma\}^{\mathrm{T}}[D](\{d\varepsilon\} - d\overline{\varepsilon}^{\mathrm{p}}\{\partial\overline{F}/\partial\sigma\}). \tag{6.2.2.19}$$

From Eq. (6.2.2.15c), it follows that

$$\{\partial\overline{F}/\partial\sigma\}^{\mathrm{T}}\{d\sigma\} = d\overline{\sigma} = S\,d\overline{\varepsilon}^{\mathrm{p}}, \tag{6.2.2.20}$$

where S is the slope of the effective stress–effective plastic strain curve depicted in Fig. 6.2b.

Substituting Eq. (6.2.2.20) into (6.2.2.19) and solving for $d\bar{\varepsilon}^{p}$, we obtain

$$d\bar{\varepsilon}^{p} = \frac{\{\partial\bar{F}/\partial\sigma\}^{T}[D]}{S + \{\partial\bar{F}/\partial\sigma\}^{T}[D]\{\partial\bar{F}/\partial\sigma\}}\{d\varepsilon\} = \{W\}^{T}\{d\varepsilon\}, \quad (6.2.2.21a)$$

where $\{W\}^{T}$ is a row vector given by

$$\{W\}^{T} = \frac{\{\partial\bar{F}/\partial\sigma\}^{T}[D]}{S + \{\partial\bar{F}/\partial\sigma\}^{T}[D]\{\partial\bar{F}/\partial\sigma\}}. \quad (6.2.2.21b)$$

Combining Eqs. (6.2.2.18) and (6.2.2.21a) yields the required relation,

$$\{d\sigma\} = ([D] - [D]\{\partial\bar{F}/\partial\sigma\}\{W\}^{T})\{d\varepsilon\} = [D_{ep}]\{d\varepsilon\}, \quad (6.2.2.22a)$$

where

$$[D_{ep}] = [D] - [D]\{\partial\bar{F}/\partial\sigma\}\{W\}^{T} \quad (6.2.2.22b)$$

is the elastoplasticity matrix. Examination of the matrices involved shows that $[D_{ep}]$ is symmetric. Note that Eqs. (6.2.2.21) and (6.2.2.22) are valid even for elastic–ideally plastic materials as the values of the elements of $\{W\}^{T}$ and $[D_{ep}]$ remain finite when $S = 0$. A similar derivation can be performed using a nonassociated flow rule. In this case, the matrix $[D_{ep}]$ becomes asymmetric.

Evaluation of Derivatives of the Yield Function

To calculate the elastoplasticity matrix, we need to evaluate $\partial\bar{F}/\partial\sigma_{ij}$, usually called the flow vector. First, we write the yield function $\bar{F}$ in the form

$$\bar{F}(J_1, (J_2')^{1/2}, \theta_0) = 0. \quad (6.2.2.22c)$$

Differentiating $\bar{F}$ with respect to σ_{ij} and applying the chain rule, we obtain

$$\frac{\partial\bar{F}}{\partial\sigma_{ij}} = \frac{\partial\bar{F}}{\partial J_1}\frac{\partial J_1}{\partial\sigma_{ij}} + \frac{\partial\bar{F}}{\partial(J_2')^{1/2}}\frac{\partial(J_2')^{1/2}}{\partial\sigma_{ij}} + \frac{\partial\bar{F}}{\partial\theta_0}\frac{\partial\theta_0}{\partial\sigma_{ij}}. \quad (6.2.2.22d)$$

Differentiating θ_0 from Eq. (6.2.2.13e) with respect to σ_{ij}, we obtain

$$\frac{\partial\theta_0}{\partial\sigma_{ij}} = \frac{-\sqrt{3}}{2\cos 3\theta_0}\left(\frac{1}{(J_2')^{3/2}}\frac{\partial J_3'}{\partial\sigma_{ij}} - \frac{3J_3'}{(J_2')^{2}}\frac{\partial(J_2')^{1/2}}{\partial\sigma_{ij}}\right). \quad (6.2.2.22e)$$

Equations (6.2.2.22d) and (6.2.2.22e) can be combined and J_3' eliminated using Eq. (6.2.2.13e):

$$\frac{\partial\bar{F}}{\partial\sigma_{ij}} = \zeta_1\frac{\partial J_1}{\partial\sigma_{ij}} + \zeta_2\frac{\partial(J_2')^{1/2}}{\partial\sigma_{ij}} + \zeta_3\frac{\partial J_3'}{\partial\sigma_{ij}}, \quad (6.2.2.22f)$$

where

$$\zeta_1 = \frac{\partial \overline{F}}{\partial J_1}, \qquad \zeta_2 = \frac{\partial \overline{F}}{\partial (J_2')^{1/2}} - \frac{\tan 3\theta_0}{(J_2')^{1/2}} \frac{\partial \overline{F}}{\partial \theta_0},$$

$$\zeta_3 = \left(\frac{-\sqrt{3}}{2 \cos 3\theta_0}\right) \frac{1}{(J_2')^{3/2}} \frac{\partial \overline{F}}{\partial \theta_0}.$$

Next we note that J_1, J_2', and J_3' are defined as

$$J_1 = \sigma_{ii}, \tag{6.2.2.22g}$$

$$(J_2')^{1/2} = (\tfrac{1}{2}\sigma_{ij}' \sigma_{ij}')^{1/2}, \tag{6.2.2.22h}$$

$$J_3' = \tfrac{1}{3}\sigma_{ij}'\sigma_{jk}'\sigma_{ki}', \tag{6.2.2.22i}$$

where

$$\sigma_{ij}' = \sigma_{ij} - \tfrac{1}{3}J_1.$$

Differentiating these equations with respect to σ_{ij} one obtains

$$\frac{\partial J_1}{\partial \sigma_{ij}} = \delta_{ij}, \tag{6.2.2.22j}$$

$$\frac{\partial (J_2')^{1/2}}{\partial \sigma_{ij}} = \frac{1}{2(J_2')^{1/2}} S_{ij}, \tag{6.2.2.22k}$$

$$\frac{\partial J_3'}{\partial \sigma_{ij}} = \sigma_{ik}' \sigma_{kj}' - \frac{2J_2'}{3} \delta_{ij}, \tag{6.2.2.22l}$$

where

$$S_{ij} = \begin{cases} \sigma_{ij}' & \text{if} \quad i = j \\ 2\sigma_{ij}' & \text{if} \quad i \neq j. \end{cases}$$

Combining Eqs. (6.2.2.22f) and (6.2.2.22j)–(6.2.2.22l) gives

$$\frac{\partial \overline{F}}{\partial \sigma_{ij}} = \zeta_1 \delta_{ij} + \zeta_2 \left(\frac{S_{ij}}{2(J_2')^{1/2}}\right) + \zeta_3 \left(\sigma_{ik}' \sigma_{kj}' - \tfrac{2}{3} J_2' \delta_{ij}\right). \tag{6.2.2.22m}$$

Equation (6.2.2.22m) is generally valid irrespective of the type of yield criterion used. Table 6.1 presents the list of constants ζ_i for the Drucker–Prager and the Mohr–Coulomb criteria.

Once the coefficients ζ_1, ζ_2, and ζ_3 are evaluated, the components of the flow vector $\partial \overline{F}/\partial \sigma_{ij}$ can be evaluated from Eq. (6.2.2.22m) and is expressed in matrix form as

$$\left\{\frac{\partial \overline{F}}{\partial \sigma}\right\}^{\mathrm{T}} = \left\{\frac{\partial \overline{F}}{\partial \sigma_{11}}, \frac{\partial \overline{F}}{\partial \sigma_{22}}, \frac{\partial \overline{F}}{\partial \sigma_{33}}, \frac{\partial \overline{F}}{\partial \sigma_{12}}, \frac{\partial \overline{F}}{\partial \sigma_{23}}, \frac{\partial \overline{F}}{\partial \sigma_{31}}\right\}. \tag{6.2.2.22n}$$

Table 6.1
Yield Criteria and Associated Yield Functions and Constants of the Flow Vector

Yield criterion	Yield function	Coefficients of the flow vector		
		ζ_1	ζ_2	ζ_3
Drucker–Prager	$\overline{F} = \alpha J_1 + (J_2')^{1/2} - \overline{\sigma}_0$	α	1	0
Mohr–Coulomb	$\overline{F} = \left(\frac{\sin\phi}{3}\right)J_1 + \left[\cos\theta_0 - \left(\frac{\sin\theta_0 \sin\phi}{\sqrt{3}}\right)\right] \times (J_2')^{1/2} - c(\cos\phi)$	$\frac{\sin\phi}{3}$	$\cos\theta_0[(1 + \tan\theta_0 \tan 3\theta_0) + \frac{\sin\phi}{\sqrt{3}}(\tan 3\theta_0 - \tan\theta_0)]$	$\frac{(\sqrt{3}\sin\theta_0 + \cos\theta_0 \sin\phi)}{(2J_2' \cos 3\theta_0)}$

Treatment of Nonlinear Equations

Using $[D_{ep}]$ instead of $[D]$ in the case of plastically deforming elements, the coefficient matrix $[H]^e$ of Eq. (6.2.2.10) can be computed using $[D_{ep}]$ instead of $[D]$. One first constructs the stiffness submatrix $[K]^e$ using the formula given by Eq. (6.2.2.11b), and then generates the remaining submatrices in the same manner as for the elastic case. The resulting global system of algebraic equations is nonlinear because $[D_{ep}]$ depends on the solution. This nonlinear system can be solved using one or a combination of the following methods:

(a) NONITERATIVE INCREMENTAL METHOD In this method, we write the global matrix equation in an incremental form as

$$[H]_t\{\Delta\chi\} = \{\Delta R\}, \tag{6.2.2.23a}$$

where

$$\{\Delta\chi\} = \{\chi\}_{t+\Delta t} - \{\chi\}_t \tag{6.2.2.23b}$$

and

$$\{\Delta R\} = \{R\}_{t+\Delta t} - [H]_t\{\chi\}_t. \tag{6.2.2.23c}$$

A typical step in the solution is described with the aid of Fig. 6.4, which is a single-degree-of-freedom representation of the actual multiple-degrees-of-freedom problem. First we compute the coefficient matrix $[H]_t$ using known displacement values at time t. This matrix corresponds to the

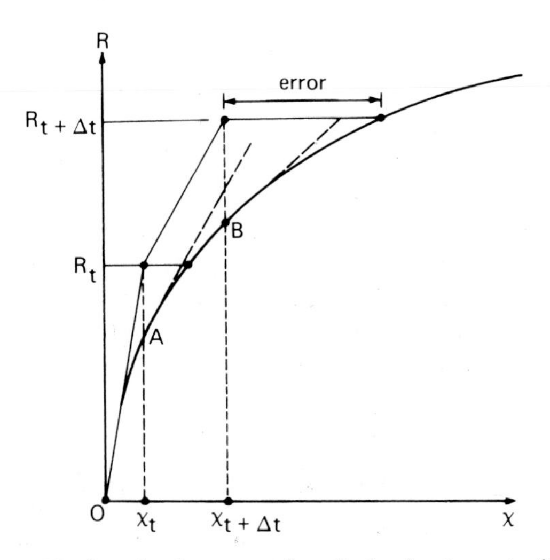

Fig. 6.4. Noniterative incremental method using tangent stiffness.

tangent at point A. The nodal values $\{\chi\}_{t+\Delta t}$ produced by the next load increment $\{\Delta R\}$ are then computed from Eqs. (6.2.2.23a) and (6.2.2.23b). Stresses, strains, and tangent element stiffness submatrices must now be updated. Incremental displacements $\{\Delta u\}$ of an element are extracted from $\{\Delta\chi\}$. For each sampling (Gaussian) point in the element, we evaluate $\{\Delta\varepsilon\}$ from

$$\{\Delta\varepsilon\} = [B]\,\{\Delta u\}, \tag{6.2.2.24a}$$

where, for the plane strain and axisymmetric problems, typical elements of $[B]$ are given in Eqs. (6.2.2.11c) and (6.2.2.12e). Knowing $\{\Delta\varepsilon\}$, we can then evaluate the incremental plastic strain and stress components from

$$\Delta\bar{\varepsilon}^{\mathrm{p}} = \int_t^{t+\Delta t} \{W\}^{\mathrm{T}}\{d\varepsilon\} \simeq \{W\}_t^{\mathrm{T}}\{\Delta\varepsilon\}, \tag{6.2.2.24b}$$

$$\{\Delta\varepsilon^{\mathrm{p}}\} = \int_t^{t+\Delta t} \{\partial\bar{F}/\partial\sigma\}\, d\bar{\varepsilon}^{\mathrm{p}} \simeq \{\partial\bar{F}/\partial\sigma\}_t\, \Delta\bar{\varepsilon}^{\mathrm{p}}, \tag{6.2.2.24c}$$

$$\Delta\bar{\sigma} = \int_t^{t+\Delta t} S\, d\bar{\varepsilon}^{\mathrm{p}} \simeq S_t\, \Delta\bar{\varepsilon}^{\mathrm{p}}, \tag{6.2.2.24d}$$

$$\{\Delta\sigma\} = [D](\{\Delta\varepsilon\} - \{\Delta\varepsilon^{\mathrm{p}}\}). \tag{6.2.2.24e}$$

Equations (6.2.2.24b)–(6.2.2.24e) are simply first-order forward difference approximations of Eqs. (6.2.2.21a), (6.2.2.15a), (6.2.2.20), and (6.2.2.17), respectively. Next we compute the stress and strain quantities at time $t+\Delta t$ and use these to update the elastoplasticity submatrix $[D_{ep}]$, the stiffness submatrix $[K]^e$, and finally the coefficient matrix $[H]^e$. Elements are assembled in the usual way to obtain the global matrix $[H]$. As soon as $[H]$ is produced, another load increment $\{\Delta R\}$ can be applied and the time-marching procedure can be repeated.

Frequently, we encounter "transitional elements" with strain conditions that are elastic prior to the application of the load increment and elastic–plastic afterward. An approximate way to account for such a transitional behavior is illustrated in Fig. 6.5. First, we estimate the next strain increment from results of the previous load increment and compute the expected change in $\bar{\sigma}$ using the elastic coefficients. This is equivalent to moving from A to B on the stress–strain curve. Thus, one can write

$$\begin{aligned}\{\Delta\sigma\} &= [D]\{\Delta\varepsilon^{e}\} + [D_{\mathrm{ep}}]\{\Delta\varepsilon^{\mathrm{p}}\} \\ &= m[D]\{\Delta\varepsilon\} + (1 - m)[D_{\mathrm{ep}}]\{\Delta\varepsilon\},\end{aligned} \tag{6.2.2.25}$$

where m is the fraction of the total strain increment $\Delta\bar{\varepsilon}$ required to cause

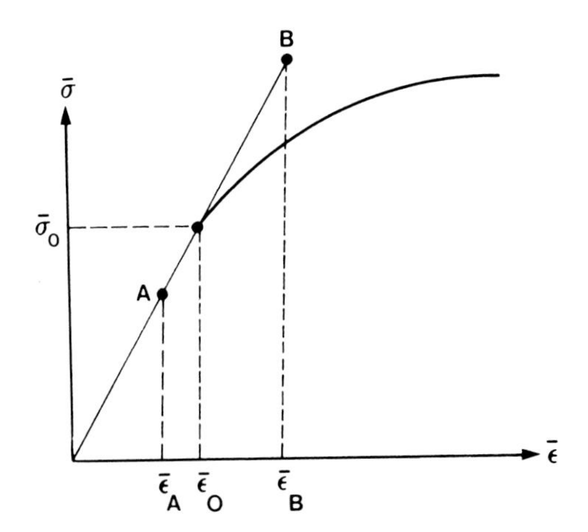

Fig. 6.5. Elastic–plastic transition. $m = (\bar{\varepsilon}_0 - \varepsilon_A)/(\bar{\varepsilon}_B - \bar{\varepsilon}_A)$.

yield. Equation (6.2.2.25) can be put in the form

$$\{\Delta\sigma\} = [D^*_{\rm ep}]\{\Delta\varepsilon\}, \tag{6.2.2.26}$$

where

$$\begin{aligned}[D^*_{\rm ep}] &= m[D] + (1 - m)[D_{\rm ep}] \\ &= [D] - (1 - m)[D]\{\partial\overline{F}/\partial\sigma\}\{W\}^{\rm T}.\end{aligned}$$

Hence, for the transitional elements, we would use $[D^*_{ep}]$ instead of $[D_{ep}]$ to compute the stiffness submatrix $[K]$.

It should be noted that the incremental method is based on a straightforward approximation of the nonlinear equations using linear increments. The cumulative error arising at the end of a particular increment is depicted in Fig. 6.4. Unless the increments (and hence the time steps) are kept sufficiently small, there will be a considerable drift from the true solution. This type of error may be corrected by proper combination of the preceding procedure with the Newton–Raphson technique.

(b) NEWTON–RAPHSON CORRECTION SCHEMES Two possible combinations of the Newton–Raphson and the incremental procedures are illustrated in Figs. 6.6a and 6.6b. In the first scheme, the tangent stiffness matrices are updated at each iteration of a particular load increment. To describe the iterative procedure, we write the global matrix equation in the form

$$[H]^r\,\{\Delta\chi\}^{r+1} = \{\Delta R\}^r, \tag{6.2.2.27}$$

where r and $r + 1$ denote the old and new iterations at $t + \Delta t$,

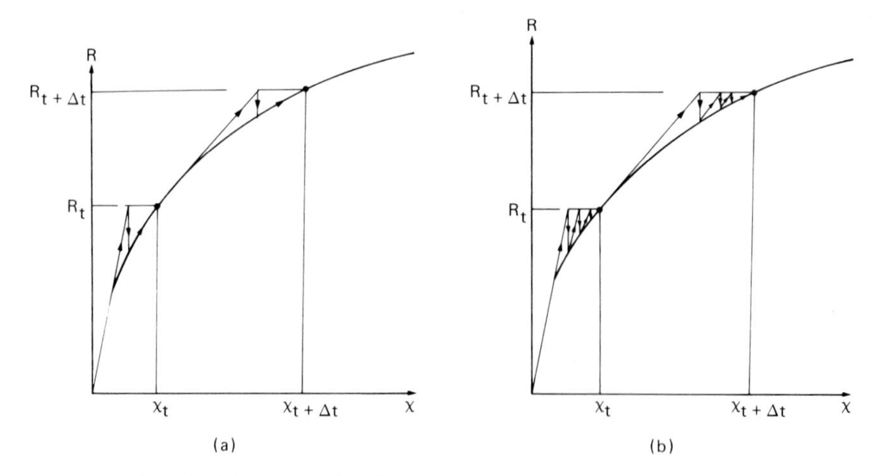

Fig. 6.6. Combined incremental and Newton–Raphson correction method: (a) variable slope; (b) constant (initial) slope.

$\{\Delta\chi\}^{r+1} = \{\chi\}^{r+1} - \{\chi\}_t$ and $\{\Delta R\}^r = \{R\}^r - [H]^r\{\chi\}_t$. For the first solution we calculate $[H]^r$ using the nodal values at the previous time level and solve Eq. (6.2.2.27) for $\{\Delta\chi\}^{r+1}$. These newly obtained nodal values are then used to recalculate $[H]^r$ and $\{\Delta R\}^r$ for the next iteration. The iterative cycle is repeated until satisfactory convergence of displacements is achieved.

In the second scheme, the initial tangent stiffness matrices are used in the first solution and subsequent iterations. To describe this iterative procedure we write the global matrix equation in the form

$$[H]_t\{\Delta\chi\}^{r+1} = \{\Delta R\}^r, \tag{6.2.2.28}$$

where $[H]_t$ represents $[H]$ evaluated at the last known time level. For the first solution, we calculate $[H]_t$ and solve Eq. (6.2.2.28) for $\{\Delta\chi\}^{r+1}$. For the subsequent iterations only the right-hand-side vector $\{\Delta R\}^r$ is updated using the most current nodal values of displacements. The iterative cycle is repeated until satisfactory convergence is achieved.

From the comparison of Figs. 6.6a and 6.6b it is seen that the initial slope scheme takes more iterations than the variable slope scheme to converge. However, the latter requires more computational time per iteration because the matrix $[H]$ needs to be generated and decomposed at each iteration. With the initial slope scheme, the solutions in the subsequent iterations can be obtained by performing only backward substitutions.

6.2.3 *Formulation of Areal Subsidence Problems*

The phenomenon of land subsidence is usually associated with the withdrawal of fluids from underground reservoirs by pumping wells. When well interference is encountered, the analysis based upon the rigorous Biot consolidation theory requires solution of the three-dimensional displacement and fluid flow equations. For single-phase flow, one has to solve four coupled partial differential equations. Despite the versatility of the finite element method and the generality of the formulation presented earlier, such an analysis involves considerable computational effort and is extremely costly. The discussion is restricted to linear electric solids.

In this section, we present an approach that leads to a significant reduction in computational effort. For many situations, this approach yields results consistent with field observations. It is convenient once again to formulate the problem in terms of the increments of the dependent variables. The increment of a given variable is defined as the difference between its value at time t and its initial value. Thus, if we let $\bar{\tau}_{ij}$ denote the incremental stress, it follows that

$$\tau_{ij} = \tau_{ij}^0 + \bar{\tau}_{ij}, \tag{6.2.3.1}$$

where τ_{ij}^0 is the initial total stress. By using Eq. (6.2.3.1), the equilibrium equation becomes

$$\frac{\partial \tau_{ij}^0}{\partial x_j} + F_i + \frac{\partial \bar{\tau}_{ij}}{\partial x_j} = 0. \tag{6.2.3.2}$$

Because the body is initially in equilibrium, Eq. (6.2.3.2) reduces to

$$\partial \bar{\tau}_{ij}/\partial x_j = 0. \tag{6.2.3.3}$$

If we start with Eq. (6.2.3.3) and follow the procedure described in Section 6.2.1 for formulating the displacement equations, the following result will be obtained:

$$G \frac{\partial^2 \bar{u}_1}{\partial x_j \, \partial x_j} + (\lambda + G) \frac{\partial \bar{\varepsilon}^*}{\partial x_1} - \frac{\partial \bar{p}}{\partial x_1} = 0, \tag{6.2.3.4a}$$

$$G \frac{\partial^2 \bar{u}_2}{\partial x_j \, \partial x_j} + (\lambda + G) \frac{\partial \bar{\varepsilon}^*}{\partial x_2} - \frac{\partial \bar{p}}{\partial x_2} = 0, \tag{6.2.3.4b}$$

$$G \frac{\partial^2 \bar{u}_3}{\partial x_j \, \partial x_j} + (\lambda + G) \frac{\partial \bar{\varepsilon}^*}{\partial x_3} - \frac{\partial \bar{p}}{\partial x_3} = 0, \tag{6.2.3.4c}$$

where $\bar{u}_i$ is the incremental displacement, $\bar{p}$ the incremental fluid pressure, and $\bar{\varepsilon}^*$ the incremental dilation or volumetric strain equal to $\partial \bar{u}_j/\partial x_j$.

Similarly, the governing equation for fluid flow can be derived by writing Darcy's law in terms of incremental pressure and combining this with the continuity equation. The result is

$$\frac{\partial}{\partial x_i}\left(\frac{k}{\mu}\frac{\partial \bar{p}}{\partial x_i}\right) = \phi\beta\frac{\partial \bar{p}}{\partial t} + \frac{\partial}{\partial t}\left(\frac{\partial \bar{u}_i}{\partial x_i}\right), \tag{6.2.3.5}$$

assuming the initial velocity field to be divergence free. Differentiating Eq. (6.2.3.4a) with respect to x_1, Eq. (6.2.3.4b) with respect to x_2, and Eq. (6.2.3.4c) with respect to x_3, and then adding all three, we obtain (Verruijt, 1969)

$$(\lambda + 2G)\frac{\partial^2 \bar{\varepsilon}^*}{\partial x_j\, \partial x_j} = \frac{\partial^2 \bar{p}}{\partial x_j\, \partial x_j}. \tag{6.2.3.6}$$

Integration of Eq. (6.2.3.6) gives

$$(\lambda + 2G)\bar{\varepsilon}^* = \bar{p} + f,$$

or

$$\bar{\varepsilon}^* = \alpha(\bar{p} + f), \tag{6.2.3.7}$$

where $\alpha = 1/(\lambda + 2G)$ and f is an unknown function of x_1, x_2, x_3, and t which must satisfy Laplace's equation, $\partial^2 f/\partial x_j\, \partial x_j = 0$, for all time t.

A simplification of the problem can be made by assuming that f is negligible compared with $\bar{p}$. Then Eq. (6.2.3.7) reduces to

$$\bar{\varepsilon}^* = \partial \bar{u}_j/\partial x_j = \alpha\bar{p}. \tag{6.2.3.8}$$

Substituting Eq. (6.2.3.8) into Eq. (6.2.3.5), we obtain

$$\frac{\partial}{\partial x_i}\left[\frac{k}{\mu}\frac{\partial \bar{p}}{\partial x_i}\right] = \left(\alpha + \phi\beta\right)\frac{\partial \bar{p}}{\partial t}. \tag{6.2.3.9}$$

Equation (6.2.3.9) is the governing equation of Terzaghi's (1943) consolidation theory in soil mechanics. This equation also corresponds to the flow equation [see Eq. (4.2.1.9b)] commonly recognized by groundwater hydrologists. It is thus evident that the assumption of vanishing f is necessary to enable us to reduce the complex system of partial differential equations describing Biot's consolidation theory to the simple flow equation underlying both Terzaghi's consolidation theory and the theory of groundwater hydrology. For a detailed discussion on the function f and the conditions required for the vanishing of f, the reader is referred to De Leeuw (1965) and Verruijt (1969).

Granted that Eq. (6.2.3.9) is valid, we can solve it for the pore pressure distribution. Once $\bar{p}$ is known, the dilation $\bar{\varepsilon}^*$ can be obtained from Eq. (6.2.3.8). The displacement components can then be determined by introducing a scalar function called the displacement potential Φ and solving the Poisson equation

$$\partial^2\Phi/\partial x_j\,\partial x_j = \bar{\varepsilon}^* = \alpha\bar{p}, \tag{6.2.3.10}$$

where we have employed the relation

$$\bar{u}_j = \partial\Phi/\partial x_j. \tag{6.2.3.11}$$

Based on this approach, the solution of the subsidence problem reduces to (1) solution of the fluid flow equation (6.2.3.9) for $\bar{p}$, (2) use of this result to solve Poisson's equation (6.2.3.10) for Φ and (3) calculation of the displacements using Eq. (6.2.3.11) and the known function Φ.

Where the land subsidence occurs over a large area, the horizontal displacements are usually small in comparison with the vertical displacement. Because the lateral dimensions of such a reservoir are large compared with its thickness, it is also reasonable to assume that the total vertical stress does not change during the consolidation of the reservoir. According to this assumption, the incremental vertical total stress is zero. Thus,

$$\bar{\tau}_{33} = \bar{\sigma}_{33} - \bar{p} = 0,$$

or

$$\bar{\tau}_{33} = 2G\bar{\varepsilon}_{33} + \lambda\bar{\varepsilon}^* - \bar{p} = 0. \tag{6.2.3.12}$$

Because $\bar{u}_1$ and $\bar{u}_2$ are negligible, $\bar{\varepsilon}^* \simeq \partial\bar{u}_3/\partial x_3$ and Eq. (6.2.3.12) becomes

$$(2G + \lambda)\bar{\varepsilon}^* = \bar{p},$$

or, with the introduction of $\alpha = 1/(2G + \lambda)$,

$$\bar{\varepsilon}^* = \alpha\bar{p}. \tag{6.2.3.13}$$

Comparing this equation with Eq. (6.2.3.7) we immeditely deduce that $f = 0$, and the flow equation reduces to Eq. (6.2.3.9).

As a result of these assumptions, the reservoir compaction reduces to vertical deformation, which can be evaluated from

$$\partial\bar{u}_3/\partial x_3 = \alpha\bar{p}, \tag{6.2.3.14}$$

where it should be noted that α corresponds to the vertical compressibility and the incremental pressure $\bar{p}$ corresponds to the change in pore pressure.

From Eq. (6.2.3.14), it follows that the reduction in reservoir thickness is given by

$$\Delta b = \Delta\bar{u}_3 = \int_{z_1}^{z_2} \alpha\bar{p}(z)\,dz, \qquad z = x_3, \tag{6.2.3.15}$$

where b is the thickness of the reservoir.

6.3 Discrete Fracture Flow Deformation Model

6.3.1 Governing Equations

The theory and finite element formulation presented thus far are strictly applicable to conventional unfractured or granular porous media. To deal with fractured media we must take into account fractures as well as porous matrix blocks. Currently there are two theoretical approaches to fractured porous medium systems. We first present the "discrete fracture" approach, which regards the fractured medium as a continuum in which there exist discrete discontinuities. Such discontinuities represent fractures whose geometries are assumed to be known. Based on the discrete fracture approach, several analytical and numerical studies of interesting flow problems have been developed. These include works by Gringarten (1971), Gringarten *et al.* (1974), Gureghian (1975), and Heber-Cinco *et al.* (1976).

The governing equations of a discrete fracture model consist of the equations for deformation and fluid flow that take place in the porous block and fracture subregions. Using Biot's assumptions, the governing equations in the porous matrix subregion are simply Biot's equations of consolidation, Eqs. (6.2.1.8) and (6.2.1.22). These equations can be re-written as

$$G \frac{\partial^2 u_i}{\partial x_j \, \partial x_j} + (\lambda + G) \frac{\partial^2 u_j}{\partial x_i \, \partial x_j} - \alpha \frac{\partial p}{\partial x_i} = -F_i - \frac{\partial \sigma_{ij}^0}{\partial x_j}, \tag{6.3.1.1a}$$

$$\frac{\partial}{\partial x_i} \left(\frac{k_{ij}}{\mu} \frac{\partial p}{\partial x_j} \right) = \phi \beta \frac{\partial p}{\partial t} + \alpha \frac{\partial}{\partial t} \left(\frac{\partial u_i}{\partial x_i} \right), \tag{6.3.1.1b}$$

where, for convenience, we assume that the flow takes place in a horizontal plane and all of the symbols are as described in the previous section. The governing equation for displacement of the fracture (joint) can be written in the form

$$\frac{\partial}{\partial x_j} \left[\frac{D_{ijkl}}{2} \left(\frac{\partial u_k}{\partial x_l} + \frac{\partial u_l}{\partial x_k} \right) - \alpha p \delta_{ij} \right] = -F_i - \frac{\partial \sigma_{ij}^0}{\partial x_j}, \tag{6.3.1.2a}$$

where D_{ijkl} is a constitutive tensor relating stress and strain and α is a parameter discussed in Section 6.2.1. Henceforth, we take $\alpha = 1$. We postpone the discussion of D_{ijkl} until Section 6.3.2. Equation (6.3.1.2a) is obtained simply by combining Eqs. (6.2.1.1) and (6.2.1.2) with the general constitutive stress–strain relation ($\sigma_{ij} = D_{ijkl}\varepsilon_{kl}$) and the strain–displacement gradient equation (6.2.1.7).

The remaining governing equation is the equation of flow in the fractures.

This equation takes the form

$$\frac{\partial}{\partial x_i}\left(\frac{k_{ij}^{\mathrm{f}}}{\mu}\frac{\partial p}{\partial x_j}\right) = \phi_{\mathrm{f}}\beta\frac{\partial p}{\partial t} + \frac{\partial}{\partial t}\left(\frac{\partial u_i}{\partial x_i}\right), \qquad (6.3.1.2b)$$

where k_{ij}^{f} is called the "fracture intrinsic permeability" tensor and ϕ_f is the porosity in the fracture. Note that ϕ_f is the ratio of the volume of voids in the fracture to the volume of fracture and that if we consider the case of fractures containing fluid only, $\phi_f = 1$.

To derive Eq. (6.3.1.2b), consider a typical fracture joint element depicted in Fig. 6.7. It is assumed that the axes x_1 and x_2 are in a horizontal plane. If the flow is laminar and predominantly in the x_1' direction, the average velocity component $V_1' \equiv V_{X_1'}$ and the pressure gradient $\partial p/\partial x_1'$ can be related by (Wilson, 1970)

$$V_1' = -\frac{b^2}{12\mu}\frac{\partial p}{\partial x_1'} = -\frac{k_{11}'^{\mathrm{f}}}{\mu}\frac{\partial p}{\partial x_1'}, \qquad (6.3.1.3)$$

where

$$k_{11}'^{\mathrm{f}} = b^2/12. \qquad (6.3.1.4)$$

Equation (6.3.1.3) is the expression for the average velocity of incompressible laminar flow between two parallel plates. Its derivation may

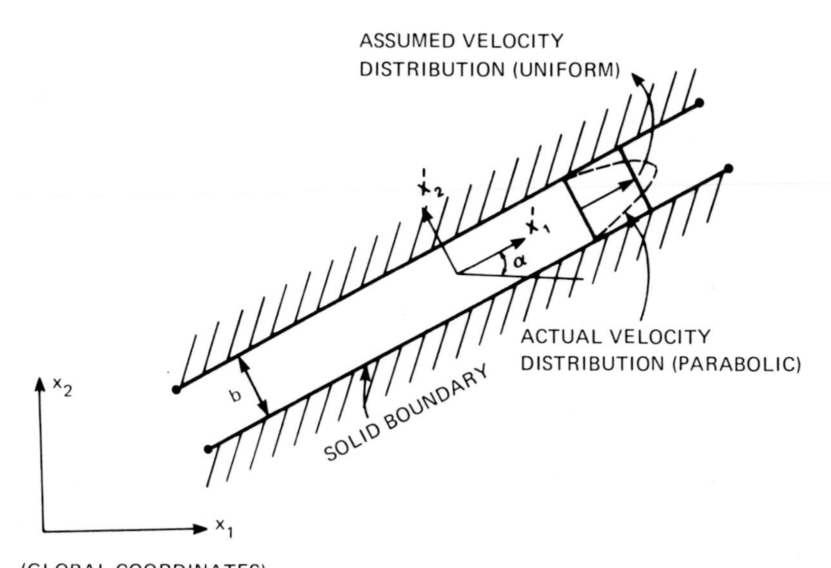

Fig. 6.7. Laminar flow in a fracture element bounded by two parallel planes.

be found in a standard text book of fluid mechanics, e.g., Streeter and Wiley (1979).

In a similar manner, we write the relation between the normal velocity and pressure gradient as

$$V_2' = -\frac{k_{22}'^{\mathrm{f}}}{\mu}\frac{\partial p}{\partial x_2'}, \tag{6.3.1.5}$$

where $k_{22}'^{\mathrm{f}}$ is the fracture permeability in the direction normal to the fracture plane. Equations (6.3.1.3) and (6.3.1.5) can be regarded as Darcy's law for fracture flow. They can be written

$$V_i' = -\frac{k_{ij}'^{\mathrm{f}}}{\mu}\frac{\partial p}{\partial x_j'}, \tag{6.3.1.6}$$

or, in global coordinates,

$$V_i = -\frac{k_{ij}^{\mathrm{f}}}{\mu}\frac{\partial p}{\partial x_j}, \tag{6.3.1.7}$$

where the transformation relation between the tensor components in global and local coordinate systems can be expressed in matrix form as

$$[k^{\mathrm{f}}] = [T]^{\mathrm{T}}[k'^{\mathrm{f}}][T], \tag{6.3.1.8}$$

in which

$$[k'^{\mathrm{f}}] = \begin{bmatrix} k_{11}'^{\mathrm{f}} & 0 \\ 0 & k_{22}'^{\mathrm{f}} \end{bmatrix} \tag{6.3.1.9a}$$

and $[T]$ is the coordinate transformation matrix given by

$$[T] = \begin{bmatrix} \cos\alpha & \sin\alpha \\ -\sin\alpha & \cos\alpha \end{bmatrix}. \tag{6.3.1.9b}$$

Substituting Eqs. (6.3.1.9a) and (6.3.1.9b) into (6.3.1.8) and multiplying, we obtain

$$k_{11}^{\mathrm{f}} = k_{11}'^{\mathrm{f}}\cos^2\alpha + k_{22}'^{\mathrm{f}}\sin^2\alpha, \tag{6.3.1.10a}$$

$$k_{12}^{\mathrm{f}} = (k_{11}'^{\mathrm{f}} - k_{22}'^{\mathrm{f}})\sin\alpha\cos\alpha, \tag{6.3.1.10b}$$

$$k_{21}^{\mathrm{f}} = k_{12}^{\mathrm{f}}, \tag{6.3.1.10c}$$

$$k_{22}^{\mathrm{f}} = k_{11}'^{\mathrm{f}}\sin^2\alpha + k_{22}'^{\mathrm{f}}\cos^2\alpha. \tag{6.3.1.10d}$$

The continuity of flow of a homogeneous, slightly compressible fluid in a deformable fracture can be written in the form

$$-\partial V_i/\partial x_i = \partial \zeta^*/\partial t, \tag{6.3.1.11}$$

where ζ^* is the volumetric storage due to both fluid compressibility and volumetric strain contributed by the change in displacement at the solid boundaries. Thus $\partial\zeta^*/\partial t$ is given by

$$\frac{\partial \zeta^*}{\partial t} = \phi_{\mathrm{f}}\beta \frac{\partial p}{\partial t} + \frac{\partial \varepsilon_{ii}}{\partial t}, \tag{6.3.1.12}$$

where $\varepsilon_{ii} = \partial u_i/\partial x_i$.

Combining Eqs. (6.3.1.11), (6.3.1.12), and (6.3.1.7), we obtain

$$\frac{\partial}{\partial x_i}\left(\frac{k_{ij}^{\mathrm{f}}}{\mu}\frac{\partial p}{\partial x_j}\right) = \phi_{\mathrm{f}}\beta \frac{\partial p}{\partial t} + \frac{\partial \varepsilon_{ii}}{\partial t}, \tag{6.3.1.13}$$

which is the required governing equation (6.3.1.2b).

In the next section, we describe the finite element approximation of Eqs. (6.3.1.1a), (6.3.1.1b), (6.3.1.2a), and (6.3.1.2b). Initial and boundary conditions required for the finite element solution are

$$u_i(x_1, x_2, t = 0) = u_i^0(x_1, x_2) \quad \text{on} \quad R, \tag{6.3.1.14a}$$

$$p(x_1, x_2, t = 0) = p^0(x_1, x_2) \quad \text{on} \quad R, \tag{6.3.1.14b}$$

$$u_i = U_i \quad \text{on} \quad B_u, \tag{6.3.1.14c}$$

$$\tau_{ij}n_j = S_i \quad \text{on} \quad B_s, \tag{6.3.1.14d}$$

$$p = \overline{p} \quad \text{on} \quad B_1, \tag{6.3.1.14e}$$

$$V_i n_i = q \quad \text{on} \quad B_2, \tag{6.3.1.14f}$$

in which R is the solution region, B_u and B_s are boundary portions where the displacement and surface traction are prescribed, and B_1 and B_2 are boundary portions where fluid pressure and flux are prescribed, respectively.

6.3.2 *Finite Element Discretization*

Earlier finite element formulations of discrete fracture flow and deformation were presented by Ayatollahi (1978), Hilber and Taylor (1976), and Noorishad *et al.* (1971). We revise these formulations and present here a general formulation that is not only straightforward but also reducible to the previous works.

Consider a typical region depicted in Fig. 6.8a. We discretize this region into a network of porous medium and joint elements as illustrated in Fig. 6.8b. It is convenient to present the derivation of element matrices separately for porous medium and joint elements.

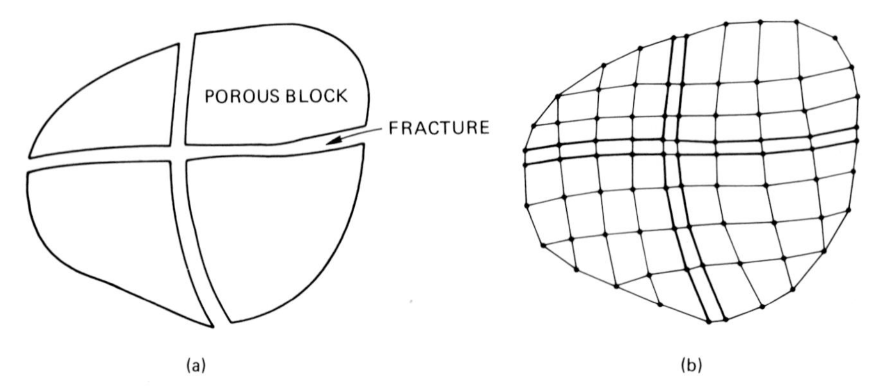

Fig. 6.8. An idealized fracture porous medium system: (a) two-dimensional region; (b) finite element discretization.

Derivation of Porous Medium Element Matrix

Derivation of the element matrix equation for the porous medium element is identical to that presented in Section 6.2.2 for the plane strain consolidation problem. The result can be written as

$$[H]^e\{\chi\}^e_{t+\Delta t} = \{R\}^e_t,$$

or in the partitioned form

$$\begin{bmatrix} [K] & [A] \\ [A]^T & -[f] \end{bmatrix}^e \begin{Bmatrix} \underset{\sim}{u} \\ \underset{\sim}{p} \end{Bmatrix}^e_{t+\Delta t} = \begin{Bmatrix} \underset{\sim}{R}_u \\ \underset{\sim}{R}_p \end{Bmatrix}^e_t, \tag{6.3.2.1}$$

where

$$[H]^e = \begin{bmatrix} [K] & [A] \\ [A]^{\mathrm{T}} & -[f] \end{bmatrix}^e$$

and typical submatrix elements can be obtained from Section 6.2.2.

Derivation of Fracture Element Matrix

For the sake of simplicity, we consider a linear fracture (joint) element consisting of four nodes as shown in Fig. 6.9. There exist two interfaces with the porous blocks adjacent to the joint. These interfaces are referred to as the top and bottom planes. The element matrix for the joint element, $[H]^e$, can be partitioned as (see above)

$$[H]^e = \begin{bmatrix} [K] & [A] \\ [A]^{\mathrm{T}} & -[f] \end{bmatrix}^e, \tag{6.3.2.2}$$

where, as before, $[K]^e$ is the conventional stiffness matrix of structural

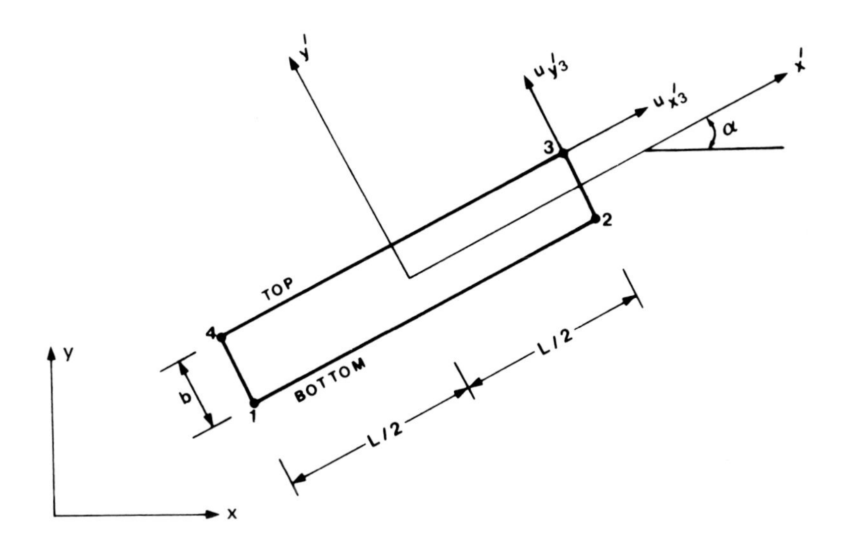

Fig. 6.9. Geometry and coordinate systems of the undeformed joint element.

analysis. We determine $[H]^e$ using the following procedure:

(1) Derive the local stiffness submatrix from the formula

$$[K']^e = \int_{R^e} [B]^{\mathrm{T}}[D][B]\, dR$$

$$= \int_{-L/2}^{L/2} [B]^{\mathrm{T}}[D][B] b\, dx', \tag{6.3.2.3a}$$

where $[D]$ is the elasticity matrix, $[B]$ the matrix relating strain to nodal displacements along the local coordinates, and it is assumed that the joint element has a unit thickness in the direction normal to the (x', y') plane.

(2) Having obtained $[K']^e$, we determine $[K]^e$ by the congruent coordinate transformation [see Eq. (6.3.1.8)]

$$[K]^e = [T]^{\mathrm{T}}[K']^e[T], \tag{6.3.2.3b}$$

where $[T]$ is the coordinate transformation matrix relating the local and global coordinates.

(3) Once $[K]^e$ has been determined, we can proceed to derive $[A]^{\mathrm{T}}$ and $[f]$ by performing the Galerkin approximation of the equation describing fluid flow in the fracture, Eq. (6.3.1.13).

To derive $[K']^e$ we first evaluate the three strain components, $\varepsilon_{x'x'}$, $\varepsilon_{y'y'}$, and $\varepsilon_{x'y'}$, in terms of the nodal displacements along the x' and y' axes. Because the fracture width is very small compared to the length L, we assume that the strain does not vary in the y' direction. Thus the three components of strain at position x' within the joint element are given by (cf. Wilson, 1977)

$$\varepsilon_{x'x'} = \frac{\partial u_{x'}}{\partial x'} \simeq \frac{1}{L}(u_{x'2} - u_{x'1}),$$

$$\varepsilon_{y'y'} = \frac{\partial u_{y'}}{\partial y'} \simeq \frac{1}{b}(u_{y'}^{\text{top}} - u_{y'}^{\text{bot}}),$$

$$= (1/b)\,[f(x')u_{y'4} + g(x')u_{y'3} - f(x')u_{y'1} - g(x')u_{y'2}]$$

$$= (1/b)\,[f(x')(u_{y'4} - u_{y'1}) + g(x')(u_{y'3} - u_{y'2})],$$

$$2\varepsilon_{x'y'} = \frac{\partial u_{x'}}{\partial y'} + \frac{\partial u_{y'}}{\partial x'}$$

$$\simeq (1/b)\,[f(x')(u_{x'4} - u_{x'1}) + g(x')(u_{x'3} - u_{x'2})] + (1/L)\,(u_{y'2} - u_{y'1}),$$

where $f(x')$ and $g(x')$ are one-dimensional interpolation functions given by

$$f(x') = \tfrac{1}{2}(1 - 2x'/L), \qquad g(x') = \tfrac{1}{2}(1 + 2x'/L).$$

In matrix notation the equations for the three strain components can be written as

$$\begin{Bmatrix} \varepsilon_{x'x'} \\ \varepsilon_{y'y'} \\ 2\varepsilon_{x'y'} \end{Bmatrix} = \frac{1}{b}\begin{bmatrix} -b/L & 0 & b/L & 0 & 0 & 0 & 0 & 0 \\ 0 & -f(x') & 0 & -g(x') & 0 & g(x') & 0 & f(x') \\ -f(x') & -b/L & -g(x') & b/L & g(x') & 0 & f(x') & 0 \end{bmatrix}$$

$$\times \begin{Bmatrix} u_{x'1} \\ u_{y'1} \\ u_{x'2} \\ u_{y'2} \\ u_{x'3} \\ u_{y'3} \\ u_{x'4} \\ u_{y'4} \end{Bmatrix} \qquad (6.3.2.4a)$$

or

$$\{\varepsilon'\} = [B]\{u'\}, \qquad (6.3.2.4b)$$

where

$$[B] = \frac{1}{b}\begin{bmatrix} -b/L & 0 & b/L & 0 & 0 & 0 & 0 & 0 \\ 0 & -f(x') & 0 & -g(x') & 0 & g(x') & 0 & f(x') \\ -f(x') & -b/L & -g(x') & b/L & g(x') & 0 & f(x') & 0 \end{bmatrix}. \tag{6.3.2.4c}$$

Next, we assume that the joint element does not undergo any dilation (volume change) due to shearing strains. Because of this nondilatancy assumption the stress–strain matrix relation takes the form

$$\begin{Bmatrix} \sigma_{x'x'} \\ \sigma_{y'y'} \\ \sigma_{x'y'} \end{Bmatrix} = \begin{bmatrix} D_{x'x'} & 0 & 0 \\ 0 & D_{y'y'} & 0 \\ 0 & 0 & D_{x'y'} \end{bmatrix} \begin{Bmatrix} \varepsilon_{x'x'} \\ \varepsilon_{y'y'} \\ 2\varepsilon_{x'y'} \end{Bmatrix}, \tag{6.3.2.5}$$

where $D_{x'x'}$ and $D_{y'y'}$ are moduli of normal deformations and $D_{x'y'}$ is the modulus of shear deformation.

Performing the triple matrix product $[B]^{\mathrm{T}}[D][B]$ and integrating term by term, we obtain the required expression for $[K']^e$

$$[K']^e = (L/6b)$$

$$\times \begin{bmatrix} 6D_{x'}\zeta^2+2D_s & 3D_s\zeta & -6D_{x'}\zeta^2+D_s & -3D_s\zeta & -D_s & 0 & -2D_s & 0 \\ & 6D_s\zeta^2+2D_n & 3D_s\zeta & -6D_s\zeta^2+D_n & -3D_s\zeta & -D_n & -3D_n\zeta & -2D_n \\ & & 6D_{x'}\zeta^2+2D_s & -3D_s\zeta & -2D_s & 0 & -D_s & 0 \\ & & & 6D_s\zeta^2+2D_n & 3D_s\zeta & -2D_n & 3D_s\zeta & -D_n \\ & & & & 2D_s & 0 & D_s & 0 \\ & & \text{sym} & & & 2D_n & 0 & D_n \\ & & & & & & 2D_s & 0 \\ & & & & & & & 2D_n \end{bmatrix}, \tag{6.3.2.6}$$

where sym means the matrix is symmetric. Note that for convenience we let $D_{x'} = D_{x'x'}$, $D_n = D_{y'y'}$, $D_s = D_{x'y'}$, and $\zeta = b/L$.

A simpler expression for $[K']^e$ was derived in the original paper on joint elements by Goodman *et al.* (1968). It is based on the assumption that the ratio b/L is so small that $\varepsilon_{x'x'}$ is negligible compared to $\varepsilon_{y'y'}$ and $\varepsilon_{x'y'}$. By this assumption, the matrices $[B]$ and $[D]$ reduce to

$$[B] = \frac{1}{b}\begin{bmatrix} 0 & -f(x') & 0 & -g(x') & 0 & g(x') & 0 & f(x') \\ -f(x') & 0 & -g(x') & 0 & g(x') & 0 & f(x') & 0 \end{bmatrix}, \tag{6.3.2.7a}$$

$$[D] = \begin{bmatrix} D_{y'y'} & 0 \\ 0 & D_{x'y'} \end{bmatrix}. \tag{6.3.2.7b}$$

Using these expressions for $[B]$ and $[D]$ leads to an expression for $[K']^e$ that is obtainable from Eq. (6.3.2.6) by setting $\zeta = 0$.

Once $[K']^e$ has been determined it is a simple matter to evaluate $[K]^e$. We only need to use Eq. (6.3.2.3b) with $[T]$ being an 8×8 matrix given by

$$[T] = \begin{bmatrix} \mathbf{A} & \mathbf{0} & \mathbf{0} & \mathbf{0} \\ \mathbf{0} & \mathbf{A} & \mathbf{0} & \mathbf{0} \\ \mathbf{0} & \mathbf{0} & \mathbf{A} & \mathbf{0} \\ \mathbf{0} & \mathbf{0} & \mathbf{0} & \mathbf{A} \end{bmatrix}, \tag{6.3.2.8}$$

where

$$\mathbf{A} = \begin{bmatrix} \cos\alpha & \sin\alpha \\ -\sin\alpha & \cos\alpha \end{bmatrix}, \quad \mathbf{0} = \begin{bmatrix} 0 & 0 \\ 0 & 0 \end{bmatrix}. \tag{6.3.2.9}$$

Having obtained $[K]^e$, our remaining task is to derive submatrices $([A]^e)^{\mathrm{T}}$ and $[f]^e$ of Eq. (6.3.2.1). This can be achieved by applying the Galerkin finite element procedure to the fracture flow equation (6.3.1.13). If we adopt the sign convention that p is positive for tension and negative for compression, the resulting discretized equation takes the form

$$\sum_{e=1}^{m} \left[\int_{R^e} \left(N_I \varepsilon_{ii}^{t+\Delta t} - \phi_{\mathrm{f}} \beta N_I N_J \, p_J^{t+\Delta t} - \theta \, \Delta t \frac{k_{ij}^{\mathrm{f}}}{\mu} \frac{\partial N_I}{\partial x_i} \frac{\partial N_J}{\partial x_j} \, p_J^{t+\Delta t} \right) dR \right]$$

$$= \sum_{e=1}^{m} \left[\int_{R^e} \left(N_I \varepsilon_{ii}^{t} - \phi_{\mathrm{f}} \, \beta N_I N_J p_J^t + (1 - \theta) \, \Delta t \frac{k_{ij}^{\mathrm{f}}}{\mu} \frac{\partial N_I}{\partial x_i} \frac{\partial N_J}{\partial x_j} \, p_J^t \right) dR \right.$$

$$\left. - \int_{B_2^e} \Delta t \, N_I q \, dB \right], \qquad I, J = 1, 2, \ldots, n. \tag{6.3.2.10}$$

By the use of Eq. (6.3.2.4a), the volumetric strain in the joint element in Fig. 6.8 can be expressed as

$$\begin{aligned} \varepsilon_{ii} &= \varepsilon_{x'x'} + \varepsilon_{y'y'} \\ &= \frac{1}{L}(u_{x'2} - u_{x'1}) + \frac{1}{b}[(u_{y'4} - u_{y'1}) f(x') + (u_{y'3} - u_{y'2}) \, g(x')] \end{aligned}$$

(6.3.2.11a)

where the second subscripts of u refer to local node numbers and $f(x')$ and $g(x')$ are the linear interpolation functions given earlier. Let $\bar{\varepsilon}_{ii}$ be the value of ε_{ii} at the centroid of the element. It is easy to see that

$$\bar{\varepsilon}_{ii} = \frac{1}{L}(u_{x'2} - u_{x'1}) + \frac{1}{2b}(u_{y'4} - u_{y'1} + u_{y'3} - u_{y'2}),$$

or, in terms of nodal displacements along the global axes,

$$\bar{\varepsilon}_{ii} = \frac{1}{L}(u_{x2}\cos\alpha + u_{y2}\sin\alpha - u_{x1}\cos\alpha - u_{y1}\sin\alpha)$$
$$+ \frac{1}{2b}[(u_{x1} - u_{x4} + u_{x2} - u_{x3})\sin\alpha$$
$$+ (u_{y4} - u_{y1} + u_{y3} - u_{y2})\cos\alpha]. \qquad (6.3.2.11b)$$

The integral involving ε_{ii} can be simply evaluated if we use one-point Gauss quadrature. We thus obtain

$$\int_{R^e} N_I \varepsilon_{ii}\, dR = \bar{\varepsilon}_{ii}\frac{bL}{4}, \qquad (6.3.2.12a)$$

which, upon substitution for $\bar{\varepsilon}_{ii}$, becomes

$$\int_{R^e} N_I \varepsilon_{ii}\, dR$$
$$= \tfrac{1}{4}[(-b\cos\alpha + \tfrac{1}{2}L\sin\alpha)u_{x1} + (b\cos\alpha + \tfrac{1}{2}L\sin\alpha)u_{x2}$$
$$- (\tfrac{1}{2}L\sin\alpha)u_{x3} - (\tfrac{1}{2}L\sin\alpha)u_{x4}]$$
$$+ \tfrac{1}{4}[(-b\sin\alpha - \tfrac{1}{2}L\cos\alpha)u_{y1} + (b\sin\alpha - \tfrac{1}{2}L\cos\alpha)u_{y2}$$
$$+ (\tfrac{1}{2}L\cos\alpha)\, u_{y3} + (\tfrac{1}{2}L\cos\alpha)u_{y4}]. \qquad (6.3.2.12b)$$

By using our usual subscript notation, Eq. (6.3.2.12b) can be written in a compact form as

$$\int_{R^e} N_I \varepsilon_{ii}\, dR = U_I W_{iJ} u_{iJ}, \qquad i = 1, 2, \qquad (6.3.2.12c)$$

where

$$U_I = 1, \qquad u_{1J} \equiv u_{xJ}, \qquad u_{2J} \equiv u_{yJ},$$

and W_{1J} and W_{2J} are weighting factors that can be extracted from Eq. (6.3.2.12b).

Combination of Eqs. (6.3.2.10) and (6.3.2.12c) gives

$$\sum_{e=1}^{m} \Bigg[U_I W_{iJ} u_{iJ}^{t+\Delta t} - \int_{R^e} \phi_f \beta N_I N_J \, dR \; p_J^{t+\Delta t} - \int_{R^e} \theta \, \Delta t \, \frac{k_{ij}^f}{\mu} \frac{\partial N_I}{\partial x_i} \frac{\partial N_J}{\partial x_j} \, dR \; p_J^{t+\Delta t} \Bigg]$$

$$= \sum_{e=1}^{m} \Bigg[U_I W_{iJ} u_{iJ}^{t} - \int_{R^e} \phi_f \beta N_I N_J \, dR \; p_J^{t} + \int_{R^e} (1-\theta) \, \Delta t \, \frac{k_{ij}^f}{\mu} \frac{\partial N_I}{\partial x_i} \frac{\partial N_J}{\partial x_j} \, dR \; p_J^{t} - \int_{B_2^e} \Delta t \, N_I q \, dB \Bigg], \qquad I, J = 1, 2, \ldots, n. \tag{6.3.2.13}$$

A typical element equation can be written in the form

$$([A]^{\mathrm{T}})^e \{u\}_{t+\Delta t}^e + [f]^e \{p\}_{t+\Delta t}^e = \{R_p\}_t^e, \tag{6.3.2.14}$$

where, according to our previous notation, a typical element of $([A]^{\mathrm{T}})^e$ and $\{u\}_{t+\Delta t}^e$ are defined as

$$\{u_J\}_{t+\Delta t}^e = \begin{Bmatrix} u_{1J} \\ u_{2J} \end{Bmatrix}_{t+\Delta t}^e, \qquad ([A_{IJ}]^{\mathrm{T}})^e = [(c_{IJ})^{\mathrm{T}}, (e_{IJ})^{\mathrm{T}}]^e.$$

The expressions for $(c_{IJ})^{\mathrm{T}}$, $(e_{IJ})^{\mathrm{T}}$, and f_{IJ} result from Eq. (6.3.2.13). These are given by

$$(c_{IJ})^{\mathrm{T}} = U_I W_{1J},$$

$$(e_{IJ})^{\mathrm{T}} = U_I W_{2J},$$

$$f_{IJ} = -\int_{R^e} \phi_f \beta N_I N_J \, dR - \int_{R^e} \theta \, \Delta t \, \frac{k_{ij}^f}{\mu} \frac{\partial N_I}{\partial x_i} \frac{\partial N_J}{\partial x_j} \, dR.$$

Material Properties of Fractures

The material properties of fractures have recently been the subject of detailed laboratory investigations (Gale, 1975; Goodman, 1974; Iwai, 1976; Ohnishi and Goodman, 1974). In accordance with the experimental data reviewed by Goodman and Dubois (1972), the basic characteristics of fractures that should be modeled are as follows:

(1) A fracture is tabular as its dimension in the direction normal to the fracture plane is usually much smaller than the other two dimensions. Thus, a typical fracture resembles an irregular line rather than a zone of appreciable thickness; i.e., we may assume that the fracture width b is negligible compared with the fracture length L.
(2) Fractures have no strength in tension but offer high resistance to compression. Normal strains occur under normal compressive stresses and may include elastic and plastic strains.
(3) Shear displacement along a fracture is a function of the normal stress. At low normal stresses, the shear strength of the fracture is mainly due to cohesion. At high normal stresses, shearing induced by shear stresses occurs through the irregularities in the fracture plane, and thus the shear strength is a function of both cohesion and friction.

Typical constitutive stress-strain relations for a nondilatant fracture joint element with zero initial stresses are illustrated in Fig. 6.10. More complex relations for dilatant joints can be found in Goodman (1975) and Goodman and St. John (1975). The modeled normal stress–strain relation, Fig. 6.10a, consists of three stages (cf. Ghaboussi *et al.*, 1973):

(i) Joint separation:

$$D_{\mathrm{n}} = 0 \qquad \text{when} \quad \varepsilon_{\mathrm{n}} > 0;$$

(ii) Compression of the material in the joint or crushing of surface irregularities:

$$D_{\mathrm{n}} = E_{\mathrm{n}}^{\mathrm{i}} \qquad \text{when} \quad \varepsilon_{\mathrm{n}}^{\mathrm{c}} < \varepsilon_{\mathrm{n}} < 0;$$

(iii) Contact:

$$D_{\mathrm{n}} = E_{\mathrm{n}}^{\mathrm{I}} \qquad \text{when} \quad \varepsilon_{\mathrm{n}} < \varepsilon_{\mathrm{n}}^{\mathrm{c}}.$$

The shear stress–shear strain relation, Fig. 6.10b, is assumed to be elastic–perfectly plastic. Failure is specified according to the linearized Mohr–Coulomb yield criterion:

$$\sigma_{\mathrm{sy}} = \begin{cases} c - \sigma_{\mathrm{n}} \tan \phi & \text{if} \quad \sigma_{\mathrm{n}} < 0 \\ c & \text{if} \quad \sigma_{\mathrm{n}} \geqslant 0, \end{cases}$$

where c and ϕ are the cohesion and angle of friction, respectively. Shear failure occurs when $|\sigma_{\mathrm{s}}| \geqslant \sigma_{\mathrm{sy}}$. Thus,

$$D_{\mathrm{s}} = \begin{cases} E_{\mathrm{s}} & \text{when} \quad |\sigma_{\mathrm{s}}| < \sigma_{\mathrm{sy}} \\ 0 & \text{when} \quad |\sigma_{\mathrm{s}}| \geqslant \sigma_{\mathrm{sy}}. \end{cases}$$

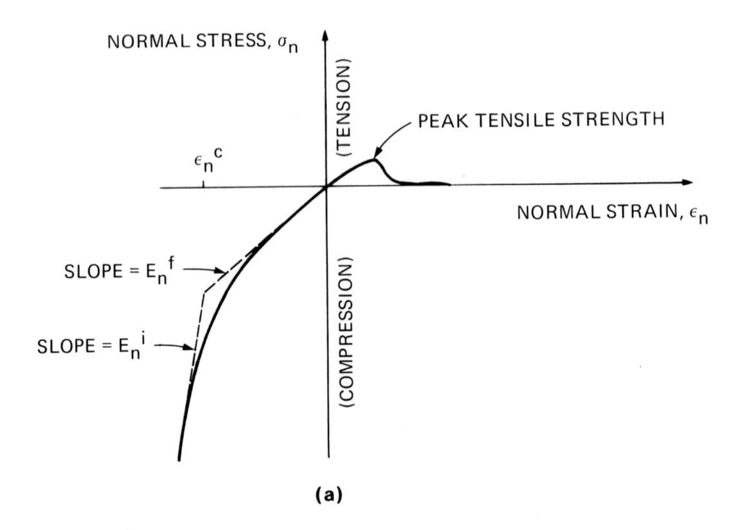

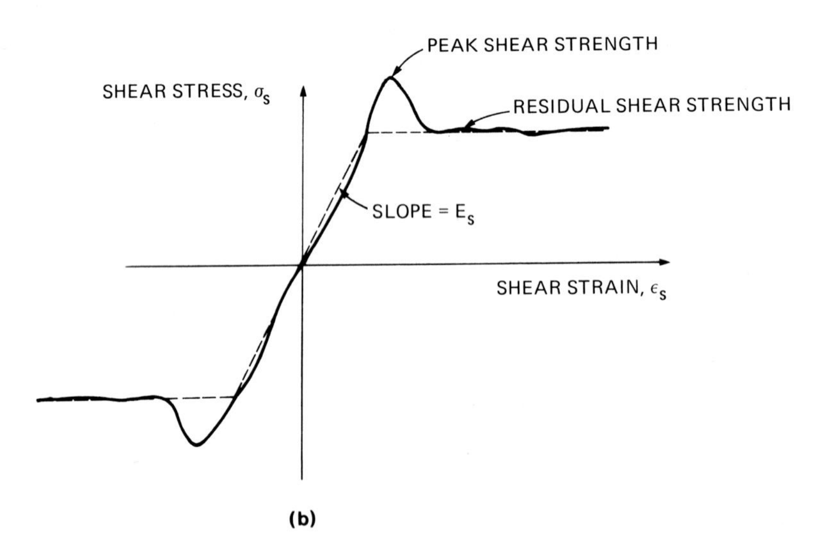

Fig. 6.10. Typical constitutive relations for a joint element without initial stresses: (a) normal stress versus normal strain: (b) shear stress versus shear strain. Legend: ——, actual behavior; - - -, modeled behavior.

6.3.3 *Solution Procedures*

Two solution methods are applicable to a nonlinear set of algebraic equations resulting from the finite element formulation of the discrete fracture flow deformation model. These are referred to as simultaneous and sequential algorithms. We shall assume in the calculation of element matrices that the nodal coordinates are fixed.

Simultaneous Solution Algorithm

In the simultaneous solution algorithm, all the nodal values of displacement and fluid pressure at the current time level are solved simultaneously. The solution procedure is similar to that described earlier for the consolidation problem. Owing to the nonlinearity of the problem, we need to perform iterations within each time step unless the time step is kept sufficiently small. At the beginning of a new time level, we form the element matrices using the most current values of the fracture aperture and stress and strain values that correspond to the last iteration of the previous time level. We then assemble all element matrices into a global matrix and solve the resulting system of equations using a standard Gauss algorithm for a symmetric banded matrix. Once the new set of nodal displacement and pressure values are obtained we can recalculate each element matrix by updating the values of the fracture width b and the joint stiffness constants D_n and D_s. The procedure for updating b is straightforward. For a typical joint element depicted in Fig. 6.9 we calculate the change in aperture width from

$$\Delta b = \tfrac{1}{2}[(u_{y'3} - u_{y'2}) + (u_{y'4} - u_{y'1})]. \tag{6.3.3.1}$$

The procedure for determining the normal and shear moduli has been described in the foregoing section.

Once all of the element matrices have been calculated, they can be assembled into the global matrix, and the resulting system of equations can be resolved. Iterations are performed until the changes in successive values of displacement and pressure lie within a prescribed tolerance.

Sequential Solution Algorithm

In the sequential solution algorithm, the nodal values of displacement and pressure are solved sequentially. This is achieved by writing the global matrix equation in the form

$$\begin{bmatrix} [K] & [A] \\ [A]^{\mathrm{T}} & [f] \end{bmatrix} \begin{Bmatrix} \mathbf{u} \\ \mathbf{p} \end{Bmatrix}_{t+\Delta t} = \begin{Bmatrix} \mathbf{R}_u \\ \mathbf{R}_p \end{Bmatrix}_t . \tag{6.3.3.2}$$

To enable the unknown vectors $\{\mathbf{u}\}_{t+\Delta t}$ and $\{\mathbf{p}\}_{t+\Delta t}$ to be solved independently, we rewrite Eq. (6.3.3.2) as

$$[K]\{\mathbf{u}\}_{t+\Delta t} = \{\mathbf{R}_u^*\} \tag{6.3.3.3a}$$

and

$$[f]\{\mathbf{p}\}_{t+\Delta t} = \{\mathbf{R}_p^*\}, \tag{6.3.3.3b}$$

where

$$\{\mathbf{R}_u^*\} = \{\mathbf{R}_u\} - [A]\{\mathbf{p}\}_{t+\Delta t}$$

and

$$\{\mathbf{R}_p^*\} = \{\mathbf{R}_p\} - [A]^{\mathrm{T}}\{\mathbf{u}\}_{t+\Delta t}.$$

It is evident that Eqs. (6.3.3.3a) and (6.3.3.3b) are the subsets of Eq. (6.3.3.2). Because these subsets of equations are coupled, they must be solved iteratively. The iteration procedure begins with an initial estimate for $\{\mathbf{p}\}_{t+\Delta t}$.

To enhance stability of the numerical solution we use the final pressure values at the previous time step as the initial estimate. By solving Eq. (6.3.3.3a), we obtain the first approximation for $\{\mathbf{u}\}_{t+\Delta t}$, which can then be used to determine the fracture width, and form the flow matrix and the right-hand-side vector of Eq. (6.3.3.3b). Once this is done, Eq. (6.3.3.3b) can be solved to obtain the first approximation for $\{\mathbf{p}\}_{t+\Delta t}$.

In the second iteration, we use the improved estimate of $\{\mathbf{p}\}_{t+\Delta t}$ to recalculate the matrix and right-hand-side vector of Eq. (6.3.3.3a) and solve the resulting system of equations for the second approximation of $\{\mathbf{u}\}_{t+\Delta t}$, which is then used to obtain the second approximation of $\{\mathbf{p}\}_{t+\Delta t}$. The sequential process is continued until the change in successive values of displacement and pressure lies within a prescribed tolerance.

For a steady-state problem, Eqs. (6.3.3.3a) and (6.3.3.3b) become

$$[K]\{\mathbf{u}\} = \{\mathbf{R}_u\} - [A]\{\mathbf{p}\} \tag{6.3.3.4a}$$

and

$$[f]\{\mathbf{p}\} = \{\mathbf{R}_p\}. \tag{6.3.3.4b}$$

The sequential solution procedure described here corresponds to that employed by Noorishad *et al.* (1971).

6.3.4 Calculation of Stresses and Velocities

Once the nodal displacement and pressure values are determined for the entire region, the components of stress and velocity can be readily obtained. The stress components are simply calculated from the equation

$$\{\sigma\} = \{\sigma^0\} + [D]\{\varepsilon\} = \{\sigma^0\} + [D][B]\{\mathbf{u}\}. \tag{6.3.4.1}$$

It is a standard practice to evaluate stress and strain at the Gauss points of the element. Once the stress components are determined the principle stresses and their orientation can be computed. If we let σ_x, σ_y, and σ_{xy} be the stress components in the (x, y) coordinate system, it can be shown that the principle stresses and their orientation with respect to the x axis are given by

$$\sigma_1 = \tfrac{1}{2}(\sigma_x + \sigma_y) + [\sigma_{xy}^2 + \tfrac{1}{4}(\sigma_x - \sigma_y)^2]^{1/2}, \tag{6.3.4.2a}$$

$$\sigma_2 = \tfrac{1}{2}(\sigma_x + \sigma_y) - [\sigma_{xy}^2 + \tfrac{1}{4}(\sigma_x - \sigma_y)^2]^{1/2}, \tag{6.3.4.2b}$$

$$\alpha = \tan^{-1}[2\sigma_{xy}/(\sigma_y - \sigma_x)]. \tag{6.3.4.2c}$$

As shown in Fig. 6.11, Eqs. (6.3.4.2a)–(6.3.4.2c) can be represented by Mohr's circle.

The calculation of velocities at Gauss points from nodal values of pressure is straightforward. We simply use Darcy's law for either porous block or fracture flow elements. For this reason, no detail is given here.

6.4 Double-Porosity Flow Models

As an alternative to the discrete fracture modeling of fractured porous media, we present a second approach, called the double-porosity modeling

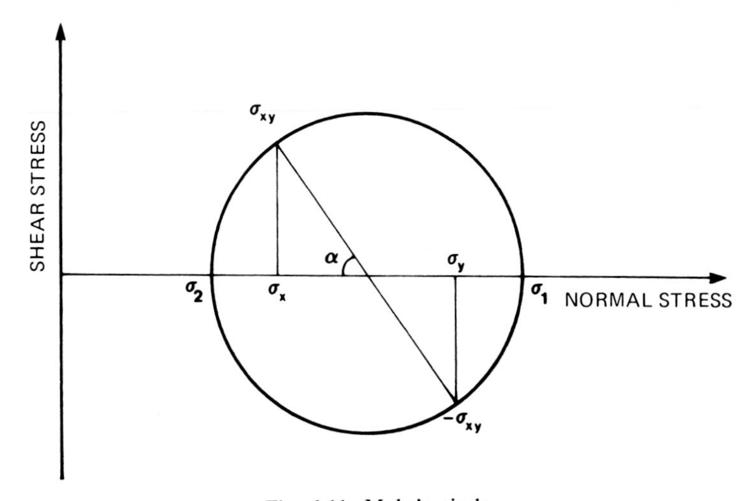

Fig. 6.11. Mohr's circle.

approach. In deriving double-porosity models, we rely on the use of the concepts of statistical averaging, volume averaging, or the theory of mixtures. The fractured medium is represented by two completely overlapping continua, one representing the porous matrix and the other representing the fractures. This type of flow model was introduced by Barenblatt *et al.* (1960) and is expounded by Closmann (1975), Kazemi (1969), Odeh (1965), van Golf-Racht (1982), and Warren and Root (1963).

In this section we present a derivation of the governing equations for saturated, single-phase flow and the finite element approximation for the resulting fracture flow model. An extension that includes the coupling of flow and deformation will be presented in the next section. For a more detailed description of the formulation of this type of problem, the reader is referred to Huyakorn and Lester (1983b).

6.4.1 Governing Equations

Barenblatt's Quasi-Steady Leakage Model

To derive the governing equations of the flow model developed by Barenblatt *et al.* (1960), we consider a representative elementary volume (REV) of a fractured porous medium, shown in Fig. 6.12. Such an REV is a volume including porous matrix and fractures which is large enough compared to the dimensions of the individual blocks and fractures to allow meaningful spatial averaging, yet small enough to characterize variations in reservoir properties at scales of interest.

Let m be a subscript used to identify the two components (porous blocks and fractures) of the medium, with $m = 1$ denoting porous blocks and $m = 2$ denoting fractures. Darcy's law for flow in the porous blocks and fractures can be expressed as

$$V_{im} = -\frac{k_{ijm}}{\mu}\left(\frac{\partial p_m}{\partial x_j} + \rho g_j\right), \qquad m = 1, 2, \tag{6.4.1.1}$$

where p_1 represents the average pressure of the fluid in the porous blocks in the neighborhood of a given point and p_2 represents the average

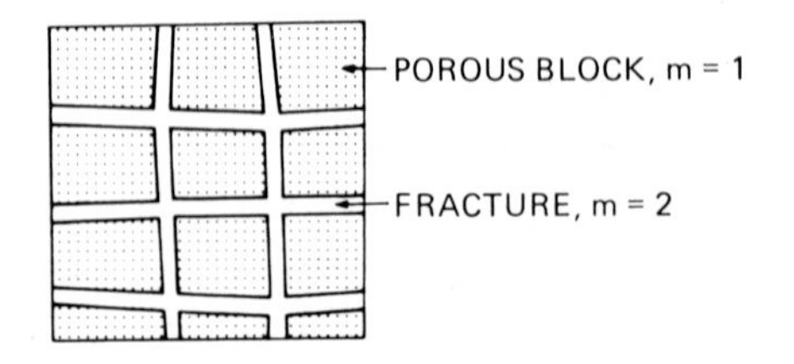

Fig. 6.12. Representative elementary volume of a conceptualized fractured porous medium.

pressure of the fluid in the fractures in the neighborhood of the given point. Similar meanings are given to V_{i1}, V_{i2}, k_{ij1}, and k_{ij2}.

Next, we consider the continuity of fluid flow in the porous blocks and the fractures. In writing the continuity equations we must take into account fluid transfer between the porous blocks and fractures. Let Γ denote the rate of fluid mass transfer from the porous blocks to the fractures per unit bulk volume of the medium. The continuity equation for flow in the porous blocks can be written as

$$-\frac{\partial}{\partial x_i}\rho V_{i1} = \frac{\partial}{\partial t}\phi_1\rho + \Gamma, \qquad (6.4.1.2)$$

where ϕ_1 is the primary porosity, defined as the ratio of the volume of fluid in the porous blocks to the bulk volume of the porous medium.

Similarly, the continuity equation for flow in the fractures can be written as

$$-\frac{\partial}{\partial x_i}\rho V_{i2} = \frac{\partial}{\partial t}\phi_2\rho - \Gamma, \qquad (6.4.1.3)$$

where ϕ_2 is the secondary porosity, defined as the ratio of the volume of fluid in the fractures to the bulk volume of the porous medium. The reader should not confuse ϕ_2 with ϕ_f, which denotes the porosity within the fracture. Equations (6.4.1.2) and (6.4.1.3) can be represented by

$$\frac{\partial}{\partial x_i}\left[\rho\frac{k_{ijm}}{\mu}\left(\frac{\partial p_m}{\partial x_j} + \rho g_j\right)\right] = \frac{\partial}{\partial t}\phi_m\rho + (-1)^{m+1}\Gamma, \qquad m = 1, 2, \qquad (6.4.1.4)$$

where Γ is the rate of mass transfer of fluid from porous blocks to fractures. If it is assumed that the fractured medium is isotropic and homogeneous, and the fluid is slightly compressible, Eq. (6.4.1.4) may be approximated by

$$\frac{\partial}{\partial x_i}\left(\rho_0\frac{k_m}{\mu}\frac{\partial p_m}{\partial x_i}\right) = \frac{\partial}{\partial t}\phi_m\rho + (-1)^{m+1}\Gamma, \qquad m = 1, 2, \qquad (6.4.1.5)$$

where ρ_0 is the fluid density at a standard reference pressure, and k_1 and k_2 are the porous block permeability and fracture permeability, respectively.

Based upon dimensional analysis and the assumption of quasi-steady flow, the rate of fluid mass transfer from porous matrix blocks to fractures can be deduced as (see Barenblatt *et al.*, 1960, pp. 1289–1292)

$$\Gamma = \Pi\,(k_1\sigma^2/\mu)\,\rho_0(p_1 - p_2) = (\Lambda/\mu)\,\rho_0\,(p_1 - p_2), \qquad (6.4.1.6)$$

where Π is a dimensionless parameter assumed to be constant, σ is the specific surface of the fractures, i.e., the surface area of fractures per unit volume of the porous medium, and $\Lambda = \Pi k_1 \sigma^2$.

To arrive at the final governing equations, Barenblatt *et al.* (1960) assumed that the compressibility of fluid in the porous blocks and fractures may be described by

$$(d\rho)_{\text{pore}} = \rho_0 \beta \, dp_1, \tag{6.4.1.7a}$$

$$(d\rho)_{\text{fracture}} = \rho_0 \beta \, dp_2. \tag{6.4.1.7b}$$

They also assumed that the change in primary and secondary porosities may be expressed as

$$d\phi_1 = \alpha_1 \, dp_1 - \alpha_1^* \, dp_2, \tag{6.4.1.8a}$$

$$d\phi_2 = \alpha_2 \, dp_2 - \alpha_2^* \, dp_1, \tag{6.4.1.8b}$$

where α_1, α_1^*, α_2, and α_2^* are compressibility coefficients of the medium. Combining Eqs. (6.4.1.6), (6.4.1.7a), (6.4.1.8a), and (6.4.1.5) with $m = 1$, and combining Eqs. (6.4.1.6), (6.4.1.7b), (6.4.1.8b), and (6.4.1.5) with $m = 2$, we obtain

$$\frac{\partial}{\partial x_i}\left(\frac{k_2}{\mu}\frac{\partial p_2}{\partial x_i}\right) = (\alpha_2 + \phi_2 \beta)\frac{\partial p_2}{\partial t} + \frac{\Lambda}{\mu}\int_0^t \frac{\partial p_2}{\partial \tau} \exp\left[-\omega_1(t - \tau)\right] d\tau, \tag{6.4.1.9a}$$

$$\frac{\partial}{\partial x_i}\left(\frac{k_2}{\mu}\frac{\partial p_2}{\partial x_i}\right) = (\alpha_2 + \phi_2 \beta)\frac{\partial p_2}{\partial t} - \alpha_2^* \frac{\partial p_1}{\partial t} - \frac{\Lambda}{\mu}(p_1 - p_2). \tag{6.4.1.9b}$$

Equations (6.4.1.9a) and (6.4.1.9b) are the general governing equations for fluid flow in a double-porosity medium. Several simplified forms of these equations can be obtained by introducing simplifying approximations. This enables analytical solutions to be obtained. The first form is obtained assuming that the terms associated with coefficients α_1^* and α_2^* are negligible and that all material coefficients are constant. Equations (6.4.1.9a) and (6.4.1.9b) can thus be reduced to

$$\frac{\partial}{\partial x_i}\left(\frac{k_1}{\mu}\frac{\partial p_1}{\partial x_i}\right) = (\alpha_1 + \phi_1 \beta)\frac{\partial p_1}{\partial t} + \frac{\Lambda}{\mu}(p_1 - p_2), \tag{6.4.1.10a}$$

$$\frac{\partial}{\partial x_i}\left(\frac{k_2}{\mu}\frac{\partial p_2}{\partial x_i}\right) = (\alpha_2 + \phi_2 \beta)\frac{\partial p_2}{\partial t} - \frac{\Lambda}{\mu}(p_1 - p_2). \tag{6.4.1.10b}$$

Despite this simplification, an analytical solution of the preceding equations is still difficult to obtain. For this reason it is common practice to assume further that the permeability of the blocks is small, so that the left-hand term of Eq. (6.4.1.10a) is negligible compared to the right-hand terms. Using this assumption, Eq. (6.4.1.10a) reduces to

$$(\alpha_1 + \phi_1\beta)\frac{\partial p_1}{\partial t} = -\frac{\Lambda}{\mu}(p_1 - p_2), \tag{6.4.1.11a}$$

which upon combining with Eq. (6.4.1.10b) gives

$$\frac{\partial}{\partial x_i}\left(\frac{k_2}{\mu}\frac{\partial p_2}{\partial x_i}\right) = (\alpha_2 + \phi_2\beta)\frac{\partial p_2}{\partial t} + (\alpha_1 + \phi_1\beta)\frac{\partial p_1}{\partial t}. \tag{6.4.1.11b}$$

Several analytical solutions of Eqs. (6.4.1.11a) and (6.4.1.11b) are available in the literature for various cases of the well flow problem (Warren and Root, 1963; Closmann, 1975).

An equivalent form of Eq. (6.4.1.11a) is also obtainable by solving it using the Laplace transform method and substituting the solution into Eq. (6.4.1.11b). This is the procedure used by Streltsova-Adams (1978), except that here the problem is cast in terms of fluid pressures instead of drawdowns. Application of such a procedure to Eqs. (6.4.1.11a) and (6.4.1.11b) yields

$$p_1 = p_2 - \int_0^t \frac{\partial p_2}{\partial \tau}\exp\left[-\omega_1(t - \tau)\right] d\tau, \tag{6.4.1.12a}$$

$$\frac{\partial}{\partial x_i}\left(\frac{k_2}{\mu}\frac{\partial p_2}{\partial x_i}\right) = (\alpha_2 + \phi_2\beta)\frac{\partial p_2}{\partial t} + \frac{\Lambda}{\mu}\int_0^t \frac{\partial p_2}{\partial \tau}\exp\left[-\omega_1(t - \tau)\right] d\tau, \tag{6.4.1.12b}$$

where $\omega_1 = \Lambda/\mu(\alpha_1 + \phi_1\beta)$ and it is assumed that the initial values of p_1 and p_2 are zero.

An analytical solution of Eqs. (6.4.1.12a) and (6.4.1.12b) is given in Streltsova-Adams (1978) for a case of transient flow to a well fully penetrating a confined fractured reservoir. It is interesting to note that Eq. (6.4.1.12b) is of the same form as the equation commonly employed to describe transient flow to a well in a conventional unconfined aquifer with a delayed yield effect (Boulton, 1963).

Finally, we derive the simplified fracture flow equation solved analytically by Barenblatt *et al.* (1960). This equation is obtained from Eqs. (6.4.1.11a) and (6.4.1.11b) by assuming that the first compressibility term in Eq. (6.4.1.11b) is negligible. Such an assumption is valid provided that the rate of volume change due to compressibility of the fractures is small compared with the fluid flux transferred from the porous blocks to the fractures. By the stated assumption, Eqs. (6.4.1.11a) and (6.4.1.11b) become

$$(\alpha_1 + \phi_1\beta)\frac{\partial p_1}{\partial t} = -\frac{\Lambda}{\mu}(p_1 - p_2), \tag{6.4.1.13a}$$

$$\frac{\partial}{\partial x_i}\left(\frac{k_2}{\mu}\frac{\partial p_2}{\partial x_i}\right) = -\frac{\Lambda}{\mu}(p_1 - p_2). \tag{6.4.1.13b}$$

If the material coefficients are constant, we can eliminate p_1 from the above equations and obtain

$$\frac{k_2}{\mu}\frac{\partial^2 p_2}{\partial x_i\, \partial x_i} = (\alpha_1 + \phi_1\beta)\frac{\partial p_2}{\partial t} - \frac{k_2}{\mu\omega_1}\frac{\partial}{\partial t}\left(\frac{\partial^2 p_2}{\partial x_i\, \partial x_i}\right), \tag{6.4.1.14}$$

or

$$\chi\frac{\partial^2 p_2}{\partial x_i\, \partial x_i} = \frac{\partial p_2}{\partial t} - \frac{k_2}{\Lambda}\frac{\partial}{\partial t}\left(\frac{\partial^2 p_2}{\partial x_i\, \partial x_i}\right), \tag{6.4.1.15}$$

where $\chi = k_2/\mu(\alpha_1 + \phi_1\beta)$ is called the "piezo-conductivity" of the fractured medium. Equation (6.4.1.15) is the governing equation solved analytically by Barenblatt *et al.* (1960) for two cases involving one-dimensional and radial well flow.

Transient Leakage Models

Perhaps the main drawback of Barenblatt's flow model is that it is based on a simple but rather ad hoc formula for describing fluid transfer between the porous matrix and fractures. This formula is expressed by Eq. (6.4.1.6). In essence, it is based on the assumption of quasi-steady state leakage. Such an assumption may not be accurate during early and intermediate pumping periods.

As an alternative to the approach of Barenblatt *et al.* (1960), a more elaborate approach has recently been advanced for the evaluation of transient fluid transfer rates. In this approach, the porous matrix blocks are idealized as simple geometrical shapes, e.g., parallel rectangular slabs or spheres of equal size. Fluid flow in an individual block is then considered using a one-dimensional, second-order transient flow equation. The flux of fluid into the fracture is derived by obtaining the solution of this one-dimensional equation and applying Darcy's law at the interface of the matrix block and the fracture. Derivation of the governing equations can be found in Huyakorn *et al.* (1983) and Streltsova-Adams (1978).

In essence, the governing equations for flow in the fracture domain of transient leakage models take the same form as Eq. (6.4.1.12b) of Barenblatt's quasi-steady leakage model. The only difference is that the convolution integral term that accounts for quasi-steady state leakage is now replaced by a more elaborate analytical expression that better accounts for transient storage effects in the porous matrix blocks. Without loss of generality, we now modify Eq. (6.4.1.12b) and rewrite it in a general form

$$\frac{\partial}{\partial x_i}\left(\frac{k_2}{\mu}\frac{\partial p_2}{\partial x_i}\right) = (\alpha_2 + \phi_2\beta)\frac{\partial p_2}{\partial t} - \Gamma, \qquad (6.4.1.16)$$

where Γ now represents the leakage term, the expression that depends on the type of the dual-porosity medium.

In the case where the matrix blocks of the dual-porosity system are idealized as parallel rectangular slabs of uniform size, the expression for Γ is given by [for derivation see Huyakorn *et al.* (1983)]

$$\Gamma = \frac{-2K'}{a(a+b)}\sum_{n=0}^{\infty}\int_0^t \frac{\partial p_2}{\partial \tau}\exp\left[-\alpha_n(t-\tau)\right]d\tau, \qquad (6.4.1.17)$$

where

$$\alpha_n = \pi^2(2n+1)^2K'/4S_s'a^2,$$

K' is the hydraulic conductivity of the matrix block, a and b are one half of the block thickness and one half of the fracture aperture, respectively, and S_s' is the specific storage of the matrix blocks. Note that by definition $K' = k_1\rho g/\mu$, and $S_s' = (\alpha_1 + \phi_1\beta)\rho g$.

In the case where the matrix blocks are idealized as uniform spheres, expression for the leakage term $-\Gamma$ is given by

$$\Gamma = \frac{-6K'}{a(a+b)}\sum_{n=1}^{\infty}\int_0^t \frac{\partial p_2}{\partial \tau}\exp\left[-\hat{\alpha}_n(t-\tau)\right]d\tau \qquad (6.4.1.18)$$

where a is the radius of each sphere and

$$\hat{\alpha}_n = \pi^2n^2K'/S_s'a^2.$$

6.4.2 Finite Element Formulation

To model transient flow in double-porosity media, let us first consider Eqs. (6.4.1.11b) and (6.4.1.11a) of Barenblatt's model and rewrite these in terms of the piezometric heads in the fractures and the matrix blocks, h and h',

$$\frac{\partial}{\partial x_i}\left(K\frac{\partial h}{\partial x_i}\right) = S_s\frac{\partial h}{\partial t} + S_s'\frac{\partial h'}{\partial t}, \qquad (6.4.2.1)$$

$$S_s'\frac{\partial h'}{\partial t} + \zeta(h' - h) = 0, \qquad (6.4.2.2)$$

where the variables h and h' are now defined as

$$h = (p_2/\gamma) + z, \qquad h' = (p_1/\gamma) + z, \qquad \gamma = \text{specific weight of fluid},$$

and the physical parameters K, ζ, S_s, and S'_s are given by

$$K = k_2\gamma/\mu, \qquad \zeta = \Lambda/\mu,$$

$$S_s = \gamma(\alpha_2 + \phi_2\beta), \qquad S'_s = \gamma(\alpha_1 + \phi_1\beta).$$

An efficient solution method can be derived using the Galerkin finite element procedure in conjunction with a simple finite difference method to approximate (6.4.2.1) and (6.4.2.2), respectively. The Galerkin approximation of (6.4.2.1) is obtained in the same manner as described previously. If one adopts trial functions of the form

$$\begin{aligned} \hat{h}(x_1, x_2, t) &= N_I(x_1, x_2)h_I(t), \qquad I = 1, 2, \ldots, n, \\ \hat{h}'(x_1, x_2, t) &= N_I(x_1, x_2)h'_I(t), \qquad I = 1, 2, \ldots, n, \end{aligned} \tag{6.4.2.3}$$

and follows the standard procedure described in Sections 2.5 and 4.3, the following set of algebraic equations will result:

$$\left(\theta A_{IJ} + \frac{B_{IJ}}{\Delta t}\right) h_J^{t+\Delta t} + \left(\frac{B'_{IJ}}{\Delta t}\right) h_J'^{t+\Delta t} = \left((\theta - 1)A_{IJ} + \frac{B_{IJ}}{\Delta t}\right) h_J^t + \left(\frac{B'_{IJ}}{\Delta t}\right) h_J'^t$$

$$+ (\theta F_I^{t+\Delta t} + (1 - \theta)\, F_I^t), \tag{6.4.2.4}$$

where the superscripts t and $t+\Delta t$ denote the previous and current time values, θ is the time-weighting factor, and

$$A_{IJ} = \sum_{e=1}^{m} \int_{R^e} K \frac{\partial N_I}{\partial x_i} \frac{\partial N_J}{\partial x_i}\, dR,$$

$$B_{IJ} = \sum_{e=1}^{m} \int_{R^e} S_s N_I N_J\, dR,$$

$$B'_{IJ} = \sum_{e=1}^{m} \int_{R^e} S'_s N_I N_J\, dR,$$

$$F_I = \sum_{e=1}^{m} \int_{B_2^e} N_I q\, dB.$$

To eliminate the unknown $h_J'^{t+\Delta t}$ from (6.4.2.4), we make use of Eq. (6.4.2.2) and approximate it by

$$h_J'^{t+\Delta t} - h_J'^t = \omega(h_J^{t+\Delta t} - h_J'^t), \tag{6.4.2.5}$$

where

$$\omega = (\zeta\Delta t)/(\zeta\,\Delta t + S_s').$$

Combination of Eqs. (6.4.2.4) and (6.4.2.5) and rearrangement of the terms yield

$$\left(\theta A_{IJ} + \frac{B_{IJ} + \omega B_{IJ}'}{\Delta t}\right) h_J^{t+\Delta t} = \left((\theta - 1)A_{IJ} + \frac{B_{IJ}}{\Delta t}\right) h_J^t + \left(\frac{\omega B_{IJ}'}{\Delta t}\right) h_J'^t$$

$$+ (\theta F_I^{t+\Delta t} + (1 - \theta)\, F_I^t) \qquad (6.4.2.6)$$

and

$$h_J'^{t+\Delta t} = h_J'^t + \omega(h_J^{t+\Delta t} - h_J'^t). \qquad (6.4.2.7)$$

A simple sequential solution procedure can now be employed. At a given time level, Eq. (6.4.2.6) is first solved for the nodal values of piezometric head in the fractures $h_J^{t+\Delta t}$ using known values at the previous time level. After $h_J^{t+\Delta t}$ has been determined, Eq. (6.4.2.7) is used to compute $h_J'^{t+\Delta t}$, the nodal values of piezometric head in the matrix blocks. The solution algorithm is repeated until the final time level is reached.

As an alternative to the preceding formulation, one can also model flow in double-porosity media by applying the Galerkin finite element method to (6.4.1.12a) and (6.4.1.12b), which are integro-differential equations. The convolution integrals can be evaluated simply by using the recurrence formulas and the solution technique given in Section 4.3.2. This technique has recently been shown by Huyakorn *et al.* (1983) to be suitable for handling the more complex Eqs. (6.4.1.16)–(6.4.1.18) of the two transient leakage models.

6.5 Double-Porosity Flow Deformation Model

An extension of the theoretical development for the double-porosity model was presented by Duguid (1973), Duguid and Abel (1974), and Duguid and Lee (1977). Duguid derived a system of governing equations taking into account the coupled effects of flow in the porous blocks and fractures as well as solid displacement. To simplify the problem he uncoupled the flow and displacement equations and employed the Galerkin finite element method to solve an example problem using the reduced set of flow equations.

In this section, we derive the governing equations for solid displacements and fluid pressures in the porous matrix and the fractures using a more simplistic approach than that presented in Duguid (1973).

6.5.1 Governing Equations

The governing equations of a double-porosity flow deformation model can be derived in a similar manner to the derivation of the equations describing consolidation in a conventional (unfractured) porous medium (see Section 6.2). However, it should be recognized that in this case we have two coexisting fluid pressures and hence two associated governing flow equations.

Displacement Equations

Once again consider the representative elementary volume shown in Fig. 6.12. For convenience we let ϕ be the total porosity of the fractured porous medium, i.e., $\phi = \phi_1 + \phi_2$. If it is assumed that the volume of the fluid in the primary pores is large compared to the volume of fluid in the fractures (i.e., $\phi_1 >> \phi_2$), then the total stress consists of the effective stress and the fluid pressure in the primary pores. Consequently, the governing displacement equations for the elastic and isotropic fractured medium are identical to those derived in Section 6.2 for the conventional porous medium. These equations can be rewritten as

$$G \frac{\partial^2 u_j}{\partial x_j \, \partial x_j} + (\lambda + G) \frac{\partial}{\partial x_i} \left(\frac{\partial u_i}{\partial x_j} \right) - \frac{\partial p}{\partial x_i} = -F_i - \frac{\partial \sigma_{ij}^0}{\partial x_j}, \quad (6.5.1.1)$$

where $p = p_1$ if $\phi_1 >> \phi_2$. In the case where ϕ_1 and ϕ_2 are of the same order of magnitude, we suggest that p be replaced by the weighted average of p_1 and p_2, i.e., $p = (\phi_1 p_1 + \phi_2 p_2)/\phi$.

Flow Equations

We begin the development by considering mass conservation of solid and fluid in the porous blocks. The continuity equation for solid mass takes the form [cf. Eq. (6.2.1.12)]

$$-\frac{\partial}{\partial x_i}(1 - \phi)\rho_s V_{is} = \frac{\partial}{\partial t}(1 - \phi)\rho_s, \quad (6.5.1.2)$$

where $\phi = \phi_1 + \phi_2$ and the rest of the symbols have the same meaning as before. By the introduction of the substantial time differential operator,

$$\frac{D}{Dt} = \frac{\partial}{\partial t} + V_{is} \frac{\partial}{\partial x_i},$$

and the assumption that $D\rho_s/Dt = 0$, Eq. (6.5.1.2) can be reduced to

$$\frac{D\phi}{Dt} = (1 - \phi) \frac{\partial V_{is}}{\partial x_i},$$

or

$$\frac{D\phi_1}{Dt} + \frac{D\phi_2}{Dt} = (1 - \phi_1 - \phi_2)\frac{\partial V_{is}}{\partial x_i}. \tag{6.5.1.3}$$

Now consider the continuity of fluid flow in the primary pores of the compacting fractured medium. Let V_{i1} be the relative Darcy velocity of the fluid in the porous blocks. Note that according to our previous notation in Section 6.2, the relative velocity $V_{i1} = \phi_1(V_{if1} - V_{is})$. The continuity equation of flow in the porous blocks can now be written in the form [cf. Eqs. (6.2.1.16) and (6.4.1.2)]

$$-\frac{\partial}{\partial x_i}(\rho V_{i1}) - \frac{\partial}{\partial x_i}(\rho\phi_1 V_{is}) = \frac{\partial}{\partial t}(\phi_1\rho) + \Gamma, \tag{6.5.1.4}$$

where Γ is the rate of fluid mass transferred from the porous blocks to the fractures per unit bulk volume of the medium. Once again, by introducing the definition of the substantial time derivative, Eq. (6.5.1.4) can be reduced to

$$-\frac{\partial}{\partial x_i}(\rho V_{i1}) = \phi_1\frac{D\rho}{Dt} + \rho\frac{D\phi_1}{Dt} + \rho\phi_1\frac{\partial V_{is}}{\partial x_i} + \Gamma. \tag{6.5.1.5}$$

Combining Eqs. (6.5.1.5) and (6.5.1.3) and using the fluid compressibility relation, $d\rho = \rho_0\beta\, dp_1$, we obtain

$$-\frac{\partial}{\partial x_i}(\rho V_{i1}) = \phi_1\rho_0\beta\frac{Dp_1}{Dt} - \rho\frac{D\phi_2}{Dt} + \rho(1 - \phi_2)\frac{\partial V_{is}}{\partial x_i} + \Gamma. \tag{6.5.1.6}$$

We require an equation relating the change in secondary porosity to the change in fluid pressures. This equation can be derived using the definition of ϕ_2,

$$\phi_2 = \mathbf{V}_2/\mathbf{V}, \tag{6.5.1.7}$$

where $\mathbf{V}_2$ and $\mathbf{V}$ denote the volume of fluid in the fractures and the bulk volume of a given mass of the porous medium, respectively. Differentiation of Eq. (6.5.1.7) with respect to t gives

$$\frac{D\phi_2}{Dt} = \frac{1}{\mathbf{V}}\left(\frac{D\mathbf{V}_2}{Dt} - \phi_2\frac{D\mathbf{V}}{Dt}\right). \tag{6.5.1.8}$$

By definition,

$$\mathbf{V} = \mathbf{V}_1 + \mathbf{V}_2 + \mathbf{V}_s, \tag{6.5.1.9a}$$

where $\mathbf{V}_1$ and $\mathbf{V}_s$ are the volume of fluid in the fractures and the volume of solid, respectively. Assuming that the solid material is incompressible, we obtain

$$\frac{D\mathbf{V}}{Dt} = \frac{D\mathbf{V}_1}{Dt} + \frac{D\mathbf{V}_2}{Dt}. \tag{6.5.1.9b}$$

Combination of Eqs. (6.5.1.8) and (6.5.1.9b) gives

$$\frac{D\phi_2}{Dt} = \frac{1}{\mathbf{V}}\left[(1 - \phi_2)\frac{D\mathbf{V}_2}{Dt} - \phi_2\frac{D\mathbf{V}_1}{Dt}\right]. \tag{6.5.1.10}$$

From the definition of compressibility of fluid [see Eq. (6.2.1.19)] and the definition of porosity, it follows that

$$-\beta\frac{Dp_m}{Dt} = \frac{1}{\mathbf{V}_m}\frac{D\mathbf{V}_m}{Dt},$$

or

$$-\beta\phi_m\mathbf{V}\frac{Dp_m}{Dt} = \frac{D\mathbf{V}_m}{Dt}, \qquad m = 1, 2. \tag{6.5.1.11}$$

Combination of Eqs. (6.5.1.10) and (6.5.1.11) yields

$$\frac{D\phi_2}{Dt} = -\phi_2(1 - \phi_2)\beta\frac{Dp_2}{Dt} + \phi_1\phi_2\beta\frac{Dp_1}{Dt}. \tag{6.5.1.12}$$

Equation (6.5.1.12) is the required expression for $D\phi_2/Dt$, and this can now be substituted into Eq. (6.5.1.6) to yield

$$-\frac{\partial V_{i1}}{\partial x_i} = (1 - \phi_2)\phi_1\beta\frac{\partial p_1}{\partial t} + (1 - \phi_2)\phi_2\beta\frac{\partial p_2}{\partial t} + (1 - \phi_2)\frac{\partial V_{is}}{\partial x_i} + \frac{\Gamma}{\rho}, \tag{6.5.1.13a}$$

where it is assumed that the fluid is slightly compressible and that $Dp_1/Dt \simeq \partial p_1/\partial t$ and $Dp_2/Dt \simeq \partial p_2/\partial t$. Equation (6.5.1.13a) is the required continuity equation for fluid flowing in the porous blocks. Its counterpart is the continuity equation for fluid flowing in the fractures. The latter can be derived using a similar procedure to that just given. The result is given by

$$-\frac{\partial V_{i2}}{\partial x_i} = (1 - \phi_1)\phi_1\beta\frac{\partial p_1}{\partial t} + (1 - \phi_1)\phi_2\beta\frac{\partial p_2}{\partial t} + (1 - \phi_1)\frac{\partial V_{is}}{\partial x_i} - \frac{\Gamma}{\rho}. \tag{6.5.1.13b}$$

To obtain the final form of the two flow equations we combine Eqs. (6.5.1.13a) and (6.5.1.13b) with Darcy's law:

$$V_{i1} = -\frac{k_{ij1}}{\mu}\left(\frac{\partial p_1}{\partial x_j} + \rho g_j\right), \tag{6.5.1.14a}$$

$$V_{i2} = -\frac{k_{ij2}}{\mu}\left(\frac{\partial p_2}{\partial x_j} + \rho g_j\right). \tag{6.5.1.14b}$$

The required flow equations take the form

$$\frac{\partial}{\partial x_i}\left[\frac{k_{ij1}}{\mu}\left(\frac{\partial p_1}{\partial x_j}+\rho g_j\right)\right] = (1-\phi_2)\phi_1\beta\frac{\partial p_1}{\partial t}+(1-\phi_2)\phi_2\beta\frac{\partial p_2}{\partial t}+(1-\phi_2)\frac{\partial V_{is}}{\partial x_i}+\frac{\Gamma}{\rho}, \tag{6.5.1.15a}$$

$$\frac{\partial}{\partial x_i}\left[\frac{k_{ij2}}{\mu}\left(\frac{\partial p_2}{\partial x_i}+\rho g_j\right)\right] = (1-\phi_1)\phi_1\beta\frac{\partial p_1}{\partial t}+(1-\phi_1)\phi_2\beta\frac{\partial p_2}{\partial t}+(1-\phi_1)\frac{\partial V_{is}}{\partial x_i}-\frac{\Gamma}{\rho}. \tag{6.5.1.15b}$$

6.5.2 Finite Element Formulation

Equations (6.5.1.1) and (6.5.1.15) constitute the required system of governing equations for the coupled processes of flow and deformation in dual-porosity (fractured) media. The Galerkin finite element solution of such a system of equations is a relatively straightforward extension of the formulation presented in Section 6.2.2 for consolidation in unfractured (or granular) porous media.

6.6 Summary

This concludes our discussion of the simulation of flow in deforming fractured and unfractured media. Although we have not considered the simulation of solute transport in fractured systems, it should be realized that this subject is becoming increasingly important and is presently being studied by several researchers. Interested readers can refer to the recent works by Huyakorn and Lester (1983a) and Rasmuson *et al.* (1982) for details pertaining to numerical models and computer codes for simulating single and multiple species transport in fractured porous media.

We have presented several approaches to flow and deformation problems. Each approach requires a data base for the determination of the constitutive coefficients. It remains to be seen which, if any, of various conceptual models for fractured systems evolves into a widely utilized tool for reservoir analyses.

References

Ayatollahi, M. S. (1978). Stress and flow in fractured porous media, Ph.D. thesis, Department of Material Science and Mineral Engineering, University of California, Berkeley.

Barenblatt, G. I., Zheltov, Iu. P. and Kochina, N. (1960). Basics concepts in the theory of seepage of homogeneous liquids in fissured rocks. *Prikl. Mat. Mekh.* **24** (5), 852–864.

Bear, J., and Braester, C. (1972). On the flow of two immiscible fluids in fractured porous media. *In* "Development in Soil Science: Vol. 2, Fundamentals of Transport Phenomena in Porous Media," pp. 177–202. Elsevier, Amsterdam.

Biot, M. A. (1941). General theory of three-dimensional consolidation. *J. Appl. Phys.* **12,** 155–164.

Biot, M. A. (1955). General theory of three-dimensional consolidation of a porous anisotropic solid. *J. Appl. Phys.* **26,** 182–185.

Boonlualohr, P. (1977). Numerical analyses of continua. Ph.D. thesis, Department of Civil Engineering, University of New South Wales, Australia.

Boulton, N. S. (1963). Analysis of data from non-equilibrium pumping tests allowing for delayed yield from storage. *Proc. Inst. Civ. Eng.* **26,** 469–482.

Callahan, G. D., and Fossum, A. F. (1980). Quality assurance study of the special purpose finite element program SPECTROM: II. Plasticity problems. Battelle Office of Nuclear Waste Isolation, Technical Report No. RSI-0111.

Closmann, P. J. (1975). An aquifer model for fissured reservoirs. *Soc. Pet. Eng. J.* **15** (4), 385–398.

Cook, R. D. (1974). "Concepts and Applications of Finite Element Analysis." Wiley, New York.

Davies, E. H. (1968). Theories of plasticity and the failure of soil masses. *In* "Soil Mechanics, Selected Topics," I. K. Lee, ed., Chapter 6. Butterworth, London.

De Leeuw, E. H. (1965). The theory of three-dimensional consolidation applied to cylindrical bodies. *Proc. 6th Int. Conf. Soil Mech. Found. Eng.* **1,** 287–290.

Drucker, D. C., and Prager, W. (1952). Soil mechanics and plastic analysis of limit design. *Q. J. Appl. Math.* **10,** (2), 157–165.

Duguid, J. (1973). Flow in fractured porous media. Ph.D. thesis, Princeton University, Princeton, New Jersey.

Duguid, J., and Abel, J. (1974). Finite element Galerkin method for flow in fractured media. *In* "Finite Element Methods in Flow Problems," J. T. Oden *et al.*, eds., pp. 599–615. University of Alabama Press, University, Alabama.

Duguid, J., and Lee, P. C. Y. (1977). Flow in fractured porous media. *Water Resour. Res.* **13** (3), 558–566.

Fung, Y. C. (1965). "Foundations of Solid Mechanics." Prentice-Hall, Englewood Cliffs, New Jersey.

Gale, J. E. (1975). A numerical, field, and laboratory study of flow in rocks with deformable fractures. Ph.D. thesis, University of California, Berkeley.

Ghaboussi, J., and Wilson, E. L. (1973). Flow of compressible fluids in elastic media. *Int. J. Numer. Meth. Eng.* **5,** 419–442.

Ghaboussi, J., Wilson, E. L., and Isenberg, J. (1973). Finite element for rock joints and interfaces. *ASCE J. Soil Mech. Found. Div.* **99** (SM10), 833–848.

Gibson, R. E., Schiffman, R. L., and Pu, S. L. (1970). Plane strain and axially symmetric consolidation of a clay layer on a smooth impervious base. *Q. J. Mech. Appl. Math.* **23** (4), 505–520.

Goodman, R. E. (1974). The mechanical properties of joints. *Proc. 3rd Congr. Int. Soc. Rock Mech.* **1,** 127–140.

Goodman, R. E. (1975). "Methods of Geological Engineering in Discontinuous Rocks." West Publishing Co., St. Paul, Minnesota.
Goodman, R. E., and Dubois, J. (1972). Duplication of dilatancy of jointed rocks. *ASCE J. Soil Mech. Found. Div.* **98** (SM4), 399–422.
Goodman, R. E., and St. John, C. (1975). Finite element analysis for discontinuous rocks. *In* "Numerical Methods in Geotechnical Engineering," C. S. Desai and J. T. Christian, eds., Ch. 4, 148–175. McGraw-Hill, New York.
Goodman, R. E., Taylor, R. L., and Brekke, T. L. (1968). A model for the mechanics of jointed rock. *ASCE J. Soil Mech. Found. Div.* **94** (SM3), 637–659.
Gringarten, A. C. (1971). Unsteady-state pressure distributions created by a well with a single horizontal fracture, partial penetration, or restricted entry. Ph.D. thesis, Stanford University.
Gringarten, A. C., Ramey, H. J., and Raghavan, R. (1974). Unsteady-state pressure distributions created by a well with a single infinite-conductivity vertical fracture. *Soc. Pet. Eng. J.* **14** (4), 347–360.
Gureghian, A. B. (1975). A study by the finite-element method of the influence of fractures in confined aquifers. *Soc. Pet. Eng. J.* **15** (2), 181–191.
Hagmann, A. J. (1971). Prediction of stress and strain under drained loading conditions. Report No. R70-18, Department of Civil Engineering, M.I.T., Cambridge, Massachusetts.
Heber-Cinco, L., Sameniego, V. F., and Dominquez, A. N. (1976). Transient pressure behavior for a well with a finite conductivity vertical fracture. *Soc. Pet. Eng.* 51st Conference, paper no. SPE 6019.
Hilber, H. M., and Taylor, R. L. (1976). A finite element model of fluid flow in systems of deformable fractured rock. Report No. U.C. SESM 76-5, Department of Civil Engineering, University of California, Berkeley.
Hill, R. (1950). "The Mathematical Theory of Plasticity." Oxford University Press, London.
Huyakorn, P. S., Lester, B. H., and Faust, C. R. (1983). Finite element techniques for modeling groundwater flow in fractured aquifers. *Water Resour. Res.* **19** (in press).
Huyakorn, P. S. and Lester, B. H. (1983a). FTRANS: A two-dimensional code for simulating fluid flow and transport of radioactive nuclides in fractured rock for repository performance assessment. Battelle Office of Nuclear Waste Isolation, Report No. 426.
Huyakorn, P. S., and Lester, B. H. (1983b). STAFAN: A two-dimensional code for simulating fluid flow and interaction of fluid pressure and stress in fractured rock for repository performance assessment. Battelle Office of Nuclear Waste Isolation, Report No. 427.
Iwai, K. (1976). Fundamental studies of fluid flow through a single fracture. Ph.D. thesis, Department of Civil Engineering, University of California, Berkeley.
Kazemi, H. (1969). Pressure transient analysis of naturally fractured reservoirs with uniform fracture distribution. *Soc. Pet. Eng. J.* **9,** 451–462.
Nayak, G. C. (1971). Plasticity and large deformation problems by the finite element method. Ph.D. thesis, Department of Civil Engineering, University of Wales, Swansea, United Kingdom.
Noorishad, J., Witherspoon, P. A., and Brekke, T. L. (1971). A method for coupled stress and flow analysis of fractured rock masses. Geotechnical Engineering Publication No. 71-6, University of California, Berkeley.
Odeh, A. S. (1965). Unsteady-state behavior of naturally fractured reservoirs. *Soc. Pet. Eng. J.* **5,** 60–66.
Ohnishi, Y., and Goodman, R. E. (1974). Results of laboratory tests on water pressure and flow in joints. *Proc. 3rd Cong. Int. Soc. Rock Mech.* **II-A,** 660–666.
Owen, D. R. J., and Hinton, E. (1980). "Finite Elements in Plasticity: Theory and Practice." Pineridge Press, Swansea, United Kingdom, 594 pp.

Prager, W. (1959). "An Introduction to Plasticity." Addison-Wesley, Reading, Massachusetts.
Rasmuson, A., Narasimhan, T. N., and Neretnieks, I. (1982). Chemical transport in a fissured rock: Verification of a numerical model. *Water Resour. Res.* **18** (5), 1479–1492.
Schiffman, R. L., Chen, A. T., and Jordan, J. C. (1969). An analysis of consolidation theories. ASCE *J. Soil Mech. Found. Div.* **95** (1), 285–312.
Snow, D. T. (1965). A parallel plate model of fractured permeable media. Ph.D. thesis, University of California, Berkeley.
Streeter, V. L., and Wiley, E. B. (1979). "Fluid Mechanics." 7th ed. McGraw-Hill, New York.
Streltsova-Adams, T. D. (1978). Well hydraulics in heterogeneous aquifer formations. *In* "Advances in Hydroscience," Vol. 11, 357–423, V. T. Chow, ed. Academic Press, New York.
Suh, N. P., and Turner, A. P. L. (1975). "Elements of a Mechanical Behavior of Solids," McGraw-Hill, New York.
Terzaghi, K. (1943). "Theoretical Soil Mechanics." Wiley, New York.
van Golf-Racht, (1982). "Fundamentals of Fractured Reservoir Engineering." Elsevier, Amsterdam.
Verruijt, A. (1969). Elastic storage of aquifers. *In* "Flow through Porous Media", Ch. 8, 331–376, R. J. M. DeWiest, ed. Academic Press, New York.
Warren, J. E., and Root, P. J. (1963). The behavior of naturally fractured reservoirs. *Soc. Pet. Eng. J.* **3,** 245–255.
Wilson, C. R. (1970). An investigation of laminar flow in fractured porous rocks. Ph.D. thesis, Department of Civil Engineering, University of California, Berkeley.
Wilson, E. L. (1977). Finite elements for foundations, joints and fluids. *In* "Finite Elements in Geomechanics", G. Gudheus, ed., Ch. 10, 319–350. Wiley, New York.
Zienkiewicz, O. C., and Pande, G. N. (1977). Some useful forms of isotropic yield surfaces for soil and rock mechanics. *In* "Finite Elements in Geomechanics," G. Gudheus, ed., Ch. 5, 179–190. Wiley, New York.

7

Alternative Finite Element Techniques and Applications

7.1 Introduction

In this chapter we describe the three numerical techniques referred to as subdomain collocation, point collocation, and boundary element methods. These methods can be considered generally as variants of the standard Galerkin weighted residual finite element method.

Collocation methods hold considerable promise in providing an efficient algorithm for solving problems such as those encountered in subsurface simulation. In particular, the point collocation method has the advantage of avoiding the integration and assembly steps normally encountered in finite element formulations. This is, of course, of particular significance in nonlinear problems.

Recently, there has been considerable interest in the engineering application of the boundary element method. The attractive feature of this approach is the resultant reduction in the dimensionality of the problems considered. The method is particularly attractive when one considers problems governed by elliptic equations with constant coefficients. Problems governed by parabolic equations are also tractable, but the effectiveness of the scheme is significantly compromised. In the ensuing discussion, we shall describe the theory behind this relatively new technique and subsequently demonstrate its application to problems in subsurface hydrology. For more detailed information related to theoretical development and various practical applications of the boundary integral and boundary element methods, the reader is referred to Brebbia and Walker (1980), Jaswon and Symm (1977), Lennon *et al.* (1979, 1980), Ligget and Liu (1979, 1983), and Liu and Ligget (1978).

7.2 The Point Collocation Technique

7.2.1 General Formulation

The point collocation technique is a conceptually simple and computationally efficient scheme for the solution of many groundwater flow problems. In its most sophisticated form it can be used with isoparametric elements so that irregular geometry is readily accommodated. In this section we solve an elementary groundwater flow problem using this method. For a detailed theoretical development and application of the point collocation method, the reader is referred to Finlayson (1972) and Frind and Pinder (1979).

The equation describing steady groundwater flow in a homogeneous isotropic aquifer is Laplace's equation

$$\frac{\partial^2 h}{\partial x^2} + \frac{\partial^2 h}{\partial y^2} = 0, \tag{7.2.1}$$

where $h(x, y)$ is the potentiometric head and the Cartesian coordinates x and y are presented in Fig. 7.1. The problem of interest is illustrated in this figure, and the associated boundary conditions are

$$\partial h(0, y)/\partial x = 0, \qquad 0 < y < 2, \tag{7.2.2a}$$

$$\partial h(2, y)/\partial x = 0, \qquad 0 < y < 2, \tag{7.2.2b}$$

$$h(x, 0) = 3 - x, \qquad 0 < x < 2; \tag{7.2.2c}$$

$$h(x, 2) = 3 - x, \qquad 0 < x < 2. \tag{7.2.2d}$$

These conditions impose no-flow constraints across the vertical boundaries of the rectangular domain and require that the head decrease linearily from a value of three to one along the horizontal boundaries in the positive x direction.

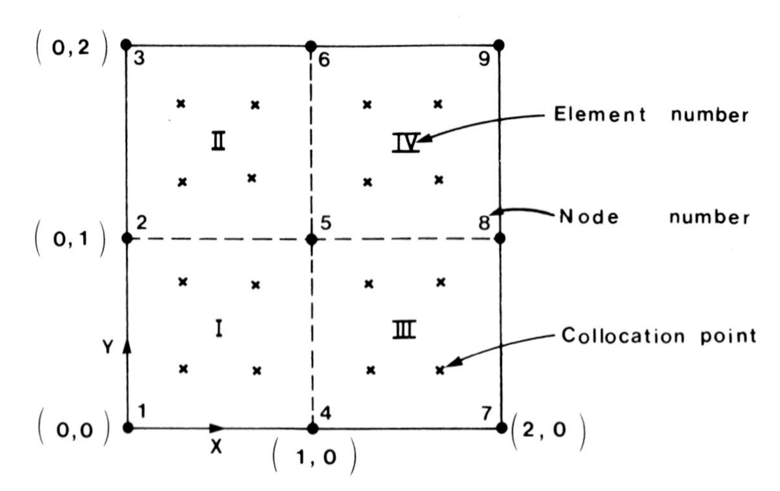

Fig. 7.1. Discretized domain for the collocation example considered.

The governing equation is second order in x and y. Consequently, it is necessary to select basis functions that are piecewise C^1 continuous to assure the existence of their second derivatives within an element. The cubic Hermite polynomials (see Chapter 3) satisfy this criteria. Thus, we introduce a trial function of the form

$$h(x, y) \simeq \hat{h}(x, y) = \sum_{J=1}^{4} \left(h_J N_{00J} + \frac{\partial h_J}{\partial x} N_{xJ} + \frac{\partial h_J}{\partial y} N_{yJ} + \frac{\partial^2 h_J}{\partial x\, \partial y} N_{xyJ}\right), \quad (7.2.3)$$

where h_J, $\partial h_J/\partial x$, $\partial h_J/\partial y$, and $\partial^2 h_J/(\partial x\, \partial y)$ denote values of h, $\partial h/\partial x$, $\partial h/\partial y$, and $\partial^2 h/(\partial x\, \partial y)$ at node J, the N_{mJ} are given by [see Fig. 7.2 for a definition of the local (ξ, η) coordinate system and Lapidus and Pinder (1982) for further development]

$$N_{xJ} = N_{10J} \frac{\partial x_J}{\partial \xi} + N_{01J} \frac{\partial x_J}{\partial \eta} + N_{11J} \frac{\partial^2 x_J}{\partial \xi\, \partial \eta}, \quad (7.2.4a)$$

$$N_{yJ} = N_{10J} \frac{\partial y_J}{\partial \xi} + N_{01J} \frac{\partial y_J}{\partial \eta} + N_{11J} \frac{\partial^2 y_J}{\partial \xi\, \partial \eta}, \quad (7.2.4b)$$

$$N_{xyJ} = N_{11J} \left[\frac{\partial x_J}{\partial \eta} \frac{\partial y_J}{\partial \xi} + \frac{\partial y_J}{\partial \eta} \frac{\partial x_J}{\partial \xi}\right], \quad (7.2.4c)$$

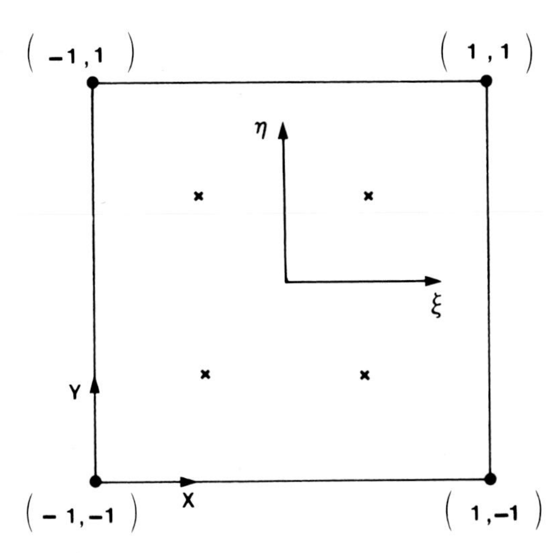

Fig. 7.2. Typical element defined in local (ξ, η) coordinate system. Legend: ●, nodal point; ×, collocation point.

and the Hermite polynomials N_{mnJ} are defined by

$$N_{00J} = \tfrac{1}{16}(\xi + \xi_J)^2(\xi\xi_J - 2)(\eta + \eta_J)^2(\eta\eta_J - 2), \tag{7.2.5a}$$

$$N_{10J} = -\tfrac{1}{16}\xi_J(\xi + \xi_J)^2(\xi\xi_J - 1)(\eta + \eta_J)^2(\eta\eta_J - 2), \tag{7.2.5b}$$

$$N_{01J} = -\tfrac{1}{16}(\xi + \xi_J)^2(\xi\xi_J - 2)\eta_J(\eta + \eta_J)^2(\eta\eta_J - 1), \tag{7.2.5c}$$

$$N_{11J} = \tfrac{1}{16}\xi_J(\xi + \xi_J)^2(\xi\xi_J - 1)\eta_J(\eta + \eta_J)^2(\eta\eta_J - 1), \tag{7.2.5d}$$

where ξ_J and η_J are the nodal values of ξ and η, respectively.

The algebra is simplified because we have selected rectangular elements. Specifically, we can write

$$\partial x_J/\partial\eta = \partial y_J/\partial\xi = 0 \tag{7.2.6a}$$

and

$$\partial x_J/\partial\xi = \partial y_J/\partial\eta = \tfrac{1}{2}. \tag{7.2.6b}$$

By employing (7.2.6) in (7.2.4), we obtain

$$N_{xJ} = N_{10J}/2, \tag{7.2.7a}$$

$$N_{yJ} = N_{01J}/2, \tag{7.2.7b}$$

$$N_{xyJ} = N_{11J}/4. \tag{7.2.7c}$$

Now that we have established an appropriate trial function it is necessary to develop the collocation relation. This is achieved by writing a weighted residual equation with the Dirac delta function as the weighting function, i.e.,

$$\int_x\int_y\left[\frac{\partial^2\hat{h}}{\partial x^2} + \frac{\partial^2\hat{h}}{\partial y^2}\right]\delta(x - x_I, y - y_I)\,dy\,dx = 0, \qquad I = 1, 2, \ldots, 16, \tag{7.2.8}$$

where I denotes the collocation points.

In our specific example, we could proceed directly from (7.2.8) to develop our approximating algebraic equations. Generally, however, we are faced with the problem of choosing suitable points (x_I, y_I). We know that high accuracy is achieved when these points are selected to be Gaussian points (DeBoor and Swartz, 1973). Although these are not easily identified in (x, y) space, they are known in the local (ξ, η) coordinate system. Thus, it is advantageous to express (7.2.8) in local (ξ, η) coordinates. In this coordinate system (7.2.8) becomes, for each element,

$$\int_\xi\int_\eta\left[\frac{\partial^2\hat{h}}{\partial x^2} + \frac{\partial^2\hat{h}}{\partial y^2}\right]\delta(\xi - \xi_K, \eta - \eta_K)\det[J]\,d\eta\,d\xi = 0,$$

$$K = 1, 2, 3, 4, \tag{7.2.9}$$

where $[J]$ is the Jacobian matrix defining the transformation from (x, y) to (ξ, η) and K denotes collocation points in a typical element. From the definition of the Dirac delta function, we can write (7.2.9) as

$$\left(\frac{\partial^2 \hat{h}}{\partial x^2} + \frac{\partial^2 \hat{h}}{\partial y^2}\right) \det[J]\bigg|_{\xi_K, \eta_K} = 0, \qquad K = 1, 2, 3, 4. \tag{7.2.10}$$

At a point (ξ_K, η_K), $\det[J]$ is simply a number and we can divide both sides of (7.2.10) by it to give

$$\left(\frac{\partial^2 \hat{h}}{\partial x^2} + \frac{\partial^2 \hat{h}}{\partial y^2}\right)\bigg|_{\xi_K, \eta_K} = 0, \qquad K = 1, 2, 3, 4. \tag{7.2.11}$$

Thus, we shall have four equations per element, one identified with each of the four Gauss points illustrated in Fig. 7.2. It turns out that we shall also have four unknown parameters per element after accommodating the boundary conditions. The system of equations can subsequently be solved using either a direct or fractional step algorithm.

Substitution of the trial function into (7.2.11) yields for each rectangular element

$$\begin{aligned}\sum_{J=1}^{4} \Bigg[& h_J \frac{\partial^2}{\partial x^2} N_{00J} + \frac{\partial h_J}{\partial x} \frac{\partial^2}{\partial x^2} \frac{N_{10J}}{2} + \frac{\partial h_J}{\partial y} \frac{\partial^2}{\partial x^2} \frac{N_{01J}}{2} \\ & + \frac{\partial^2 h_J}{\partial x\, \partial y} \frac{\partial^2}{\partial x^2} \frac{N_{11J}}{4} + h_J \frac{\partial^2}{\partial y^2} N_{00J} + \frac{\partial h_J}{\partial x} \frac{\partial^2}{\partial y^2} \frac{N_{10J}}{2} \\ & + \frac{\partial h_J}{\partial y} \frac{\partial^2}{\partial y^2} \frac{N_{01J}}{2} + \frac{\partial^2 h_J}{\partial x\, \partial y} \frac{\partial^2}{\partial y^2} \frac{N_{11J}}{4} \Bigg]\Bigg|_{\xi_K, \eta_K} = 0, \qquad K = 1, 2, 3, 4.\end{aligned} \tag{7.2.12}$$

To evaluate (7.2.12) it is necessary to transform the derivatives from (x, y) to (ξ, η). This is the same problem encountered in finite element integration. In fact, for selected numerical integration schemes, an equivalence between the two methods can be established (see Prenter, 1975). The transformation is readily achieved using the chain rule. Whereas the expansion for the general case is cumbersome, for this simplified geometry we obtain

$$\frac{\partial^2 \hat{h}}{\partial x^2} = \frac{\partial^2 \hat{h}}{\partial \xi^2}\left(\frac{\partial \xi}{\partial x}\right)^2 = \frac{\partial^2 \hat{h}}{\partial \xi^2}(4) \tag{7.2.13a}$$

and, similarly,

$$\frac{\partial^2 \hat{h}}{\partial y^2} = \frac{\partial^2 \hat{h}}{\partial \eta^2}\left(\frac{\partial \eta}{\partial y}\right)^2 = \frac{\partial^2 \hat{h}}{\partial \eta^2}(4). \tag{7.2.13b}$$

Equations (7.2.13a), (7.2.13b), and (7.2.12) can now be combined to yield

$$\sum_{J=1}^{4}\left[h_J\left(\frac{\partial^2}{\partial\xi^2}+\frac{\partial^2}{\partial\eta^2}\right)N_{00J}+\frac{\partial h_J}{\partial x}\left(\frac{\partial^2}{\partial\xi^2}+\frac{\partial^2}{\partial\eta^2}\right)\frac{N_{10J}}{2}\right.$$
$$\left.+\frac{\partial h_J}{\partial y}\left(\frac{\partial^2}{\partial\xi^2}+\frac{\partial^2}{\partial\eta^2}\right)\frac{N_{01J}}{2}+\frac{\partial^2 h_J}{\partial x\,\partial y}\left(\frac{\partial^2}{\partial\xi^2}+\frac{\partial^2}{\partial\eta^2}\right)\frac{N_{11J}}{4}\right]\Bigg|_{\xi_K,\eta_K}=0,$$
$$K = 1, 2, 3, 4, \tag{7.2.14}$$

where the derivatives of the basis functions are simply evaluated at the point ξ_K, η_K in a straightforward way. This equation can be written in matrix form as

$$\begin{bmatrix} A_{11} & A_{12} & A_{13} & A_{14} \\ A_{21} & A_{22} & A_{23} & A_{24} \\ A_{31} & A_{32} & A_{33} & A_{34} \\ A_{41} & A_{42} & A_{43} & A_{44} \end{bmatrix}\begin{Bmatrix} x_1 \\ x_2 \\ x_3 \\ x_4 \end{Bmatrix}=\begin{Bmatrix} 0 \\ 0 \\ 0 \\ 0 \end{Bmatrix}. \tag{7.2.15}$$

Each matrix element A_{JK} and x_J is itself a matrix with elements

$$A_{JK} = \left[\left(\frac{\partial^2}{\partial\xi^2}+\frac{\partial^2}{\partial\eta^2}\right)N_{00J},\ \left(\frac{\partial^2}{\partial\xi^2}+\frac{\partial^2}{\partial\eta^2}\right)\frac{N_{10J}}{2},\right. \tag{7.2.16a}$$
$$\left.\left(\frac{\partial^2}{\partial\xi^2}+\frac{\partial^2}{\partial\eta^2}\right)\frac{N_{01J}}{2},\ \left(\frac{\partial^2}{\partial\xi^2}+\frac{\partial^2}{\partial\eta^2}\right)\frac{N_{11J}}{4}\right]\Bigg|_{\xi_K,\eta_K} \tag{7.2.16b}$$

and

$$x_J = \left[h_J,\ \frac{\partial h_J}{\partial x},\ \frac{\partial h_J}{\partial y},\ \frac{\partial^2 h_J}{\partial x\partial y}\right]^{\mathrm{T}}.$$

The matrix $[A]$ in (7.2.15) is not square. Indeed, when the four rectangular elements in Fig. 7.1 are assembled, the result is a global matrix equation consisting of a 16×36 global matrix and a 36×1 column vector. To reduce the global matrix equation to a tractable form we must impose boundary conditions. Examination of Eq. (7.2.2) in conjunction with Fig. 7.1 reveals that the following undetermined coefficients are known:

$$h_1 = h_3 = 3, \tag{7.2.17a}$$

$$h_4 = h_6 = 2, \tag{7.2.17b}$$

$$h_7 = h_9 = 1, \tag{7.2.17c}$$

$$\frac{\partial h_4}{\partial x} = \frac{\partial h_6}{\partial x} = -1, \qquad (7.2.17d)$$

$$\frac{\partial h_1}{\partial x} = \frac{\partial h_2}{\partial x} = \frac{\partial h_3}{\partial x} = \frac{\partial h_7}{\partial x} = \frac{\partial h_8}{\partial x} = \frac{\partial h_9}{\partial x} = 0, \qquad (7.2.17e)$$

$$\frac{\partial^2 h_1}{\partial x\, \partial y} = \frac{\partial^2 h_2}{\partial x\, \partial y} = \frac{\partial^2 h_3}{\partial x\, \partial y} = \frac{\partial^2 h_7}{\partial x\, \partial y} = \frac{\partial^2 h_8}{\partial x\, \partial y} = \frac{\partial^2 h_9}{\partial x\, \partial y} = 0. \qquad (7.2.17f)$$

The substitution of the 20 constraints (7.2.17) into the global matrix equation obtained by assembling the element matrix equations given in (7.2.15), yields a square 16 × 16 global coefficient matrix and 16 × 1 unknown column vector. The solution of the resulting set of equations is given in Table 7.1. In this table the notation U, U_x, U_y, and U_{xy} correspond to h, $\partial h/\partial x$, $\partial h/\partial y$, and $\partial^2 h/\partial x\, \partial y$, respectively.

The example presented in this subsection is a relatively simple steady-state flow example. For additional examples of collocation solutions of more complex groundwater flow problems, the reader is referred to Frind and Pinder (1979) and Lapidus and Pinder (1982).

Table 7.1
Collocation Solution Obtained from Equation (7.2.15) Modified to Accommodate Boundary Conditions[a]

Dependent variable	*Collocation solution*	*Analytical solution*	*Percentage error*
U_{y1}	-1.553		
U_2	2.384	2.325	2.5
U_{y2}	8.325×10^{-16}	0.0000	0.0
U_{y3}	1.553		
U_{y4}	3.322×10^{-16}	0.0000	0.0
U_{xy4}	5.427×10^{-2}		
U_5	1.999	2.000	0.0
U_{x5}	-5.715×10^{-1}	-0.5000	14.0
U_{y5}	-8.240×10^{-17}	0.0000	0.0
U_{xy5}	-1.256×10^{-15}		
U_{y6}	1.474×10^{-16}	0.0000	0.0
U_{xy6}	-5.427×10^{-2}		
U_{y7}	1.553		
U_8	1.616	1.675	3.5
U_{y8}	7.372×10^{-17}	0.0000	0.0
U_{y9}	-1.553		

[a] Calculations made by M. Celia, Dept. of Civil Engineering, Princeton University.

7.3 The Subdomain Collocation Technique

7.3.1 General Formulation

The subdomain collocation technique is another method that can be used to derive a finite element approximation. This approach also has the advantage that no formal integration is required in formulating the numerical approximation. The scheme was used by Cooley (1974) to solve groundwater flow problems. We now demonstrate its application to the equation describing areal flow in a confined aquifer, i.e.,

$$\frac{\partial}{\partial x}\left(T\frac{\partial h}{\partial x}\right) + \frac{\partial}{\partial y}\left(T\frac{\partial h}{\partial y}\right) = S\frac{\partial h}{\partial t}, \tag{7.3.1}$$

where T is transmissivity, h the hydraulic head, and S the storage coefficient. A portion of the finite element mesh to be used in our derivation is illustrated in Fig. 7.3. There are eight triangular elements with node 1 as their common vertex. The subdomain surrounding node 1 corresponds to the shaded rectangle in Fig. 7.3. The boundary of this rectangle is made up of 8 segments, Γ_1–Γ_8, belonging to elements 1–8, respectively. According to the subdomain collocation method, the weighted residual

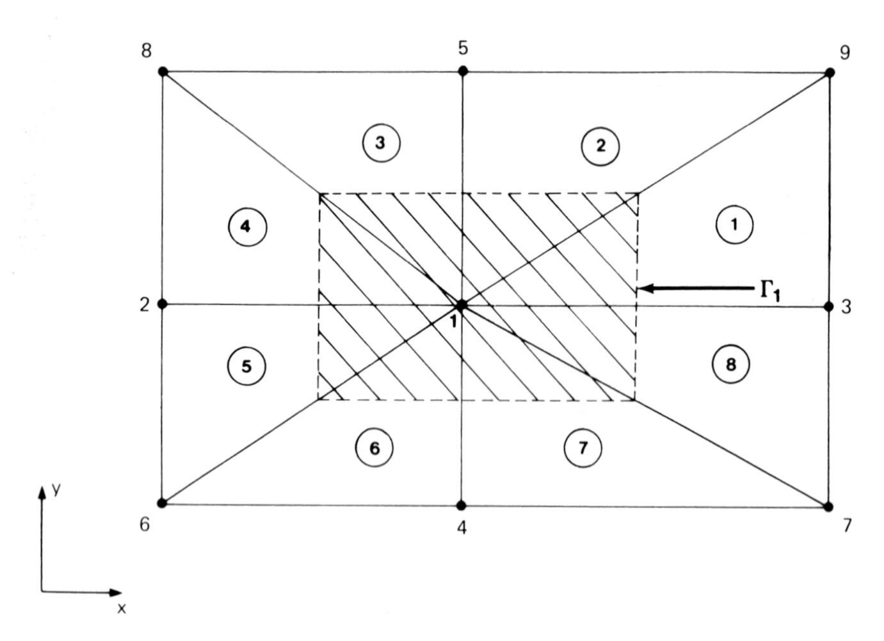

Fig. 7.3. Portion of a finite element mesh with the subdomain centered at node 1. (Adapted from Cooley, 1974.)

equation at node 1 is given by

$$\int_A W_1\left[\frac{\partial}{\partial x}\left(T\frac{\partial \hat{h}}{\partial x}\right) + \frac{\partial}{\partial y}\left(T\frac{\partial \hat{h}}{\partial y}\right) - S\frac{\partial \hat{h}}{\partial t}\right] dA = 0, \qquad (7.3.2)$$

where A is the area of the subdomain, $\hat{h}$ the trial function for the hydraulic head, and W_1 the weighting function whose value is unity in A and zero outside A. Application of the divergence theorem to the first two terms of Eq. (7.3.2) yields

$$\int_\Gamma \left[T\frac{\partial \hat{h}}{\partial x} n_x + T\frac{\partial \hat{h}}{\partial y} n_y\right] d\Gamma - \int_A S\frac{\partial \hat{h}}{\partial t} dA = 0, \qquad (7.3.3)$$

when n_x and n_y are the x and y components of the outward unit normal vector on Γ. Using $\hat{h} = N_J h_J$, and replacing the integrals by the sum of elemental contributions, Eq. (7.3.3) becomes

$$\sum_{e=1}^{8}\left[\int_{\Gamma^e} T\left(\frac{\partial N_J}{\partial x} h_J n_x + \frac{\partial N_J}{\partial y} h_J n_y\right) d\Gamma\right.$$
$$\left. - \int_{A^e}\left(SN_J\frac{dh_J}{dt}\right) dA\right] = 0. \qquad (7.3.4)$$

The typical contribution from element 1 is given by

$$I_1 = \int_{\Gamma_1} T\left(\frac{\partial N_1}{\partial x} h_1 + \frac{\partial N_3}{\partial x} h_3 + \frac{\partial N_9}{\partial x} h_9\right) dy$$
$$- \int_{A_1} S\left(N_1\frac{dh_1}{dt} + N_3\frac{dh_3}{dt} + N_9\frac{dh_9}{dt}\right) dA,$$

where the subscripts in the terms in parentheses refer to the global node numbers. Using the expressions for the shape functions for a linear triangular element, the integrals can be readily evaluated. Assuming that T and S are constant, we finally obtain

$$I_1 = \frac{T(y_9 - y_3)}{2(x_3 - x_1)}(h_3 - h_1)$$
$$- SA_1\left(\frac{2}{3}\frac{dh_1}{dt} + \frac{1}{6}\frac{dh_3}{dt} + \frac{1}{6}\frac{dh_9}{dt}\right),$$

where A_1 is the shaded area of element 1 (Fig. 7.3).

This can be written in the form

$$I_1 = G_1(h_3 - h_1) - SA_1\langle dh/dt\rangle_1,$$

where

$$G_1 = T(y_9 - y_3)/2(x_3 - x_1)$$

and

$$\langle dh/dt\rangle_1 = \frac{1}{6}\left(4\frac{dh_1}{dt} + \frac{dh_3}{dt} + \frac{dh_9}{dt}\right).$$

Expressions for the remaining seven elements may be obtained in a similar manner. It is apparent that Eq. (7.3.4) becomes

$$(G_1 + G_8)(h_3 - h_1) - (G_4 + G_5)(h_1 - h_2) + (G_2 + G_3)(h_5 - h_1)$$
$$- (G_6 + G_7)(h_1 - h_4) = \sum_{e=1}^{8} [SA_e \langle dh/dt\rangle_e], \tag{7.3.5}$$

where the expressions $(SA)_e$, G_e, and $\langle dh/dt\rangle_e$, $e = 2$–8, can be derived by inspection of $(SA)_1$, G_1, and $\langle dh/dt\rangle_1$.

The left side of Eq. (7.3.5) contains five unknowns, whereas the right side contains nine unknowns. To expedite the solution of the resulting system of algebraic equations by an iterative method, it is advantageous to reduce the number of unknowns on the right side of (7.3.5) by applying the following (lumping) approximations:

$$\sum_{e=1}^{8} [(SA)_e \langle dh/dt\rangle_e] \simeq \sum_{e=1}^{8} SA_e\, dh_1/dt = (SF)_1\, dh_1/dt, \tag{7.3.6}$$

where $(SF)_1$ is simply the product of S and the effective area at node 1. For this reason, $(SF)_1$ may be termed the storage factor at node 1.

Substituting Eq. (7.3.6) into (7.3.5), we obtain

$$(G_1 + G_8)(h_3 - h_1) - (G_4 + G_5)(h_1 - h_2) + (G_2 + G_3)(h_5 - h_1)$$
$$- (G_6 + G_7)(h_1 - h_4) = (SF)_1\, dh_1/dt. \tag{7.3.7}$$

Equation (7.3.7) is the discretized equation at node 1. It is clearly one of many integrated finite difference schemes that could be written for this nodal arrangement. The discretized equations for the remaining nodes in the finite element mesh can be obtained in an analogous way. Time integration of the resulting system of equations can be performed in the usual manner. The resulting system of algebraic equations can be solved efficiently by using either the strongly implicit procedure (SIP) or the line successive overrelaxation procedure (LSOR) (see Chapter 9). Although we choose the rectangular mesh in deriving the nodal equations, it should be emphasized that the subdomain collocation method can be used for irregular meshes (Fig. 7.4). In such a case, the contour integrals become

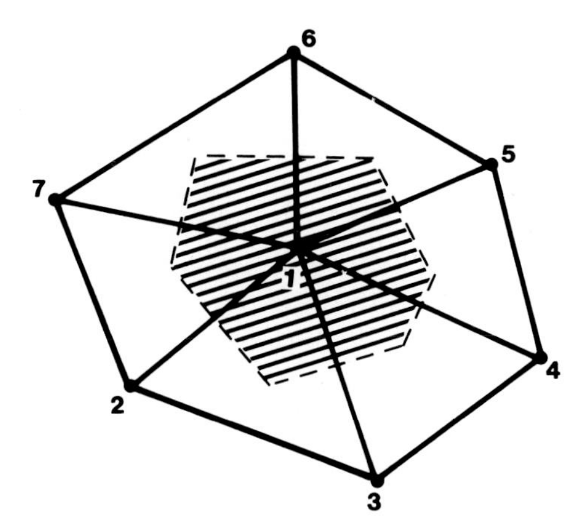

Fig. 7.4. Irregular mesh subdomain with boundary formed by joining bisectors of triangles.

more laborious to evaluate, and the structure of the resulting nodal equations will no longer resemble the conventional five-point "finite difference" approximations.

7.4 Boundary Element Method

7.4.1 General

The boundary element method seeks to exploit the efficiencies associated with the reduction of the dimensionality of a problem. Specifically, it reduces a two- or three-dimensional problem to one defined in one or two dimensions, respectively. To discuss this approach to the simulation of subsurface phenomena, we must first review some basic concepts in applied mathematics.

7.4.2 Review of Green's Functions and Identities

Green's Identities

There are many linear problems in mathematical physics in which it is possible to write the solution as an integral consisting of the product of an auxiliary function and the given information data, which may be the boundary or initial conditions or a nonhomogeneous term in the differential operator. Such an auxiliary function is called a "Green's function." It is dependent on the differential operator and on the geometry of the problem, but not on the given auxiliary information. The construction of the Green's function is often accomplished with the help of three

Green's identities, which are briefly reviewed in this section. More detailed information on Green's functions can be found in Greenberg (1971).

The three Green's identities are consequences of the divergence theorem, which may be stated as follows:

Let (F_1, F_2, F_3) be three components of a vector function $\mathbf{F}$ defined over a three-dimensional region $\mathbf{V}$. If the functions F_1, F_2, and F_3 are continuously differentiable, then the integral of the divergence of $\mathbf{F}$ over any volume is equal to the outward flux of $\mathbf{F}$ over the surface S of that volume, i.e.,

$$\int_{\mathbf{V}} \partial F_i/\partial x_i \, d\mathbf{V} = \int_S F_i n_i \, dS, \tag{6.4.2.1}$$

where $d\mathbf{V} = dx_1 dx_2 dx_3$, n_i denotes components of the outward unit normal vector to the surface S, and repeated indices again denote summation.

Green's first identity is obtained by setting $F_i = \phi \, \partial\psi/\partial x_i$ and substituting this into Eq. (7.4.2.1), i.e.,

$$\int_{\mathbf{V}} \frac{\partial}{\partial x_i}\left(\phi \frac{\partial \psi}{\partial x_i}\right) d\mathbf{V} = \int_S \phi \frac{\partial \psi}{\partial x_i} n_i \, dS. \tag{7.4.2.2}$$

Equation (7.4.2.2) may be written in the form

$$\int_{\mathbf{V}} \phi \frac{\partial^2 \psi}{\partial x_i \, \partial x_i} d\mathbf{V} = \int_S \phi \frac{\partial \psi}{\partial x_i} n_i \, dS - \int_{\mathbf{V}} \frac{\partial \psi}{\partial x_i} \frac{\partial \phi}{\partial x_i} d\mathbf{V}. \tag{7.4.2.3}$$

If the roles of ψ and ϕ in Eq. (7.4.2.3) are interchanged and the resulting equation is subtracted from (7.4.2.3), we obtain

$$\int_{\mathbf{V}} \left(\phi \frac{\partial^2 \psi}{\partial x_i \, \partial x_i} - \psi \frac{\partial^2 \phi}{\partial x_i \, \partial x_i}\right) d\mathbf{V} = \int_S \left(\phi \frac{\partial \psi}{\partial x_i} - \psi \frac{\partial \phi}{\partial x_i}\right) n_i \, dS, \tag{7.4.2.4}$$

which is known as Green's second identity. Green's third identity is derived from the second identity as follows:

Let $\psi = 1/r$, where, as depicted in Fig. 7.5, r is the distance from some fixed point $P(\xi_1, \xi_2, \xi_3)$ to any other point $Q(x_1, x_2, x_3)$. Because ψ is infinite when $r = 0$, Eq. (7.4.2.4) can only be applied to the domain $\mathbf{V}^*$, which results from $\mathbf{V}$ when the point P is excluded from $\mathbf{V}$ by a small circle with radius ε, which in the limit shrinks to zero (see Fig. 7.5). Applying (7.4.2.4) to the domain $\mathbf{V}^*$, we obtain

$$\int_{\mathbf{V}^*} \left[\phi \frac{\partial^2}{\partial x_i \, \partial x_i} \frac{1}{r} - \frac{1}{r} \frac{\partial^2 \phi}{\partial x_i \, \partial x_i}\right] d\mathbf{V}$$

$$= \oint_S \left[\phi \frac{\partial}{\partial n} \frac{1}{r} - \frac{1}{r} \frac{\partial \phi}{\partial n}\right] dS + \oint_{S_1} \left[\phi \frac{\partial}{\partial n} \frac{1}{r} - \frac{1}{r} \frac{\partial \phi}{\partial n}\right] dS. \tag{7.4.2.5}$$

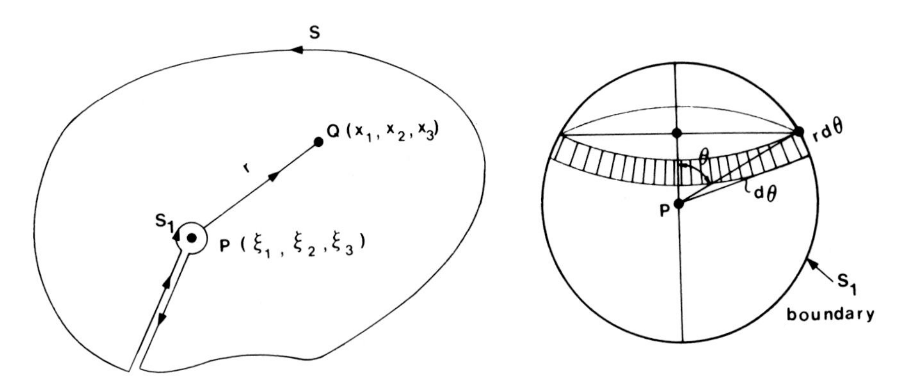

Fig. 7.5. Solution domain with a branch cut to exclude point P.

It can be shown that

$$\int_{V^*} \phi \frac{\partial^2}{\partial x_i \partial x_i} \frac{1}{r}\, d\mathbf{V} = 0 \tag{7.4.2.6}$$

and

$$\oint_{S_1} \left[\phi \frac{\partial}{\partial n} \frac{1}{r} - \frac{1}{r} \frac{\partial \phi}{\partial n} \right] dS = \lim_{\varepsilon \to 0} \int_0^{\Pi} \left[-\phi \frac{\partial}{\partial r} \frac{1}{r} + \frac{1}{r} \frac{\partial \phi}{\partial r} \right] 2\pi r^2 \sin\theta\, d\theta \bigg|_{r=\varepsilon}$$

$$= \lim_{\varepsilon \to 0} \left[\frac{\phi}{\varepsilon^2} + \frac{1}{\varepsilon} \frac{\partial \phi}{\partial r} \right] \varepsilon^2 4\pi \bigg|_{r=\varepsilon}$$

$$= 4\,\pi \phi(\xi_1, \xi_2, \xi_3). \tag{7.4.2.7}$$

Substituting Eqs. (7.4.2.6) and (7.4.2.7) into (7.4.2.5) and rearranging, we obtain

$$4\pi\phi(\xi_1, \xi_2, \xi_3) = \oint_S \left[\frac{1}{r} \frac{\partial \phi}{\partial n} - \phi \frac{\partial}{\partial n} \frac{1}{r} \right] dS - \int_{V^*} \frac{1}{r} \frac{\partial^2 \phi}{\partial x_i \partial x_i}\, d\mathbf{V},$$

which becomes

$$\phi(\xi_1, \xi_2, \xi_3) = \frac{1}{4\pi} \int_S \left[\frac{1}{r} \frac{\partial \phi}{\partial n} - \phi \frac{\partial}{\partial n} \frac{1}{r} \right] dS - \frac{1}{4\pi} \int_V \frac{1}{r} \frac{\partial^2 \phi}{\partial x_i \partial x_i}\, d\mathbf{V}. \tag{7.4.2.8}$$

Equation (7.4.2.8) is known as Green's third identity in three dimensions. Using the logarithmic function $\psi = \ln r$, the two-dimensional form of this identity is

$$2\pi\phi(\xi_1, \xi_2) = \int_\Gamma \left[\phi \frac{\partial}{\partial n}(\ln r) - \ln r \frac{\partial \phi}{\partial n} \right] d\Gamma + \int_A \ln r \frac{\partial^2 \phi}{\partial x_i \, \partial x_i} \, dA, \tag{7.4.2.9}$$

where $d\Gamma$ is the element of arc length of the curve Γ enclosing the area A. Again, the point (ξ_1, ξ_2) is held fixed in the integrations on the right-hand side.

Fundamental Solutions

The functions $\phi = 1/r$ and $\phi = \ln r$ are often called the fundamental solutions of Laplace's equation in two and three dimensions, respectively. They satisfy the Laplace equation at all points, excluding the origin (pole) P. The behavior of $\nabla^2(1/r)$ at point P can be more clearly understood by considering another function, ϕ_ε, defined as

$$\phi_\varepsilon = 1/(r + \varepsilon), \qquad \varepsilon \ll 1. \tag{7.4.2.10}$$

It follows that

$$\nabla^2 \phi_\varepsilon = \frac{1}{r^2} \frac{\partial}{\partial r}\left(r^2 \frac{\partial \phi_\varepsilon}{\partial r} \right) = \left(\frac{\partial^2}{\partial r^2} + \frac{2}{r}\frac{\partial}{\partial r} \right) \frac{1}{r + \varepsilon} = \frac{-2\varepsilon}{r(r + \varepsilon)^3}. \tag{7.4.2.11}$$

Taking the limit, we obtain

$$\lim_{\varepsilon \to 0} (\nabla^2 \phi_\varepsilon) = \begin{cases} 0 & \text{for } r \neq 0 \\ -\infty & \text{for } r = 0. \end{cases} \tag{7.4.2.12}$$

In addition, $\nabla^2 \phi_\varepsilon$ has a volume integral that is independent of r:

$$\int_0^\infty \nabla^2 \phi_\varepsilon 4\pi r^2 \, dr = \int_0^\infty \frac{-2\varepsilon}{r(r + \varepsilon)^3} 4\pi r^2 \, dr = -4\pi. \tag{7.4.2.13}$$

From the definition of the Dirac delta function,

$$\nabla^2 (1/r) = \lim_{\varepsilon \to 0} (\nabla^2 \phi_\varepsilon) = -4\pi\delta(x_1 - \xi_1)\delta(x_2 - \xi_2)\delta(x_3 - \xi_3), \tag{7.4.2.14}$$

and similarly for the two-dimensional case the function $\ln r$ satisfies

$$\nabla^2(\ln r) = 2\pi\delta(x_1 - \xi_1)\delta(x_2 - \xi_2). \tag{7.4.2.15}$$

Green's Functions for Elliptic Equations

The general solution of the equation

$$\nabla^2\phi = f \qquad \text{or} \qquad \nabla^2\phi = 0 \tag{7.4.2.16}$$

in three dimensions may be obtained by constructing a function g defined as

$$g(x_1, x_2, x_3; \xi_1, \xi_2, \xi_3) = -\frac{1}{4\pi r} + p(x_1, x_2, x_3; \xi_1, \xi_2, \xi_3). \tag{7.4.2.17}$$

The function g is called Green's function. It consists of two parts:

(i) a part that has a singularity inside the domain R at the source point or pole $P(\xi_1, \xi_2, \xi_3)$. This part corresponds to $-1/4\pi r$ and is called the principal or fundamental solution;

(ii) a part that is a regular or smooth harmonic function in the entire domain and will be constructed so as to make g satisfy certain boundary conditions. This part corresponds to $p(x_1, x_2, x_3; \xi_1, \xi_2, \xi_3)$. It is easy to show that g satisfies the equation

$$\nabla^2 g = \delta(x_1 - \xi_1)\delta(x_2 - \xi_2)\delta(x_3 - \xi_3). \tag{7.4.2.18}$$

From a physical point of view, we may say that Green's function is the response at a point Q to an influence or disturbance located at a fixed point P. If we replace ψ in Eq. (7.4.2.4) by g and repeat the previous derivation, we obtain

$$\phi(\xi_1, \xi_2, \xi_3) = \int_S \left[\phi\frac{\partial g}{\partial n} - g\frac{\partial \phi}{\partial n}\right] dS + \int_V g\frac{\partial^2\phi}{\partial x_i\,\partial x_i}\, d\mathbf{V}, \quad \mathrm{i} = 1, 2, 3. \tag{7.4.2.19}$$

In two dimensions, the function g is given by

$$g(x_1\ x_2; \xi_1, \xi_2) = (1/2\pi)\ln r + p(x_1, x_2; \xi_1, \xi_2) \tag{7.4.2.20}$$

and Green's third identity thus becomes

$$\phi(\xi_1, \xi_2) = \int_\Gamma \left[\phi\frac{\partial g}{\partial n} - g\frac{\partial \phi}{\partial n}\right] d\Gamma + \int_A g\frac{\partial^2\phi}{\partial x_i\,\partial x_i}\, dA, \quad \mathrm{i} = 1, 2, \tag{7.4.2.21}$$

where Γ is the boundary of the domain A.

For a Dirichlet problem, in which the value of ϕ is specified on the entire boundary of the region, we select the function g so that it vanishes on the boundary. This is done to eliminate the surface integral involving the unknown $\partial\phi/\partial n$. Equation (7.4.2.19) thus reduces to

$$\phi(\xi_1, \xi_2, \xi_3) = \int_S \phi \frac{\partial g}{\partial n} dS + \int_{\mathbf{V}} g \frac{\partial^2 \phi}{\partial x_i \, \partial x_i} d\mathbf{V}. \qquad (7.4.2.22)$$

The construction of g becomes quite simple when the domain $\mathbf{V}$ can be treated by the method of images in which singularities are reflected across the boundaries of the domain. The reflection process is straightforward for certain configurations of $\mathbf{V}$, such as the infinite and semiinfinite strip, the rectangle, the circle, and the sphere. In some cases, a finite number of singularities will suffice to construct Green's function, whereas in other cases an infinite number of singularities is necessary.

Example 1 Find Green's function $g(x, y; \xi, \eta)$ in a semiinfinite plane. Use this function to solve $\nabla^2\phi = 0$ subject to the boundary condition $\phi(x, 0) = h(x)$ shown in Fig. 7.6. To derive the required Green's function, a pole $P(\xi, \eta)$ is reflected across the axis and its image is $P'(\xi, -\eta)$. Hence g is obtained by superposition:

$$g(x, y; \xi, \eta) = (1/2\pi)(\ln r_1 - \ln r_2) = \frac{\ln(r_1/r_2)}{2\pi}.$$

It is easy to see that g vanishes on the boundary of the solution domain. Knowing g, we can determine the required harmonic function ϕ from

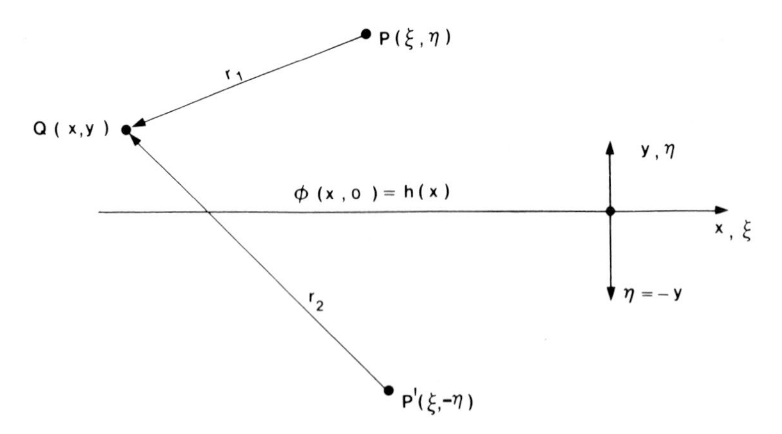

Fig. 7.6. Solution of Laplace's equation in a semiinfinite plane.

Eq. (7.4.2.22), which becomes

$$\phi(\xi, \eta) = \frac{1}{2\pi} \int_{-\infty}^{\infty} \phi(x, 0) \left\{ \frac{\partial}{\partial n} \ln \left[\frac{(x-\xi)^2 + (y-\eta)^2}{(x-\xi)^2 + (y+\eta)^2} \right]^{1/2} \right\}_{y=0} dx.$$

Because $n = -y$ and $\phi\ (x, 0) = h(x)$, this expression reduces to

$$\phi(\xi, \eta) = \frac{-1}{2\pi} \int_{-\infty}^{\infty} h(x) \left\{ \frac{\partial}{\partial y} \ln \left[\frac{(x-\xi)^2 + (y-\eta)^2}{(x-\xi)^2 + (y+\eta)^2} \right]^{1/2} \right\}_{y=0} dx$$

$$= \frac{\eta}{\pi} \int_{-\infty}^{\infty} \frac{h(x)\, dx}{(x-\xi)^2 + \eta^2}.$$

Example 2 Find Green's function g that satisfies $\nabla^2 g = 0$ in the interior of a sphere of radius R (see Fig. 7.7A). To derive the required Green's function, we adopt a spherical coordinate system and place a pole at $P(\rho, \alpha, \beta)$. It can be shown that the image of this pole lies at $P'(R^2/\rho, \alpha, \beta)$. The function g is given by

$$g(r, \theta, \psi; \rho, \alpha, \beta) = \left(\frac{R}{\rho r_2} - \frac{1}{r_1} \right) \Big/ 4\pi,$$

where r_1 and r_2 are distances from $Q(r, \theta, \psi)$ to $P(\rho, \alpha, \beta)$ and $P'(R^2/\rho, \alpha, \beta)$, respectively. Next, we outline the solution of a simple von Neumann boundary value problem by means of the Green's function technique. In the von Neumann problem only the normal derivative $\partial u/\partial n$ is given, and hence no information about the function u is known along the boundary S of the solution domain. The general solution is given by Eq. (7.4.2.19) or (7.4.2.21) depending on whether the problem is two or

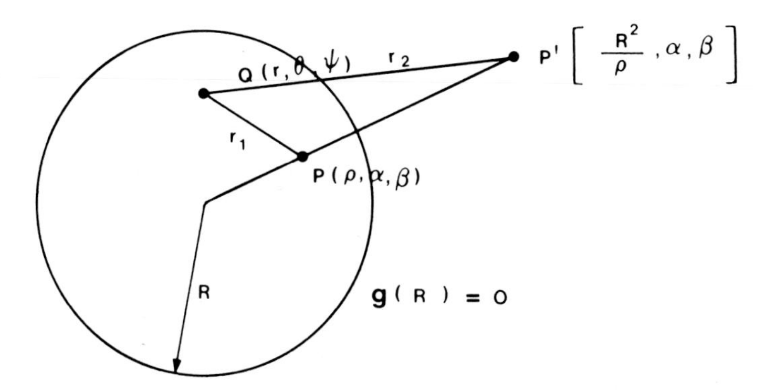

Fig. 7.7A. Green's function in the interior of a sphere.

three dimensional. If Green's function g is constructed so that $\partial g/\partial n$ vanishes on the boundary S, Eq. (7.4.2.19) becomes

$$\phi(\xi_1, \xi_2, \xi_3) = \int_{\mathbf{V}} g \frac{\partial^2 \phi}{\partial x_i\, \partial x_i}\, d\mathbf{V} - \int_S g \frac{\partial \phi}{\partial n}\, dS. \tag{7.4.2.23}$$

For the mixed boundary value problem, we have to solve the partial differential equation in domain $\mathbf{V}$ when the function u is prescribed on some portion and $\partial u/\partial n$ is prescribed on the remaining portion of the boundary S. In this case, we construct Green's function g so that it vanishes over portion S_1 of the boundary and its normal derivative vanishes over the remaining portion S_2. Such a Green's function is called a "mixed Green's function." If Eq. (7.4.2.19) is used, the solution of the problem is given by

$$\phi(\xi_1, \xi_2, \xi_3) = \int_{S_1} \phi \frac{\partial g}{\partial n}\, dS - \int_{S_2} g \frac{\partial \phi}{\partial n}\, dS + \int_{\mathbf{V}} g \frac{\partial^2 \phi}{\partial x_i\, \partial x_i}\, d\mathbf{V}. \tag{7.4.2.24}$$

Example 3 Solve $\nabla^2\phi = 0$ in the first quadrant of the (x, y) plane, given the boundary conditions $\phi(x, 0) = h(x)$ and $(\partial\phi/\partial x)(0, y) = -q$ (see Fig. 7.7B).

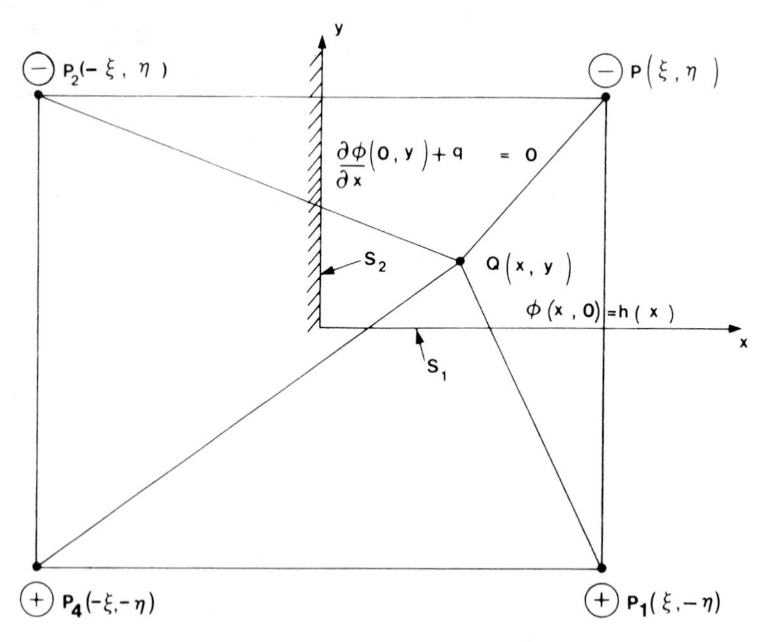

Fig. 7.7B. Solution of Laplace's equation in the first quadrant of the (x, y) plane. (Adapted from De Wiest, 1969, p. 412.)

Using the method of images, we first construct the Green's function of the form

$$g(x, y; \xi, \eta) = (1/2\pi) \ln (r_{PQ} r_{P_2Q}/r_{P_1Q} r_{P_4Q}).$$

This satisfies the given boundary conditions on the x and y axes. The solution of the problem is given by inserting this expression for g into Eq. (7.4.2.21). Nothing that $\xi_1 = \xi$ and $\xi_2 = \eta$, we obtain

$$2\pi\phi(\xi, \eta) = \int_0^\infty h(x) \left\{ \left(\frac{-\partial}{\partial y} \right) \frac{1}{2} \ln \frac{[(x-\xi)^2 + (y-\eta)^2][(x+\xi)^2 + (y-\eta)^2]}{[(x+\xi)^2 + (y+\eta)^2][(x-\xi)^2 + (y+\eta)^2]} \right\}_{y=0} dx$$

$$- \int_0^\infty \frac{1}{2} \left\{ \ln \frac{[(x-\xi)^2 + (y-\eta)^2][(x+\xi)^2 + (y-\eta)^2]}{[(x+\xi)^2 + (y+\eta)^2][(x-\xi)^2 + (y+\eta)^2]} \right\}_{x=0} q\, dy.$$

This reduces to

$$\phi(\xi, \eta) = \frac{\eta}{\pi} \int_0^\infty h(x) \left\{ \frac{1}{(x-\xi)^2 + \eta^2} + \frac{1}{(x+\xi)^2 + \eta^2} \right\} d\xi$$

$$- \frac{1}{2\pi} \int_0^\infty q \ln \left[\frac{\xi^2 + (y-\eta)^2}{\xi^2 + (y+\eta)^2} \right] dy.$$

7.4.3 Boundary Element Solution of a Steady-State Confined Flow

The classical Green's function approach to linear boundary value problems is limited to domains with simple boundary geometry. To deal with problems involving complex boundary geometry, we now present the boundary element method (BIEM or boundary integral equation method), so called because it envisions the boundary of the solution domain as being made up of a series of interconnected elements.

We begin by considering a simple steady-state confined groundwater flow problem that requires the solution of the potential equation

$$\partial^2\phi/\partial x_i\, \partial x_i = 0 \quad \text{in } R \tag{7.4.3.1a}$$

subject to the boundary conditions

$$\phi = H \quad \text{on } S_1 \tag{7.4.3.1b}$$

and

$$\partial\phi/\partial n = -q/K = f \quad \text{on } S_2, \tag{7.4.3.1c}$$

in which the function ϕ corresponds to the hydraulic head and it is assumed that the hydraulic conductivity K is constant and S_1 and S_2 are segments of the boundary enclosing the region R. The first step in the boundary element solution is to formulate the boundary integral equation. This involves discretization of the boundary into elements.

In essence, we make direct use of Green's second and third identities. If ψ is selected to be the fundamental solution and the fixed point P is placed on the boundary of the region, then Green's second identity reduces to

$$\int_S \left(\phi \frac{\partial \psi}{\partial n} - \psi \frac{\partial \phi}{\partial n}\right) dS = 0, \qquad S = S_1 \cup S_2, \qquad (7.4.3.2a)$$

or

$$\int_{S_1} \left(H \frac{\partial \psi}{\partial n} - \psi \frac{\partial \phi}{\partial n}\right) dS + \int_{S_2} \left(\phi \frac{\partial \psi}{\partial n} - \psi f\right) dS = 0, \qquad (7.4.3.2b)$$

where ψ is the fundamental solution given by

$$\psi = \begin{cases} (1/2\pi) \ln \mathrm{r} & \text{for the two-dimensional case} \\ -1/4\pi r & \text{for the three-dimensional case.} \end{cases}$$

Equation (7.4.3.2a) is the required boundary integral equation. Note that in a well-posed problem, either ϕ or its normal derivative $\partial\phi/\partial n$ is prescribed along any given boundary segment. It is desirable to know the values of ϕ and $\partial\phi/\partial n$ so that Green's third identity can be used to determine the value in the interior of the solution domain. Consequently, we have to use Eq. (7.4.3.2a) to obtain the missing boundary data. In other words, this equation serves as an equation for ϕ if $\partial\phi/\partial n$ is known and as an equation for $\partial\phi/\partial n$ if ϕ is known.

To solve for the unknown values of ϕ and $\partial\phi/\partial n$, we now proceed to discretize the boundary into a number of interconnected elements. For clarity in presentation, the two-dimensional case is considered here. Figure 7.8 illustrates the subdivision of a closed curve into m straight-line elements. The nodal points where the unknown values reside may be placed in the middle of each element (constant elements) or at the intersection between the two adjacent elements (linear elements). In general, the elements need not be straight-line segments; we can also use curved elements having three or more nodes per element.

The contour integral in Eq. (7.4.3.2) is now represented by the summation of integrals over individual elements. For a particular fixed point i on the boundary, we use the fundamental solution $\psi = (1/2\pi) \ln r$ to expand Eq. (7.4.3.2a) to give

$$\sum_{j=1}^{m} \int_{S_j} \left[\phi \frac{\partial}{\partial n} (\ln r_i) - \frac{\partial \phi}{\partial n} \ln r_i\right] dS = 0, \qquad (7.4.3.2c)$$

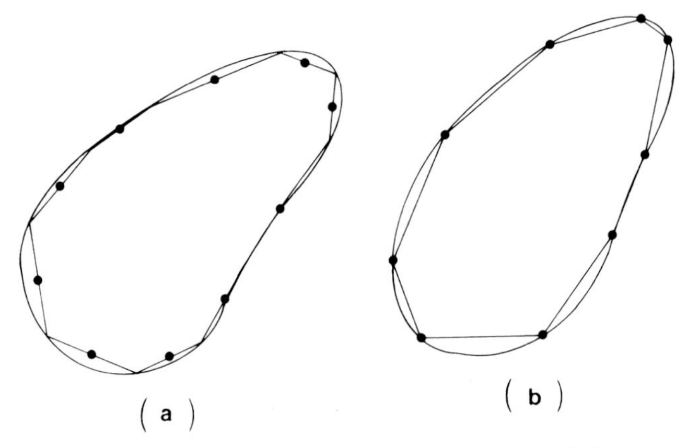

Fig. 7.8. Discretization of boundary: (a) constant elements; (b) linear elements.

where r_i is the radial distace measured from the fixed point i to a variable point on the boundary and S_j the boundary of element j. As one might expect, Eq. (7.4.3.2c) represents a system of discretized equations. Its solution enables us to determine the missing boundary data. Once the complete information on the boundary has been obtained, the final step is to determine the desired interior values of ϕ using Green's third identity.

To present the formulation of the discretized boundary equations, we first consider the use of constant elements and then of linear and quadratic elements.

Constant Elements

For this type of element, the functions ϕ and $\partial\phi/\partial n$ are assumed to be constant on each element and equal to the value at the midside node of the element. Let node i be identified with element number i. For an element j, the integral is given by

$$\int_{S_j}\left[\phi\frac{\partial}{\partial n}\left(\ln r_i\right) - \ln r_i\frac{\partial\phi}{\partial n}\right] dS = \phi_j\int_{S_j}\left(\frac{1}{r}\frac{\partial r}{\partial n}\right)_i dS - \left(\frac{\partial\phi}{\partial n}\right)_j\int_{S_j}\left(\ln r_i\right) dS \qquad (7.4.3.3a)$$

or

$$\int_{S_j}\left[\phi\frac{\partial}{\partial n}\left(\ln r_i\right) - \left(\ln r_i\right)\frac{\partial\phi}{\partial n}\right] dS = G_{ij}\phi_j + H_{ij}\left(\frac{\partial\phi}{\partial n}\right)_j,$$

$$G_{ij} = \int_{S_j}\left(\frac{1}{r}\frac{\partial r}{\partial n}\right)_i dS \qquad (7.4.3.3b)$$

and

$$H_{ij} = -\int_{S_j} \left(\ln r_i\right) dS.$$

In general, these integrals are easily evaluated. However, a minor difficulty arises in evaluating the integral for the ith element ($j = i$). Here we run into the problem of a singularity in $\ln r_i$ when $r_i = 0$. To avoid this difficulty, we replace the subdomain S_i by S_i^* (see Fig. 7.9), which results from S_i when point i is excluded from it by a small semicircle with radius ε (which in the limit shrinks to zero). Thus,

$$\int_{S_i} \left[\phi \frac{\partial}{\partial n}\left(\ln r_i\right) - \left(\ln r_i\right)\frac{\partial \phi}{\partial n}\right] dS$$
$$= \lim_{\varepsilon \to 0} \left[\int_{L_1} I\, dS + \int_C I\, dS + \int_{L_2} I\, dS\right], \tag{7.4.3.4}$$

where C is the semicircle surrounding point i, L_1 and L_2 the line segments to the left and right of C, respectively, and I is defined as

$$I = \frac{\phi}{r_i}\frac{\partial r_i}{\partial n} - (\ln r_i)\frac{\partial \phi}{\partial n}. \tag{7.4.3.5}$$

The integral around the semicircle is given by

$$\lim_{\varepsilon \to 0} \int_C I\, dS = \lim_{\varepsilon \to 0}\left[\left(-\frac{\phi}{\varepsilon} - \ln \varepsilon \frac{\partial \phi}{\partial n}\right)\pi\varepsilon\right]$$
$$= -\pi\phi_i - \pi \lim_{\varepsilon \to 0}\left(\varepsilon \ln \varepsilon \frac{\partial \phi}{\partial n}\right). \tag{7.4.3.6a}$$

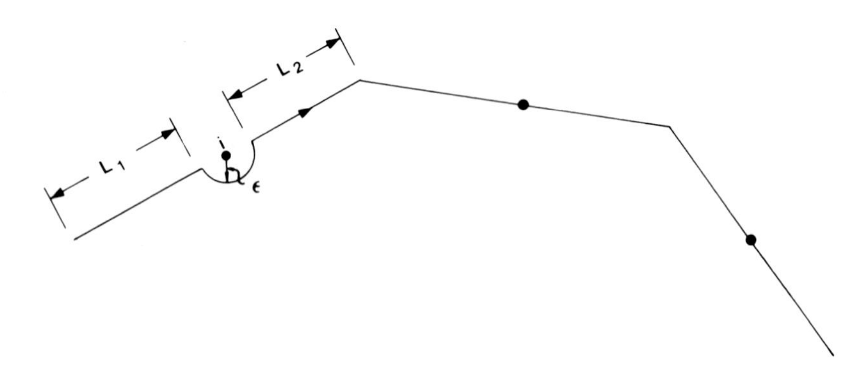

Fig. 7.9. Treatment of singularity at point i.

Because $\lim_{\varepsilon\to 0}(\varepsilon \ln \varepsilon) = 0$, Eq. (7.4.3.6a) becomes

$$\lim_{\varepsilon\to 0} \int_C I\, dS = -\pi\phi_i. \tag{7.4.3.6b}$$

Substituting Eq. (7.4.3.6b) into (7.4.3.4), we obtain

$$\int_{S_i} \left[\phi \frac{\partial}{\partial n}\left(\ln r_i\right) - \left(\ln r_i\right) \frac{\partial \phi}{\partial n} \right] dS = \lim_{\varepsilon\to 0}\left[\int_{L_1} I\, dS + \int_{L_2} I\, dS \right] - \pi\phi_i,$$

which may be written in the form

$$\int_{S_i} \left[\phi \frac{\partial}{\partial n}(\ln r_i) - (\ln r_i) \frac{\partial \phi}{\partial n} \right] dS$$

$$= \lim_{\varepsilon\to 0}\left[\int_{L_1+L_2} \phi \frac{\partial}{\partial n}(\ln r_i)\, dS - \int_{L_1+L_2} \left(\ln r_i\right) \frac{\partial \phi}{\partial n}\, dS \right] - \pi\phi_i$$

or

$$\int_{S_i} \left[\phi \frac{\partial}{\partial n}(\ln r_i) - (\ln r_i) \frac{\partial \phi}{\partial n} \right] dS = \hat{G}_{ii}\phi_i + \hat{H}_{ii}\left(\frac{\partial \phi}{\partial n}\right)_i - \pi\phi_i, \tag{7.4.3.7}$$

where

$$\hat{G}_{ii} = \int_{S_i^*} \left(\frac{1}{r}\frac{\partial r}{\partial n}\right)_i dS,$$

$$\hat{H}_{ii} = -\int_{S_i^*} (\ln r_i)\, dS,$$

and S_i^* is the subdomain S_i with point i excluded. Substituting Eqs. (7.4.3.3b) and (7.4.3.7) into (7.4.3.2c) and noting that, for $i \neq j$, $G_{ij} = \hat{G}_{ij}$ and $H_{ij} = \hat{H}_{ij}$, we obtain

$$-\pi\phi_i + \sum_{j=1}^{m} [\hat{G}_{ij}\phi_j + H_{ij}(\partial\phi/\partial n)_j] = 0. \tag{7.4.3.8}$$

Equation (7.4.3.8) is the equation for node i on the boundary. We can derive a system of simultaneous equations from it by varying i from 1 to m. Such a system may be written in matrix notation as

$$[G]\{\phi\} + [H]\{\partial\phi/\partial n\} = \{0\}, \tag{7.4.3.9}$$

where

$$[G] = [\hat{G}] - \pi[I]$$

and $[I]$ is an $m \times m$ identity matrix and $\{\phi\}$ and $\{\partial\phi/\partial n\}$ are column vectors defined as

$$\{\phi\} = \begin{Bmatrix} \phi_1 \\ \phi_2 \\ \vdots \\ \phi_m \end{Bmatrix}, \qquad \{\partial\phi/\partial n\} = \begin{Bmatrix} (\partial\phi/\partial n)_1 \\ (\partial\phi/\partial n)_2 \\ \vdots \\ (\partial\phi/\partial n)_m \end{Bmatrix}. \tag{7.4.3.10}$$

It is easy to see that the jth columns of matrices $[G]$ and $[H]$ are contributed by element j with node number j at its center. This matrix equation contains m unknowns because, in a well-posed boundary value problem, either the value of function ϕ or its normal derivative $\partial\phi/\partial n$ is specified at each boundary point. The solution of this matrix equation provides the missing boundary data. The coefficients G_{ij} and H_{ij} in Eq. (7.4.3.9) contain integrals that can be directly integrated via the use of the local coordinate system (l, n) depicted in Fig. 7.10.

For $j \neq i$, G_{ij} may be evaluated as

$$\begin{aligned} G_{ij} &= \int_{S_j} \frac{1}{r_i} \frac{\partial r_i}{\partial n}\, dS = \int_{l_1}^{l_2} \frac{1}{r_i} \frac{\partial r_i}{\partial n}\, dl \\ &= \int_{l_1}^{l_2} \frac{\cos\theta}{r_i}\, dl = \int_{l_1}^{l_2} \frac{p_j\, dl}{r_i^2} = \int_{l_1}^{l_2} \frac{p_j\, dl}{p_j^2 + l^2} \\ &= \tan^{-1}\left(\frac{l_2}{p_j}\right) - \tan^{-1}\left(\frac{l_1}{p_j}\right) = \beta_{ij}, \end{aligned} \tag{7.4.3.11}$$

where β_{ij} is the angle at point i subtended by the line element j. The

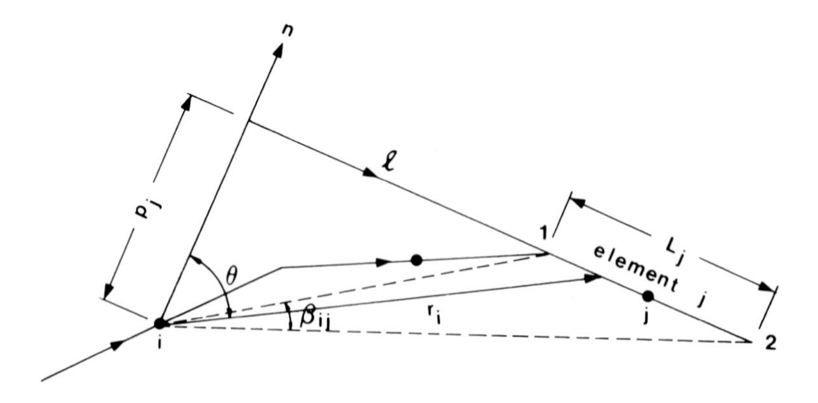

Fig. 7.10. Local coordinate system for element j.

sign of G_{ij} depends on the direction of the unit normal vector n. If n is the direction opposite to that shown in Fig. 7.10, then $G_{ij} = -\beta_{ij}$. Thus we can deduce that the general expression for G_{ij} is given by

$$G_{ij} = \pm\beta_{ij}, \qquad (7.4.3.12)$$

where the positive sign is applicable when (l, n) forms a right-handed coordinate system and the negative sign is applicable when (l, n) forms a left-handed coordinate system.

Similarly, for $j \neq i$, H_{ij} may be evaluated as

$$H_{ij} = -\int_{S_j} \ln(r_i)\, dS = -\int_{l_1}^{l_2} \ln r_i\, dl. \qquad (7.4.3.13)$$

Upon integration, we obtain

$$\begin{aligned} H_{ij} &= \{-\tfrac{1}{2}\ln(p_j^2 + l^2) + l - p_j \tan^{-1}(l/p_j)\}\big|_{l_1}^{l_2} \\ &= [(l_1 \ln r_{i1} - l_2 \ln r_{i2}) + (l_2 - l_1) - p_j(\theta_2 - \theta_1)] \\ &= [-l_2 \ln r_{i2} + l_1 \ln r_{i1} + L_j - p_j\beta_{ij}], \end{aligned} \qquad (7.4.3.14)$$

where L_j is the length of element j. For $i = j$, $G_{ii} = \hat{G}_{ii} - \pi = -\pi$ (the value of $\hat{G}_{ii}$ is zero because of the orthogonality of the vector $\mathbf{r}_i$ and $\mathbf{n}$) and H_{ii} is given by

$$H_{ii} = \hat{H}_{ii} = -\int_{S_i^*} (\ln r_i)\, dS = -2 \lim_{\varepsilon\to 0} \int_{\varepsilon}^{L_i/2} \ln l\, dl, \qquad (7.4.3.15)$$

where ε is the radius of a semicircle surrounding point i and L_i the length of element i. Upon integration, we obtain

$$H_{ii} = -2\left[\frac{L_i}{2} \ln \frac{L_i}{2} - \lim_{\varepsilon\to 0} \int_{\varepsilon}^{L_i/2} dl\right] = L_i\left[\ln\left(\frac{2}{L_i}\right) + 1\right]. \qquad (7.4.3.16)$$

In summary, we can rewrite the boundary matrix equation (7.4.3.9) as

$$[G]\{\phi\} + [H]\{\partial\phi/\partial n\} = \{0\}, \qquad (7.4.3.17)$$

where

$$\begin{aligned} G_{ij} &= \pm\beta_{ij} \quad \text{for} \quad i \neq j, \\ G_{ii} &= -\pi, \\ H_{ij} &= [-l_2 \ln r_{i2} + l_1 \ln r_{i1} + L_j - p_j\beta_{ij}] \quad \text{for} \quad i \neq j, \\ H_{ii} &= L_i[\ln(2/L_i) + 1]. \end{aligned}$$

Linear Elements

Another way to approximate boundary integrals is to discretize the boundary into a number of linear elements. Within each of these elements, we assume that the distribution of ϕ and $\partial\phi/\partial n$ is linear, as illustrated in Fig. 7.11. The nodes are located at the intersection of two adjacent elements. If the boundary is subdivided into m elements, the boundary equation becomes

$$\sum_{e=1}^{m} \int_{S_e} \left[\phi \frac{\partial}{\partial n}(\ln r_i) - \ln r_i \frac{\partial \phi}{\partial n} \right] dS = 0. \qquad (7.4.3.18a)$$

Next, we consider an element e defined by two nodes, j and $j + 1$. For convenience, we adopt a local node numbering system and call the two nodes 1 and 2, respectively. In terms of the local coordinate system (l, n) we represent ϕ and $\partial\phi/\partial n$ by

$$\phi = N_1(l)\phi_1 + N_2(l)\phi_2 \qquad (7.4.3.18b)$$

and

$$\partial\phi/\partial n = N_1(l)(\partial\phi/\partial n)_1 + N_2(l)(\partial\phi/\partial n)_2, \qquad (7.4.3.18c)$$

where $N_1(l)$ and $N_2(l)$ are linear interpolation or basis functions of the form

$$N_1(l) = (1/L_e)[l_2 - l],$$
$$N_2(l) = (1/L_e)[l - l_1], \qquad (7.4.3.18d)$$

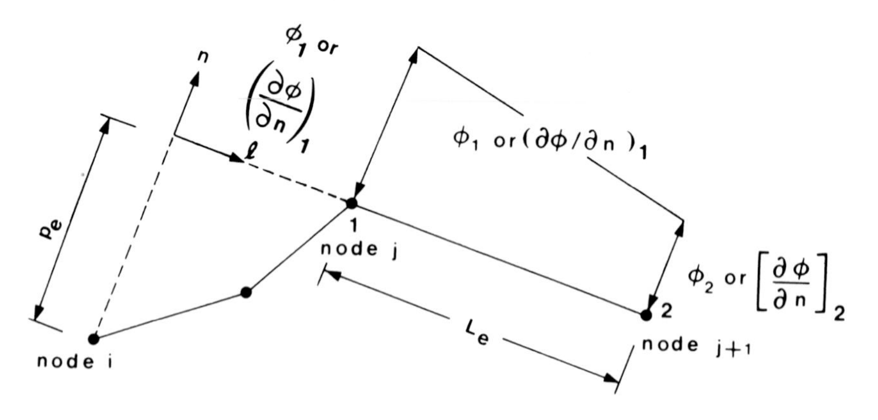

Fig. 7.11. Linear elements and local coordinates.

where $L_e = l_2 - l_1$. Using Eqs. (7.4.3.18), the boundary integral contributed by element e is given by

$$\int_{S_e}\left[\phi\frac{\partial}{\partial n}(\ln r_i) - \ln r_i\frac{\partial \phi}{\partial n}\right] dS = \int_{l_1}^{l_2}(N_1\phi_1 + N_2\phi_2)\frac{\partial}{\partial n}(\ln r_i)\, dl - \int_{l_1}^{l_2}\ln r_i\left[N_1\left(\frac{\partial \phi}{\partial n}\right)_1 + N_2\left(\frac{\partial \phi}{\partial n}\right)_2\right] dl,$$

which may be written in the form

$$\int_{S_e}\left[\phi\frac{\partial}{\partial n}(\ln r_i) - \ln r_i\frac{\partial \phi}{\partial n}\right] dS = G_{i1}^e\phi_1 + G_{i2}^e\phi_2 + H_{i1}^e\left(\frac{\partial \phi}{\partial n}\right)_1 + H_{i2}^e\left(\frac{\partial \phi}{\partial n}\right)_2, \tag{7.4.3.19}$$

where the superscript e is used to denote the contribution from element e and the coefficients G_{i1}^e, G_{i2}^e, H_{i1}^e, and H_{i2}^e are defined as

$$G_{i1}^e = \int_{i_1}^{l_2} N_1\frac{\partial}{\partial n}(\ln r_i)\, dl, \qquad H_{i1}^e = -\int_{l_1}^{l_2}\ln r_i\, N_1\, dl,$$

$$G_{i2}^e = \int_{l_1}^{l_2} N_2\frac{\partial}{\partial n}(\ln r_i)\, dl, \qquad H_{i2}^e = -\int_{l_1}^{l_2}\ln r_i\, N_2\, dl.$$

It should be pointed out that the use of superscript e was unnecessary in the previous case of constant elements because then each node on the boundary belongs to only one element. However, in the present case of linear elements, each node on the boundary is shared by two elements and so their contributions must be added up.

In matrix notation, Eq. (7.4.3.19) may be written as

$$\{Q\}^e = [G]^e\{\phi\}^e + [H]^e\{\partial\phi/\partial n\}^e, \tag{7.4.3.20}$$

where $[G]^e$ and $[H]^e$ are element coefficient matrices given by

$$[G]^e = [G_{i1}^e, G_{i2}^e], \qquad i = 1, 2, \ldots, M,$$

$$[H]^e = [H_{i1}^e, H_{i2}^e], \qquad i = 1, 2, \ldots, M,$$

where M is the number of nodes on the entire boundary and $\{\phi\}^e$ and $\{\partial\phi/\partial n\}^e$ are given by

$$\left\{\phi\right\}^e = \begin{Bmatrix}\phi_1\\ \phi_2\end{Bmatrix}^e, \qquad \left\{\partial\phi/\partial n\right\}^e = \begin{Bmatrix}(\partial\phi/\partial n)_1\\ (\partial\phi/\partial n)_2\end{Bmatrix}^e$$

and $\{Q\}^e$ is defined as

$$\{Q\}^e = \{Q_i\}^e,$$

$$Q_i^e = \int_{S_e} \left[\phi \frac{\partial}{\partial n} (\ln r_i) - \ln r_i \frac{\partial \phi}{\partial n} \right] dS.$$

Equation (7.4.3.20) is an element matrix equation. Summing up all elemental contributions and replacing the local node numbers by the global node numbers, we obtain

$$\sum_{e=1}^{m} \{Q\}^e = \sum_{e=1}^{m} [G]^e \{\phi\} + \sum_{e=1}^{m} [H]^e \{\partial\phi/\partial n\}, \qquad (7.4.3.21)$$

in which

$$\{\phi\} = \begin{Bmatrix} \phi_1 \\ \vdots \\ \phi_M \end{Bmatrix}, \qquad \{\partial\phi/\partial n\} = \begin{Bmatrix} (\partial\phi/\partial n)_1 \\ \vdots \\ (\partial\phi/\partial n)_M \end{Bmatrix}.$$

By enforcing the condition

$$\sum_{e=1}^{m} \{Q\}^e = \{0\}, \qquad (7.4.3.22)$$

Eq. (7.4.3.21) becomes the global matrix equation

$$[G]\{\phi\} + [H]\{\partial\phi/\partial n\} = \{0\}, \qquad (7.4.3.23)$$

where $[G]$ and $[H]$ are the global coefficient matrices, defined as

$$[G] = \sum_{e=1}^{m} [G]^e, \qquad [H] = \sum_{e=1}^{m} [H]^e.$$

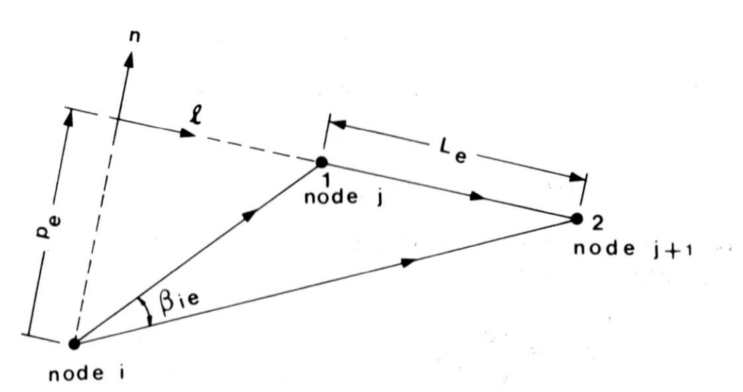

Fig. 7.12. Integration along boundary segment L_e.

As in the case of constant elements, the matrices $[G]^e$ and $[H]^e$ are integrals that can be directly evaluated via the use of local coordinate system (l, n) shown in Fig. 7.12. There are three cases that must be considered.

(1) When node i corresponds to neither node j nor $j + 1$, the matrix elements G_{i1}^e, G_{i2}^e, H_{i1}^e, and H_{i2}^e are given by

$$G_{i1}^e = \int_{l_1}^{l_2} \frac{N_1(l)}{r_i} \frac{\partial r_i}{\partial n}\, dl$$

$$= (1/L_e)\,[-p_e \ln(r_{i2}/r_{i1}) + \beta_{ie} l_2],$$

$$G_{i2}^e = \int_{l_1}^{l_2} \frac{N_2(l)}{r_i} \frac{\partial r_i}{\partial n}\, dl$$

$$= (1/L_e)\,[p_e \ln(r_{i2}/r_{i1}) - \beta_{ie} l_1]$$

$$H_{i1}^e = -\int_{l_1}^{l_2} N_1(l) \ln r_i \, dl$$

$$= (-1/4L_e)[r_{i2}^2(\ln r_{i2}^2 - 1) - r_{i1}^2 (\ln r_{i1}^2 - 1)$$

$$- 4l_2(l_2 \ln r_{i2} - l_1 \ln r_{i1} - L_e + p_e \beta_{ie})],$$

$$H_{i2}^e = -\int_{l_1}^{l_2} N_2(l) \ln r_i \, dl$$

$$= (1/4L_e)\,[r_{i2}^2(\ln r_{i2}^2 - 1) - r_{i1}^2(\ln r_{i1}^2 - 1)$$

$$- 4l_1(l_2 \ln r_{i2} - l_1 \ln r_{i1} - L_e - p_e \beta_{ie})],$$

where β_{ie} is the angle defined by the radii connecting node i to nodes j and $j + 1$.

(2) When node i corresponds to node j (or local node 1) of element e, the integration along the boundary segment $[j - 1, i] \cup [i, j + 1]$ must be divided into three parts, as shown in Fig. 7.13. First, we integrate along the line segment $[j - 1, i^*]$, then around the arc $[i^*, \hat{\imath}]$, and finally along the line segment $[\hat{\imath}, j + 1]$. It is convenient to divide the contribution of element e into two parts: around the second half of arc $[i^*, \hat{\imath}]$ and then along the line segment $[\hat{\imath}, j + 1]$. The matrix element G_{i1}^e may be evaluated as

$$G_{i1}^e = \lim_{\varepsilon \to 0} \left[\int_0^{\theta_i/2} \frac{N_1(\varepsilon)\varepsilon \, d\theta}{\varepsilon} + \int_{\varepsilon}^{l_2} \frac{N_1(l)}{l} \frac{\partial r_i}{\partial n}\, dl \right],$$

where ε is the radius of the arc and θ_i the angle subtended by the arc

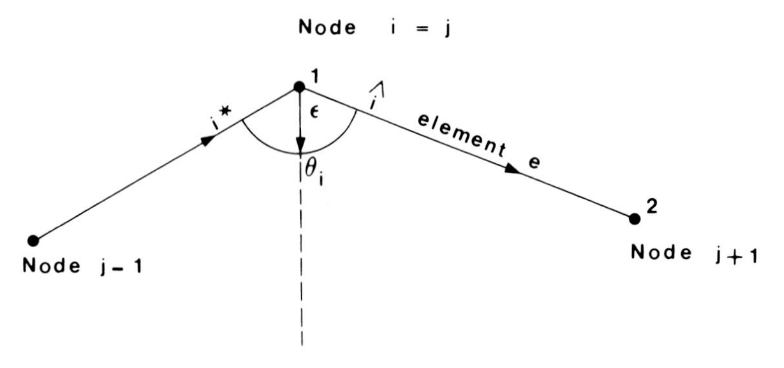

Fig. 7.13. Integration along $[j - 1, i] \cup [i, j + 1]$.

around node i. Because $\lim_{\varepsilon\to 0} N_1(\varepsilon) = 1$ along the arc and $\partial r_i/\partial n = 0$ along the line segment $[\hat{\imath}, j + 1]$, the preceding expression reduces to

$$G_{i1}^e = -\theta_i/2.$$

Similarly, we can show that G_{i2}^e is given by

$$G_{i2}^e = \lim_{\varepsilon\to 0}[-\tfrac{1}{2}\theta_i N_2] = 0.$$

The evaluation of matrix $[H]^e$ poses no problem, because the logarithmic function is a singularly integrable function. Thus, the previous expressions for H_{i1}^e and H_{i2}^e apply with $p_e = r_{i1} = l_1 = \beta_{ie} = 0$ and $r_{i2} = L_e$. These expressions reduce to

$$H_{i1}^e = (-1/4L_e)[L_e^2(\ln L_e^2 - 1) - 4L_e^2 (\ln L_e - 1)],$$

$$H_{i2}^e = (1/4L_e)[L_e^2(\ln L_e^2 - 1)].$$

(3) Finally, we consider the case when i corresponds to node $j + 1$ (or local node 2) of element e. In this case, the matrix elements are given by

$$G_{i1}^e = 0, \qquad G_{i2}^e = -\theta_i/2,$$

$$H_{i1}^e = (1/4L_e)[L_e^2(\ln L_e^2 - 1)],$$

$$H_{i2}^e = (-1/4L_e)[L_e^2 (\ln L_e^2 - 1) - 4L_e^2(\ln L_e - 1)].$$

Quadratic Elements

Quadratic elements with curved geometry can be used to provide a better representation of a complex boundary configuration. In this case, the integration becomes complicated and should be performed numerically.

It is convenient to introduce a local isoparametric coordinate system. Such a system is shown in Fig. 7.14 for a typical element e. The transformation from the global (x, y) coordinates to the local (ξ) coordinate is given by

$$x = \sum_{k=1}^{3} N_k(\xi)x_k, \qquad y = \sum_{k=1}^{3} N_k(\xi)y_k, \tag{7.4.3.24}$$

where

$$N_1 = \tfrac{1}{2}\xi\,(\xi - 1), \qquad N_2 = \tfrac{1}{2}\xi\,(1 + \xi), \qquad N_3 = (1 - \xi^2).$$

We also represent ϕ and $\partial\phi/\partial n$ by

$$\phi = \sum_{k=1}^{3} N_k\phi_k, \qquad \partial\phi/\partial n = \sum_{k=1}^{3} N_k(\partial\phi/\partial n)_k. \tag{7.4.3.25}$$

The boundary integral contributed by element e may be written in the form

$$\begin{aligned}\int_{S_e}&\left[\phi\frac{\partial}{\partial n}(\ln r_i) - \ln r_i\frac{\partial\phi}{\partial n}\right]dS \\ &= \int_{S_e}\left[\sum_{k=1}^{3} N_k\phi_k\frac{\partial}{\partial n}(\ln r_i) - \ln r_i\sum_{k=1}^{3} N_k\left(\frac{\partial\phi}{\partial n}\right)_k\right]dS \\ &= \sum_{k=1}^{3}\left[G_{ik}^e\phi_k + H_{ik}^e\left(\frac{\partial\phi}{\partial n}\right)_k\right],\end{aligned} \tag{7.4.3.26}$$

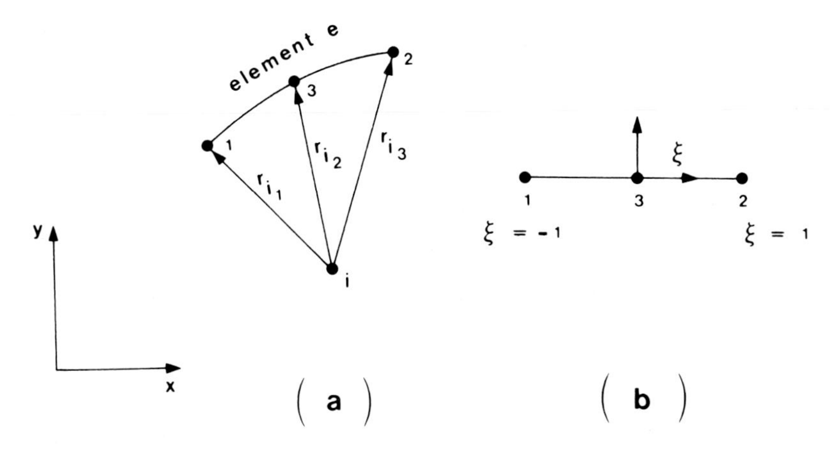

Fig. 7.14. Quadratic element: (a) global coordinates; (b) isoparametric coordinates.

where

$$G_{ik}^e = \int_{S_e} N_k \frac{\partial}{\partial n}(\ln r_i)\, dS, \qquad H_{ik}^e = -\int_{S_e} \ln r_i\, N_k\, dS, \qquad k = 1, 2, 3.$$

To evaluate the preceding integrals, we need to replace dS by

$$dS = [(dx/d\xi)^2 + (dy/d\xi)^2]^{1/2}\, d\xi = |J|\, d\xi.$$

Thus we obtain

$$G_{ik}^e = \int_{-1}^{1} N_k\left(\frac{1}{r_i}\frac{\partial r_i}{\partial n}\right)|J|\, d\xi, \qquad H_{ik}^e = -\int_{-1}^{1} \ln r_i\, N_k|J|\, d\xi, \qquad (7.4.3.27)$$

where

$$r_i = \sum_{k=1}^{3} N_k r_{ik}, \qquad \partial r_i/\partial n = \sum_{k=1}^{3} N_k(\partial r_i/\partial n)_k.$$

Numerical integration can be performed by a Gauss quadrature technique. The integral in H_{ik}^e may be efficiently evaluated using a logarithmic Gauss quadrature formula of the form

$$\int_0^1 \ln(1/\tau) f(\tau)\, d\tau = \sum_{I=1}^{n_g} W_I f(\tau_I),$$

where $\tau = (1 + \xi)/2$, n_g is the number of Gaussian points and ξ_I and W_I the coordinate and weighting factor of the Gaussian point I, respectively. A tabulation of values of the Gaussian point coordinates and weights can be found in Stroud and Secrest (1966).

Except for the element corresponding to the fixed node i under consideration, the integral in G_{ik}^e may be evaluated using a standard Gauss quadrature formula

$$\int_{-1}^{1} f(\xi)\, d\xi = \sum_{I=1}^{n_g} W_I f(\xi_I),$$

for which values of the Gaussian point coordinates and the weights are given in Table 3.1. In general, three Gaussian points are sufficient to obtain the required accuracy.

For the element corresponding to the fixed node i, the integral in G_{ik}^e cannot be accurately evaluated by the use of the standard Gauss quadrature. This difficulty can be circumvented in the following way. First, we note that when node i lies on the element e, $G_{ik}^e = 0$ for $i \neq k$. We can avoid evaluating G_{ii}^e as it is possible to calculate the diagonal element of the global matrix $[G]$ indirectly. This is achieved by applying a uniform distribution of the function ϕ to the global matrix equation

(7.4.3.23). Such a distribution is a trivial solution of the Laplace equation. Because ϕ is constant, $\partial\phi/\partial n$ must be zero:

$$[G]\{\phi\} = \{0\}, \tag{7.4.3.28}$$

which implies that the sum of all elements in any row of $[G]$ is zero. Thus,

$$G_{ii} = -\sum_{j=1}^{M} G_{ij}, \qquad j \neq i. \tag{7.4.3.29}$$

It should be emphasized that Eq. (7.4.3.29) is valid for all types of boundary elements. Using this equation, we need not evaluate the contribution of element e at the fixed node i whenever i lies on the element.

Solution of Algebraic Equations and Determination of Internal Values

It is evident that the system of algebraic equations resulting from the boundary element approximation can be written in matrix form as

$$[G]\{\phi\} + [H]\{\partial\phi/\partial n\} = \{0\}.$$

Unlike the global matrices of the finite element method, which are banded and symmetric, the matrices $[G]$ and $[H]$ in these matrix equations are not only fully populated but also usually nonsymmetric. Before the solution of the global boundary element equations can be obtained, we must separate the known and unknown components of the nodal vectors $\{\phi\}$ and $\{\partial\phi/\partial n\}$ and form the following system

$$[A]\{\chi\} = \{F\},$$

where $[A]$ is a fully populated $n \times n$ matrix derived from $[G]$ and $[H]$, $\{\chi\}$ an $n \times 1$ vector containing the unknown boundary values of ϕ and $\partial\phi/\partial n$, and $\{F\}$ an $n \times 1$ right-hand side compiled from the prescribed boundary values of ϕ and $\partial\phi/\partial n$. This system of equations can be solved simply by using the general Gaussian elimination scheme for a fully populated and asymmetric coefficient matrix.

Once all the nodal values of ϕ and $\partial\phi/\partial n$ on the boundary are determined, we can calculate the value of ϕ at any point P in the region by the linking equation (Green's third identity), which is rewritten in discretized form as

$$\begin{aligned}\phi_P &= \frac{1}{2\pi}\sum_{e=1}^{m}\int_{S^e}\left[\phi\frac{\partial}{\partial n}(\ln r_P) - \ln r_P\frac{\partial\phi}{\partial n}\right] dS \\ &= \frac{1}{2\pi}\sum_{e=1}^{m}\left([G]_P^e\{\phi\}^e + [H]_P^e\left\{\frac{\partial\phi}{\partial n}\right\}^e\right).\end{aligned}$$

The formulation and solution procedure just described can be used to provide simple solutions of problems involving two-dimensional potential flow in homogeneous and isotropic aquifers. Numerical examples can be found in Lapidus and Pinder (1982) and Liggett and Liu (1983).

Inhomogeneous Regions

If a region consists of different materials, the governing partial differential equation will depend on the material coefficient K. In such a case, we must divide the entire region into a number of subregions, each of which belongs to one material. Within each subregion, the value of K is assumed to be constant. This assumption enables us to treat individual subregions separately. The subregions can then be connected together using the continuity of the potential and the normal flux across the interface of two adjacent subregions.

As an example, consider a region consisting of three different materials as shown in Fig. 7.15. We discretize the boundary of the entire region and the interfaces between adjacent subregions using linear elements. The boundary matrix equation for subregion 1 is given by

$$K_1[G]_1\{\phi\}_1 + K_1[H]_1\{\partial\phi/\partial n\}_1 = \{0\}, \tag{7.4.3.30}$$

where the subscript is used to denote the subregion number. Equation (7.4.3.30) may be rewritten as

$$[G^*]_1\{\phi\}_1 + [H]_1\{q\}_1 = \{0\}, \tag{7.4.3.31}$$

where $\{q\}_1$ is the nodal flux column vector, defined as

$$\{q\}_1 = K_1\{\partial\phi/\partial n\}_1$$

and

$$[G^*]_1 = K_1[G]_1.$$

We can introduce the following submatrix notation:

(a) ϕ_I is the submatrix containing potential values on the external surface of subregion I;

(b) q_I the submatrix containing flux values on the external surface of subregion I;

(c) ϕ_{IJ} the submatrix containing potential values on the interface between subregions I and J; and

(d) q_{IJ} the submatrix containing flux values on the interface between subregions I and J.

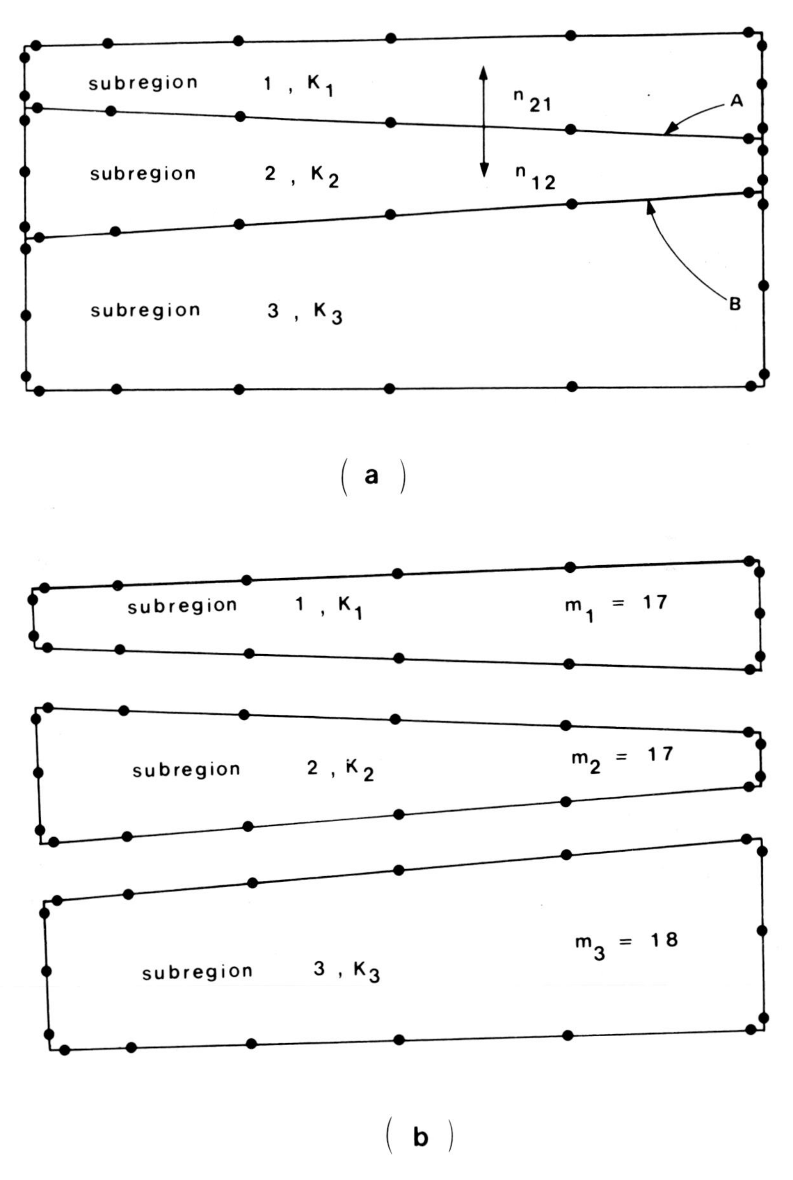

Fig. 7.15. Inhomogeneous region (a) region consisting of three distinct zones (b) subregion separation. For the line in 7.15a identified as A, $\phi_{12} = \phi_{21}$ and $K_1(\partial\phi/\partial n)_{12} = -K_2(\partial\phi/\partial n)_{21}$. For the line in 7.15a identified as B, $\phi_{23} = \phi_{32}$ and $K_2(\partial\phi/\partial n)_{23} = -K_3(\partial\phi/\partial n)_{32}$.

Using this notation, Eq. (7.4.3.31) becomes

$$[G_1^*, G_{12}^*]\begin{Bmatrix}\phi_1\\ \phi_{12}\end{Bmatrix} = [-H_1, -H_{12}]\begin{Bmatrix}q_1\\ q_{12}\end{Bmatrix}. \tag{7.4.3.32a}$$

Similarly, for subregions 2 and 3, we obtain

$$[G_2^*, G_{21}^*, G_{23}^*]\begin{Bmatrix}\phi_2\\ \phi_{21}\\ \phi_{23}\end{Bmatrix} = [-H_2, -H_{21}, -H_{23}]\begin{Bmatrix}q_2\\ q_{21}\\ q_{23}\end{Bmatrix} \tag{7.4.3.32b}$$

and

$$[G_3^*, G_{32}^*]\begin{Bmatrix}\phi_3\\ \phi_{32}\end{Bmatrix} = [-H_3, -H_{32}]\begin{Bmatrix}q_3\\ q_{32}\end{Bmatrix}. \tag{7.4.3.32c}$$

It is apparent that if m_I is the number of nodes on the boundary of subregion I, then Eqs. (7.4.3.32) represent a system of $m_1 + m_2 + m_3 + 2M$ unknowns in this system, M being the total number of nodes on the interfaces. Such a system of equations can be solved if we apply the continuity conditions for potential and flux values at the interfaces. These conditions constitute $2M$ additional simultaneous equations, which may be written as follows:

(1) at the interface between zones 1 and 2,

$$\phi_{12} = \phi_{21}, \qquad q_{12} = -q_{21}; \tag{7.4.3.33a}$$

(2) at the interface between subregions 2 and 3,

$$\phi_{23} = \phi_{32}, \qquad q_{23} = -q_{32}. \tag{7.4.3.33b}$$

Equations (7.4.3.32) and (7.4.3.33) can be combined into the following global matrix equations:

$$\begin{bmatrix} H_1 & G_{12}^* & H_{12} & 0 & 0 & 0 & 0\\ 0 & G_{21}^* & -H_{21} & H_2 & G_{23}^* & H_{23} & 0\\ 0 & 0 & 0 & 0 & G_{32}^* & -H_{32} & H_3 \end{bmatrix}\begin{Bmatrix}q_1\\ \phi_{12}\\ q_{12}\\ q_2\\ \phi_{23}\\ q_{23}\\ q_3\end{Bmatrix}$$

$$= \begin{bmatrix} -G_1^* & 0 & 0\\ 0 & -G_2^* & 0\\ 0 & 0 & -G_3^* \end{bmatrix}\begin{Bmatrix}\phi_1\\ \phi_2\\ \phi_3\end{Bmatrix}. \tag{7.4.3.34}$$

Equation (7.4.3.34) now represents a system of $m_1 + m_2 + m_3$ equations in $m_1 + m_2 + m_3$ unknowns. These unknowns may appear on both sides. To obtain the solution, we must permute the rows and columns so that all the unknowns are transferred to the left side and the known values to the right side.

Treatment of Anisotropic Material

In practice, we may encounter a situation where the flow takes place in a region of anisotropic but homogeneous material. In such a case, the hydraulic conductivity is a tensor of second rank. If the coordinate axes x_1 and x_2 coincide with the principal axes of the hydraulic conductivity tensor, the governing partial differential equation is given by

$$K_{11}\,(\partial^2\phi/\partial x_1^2) + K_{22}\,(\partial^2\phi/\partial x_2^2) = 0. \qquad (7.4.3.35)$$

In a special case, the directions of the principal axes remain the same throughout the entire region. Such a case is often called an "orthotropic" case. If we place a concentrated potential at $(x_1, x_2) = (\lambda_1, \lambda_2)$, then the fundamental solution, ψ satisfies

$$K_{11}\,(\partial^2\psi/\partial x_1^2) + K_{22}\,(\partial^2\psi/\partial x_2^2) = \delta(x_1 - \lambda_1)\delta(x_2 - \lambda_2). \qquad (7.4.3.36)$$

The simplest way to perform the boundary integral formulation is by using the transformation

$$X_i = x_i/\sqrt{K_{ii}}, \qquad i = 1, 2. \qquad (7.4.3.37)$$

Substitution of Eq. (7.4.3.37) into (7.4.3.35) and (7.4.3.36) yields

$$(\partial^2\phi/\partial X_1^2) + (\partial^2\phi/\partial X_2^2) = 0 \qquad (7.4.3.38)$$

and

$$(\partial^2\psi/\partial X_1^2) + (\partial^2\psi/\partial X_2^2) = \delta[\sqrt{K_{11}}(X_1 - \Lambda_1)]\delta[\sqrt{K_{22}}(X_2 - \Lambda_2)]. \qquad (7.4.3.39)$$

The solution to (7.4.3.39) is

$$\psi = \ln r/2\pi, \qquad (7.4.3.40)$$

where

$$r = [(X_1 - \Lambda_1)^2 + (X_2 - \Lambda_2)^2]^{1/2}.$$

With respect to the transformed region R enclosed by the boundary S, Green's second identity is given by

$$\int_R \left[\phi \frac{\partial^2\psi}{\partial X_i\,\partial X_i} - \psi \frac{\partial^2\phi}{\partial X_i\,\partial X_i}\right] dR = \int_S \left(\phi \frac{\partial\psi}{\partial n} - \psi \frac{\partial\phi}{\partial n}\right) dS. \qquad (7.4.3.41)$$

The second term in the integrand that forms the left-hand side of Eq. (7.4.3.41) vanishes as ϕ satisfies Eq. (7.4.3.38) everywhere in R. If we select a fixed point (Λ_1, Λ_2) inside R, ψ then satisfies Eq. (7.4.3.39). Substitution of (7.4.3.39) and (7.4.3.40) into (7.4.3.41) yields

$$\int_R [\phi(X_1, X_2)\delta[\sqrt{K_{11}}(X_1 - \Lambda_1)]\delta[\sqrt{K_{22}}(X_2 - \Lambda_2)]\, dR$$
$$= \frac{1}{2\pi}\int_S \left[\phi \frac{\partial}{\partial n}(\ln r) - \ln r \frac{\partial \phi}{\partial n}\right] dS. \tag{7.4.3.42}$$

Using Eq. (7.4.3.37), the left side of Eq. (7.4.3.42) may be written as

$$\int_R [\phi(X_1, X_2)\delta(\sqrt{K_{11}}(X_1 - \Lambda_1))\delta(\sqrt{K_{22}}(X_2 - \Lambda_2))]\, dR$$
$$= \lim_{\varepsilon\to 0}\left[\int_{\lambda_1-\varepsilon}^{\lambda_1+\varepsilon}\int_{\lambda_2-\varepsilon}^{\lambda_2+\varepsilon} \phi\left(\frac{x_1}{\sqrt{K_{11}}}, \frac{x_2}{\sqrt{K_{22}}}\right)\delta(x_1 - \lambda_1)\delta(x_2 - \lambda_2)\frac{dx_1\, dx_2}{\sqrt{K_{11}K_{22}}}\right]$$
$$= \frac{1}{\sqrt{K_{11}K_{22}}}\phi\left(\frac{\lambda_1}{\sqrt{K_{11}}}, \frac{\lambda_2}{\sqrt{K_{22}}}\right) = \frac{1}{\sqrt{K_{11}K_{22}}}\phi(\Lambda_1, \Lambda_2). \tag{7.4.3.43}$$

Combination of Eqs. (7.4.3.43) and (7.4.3.42) yields

$$\phi(\Lambda_1, \Lambda_2) = \frac{\sqrt{K_{11}K_{22}}}{2\pi}\int_S \left[\phi \frac{\partial}{\partial n}(\ln r) - \ln r \frac{\partial \phi}{\partial n}\right] dS. \tag{7.4.3.44}$$

If we place a fixed point i on the boundary of region R, the left side of Eq. (7.4.3.42) vanishes and we obtain the boundary integral equation

$$\int_S \left[\phi \frac{\partial}{\partial n}(\ln r_i) - \ln r_i \frac{\partial \phi}{\partial n}\right] dS = 0. \tag{7.4.3.45}$$

We can then solve Eq. (7.4.3.45) by the boundary element approximation and use Eq. (7.4.3.44) to determine the value of ϕ at any specified interior point. It should be noted that the boundary integral equation (7.4.3.45) for the orthotropic case is identical to that for the isotropic case except that now

$$r = [(X_1 - \Lambda_1)^2 + (X_2 - \Lambda_1)^2]^{1/2} = \left[\frac{(x_1 - \lambda_1)^2}{K_{11}} + \frac{(x_2 - \lambda_2)^2}{K_{22}}\right]^{1/2}. \tag{7.4.3.46}$$

Furthermore, to reduce Eq. (7.4.3.44) to the same form as the corresponding equation for the isotropic case, we only need to define the fundamental solution for the orthotropic case as

$$\psi = \frac{1}{\sqrt{K_{11}K_{22}}}\left(\frac{1}{2\pi}\ln r\right). \tag{7.4.3.47}$$

This formulation was derived for the case where the hydraulic conductivity tensor is diagonal. However, it can also be adapted to a more general case involving a nondiagnoal hydraulic conductivity tensor with components K'_{ij} defined in a coordinate system (x'_1, x'_2). The solution procedure is straightforward. We first determine the coordinate system (x_1, x_2) in which the hydraulic conductivity tensor is diagonal and then determine the two principal hydraulic conductivity values K_{11} and K_{22}. Once the directions of the x_1 and x_2 axes and the values of K_{11} and K_{22} are known, we can proceed to solve the flow problem in the manner described previously. For the sake of completeness, the formulas for computing K_{11}, K_{22}, and the angle α between x_1 and x'_1 are given here.

$$K_{11} = -\frac{K'_{11} - K'_{22}}{2} \sin 2\alpha + K'_{12} \cos 2\alpha \tag{7.4.3.48a}$$

$$K_{22} = \frac{K'_{11} + K'_{22}}{2} - \frac{K'_{11} - K'_{22}}{2} \cos \alpha - K'_{12} \sin 2\alpha \tag{7.4.3.48b}$$

$$\alpha = \frac{1}{2}\left[\tan^{-1}\left(\frac{2K'_{12}}{K'_{11} - K'_{22}}\right)\right] \tag{7.4.3.48c}$$

7.4.4 *Boundary Element Solution of Free Surface Flow*

Perhaps the most interesting application of the boundary element method in subsurface simulation involves problems of steady or transient flow in an unconfined aquifer (Banerjee *et al.* (1981) and Liggett, 1977). As described in Chapter 4, this type of problem is characterized by the presence of a free surface, defined as the surface on which the pressure is atmospheric. If the aquifer is isotropic and homogeneous and there is a negligible elastic storage effect, then the governing flow equation is simply the Laplace equation. Except on the free surface, the boundary conditions take the form of prescribed values of either ϕ or $\partial\phi/\partial n$. The elevation of the free surface is unknown a priori; its location is part of the problem. If the flow is three dimensional and time dependent, the conditions to be satisfied on the free surface are

$$\phi(x_1, x_2, x_3, t)|_{x3=z} = z(x_1, x_2, t), \tag{7.4.4.1a}$$

$$V_i n_i = \left(S_y \frac{\partial z}{\partial t} - I\right) n_3, \tag{7.4.4.1b}$$

where z is the elevation of the free surface, n_i the outward unit normal vector on the free surface, S_y the specific yield of the aquifer material, and I the rate of vertical infiltration into the free surface.

These two equations represent the atmospheric pressure condition and the continuity of flow on the free surface. To express Eq. (7.4.4.1b) in terms of ϕ and z, the normal velocity component and the unit outward normal vector on the free surface are written as

$$V_i n_i = -K \frac{\partial \phi}{\partial n}$$

and

$$n_i = [(\partial/\partial x_i)(x_3 - z)]/|\nabla(x_3 - z)|,$$

from which n_3 is given by

$$n_3 = 1/[1 + (\partial z/\partial x_1)^2 + (\partial z/\partial x_2)^2]^{1/2}.$$

Substitution of these equations into (7.4.4.1b) yields

$$\partial z/\partial t = -(K/S_y)\,[1 + (\partial z/\partial x_1)^2 + (\partial z/\partial x_2)^2]^{1/2}\,(\partial \phi/\partial n) + I. \tag{7.4.4.2a}$$

The left-hand term in Eq. (7.4.4.2a) is the local rate of change of the free surface elevation. This also corresponds to the rate of change of potential following the free surface in a vertical direction (Fig. 7.16). Thus, Eq. (7.4.4.2a) can be rewritten as

$$(\partial \phi/\partial t)_{x_1,x_2} = -(K/S_y)\,[1 + (\partial z/\partial x_1)^2 + (\partial z/\partial x_2)^2]^{1/2}\,(\partial \phi/\partial n) + I. \tag{7.4.4.2b}$$

Equations (7.4.4.1a) and (7.4.4.2b) form the required free surface equations for the boundary integral formulation. For transient two-dimensional flow

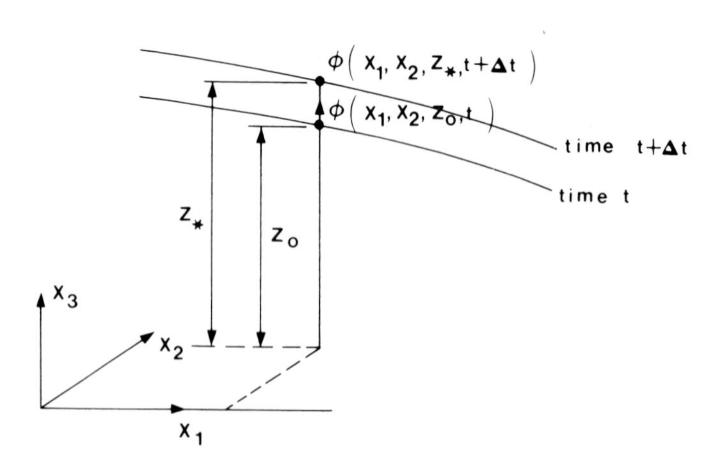

Fig. 7.16. Free surface positions at times t and $t + \Delta t$.

in the (x_1, x_3) plane, they reduce to

$$\phi(x_1, x_3, t)|_{x3=z} = z(x_1, t)$$

$$(\partial\phi/\partial t)_{x_1} = (-K/S_y)\,[1 + (\partial z/\partial x_1)^2]^{1/2}\,(\partial\phi/\partial n) + I$$

$$= (-K/S_y)\,[1 + \tan^2\alpha]^{1/2}\,(\partial\phi/\partial n) + I, \qquad (7.4.4.3a)$$

or

$$(\partial\phi/\partial t)_{x_1} = (-K \sec\alpha/S_y)(\partial\phi/\partial n) + I, \qquad (7.4.4.3b)$$

where α is the angle between the free surface and the horizontal. If the flow is steady and two dimensional and there is no vertical recharge into the free surface, Eqs. (7.4.4.3a) and (7.4.4.3b) become

$$\phi(x_1, x_3)|_{x3=z} = z(x_1), \qquad (7.4.4.4a)$$

$$\partial\phi/\partial n|_{x3=z} = 0. \qquad (7.4.4.4b)$$

Steady Two-Dimensional Flow

As an example, we consider the problem of steady flow without infiltration in a uniform unconfined aquifer located between two ditches with vertical faces. The free surface position is assumed to be as shown in Fig. 7.17. The boundary conditions are

$$\phi = h_u \quad \text{on} \quad B_1,$$

$$\partial\phi/\partial n = 0 \quad \text{on} \quad B_2,$$

$$\phi = x_3 \quad \text{on} \quad B_S,$$

$$\phi = h_d \quad \text{on} \quad B_1',$$

$$\phi = z, \quad \partial\phi/\partial n = 0 \quad \text{on} \quad B_F.$$

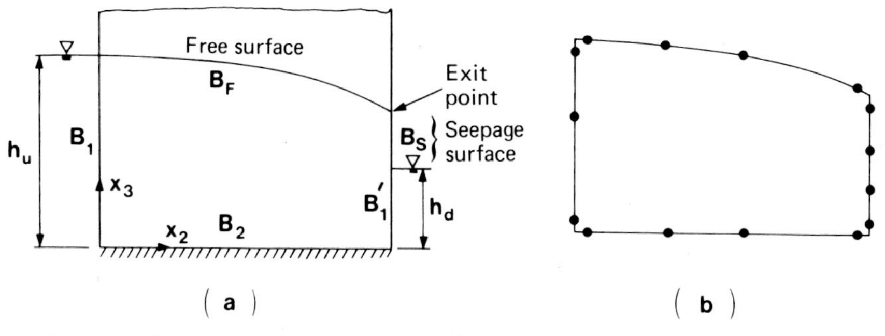

Fig. 7.17. Steady-state flow example: (a) flow region; (b) boundary discretization.

A typical discretization of the flow boundary into linear elements is depicted in Fig. 7.17b. Note that each of the four corners of the region is simulated by placing two nodes very close to each other. Two iterative approaches can be used to treat the nonlinearity owing to the free surface. Each approach starts with a trial free surface.

In the first approach, the problem is solved in the usual manner by prescribing $\partial\phi/\partial n = 0$ on the free surface and $\phi = x_3$ on the seepage face. The computed nodal values of ϕ on the free surface are then compared with the corresponding values of z. If $\max|\phi - z|$ is greater than a prescribed tolerance, the free surface position is adjusted by setting $z = \phi$, and the iteration cycle is then repeated until satisfactory convergence is achieved.

The second approach requires two computational stages per iteration cycle (see Section 4.3.2). For the first stage, the problem is solved by prescribing $\phi = z$ on the free surface and $\phi = x_3$ on the seepage face. For the second stage, the function ϕ is no longer prescribed; instead, $\partial\phi/\partial n$ is prescribed as zero on the free surface and as the $\partial\phi/\partial n$ values computed during the last iterate on the seepage face. The end of the second stage corresponds to the completion of the current iteration cycle. At this juncture, the computed values of ϕ are compared with the corresponding value of z. If need be, the free surface position is adjusted and the iteration cycle repeated until satisfactory convergence is achieved.

Transient Two-Dimensional Flow

We now consider transient flow in an unconfined aquifer such as shown in Fig. 7.18. Initially, at $t = 0$, the hydraulic head everywhere is equal to h_u. At $t > 0$, the hydraulic head at the downstream face suddenly decreases to h_d and remains constant thereafter. In addition, the aquifer is subject to a vertical infiltration rate I.

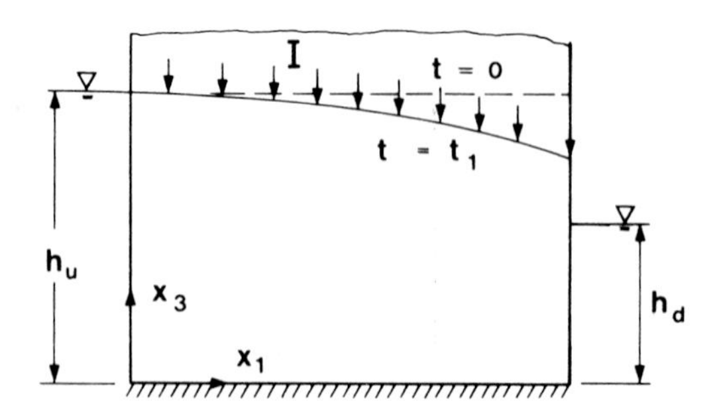

Fig. 7.18. Transient flow example.

The initial condition for this problem is given by

$$\phi(x_1, x_3, 0) = h_u. \tag{7.4.4.5a}$$

The boundary conditions on the free surface are

$$\phi|_{x_3=z} = z, \tag{7.4.4.5b}$$

$$(\partial\phi/\partial t)_{x_1} = (-K \sec\alpha/S_y)(\partial\phi/\partial n) + I. \tag{7.4.4.5c}$$

The remaining boundary conditions are the same as those previously stated for the steady flow example. We outline next two alternative schemes for locating the position of the free surface with respect to time.

Explicit Time Stepping

This scheme is explicit in the sense that there is no iteration within each time increment. This scheme may be described as follows:

(1) At a current time level $t + \Delta t$, the position of the free surface is assumed to be that calculated at the end of the previous time level t. By this assumption, we calculate the matrices $[G^t]$ and $[H^t]$ and partition them so that the boundary matrix equation takes the form

$$[G_{\mathrm{r}}^t, G_{\mathrm{f}}^t]\begin{Bmatrix}\phi_{\mathrm{r}}^{t+\Delta t}\\ \phi_{\mathrm{f}}^{t+\Delta t}\end{Bmatrix} = [H_{\mathrm{r}}^t, H_{\mathrm{f}}^t]\begin{Bmatrix}(\partial\phi/\partial n)_{\mathrm{r}}^{t+\Delta t}\\ (\partial\phi/\partial n)_{\mathrm{f}}^{t+\Delta t}\end{Bmatrix},$$

where the superscripts $t+\Delta t$ and t denote the current and previous time levels, respectively, ϕ_{f} and $(\partial\phi/\partial n)_{\mathrm{f}}$ denote the nodal values on the free surface, and ϕ_{r} and $(\partial\phi/\partial n)_{\mathrm{r}}$ denote the nodal values on the remaining parts of the boundary.

(2) The matrix equation is solved in the usual manner with $\phi_{\mathrm{f}}^{t+\Delta t}$ set equal to z^t, the free surface elevation at time t.

(3) The current position of the free surface is calculated from Eq. (7.4.4.5c), which may be written in finite difference form as

$$\frac{1}{\Delta t}(\phi_{\mathrm{f}}^{t+\Delta t} - \phi_{\mathrm{f}}^t) = \frac{1}{\Delta t}(z^{t+\Delta t} - z^t) = -\frac{K\sec\alpha}{S_y}\left(\frac{\partial\phi}{\partial n}\right)_{\mathrm{f}}^{t+\Delta t} + I^{t+\Delta t}$$

or

$$z^{t+\Delta t} = z^t - \frac{\Delta t\, K\sec\alpha}{S_y}\left(\frac{\partial\phi}{\partial n}\right)_{\mathrm{f}}^{t+\Delta t} + I^{t+\Delta t}.$$

(4) The vertical coordinates of the node are adjusted to $z^{t+\Delta t}$. We then proceed to the next time level and repeat steps 1–3.

Like other explicit time-stepping schemes for solving time dependent differential equations, one might expect a stability restriction on the size of Δt. The critical value of Δt is found to be proportional to $L^2 S_y/K$, where L is the length of the smallest element on the free surface.

Implicit Time Stepping

We now introduce an implicit time-stepping scheme that is unconditionally stable. In this scheme, we must perform a free surface iteration within each time increment. The general procedure may be described as follows:

(1) At the beginning of a current time level $t+\Delta t$, a trial position of the free surface is assumed. It is convenient to adopt the free surface at time t as the trial free surface. For the first iteration, $k = 1$, we calculate the matrices $[G^k]$ and $[H^k]$ and partition them so that the boundary matrix equation takes the form

$$[G_r^k, G_f^k]\begin{Bmatrix} \phi_r^{k+1} \\ \phi_f^{k+1} \end{Bmatrix} = [H_r^k, H_f^k]\begin{Bmatrix} (\partial\phi/\partial n)_r^{k+1} \\ (\partial\phi/\partial n)_f^{k+1} \end{Bmatrix}, \tag{7.4.4.6}$$

where $k + 1$ and k denote the current and previous iterations at time $t+\Delta t$.

(2) Next, we write Eq. (7.4.4.5c) in finite difference form as

$$\frac{\phi_f^{k+1} - \phi_f^t}{\Delta t} = -\frac{K \sec\alpha}{S_y}\left[\theta\left(\frac{\partial\phi}{\partial n}\right)_f^{k+1} + (1-\theta)\left(\frac{\partial\phi}{\partial n}\right)_f^t\right] + I, \tag{7.4.4.7}$$

where θ is in this context a time-weighting factor. To obtain unconditional stability, the value of θ should lie in the range $0.5 \leqslant \theta \leqslant 1$. Equation (7.4.4.7) may be rearranged as

$$\left(\frac{\partial\phi}{\partial n}\right)_f^{k+1} = \left(-\frac{S_y}{\theta\,\Delta t\,K\sec\alpha}\right)\phi_f^{k+1} + F_f, \tag{7.4.4.8}$$

where F_f is a known quantity given by

$$F_f = \left(\frac{S_y}{\theta\,\Delta t\,K\sec\alpha}\right)\phi_f^t + \left(\frac{\theta-1}{\theta}\right)\left(\frac{\partial\phi}{\partial n}\right)_f^t + \frac{IS_y}{\theta K\sec\alpha}.$$

Combination of Eqs. (7.4.4.6) and (7.4.4.8) yields

$$\left[G_r^k, G_f^k + \frac{S_y}{\theta\,\Delta t\,K\sec\alpha}H_f^k\right]\begin{Bmatrix} \phi_r^{k+1} \\ \phi_f^{k+1} \end{Bmatrix} = [H_r^k, H_f^k]\begin{Bmatrix} (\partial\phi/\partial n)_r^{k+1} \\ F_f \end{Bmatrix}. \tag{7.4.4.9}$$

We then solve Eq. (7.4.4.9) in the usual manner and compare the computed ϕ_f^{k+1} with the assumed elevation of the free surface z^k. If need be, the free surface is adjusted by setting $z^{k+1} = \phi_f^{k+1}$, and iteration is performed until satisfactory convergence is achieved.

The preceding implicit scheme has the advantage of being directly applicable to transient flow as well as to steady flow. In the latter situation, we need to perform the calculation for only one time increment. This calculation is achieved by setting $\Delta t = 1$, $S_y = 0$, and $\theta = 1$, thus making $F_f = 0$. It is evident that the scheme then simply reduces to the one-step iterative scheme described previously in the steady-state flow example.

7.4.5 Boundary Element Solution of Transient Confined Flow

Another interesting application of the boundary element method is in the solution of time dependent linear parabolic equations. Two approaches have been used to solve transient heat conduction and confined flow problems. The first approach (Rizzo and Shippy, 1970) is to eliminate the time dependence by a Laplace transform. This effectively reduces the parabolic equation to an elliptic equation, which can then be solved by the standard boundary element procedure described earlier. The disadvantage of such an approach is the necessity to invert the transform, which can lead to analytical and numerical difficulties. The second approach (Dubois and Buysse, 1980; Liggett and Liu, 1979) is a direct solution of the parabolic equation, using a Green's function in space and time. In doing so, however, it is necessary to perform a time integration, which can become rather lengthy.

We elect to present the second approach and demonstrate its application to a problem of areal flow in a homogeneous and isotropic confined aquifer. The governing equation for this problem is now written in the form

$$T \frac{\partial^2 \phi}{\partial x_i \, \partial x_i} = S \frac{\partial \phi}{\partial t} + Q, \qquad (7.4.5.1)$$

where the function ϕ corresponds to the hydraulic head, T and S are aquifer transmissivity and storage coefficients, respectively, and Q is the recharge $(-)$ or withdrawal rate $(+)$ per unit area. Two types of boundary conditions are associated with Eq. (7.4.5.1). These are the prescribed head and flux conditions on boundary portions Γ_1 and Γ_2, respectively. The associated initial condition is that of prescribed head, $\phi(x_1, x_2, 0) = \phi^0$. The fundamental solution for this type of problem is given in Carslaw and Jaeger (1959, p. 361). It can be written in our notation as

$$\psi(r, t-\tau) = \frac{1}{4\pi T(t-\tau)} \exp\left[\frac{-r^2 S}{4T(t-\tau)}\right], \tag{7.4.5.2}$$

where

$$r = [(x_1 - \xi_1)^2 + (x_2 - \xi_2)^2]^{1/2}$$

and ξ_1 and ξ_2 are the coordinates of the fixed point. It can be shown that the function ψ satisfies the equations

$$T \frac{\partial^2 \psi}{\partial x_i \, \partial x_i} - S \frac{\partial \psi}{\partial t} = 0, \qquad t > \tau, \tag{7.4.5.3a}$$

and

$$T \frac{\partial^2 \psi}{\partial x_i \, \partial x_i} + S \frac{\partial \psi}{\partial \tau} = 0, \qquad \tau > t. \tag{7.4.5.3b}$$

In addition, ψ satisfies the condition

$$\psi(x_1, x_2; \xi_1, \xi_2, t - \tau = 0) = (1/S)\, \delta(x_1 - \xi_1)\delta(x_2 - \xi_2)\delta(t - \tau). \tag{7.4.5.3c}$$

Physically, the function ψ represents the hydraulic head distribution at time t resulting from the instantaneous injection of a unit volume of fluid into a fully penetrating well at (ξ_1, ξ_2) at time τ. It should be noted that the distribution represented by Eq. (7.4.5.2) is simply a standard Gaussian distribution with a width proportional to $t - \tau$. As before, the boundary element solution is accomplished using Green's second identity. Because the problem is now time dependent, it is necessary to integrate Green's second identity, Eq. (7.4.2.4), with respect to time. We first obtain

$$\int_0^{t-\varepsilon} \int_A \left[\phi \frac{\partial^2 \psi}{\partial x_i \, \partial x_i} - \psi \frac{\partial^2 \phi}{\partial x_i \, \partial x_i}\right] dA \, d\tau = \int_0^{t-\varepsilon} \int_\Gamma \left[\phi \frac{\partial \psi}{\partial n} - \psi \frac{\partial \phi}{\partial n}\right] d\Gamma \, d\tau, \tag{7.4.5.4}$$

where ε is a very small positive number, Γ and A are the boundary and interior of the solution region, respectively, and $dA = dx_1 \, dx_2$. Next, we rewrite Eq. (7.4.5.1) in terms of τ as

$$T \frac{\partial^2 \phi}{\partial x_i \, \partial x_i} = S \frac{\partial \phi}{\partial \tau} + Q. \tag{7.4.5.5}$$

Using Eqs. (7.4.5.3b) and (7.4.5.5) to substitute for the left-hand terms of Eq. (7.4.5.4), we obtain

$$-\int_0^{t-\varepsilon}\int_A\left[S\left(\phi\frac{\partial\psi}{\partial\tau}+\psi\frac{\partial\phi}{\partial\tau}\right)+\psi Q\right]dA\,d\tau$$

$$=\int_0^{t-\varepsilon}\int_\Gamma T\left[\phi\frac{\partial\psi}{\partial n}-\psi\frac{\partial\phi}{\partial n}\right]d\Gamma\,d\tau. \qquad (7.4.5.6)$$

Integration of the left-hand side of Eq. (7.4.5.6) by parts yields

$$-\left[\int_A S\phi\psi\,dA\right]_{\tau=0}^{\tau=t-\varepsilon}-\int_0^{t-\varepsilon}\int_A\psi Q\,dA\,d\tau$$

$$=\int_0^{t-\varepsilon}\int_\Gamma T\left[\phi\frac{\partial\psi}{\partial n}-\psi\frac{\partial\phi}{\partial n}\right]d\Gamma\,d\tau. \qquad (7.4.5.7)$$

We now take the limit by letting ε tend to zero and then use Eq. (7.4.5.3c) to substitute for ψ at $\tau = t$. Thus, Eq. (7.4.5.7) becomes

$$-\phi(\xi_1,\xi_2,t)+\int_A S\phi^0\psi\,(r,t)\,dA-\int_0^t\int_A\psi Q\,dA\,d\tau$$

$$=\int_0^t\int_\Gamma T\left[\phi\frac{\partial\psi}{\partial n}-\psi\frac{\partial\phi}{\partial n}\right]d\Gamma\,d\tau, \qquad (7.4.5.8)$$

which may be rearranged in the form

$$\phi(\xi_1,\xi_2,t)=T\int_0^t\int_\Gamma\left[\psi\frac{\partial\phi}{\partial n}-\phi\frac{\partial\psi}{\partial n}\right]d\Gamma\,d\tau$$

$$+\int_A S\phi^0\psi\,(r,t)\,dA-\int_0^t\int_A\psi Q\,dA\,d\tau. \qquad (7.4.5.9)$$

Equation (7.4.5.9) gives the value of ϕ at a fixed point $P(\xi_1, \xi_2)$ at time t. This equation is valid as long as P lies in the interior of the solution domain. When P is on the boundary Γ, the boundary integral contains a singularity at P. This can be treated as described previously for the steady-state confined flow problem. The resulting integral equation can thus be written in a general form as

$$\lambda\phi(\xi_1,\xi_2,t)=T\int_0^t\int_\Gamma\left[\psi\frac{\partial\phi}{\partial n}-\phi\frac{\partial\psi}{\partial n}\right]d\Gamma\,d\tau$$

$$+\int_A S\phi^0\psi\,(r,t)\,dA-\int_0^t\int_A\psi Q\,dA\,d\tau, \qquad (7.4.5.10)$$

where

$$\lambda = \begin{cases} 1 & \text{if } (\xi_1, \xi_2) \text{ is in } A, \\ \frac{1}{2} & \text{if } (\xi_1, \xi_2) \text{ is on } \Gamma \text{ and } \Gamma \text{ is smooth thereupon,} \\ \theta/2\pi & \text{if } (\xi_1, \xi_2) \text{ is on } \Gamma \text{ and } \Gamma \text{ is not smooth thereupon,} \end{cases}$$

and θ is the interior angle defined as shown in Fig. 7.13. The time integration can be performed by subdividing the current time value of t_{k+1} into k small time increments. We first rewrite Eq. (7.4.5.10) in the form

$$\lambda\phi_P^{k+1} = \int_\Gamma \int_0^{t_{k+1}} \phi q^* \, d\tau \, d\Gamma - \int_\Gamma \int_0^{t_{k+1}} \psi q \, d\tau \, d\Gamma + S \int_A \phi^0 \psi(r, t_{k+1}) \, dA - \int_A \int_0^{t_{k+1}} \psi Q \, d\tau \, dA, \quad (7.4.5.11)$$

where q and q^* are called the actual and associated outward normal fluxes, respectively,

$$q = -T \, \partial\phi/\partial n, \qquad q^* = -T \, \partial\psi/\partial n, \quad (7.4.5.12)$$

and $\phi_P^{k+1} = \phi(\xi_1, \xi_2, t_{k+1})$.

Because the expression of ψ is given by Eq. (7.4.5.2), q^* can be directly obtained as

$$q^* = -T \frac{\partial\psi}{\partial x_i} n_i = \frac{S\eta}{8\pi T(t-\tau)^2} \exp\left[\frac{-r^2 S}{4T(t-\tau)}\right], \quad (7.4.5.13)$$

where $\eta = (x_i - \xi_i)n_i$.

The first time integral on the right-hand side of Eq. (7.4.5.11) can now be evaluated as

$$\begin{aligned} \int_\Gamma \int_0^{t_{k+1}} \phi q^* \, d\tau \, d\Gamma &\simeq \int_\Gamma \sum_{l=0}^{k} \langle\phi\rangle \left\{ \int_{t_l}^{t_{l+1}} \frac{S\eta}{8\pi T(t_{k+1} - \tau)^2} \times \exp\left[\frac{-r^2 S}{4T(t_{k+1} - \tau)}\right] d\tau \right\} d\Gamma \\ &= \sum_{l=0}^{k} -\frac{1}{2\pi} \int_\Gamma \langle\phi\rangle \frac{\eta}{r^2} \left\{ \exp\left[\frac{-r^2 S}{4T(t_{k+1} - t_{l+1})}\right] - \exp\left[\frac{-r^2 S}{4T(t_{k+1} - t_l)}\right] \right. d\Gamma, \end{aligned} \quad (7.4.5.14)$$

where $t_0 = 0$, $\langle\phi\rangle = (\phi^l + \phi^{l+1})/2$, and ϕ^l is the value of ϕ at time level l. The evaluation of the second time integral in Eq. (7.4.5.11) is accomplished by making the following change of variables:

$$u^* = \frac{r^2 S}{4T(t_{k+1} - \tau)}, \qquad du^* = \frac{r^2 S}{4T(t_{k+1} - \tau)^2}\, d\tau.$$

If it is assumed that q is invariant with time, the result is given by

$$\int_\Gamma \int_0^{t_{k+1}} \psi q \, d\tau \, d\Gamma = \frac{1}{4\pi T} \int_\Gamma q \mathrm{Ei}\left(\frac{r^2 S}{4T t_{k+1}}\right) d\Gamma$$

$$= \frac{1}{4\pi T} \int_\Gamma q \mathrm{Ei}(u) \, d\Gamma, \qquad (7.4.5.15)$$

where $u = r^2 S/(4T t_{k+1})$ and $\mathrm{Ei}(u)$ is the exponential integral given by

$$\mathrm{Ei}(u) = \int \frac{e^{-u}}{u} \, du = 0.5772166 + (-1)^{n-1} \frac{u^n}{n(n!)}.$$

For $u \geqslant 10$, $\mathrm{Ei}(u)$ may be evaluated using the following formula (Abramowitz and Stegun, 1970):

$$u e^u \mathrm{Ei}(u) = (u^2 + 4.0364u + 1.15198)/(u^2 + 5.03637u + 4.19160).$$

Next, we evaluate the remaining time integral in Eq. (7.4.5.11). The result is given by

$$\int_A \int_0^{t_{k+1}} \psi Q \, d\tau \, dA = \frac{1}{4\pi T} \int_A Q \, \mathrm{Ei}(u) \, dA, \qquad (7.4.5.16)$$

where, for the sake of simplicity, it is assumed that Q is invariant with time.

Substituting Eqs. (7.4.5.14)–(7.4.5.16) into (7.4.5.11), we obtain

$$\lambda \phi_P^{k+1} = \sum_{l=0}^{k} - \frac{1}{2\pi} \int_\Gamma \langle \phi \rangle \frac{\eta}{r^2} \left\{ \exp\left(\frac{-r^2 S}{4T(t_{k+1} - t_{l+1})}\right) - \exp\left(\frac{-r^2 S}{4T(t_{k+1} - t_l)}\right) \right\} d\Gamma - \frac{1}{4\pi T} \int_\Gamma q \, \mathrm{Ei}(u) \, d\Gamma$$

$$+ S \int_A \phi^0 \psi(r, t_{k+1}) \, dA - \frac{1}{4\pi T} \int_A Q \, \mathrm{Ei}(u) \, dA. \qquad (7.4.5.17)$$

The spatial discretization of Eq. (7.4.5.17) can be performed in a manner similar to that described previously for the steady-state confined flow problem. It is convenient to evaluate the boundary integrals using Gauss quadrature. The integration over the region can be performed simply by dividing the region into linear triangles or quadrilaterals. Weighting coefficients for the Gauss quadrature can be obtained from Table 3.1. Once the spatial integration has been performed, the solution of the resulting system of algebraic equations is achieved by first transferring all known information to the right-hand side and then using the standard Gauss elimination scheme. The known information includes current prescribed nodal values of ϕ and q and the information from the previous time levels.

It should be emphasized that although the preceding direct boundary element method for solving time dependent equations is straightforward, the time integration can be laborious. Because t appears both as a limit of integration and in the integrand, we cannot derive a recurrence relation between the values of the integrals at the current and previous time levels. To circumvent the problem related to the time integration, Brebbia and Wrobel (1980) employ a stepwise integration procedure. This is achieved by integrating Green's second identity over a small time interval $t_l \leq \tau \leq t_{l+1}$ and obtaining for a point P

$$\lambda\phi_P^{l+1} = \int_\Gamma \int_{t_l}^{t_{l+1}} \phi q^* \, d\tau \, d\Gamma - \int_\Gamma \int_{t_l}^{t_{l+1}} \psi \, q d\tau \, d\Gamma + S \int_A \phi^l \psi(r, t_{l+1} - t_l) \, dA - \int_A \int_{t_l}^{t_{l+1}} \psi Q \, d\tau \, dA, \quad (7.4.5.18)$$

where ψ, q^*, and $\psi(r, t_{l+1} - t_l)$ are given by

$$\psi = \frac{1}{4\pi T(t_{l+1} - \tau)} \exp\left(\frac{-r^2 S}{4T(t_{l+1} - \tau)}\right), \quad (7.4.5.19a)$$

$$q^* = \frac{S\eta}{8\pi T(t_{l+1} - \tau)^2} \exp\left(\frac{-r^2 S}{4T(t_{l+1} - \tau)}\right), \quad (7.4.5.19b)$$

and

$$\psi(r, t_{l+1} - t_l) = \frac{1}{4\pi T(t_{l+1} - t_l)} \exp\left(\frac{-r^2 S}{4T(t_{l+1} - t_l)}\right). \quad (7.4.5.19c)$$

Substituting Eqs. (7.4.5.19a)–(7.4.5.19c) into (7.4.5.17) and performing the time integration with the assumption that ϕ, q, and Q do not vary over a small time step, we obtain

$$\lambda\phi_P^{l+1} = \frac{1}{2\pi} \int_\Gamma \langle\phi\rangle \frac{\eta}{r^2} \exp(-u) \, d\Gamma - \frac{1}{4\pi T} \int_\Gamma \langle q\rangle \operatorname{Ei}(u) \, d\Gamma + S \int_A \phi^l \psi\,(r, t_{l+1} - t_l) \, dA + \frac{1}{4\pi T} \int_A \langle Q\rangle \operatorname{Ei}(u) \, dA, \quad (7.4.5.20)$$

where

$$u = r^2 S/4T(t_{l+1} - t_l), \qquad \langle\phi\rangle = \tfrac{1}{2}(\phi^l + \phi^{l+1}),$$

$$\langle q\rangle = \tfrac{1}{2}(q^l + q^{l+1}), \qquad \langle Q\rangle = \tfrac{1}{2}(Q^l + Q^{l+1}).$$

We begin the solution at the first time step by using Eq. (7.4.5.20) with $l = 0$ and ϕ^0 as the initial condition. At the end of the time step the nodal values ϕ^1 are determined and used as the initial condition for the second time step. The solution procedure is then repeated until the final time step is reached.

References

Abramowitz, M., and Stegun, A. (1970). "Handbook of Mathematical Functions." Dover, New York.

Banerjee, P. K., Butterfield, R., and Tomlin, G. R. (1981). Boundary element methods for two-dimensional problems of transient groundwater flow. *Int. J. Num. Anal. Meth. Geomech.* **5,** 15–31.

Brebbia, C. A., and Walker, S. (1980). "Boundary Element Techniques in Engineering," Newnes-Butterworths Press, London.

Brebbia, C. A., and Wrobel, L. C. (1980). "Steady and unsteady potential problems using the boundary element method," Chapter 1 *in* "Recent Advances in Numerical Methods in Fluids," C. Taylor and K. Morgan, eds. Pineridge Press, U.K.

Carslaw, H. S., and Jaeger, J. C. (1959). "Conduction of Heat in Solids," 2nd ed. Clarendon Press, Oxford.

Cooley, R. L. (1974). Finite element solutions for the equations of ground-water flow. System, Hydrology and Water Resources Publication 18, Desert Research Institute, University of Nevada.

DeBoor, C., and Swartz, B. (1973). Collocation at Gaussian Points. *SIAM J. Numer. Anal.* **10** (4), 582–606.

DeWiest, R. J. M. (1969). "Flow through Porous Media," ch. 10. Academic Press, New York.

Dubois, M., and Buysee, M. (1980). Transient heat transfer analysis by the boundary integral method. *In* "Proceedings of Second Int. Seminar on Recent Advances in Boundary Element Methods," 137–154. C. Brebbia, ed., CML Publications, University of Southampton.

Finlayson, B. M. (1972). "The Method of Weighted Residuals and Variational Principles." Academic Press, New York.

Frind, E. O., and Pinder, G. F. (1979). A collocation finite element method for potential problems in irregular domains. *Int. J. Num. Meth. Eng.* **14,** 681–701.

Greenberg, M.D. (1971). "Application of Green's Functions in Science and Engineering." Prentice-Hall, Englewood Cliffs, N.J.

Jaswon, M. A., and Symm, G. T. (1977). "Integral equation Methods in Potential Theory and Elastostatics." Academic Press, New York.

Lapidus, L., and Pinder, G. F. (1982). "Numerical Solution of Partial Differential Equations in Science and Engineering." Wiley, New York.

Lennon, G. P., Liu, P. L-F., and Liggett, J. A. (1979). Boundary integral equation solution to axisymmetric potential flows: 1. Basic formulation. *Water Resour. Res.* **15**(5), 1102–1106.

Lennon, G. P., Liu, P. L-F., and Liggett, J. A. (1980). Boundary integral solution to three-dimensional unconfined Darcy's flow. *Water Resour. Res.* **16** (4), 651–658.

Liggett, J. A. (1977). Location of Free Surface in Porous Media. *ASCE Hydraul. Div.* **103** (HY4), 353–365.

Liggett, J. A., and Liu, P. L-F. (1979). Unsteady flow in confined aquifers—A comparison of two boundary integral methods. *Water Resour. Res.* **15** (4), 861–866.

Liggett, J. A., and Liu, P. L-F. (1983). "The Boundary Integral Equation Method for Porous Media Flow." George Allen and Unwin, London.

Liu, P. L-F., and Liggett, J. A. (1978). An efficient numerical method of two-dimensional steady groundwater problems. *Water Resour. Res.* **14** (3), 385–390.

Prenter, P. M. (1975). "Splines and Variational Methods." Wiley, New York.

Rizzo, F. J., and Shippy, D. J. (1970). A method of solution for certain problems of transient heat conduction. *AIAA J.* **8** (11), 2004–2009.

Stroud, A. H., and Secrest, (1966). "Gaussian Quadrature Formulas." Prentice-Hall, Englewood Cliffs, New Jersey.

8

The Finite Difference Method

8.1 Introduction

The finite difference method has a long history and a well-established literature. There are many exhaustive treatises devoted to this subject, and we do not propose to reproduce their contents. Rather, we provide an introduction to the subject and refer the reader seeking a more exhaustive discussion to monographs dedicated to this topic [see, for example, Ames (1977), Lapidus and Pinder (1982), Mitchell and Griffiths (1980), and von Rosenberg (1969)]. We shall first introduce finite difference approximations and then consider questions of stability and convergence.

8.1.1 Finite Difference Formulas

Let us consider a continuous function $u(x)$. We shall discretize the domain of the independent variable using $(n + 1)$ points (or nodes), so that one can write

$$u(x_r) \equiv u(r\,\Delta x) \equiv u_r, \qquad r = 0, 1, 2, \ldots, n, \tag{8.1.1.1}$$

where the discretization is illustrated in Fig. 8.1. The two independent variable counterpart of $u(x)$, that is, $u(x_1, x_2)$, is similarly represented by

$$u(x_{1r}, x_{2s}) \equiv u(r\,\Delta x_1, s\,\Delta x_2) \equiv u_{r,s},$$

$$r = 0, 1, 2, \ldots, n, \quad s = 0, 1, 2, \ldots, m, \tag{8.1.1.2}$$

and this discretization appears in Fig. 8.2. Extensions to accommodate more independent variables are straightforward. The objective now is to represent the derivatives of $u(x_1, x_2)$ in terms of these discrete values at $r\,\Delta x_1$ and $s\,\Delta x_2$.

Although the approximation of derivatives can be accomplished in several ways, we focus here on the most popular approach, which is based on Taylor series expansions. Consider the Taylor series expansion

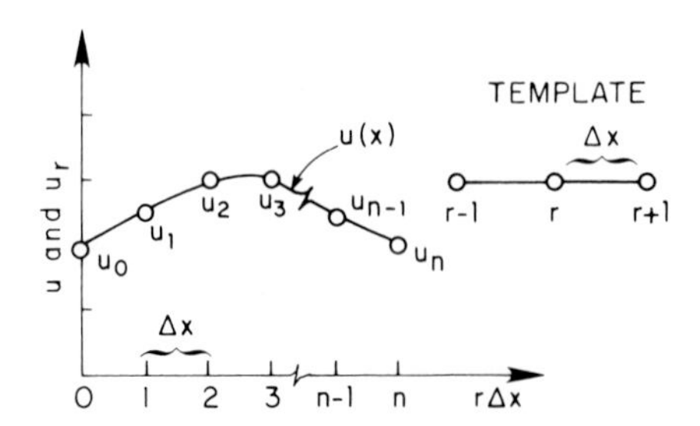

Fig. 8.1. Finite difference discretization of $u(x)$ by n increments of length Δx.

for $u(x)$ written about the point $r\,\Delta x$:

$$u((r+1)\,\Delta x) \equiv u_{r+1} = u_r + \Delta x \left.\frac{du}{dx}\right|_{r\,\Delta x} + \frac{(\Delta x)^2}{2!}\left.\frac{d^2u}{dx^2}\right|_{r\,\Delta x} + \frac{(\Delta x)^3}{3!}\left.\frac{d^3u}{dx^3}\right|_{r\,\Delta x} + \cdots \tag{8.1.1.3a}$$

and

$$u((r-1)\,\Delta x) \equiv u_{r-1} = u_r - \Delta x \left.\frac{du}{dx}\right|_{r\,\Delta x} + \frac{(\Delta x)^2}{2!}\left.\frac{d^2u}{dx^2}\right|_{r\,\Delta x} - \frac{(\Delta x)^3}{3!}\left.\frac{d^3u}{dx^3}\right|_{r\,\Delta x} + \cdots. \tag{8.1.1.3b}$$

The first derivative approximations are obtained by simple rearrangement of (8.1.1.3):

$$\left.\frac{du}{dx}\right|_{r\,\Delta x} = \frac{u_{r+1} - u_r}{\Delta x} - \frac{\Delta x}{2!}\left.\frac{d^2u}{dx^2}\right|_{r\,\Delta x} - \frac{(\Delta x)^2}{3!}\left.\frac{d^3u}{dx^3}\right|_{r\,\Delta x} - \cdots, \tag{8.1.1.4a}$$

$$\left.\frac{du}{dx}\right|_{r\,\Delta x} = \frac{u_r - u_{r-1}}{\Delta x} + \frac{\Delta x}{2!}\left.\frac{d^2u}{dx^2}\right|_{r\,\Delta x} - \frac{(\Delta x)^2}{3!}\left.\frac{d^3u}{dx^3}\right|_{r\,\Delta x} + \cdots. \tag{8.1.1.4b}$$

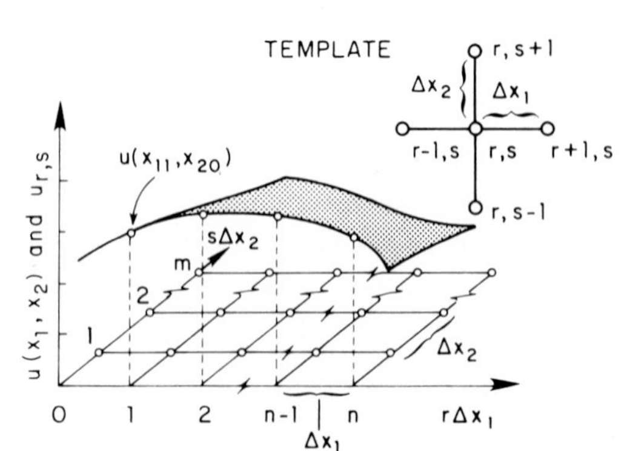

Fig. 8.2. Finite difference discretization of $u(x_1, x_2)$ by n increments of Δx_1 and m increment of Δx_2.

When these series are truncated after the first term, an error of order Δx is committed. We write this as

$$\left.\frac{du}{dx}\right|_{r\,\Delta x} = \frac{u_{r+1} - u_r}{\Delta x} + O(\Delta x), \qquad (8.1.1.5a)$$

$$\left.\frac{du}{dx}\right|_{r\,\Delta x} = \frac{u_r - u_{r-1}}{\Delta x} + O(\Delta x), \qquad (8.1.1.5b)$$

where the $O(\Delta x)$ error is in absolute value smaller than $C\,\Delta x$ (where C is an arbitrary constant) for sufficiently small Δx.

The question now arises as to how one might devise a scheme with a smaller truncation error. This is readily achieved by adding Eqs. (8.1.1.4) and dividing the result by two. There results

$$\left.\frac{du}{dx}\right|_{r\,\Delta x} = \frac{u_{r+1} - u_{r-1}}{2\,\Delta x} - \left.\frac{(\Delta x)^2}{6}\frac{d^3u}{dx^3}\right|_{r\,\Delta x} - \cdots, \qquad (8.1.1.6a)$$

or

$$\left.\frac{du}{dx}\right|_{r\,\Delta x} = \frac{u_{r+1} - u_{r-1}}{2\,\Delta x} + O((\Delta x)^2). \qquad (8.1.1.6b)$$

For sufficiently small Δx, (8.1.1.6) is considered more accurate than (8.1.1.5).

Higher-order derivative approximations may be derived in an analogous fashion. A second-order-accurate approximation to $d^2u/dx^2\,|_{r\,\Delta x}$ can be obtained by the subtraction of Eqs. (8.1.1.4) and rearrangement of the result, that is,

$$\left.\frac{d^2u}{dx^2}\right|_{r\,\Delta x} = \frac{u_{r+1} - 2u_r + u_{r-1}}{(\Delta x)^2} - \left.\frac{(\Delta x)^2}{12}\frac{d^4u}{dx^4}\right|_{r\,\Delta x} - \cdots, \qquad (8.1.1.7a)$$

or

$$\left.\frac{d^2u}{dx^2}\right|_{r\,\Delta x} = \frac{u_{r+1} - 2u_r + u_{r-1}}{(\Delta x)^2} + O((\Delta x)^2). \qquad (8.1.1.7b)$$

When a mesh-centered grid with variable spacing is encountered, such as illustrated in Fig. 8.3a, the coefficients are normally computed using mean values. Consider, for example, a second-order term containing a spatially variable coefficient:

$$d/[a(x)\,du/dx]dx$$

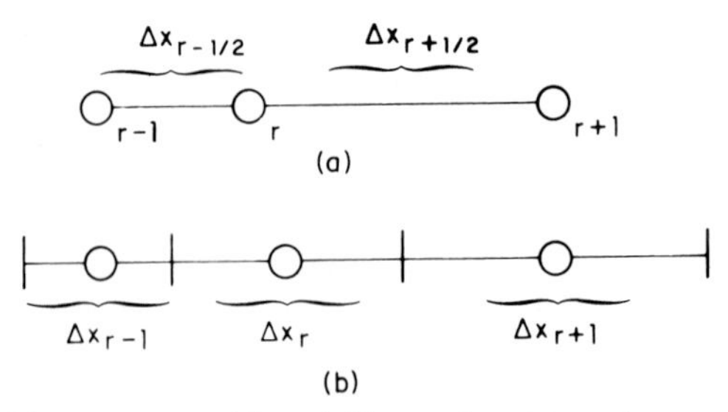

Fig. 8.3. Finite difference nets with variable spacing: (a) mesh-centered grid; (b) block-centered grid.

to be approximated at x_r. A logical difference formula would be

$$\frac{d}{dx}\left[a(x)\frac{du}{dx}\right]\bigg|_{x_r} = \frac{1}{\Delta x_r}\left[a(x_{r+1/2})\left(\frac{u_{r+1}-u_r}{\Delta x_{r+1/2}}\right) - a(x_{r-1/2})\left(\frac{u_r-u_{r-1}}{\Delta x_{r-1/2}}\right)\right], \tag{8.1.1.8}$$

with

$$\Delta x_r = \tfrac{1}{2}(\Delta x_{r+1/2} + \Delta x_{r-1/2})$$

and

$$a(x_{r\pm 1/2}) = \tfrac{1}{2}(a(x_r) + a(x_{r\pm 1})).$$

When the nodal value is assumed to be located at the center of the space increment such as shown in Fig. 8.3b (block-centered grid) one may use (8.1.1.8) with the following definitions:

$$\begin{aligned}\frac{a(x_{r\pm 1/2})}{\Delta x_{r\pm 1/2}} &= \text{harmonic mean of}\left(\frac{a(x_{r\pm 1})}{\Delta x_{r\pm 1}}, \frac{a(x_r)}{\Delta x_r}\right)\\ &= \left[\frac{1}{2}\left(\frac{\Delta x_{r\pm 1}}{a(x_{r\pm 1})} + \frac{\Delta x_r}{a(x_r)}\right)\right]^{-1}\\ &= \frac{2a(x_r)a(x_{r\pm 1})}{a(x_r)\Delta x_{r\pm 1} + a(x_{r\pm 1})\,\Delta x_r}.\end{aligned}$$

Notice that using the block-centered formulation, the harmonic mean representing the coefficient $a(x_{r\pm 1/2})/\Delta x_{r\pm 1/2}$ vanishes when either $a(x_{r\pm 1})$ or $a(x_r)$ is zero. This feature is useful in efficient code development because it automatically accommodates zero-flux boundary conditions.

8.1.2 Operator Notation

It is convenient to introduce a shorthand notation to describe the discrete formulas introduced previously and others that can be developed by extending these concepts. We present this notation in Table 8.1. Consider as an example of operator representation the central difference approximation of the first-order derivative as given by (8.1.1.6b), that is,

$$\left.\frac{du}{dx}\right|_{r\,\Delta x} \simeq \frac{u_{r+1} - u_{r-1}}{2\,\Delta x} = \frac{(u_{r+1} - u_r) + (u_r - u_{r-1})}{2\,\Delta x}$$

$$= \frac{\delta u_{r+1/2} + \delta u_{r-1/2}}{2\,\Delta x} = \frac{\mu\delta u_r}{\Delta x}. \tag{8.1.2.1}$$

Similarly, we can write the central difference approximation for the second-order derivative given by (8.1.1.7a) as

$$\left.\frac{d^2u}{dx^2}\right|_{r\,\Delta x} \simeq \frac{u_{r+1} - 2u_r + u_{r-1}}{(\Delta x^2)} = \frac{(u_{r+1} - u_r) - (u_r - u_{r-1})}{(\Delta x)^2}$$

$$= \frac{\delta u_{r+1/2} - \delta u_{r-1/2}}{(\Delta x)^2} = \frac{\delta(u_{r+1/2} - u_{r-1/2})}{(\Delta x)^2}$$

$$= \frac{\delta(\delta u_r)}{(\Delta x)^2} = \frac{\delta^2 u_r}{(\Delta x)^2}. \tag{8.1.2.2}$$

Thus, we see that the difference operators are similar to standard differential operators in their appearance and application. We shall find this shorthand very helpful in our later discussions of complicated porous media phenomena.

8.1.3 Finite Difference Formulas in Two Independent Variables

The representation of partial derivatives is generally a straightforward extension of the approximation of ordinary derivatives. Keeping in mind the discretization and accompanying template of Fig. 8.2, we write directly,

Table 8.1
Finite Difference Operators

Operator	*Symbol*	*Difference representation*
Forward difference	Δ	$\Delta u_r = u_{r+1} - u_r$
Backward difference	∇	$\nabla u_r = u_r - u_{r-1}$
Central difference	δ	$\delta u_r = u_{r+1/2} - u_{r-1/2}$
Average	μ	$\mu u_r = \dfrac{u_{r+1/2} + u_{r-1/2}}{2}$

as examples,

$$\left.\frac{\partial}{\partial x_1} u(x_1, x_2)\right|_{r\,\Delta x_1, s\Delta x_2} = \frac{u_{r+1,s} - u_{r-1,s}}{2\,\Delta x_1} + O((\Delta x_1)^2) \quad (8.1.3.1a)$$

and

$$\left.\frac{\partial^2}{\partial x_2^2} u(x_1, x_2)\right|_{r\,\Delta x_1, s\Delta x_2} = \frac{u_{r,s+1} - 2u_{r,s} + u_{r,s-1}}{(\Delta x_2)^2} + O((\Delta x_2)^2). \quad (8.1.3.1b)$$

Although the two approximations of (8.1.3.1a) and (8.1.3.1b) are analogous to their one-dimensional counterparts, the cross derivative $(\partial^2/\partial x_1\,\partial x_2)$ $u(x_1, x_2)$ has not been encountered previously. It can be shown that a second-order correct approximations is given by

$$\left.\frac{\partial^2}{\partial x_1\,\partial x_2} u(x_1, x_2)\right|_{r\Delta x_1, s\Delta x_2} = \frac{1}{2\,\Delta x_1\, 2\,\Delta x_2}[u_{r+1,s+1} - u_{r-1,s+1} - u_{r+1,s-1} + u_{r-1,s-1}] + O((\Delta x_1)^2) + O((\Delta x_2)^2).$$

The nodes used in this approximation appear as the shaded points in Fig. 8.4. Notice that these are not the same as those employed in the derivative approximations given in Eqs. (8.1.3.1a) and (8.1.3.1b).

The approximation of boundary conditions in two and three space dimensions depends upon the choice of discretization. One can employ either a block-centered or mesh-centered grid. These two mesh types are illustrated in Fig. 8.5a and 8.5b. In the case of Dirichlet or first-type boundary conditions one simply replaces the function value at the node with a specified value. This procedure is fine if the location of the simulated physical boundary corresponds to the nodal location, which is the case when a mesh-centered grid is used. However, if a block-centered grid is used and the simulated boundary corresponds to the edge of a block, it is more accurate to apply the Dirichlet condition by introducing a fictitious node located a half interval outside the solution region.

For illustrative purposes, suppose that the simulated boundary is located at $x_1 = r\,\Delta x_1$ (Figure 8.5a). The Dirichlet condition $u = \tilde{u}$ at $(r\,\Delta x_1,$

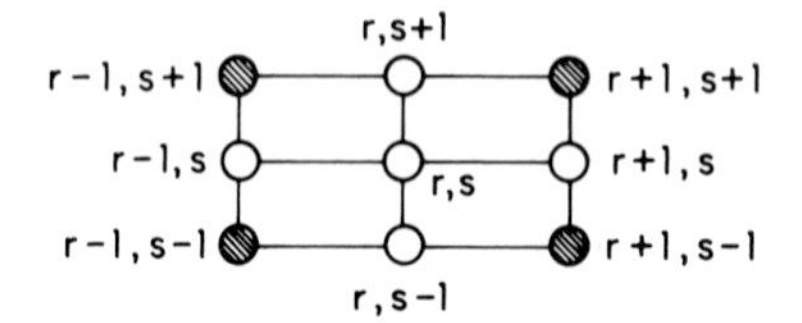

Fig. 8.4. Finite difference discretization for approximation of the cross derivative. Only the shaded points are used.

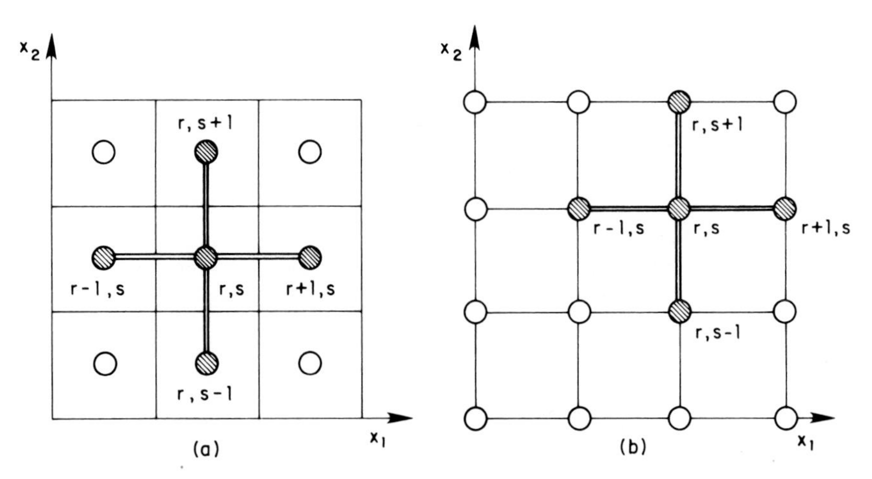

Fig. 8.5. Block-centered (a) and mesh-centered (b) finite difference approximations. Shaded nodes form a template or computational molecule.

$s\Delta x_2$) is first approximated by

$$\tfrac{1}{2}(u_{r+1,s} + u_{r,s}) = \tilde{u}, \tag{8.1.3.2}$$

where $\tilde{u}$ is the specified value of u and $u_{r+1,s}$ is the value of u at the fictitious node $(r + 1, s)$.

Although Eq. (8.1.3.2) contains the fictitious unknown $u_{r+1,s}$, this unknown can be eliminated by combining (8.1.3.2) with the finite difference representation of the governing partial differential equation for node (r, s).

In addition to the Dirichlet boundary conditions, we also deal with the von Neumann or second-type and the mixed or third-type boundary conditions. Consider the mixed condition (a von Neumann boundary is obtained by setting a and c to zero)

$$\partial u / \partial x_1 + au = c, \qquad a \text{ and } c \text{ constant.} \tag{8.1.3.3a}$$

If the simulated physical boundary is located at $r\,\Delta x_1$, the appropriate approximation of (8.1.3.2) using the block-centered grid would be

$$\frac{u_{r+1,s} - u_{r,s}}{\Delta x_1} + a\left(\frac{u_{r+1,s} + u_{r,s}}{2}\right) = c, \tag{8.1.3.3b}$$

which is, once again, a straightforward algebraic calculation utilizing the fictitious nodal point $(r + 1, s)$, located a half interval outside the solution region.

For the mesh-centered grid in Fig. 8.5b, one would naturally employ the approximation

$$\frac{u_{r+1,s} - u_{r-1,s}}{2\,\Delta x_1} + a u_{r,s} = c, \tag{8.1.3.3c}$$

which utilizes the fictitious nodal point $(r+1,\ s)$ located one interval outside the solution region.

In practice, the block-centered scheme is generally preferred because it can be programmed in a slightly more efficient manner, and it is intuitively attractive to think of porous media parameters and fluxes as identified with a grid element surrounding a node.

8.1.4 Finite Difference Representation of Partial Differential Equations

Let us now examine the use of the aforementioned methods in the approximation of partial differential equations. We shall consider one equation from each of the three types: elliptic, parabolic, and hyperbolic. The heat flow equation is a well-known equation of parabolic type. It is written for a function $u(x,\ t)$ as

$$\partial u/\partial t = \partial^2 u/\partial x^2, \tag{8.1.4.1}$$

where t is time. Employing a second-order-accurate central difference approximation in space and a first-order-accurate backward difference approximation in time, we write the difference formula for (8.1.4.1) at the point $(r\,\Delta x,\ s\,\Delta t)$ as

$$\frac{u_{r,s+1} - u_{r,s}}{\Delta t} = \frac{u_{r+1,s+1} - 2u_{r,s+1} + u_{r-1,s+1}}{(\Delta x)^2} + O(\Delta t, (\Delta x)^2). \tag{8.1.4.2a}$$

This can be written more compactly as

$$\nabla_t u_{r,s+1} = \rho \delta_x^2 u_{r,s+1} + O(\Delta t, (\Delta x)^2)\,\Delta t, \tag{8.1.4.2b}$$

where $\rho = \Delta t/(\Delta x)^2$. This approximation is stable for any value of ρ; that is, it is unconditionally stable. This is not true of all difference equations, as we shall see in Section 8.2.

Laplace's equation is a well-known equation of elliptic type. For the function $u(x_1,\ x_2)$, it is written

$$(\partial^2 u/\partial x_1^2) + (\partial^2 u/\partial x_2^2) = 0. \tag{8.1.4.3}$$

A second-order accurate central difference approximation in x_1 and x_2

yields the following discretized form of (8.1.4.3) written at the point $(r\,\Delta x_1,\ s\,\Delta x_2)$:

$$\frac{u_{r+1,s} - 2u_{r,s} + u_{r-1,s}}{(\Delta x_1)^2} + \frac{u_{r,s+1} - 2u_{r,s} + u_{r,s-1}}{(\Delta x_2)^2}$$

$$+ O((\Delta x_1)^2 + (\Delta x_2)^2) = 0, \qquad (8.1.4.4a)$$

or, using operator notation,

$$\frac{1}{(\Delta x_1)^2}\delta^2_{x_1}u_{r,s} + \frac{1}{(\Delta x_2)^2}\delta^2_{x_2}u_{r,s} + O((\Delta x_1)^2 + (\Delta x_2)^2) = 0. \qquad (8.1.4.4b)$$

A well-known equation of hyperbolic type is the wave equation. It is written for the function $u(x, t)$ as

$$\partial^2 u/\partial t^2 = \partial^2 u/\partial x^2. \qquad (8.1.4.5)$$

An obvious central difference approximation to this equation is

$$\frac{u_{r,s+1} - 2u_{r,s} + u_{r,s-1}}{(\Delta t)^2} = \frac{u_{r+1,s} - 2u_{r,s} + u_{r-1,s}}{(\Delta x)^2}, \qquad (8.1.4.6a)$$

or, employing operator notation,

$$\delta^2_t u_{r,s} = \bar{\rho}\delta^2_x u_{r,s} + O((\Delta x)^2 + (\Delta t)^2)(\Delta t)^2 = 0, \qquad (8.1.4.6b)$$

where $\bar{\rho} \equiv (\Delta t)^2/(\Delta x)^2$. Unfortunately, this equation is conditionally stable; more specifically, it is stable only when $\bar{\rho} \leqslant 1$.

The preceding development generates a set of algebraic equations, one for each node of a discretized region. These equations must be augmented with appropriate initial and boundary conditions to close the system. This information may influence the accuracy of simulation if the approximation of these auxiliary conditions is of lower order than the partial differential equation approximation.

8.2 Stability of Finite Difference Approximations

Stability refers to the behavior of numerical errors that inadvertently enter into a simulation during the computational procedure. When such errors are damped as we proceed stepwise through time, the scheme is denoted as stable; when the errors grow unboundedly we call it unstable. There are at least three ways to establish stability. One can conduct a numerical experiment (heuristic stability), examine the eigenvalues of the approximating algebraic equations [matrix stability; see Smith (1978)], or employ a Fourier expansion (von Neumann stability). We consider here only the von Neumann approach.

8.2.1 von Neumann Stability Analysis

The von Neumann analysis provides a necessary bound on the values of ρ and $\bar{\rho}$ that will ensure stability for linear equations. Variants on this basic approach can be used to extract other valuable information on numerical performance, but here we restrict our attention only to the consideration of stability.

Consider a numerical error distributed spatially at a particular time level $s\,\Delta t$. Let us represent this error by a series of the form

$$E(r\,\Delta x) = \sum_{p=0}^{n} A_p e^{\hat{\imath}\beta_p r\,\Delta x}, \qquad r = 0, 1, 2, \ldots, n, \quad \hat{\imath} \equiv \sqrt{-1}, \tag{8.2.1.1}$$

where $(n + 1)$ is the number of nodes along x, $|\beta_p|$ are the spatial frequencies of the error, and A_p are constants that can be determined from the $(n + 1)$ initial values of the error at the $(n + 1)$ nodes.

Our objective is to determine the behavior of $E(r\,\Delta x)$ as time evolves. To do this we assume an error solution to the difference equation of the form

$$E(r\,\Delta x, s\,\Delta t) = \sum_{q=0}^{m} \sum_{p=0}^{n} A_p e^{\hat{\imath}\beta_p r\,\Delta x} B_q e^{\hat{\imath}\sigma_q s\,\Delta t}, \tag{8.2.1.2}$$

which means $E(s\,\Delta t) = \sum_{q=0}^{m} B_q e^{\hat{\imath}\sigma_q s\,\Delta t}$. Because the difference equations considered here are linear, the solutions are additive. Thus, we can examine one term of the series given by (8.2.1.2), that is,

$$E^{p,q}(r\,\Delta x, s\,\Delta t) = A_p e^{\hat{\imath}\beta_p r\,\Delta x} B_q e^{\hat{\imath}\sigma_q s\,\Delta t} \tag{8.2.1.3}$$

Let $\xi \equiv B_q e^{\hat{\imath}\sigma_q \Delta t}$ such that

$$E^{p,q}(r\,\Delta x, s\,\Delta t) = A_p e^{\hat{\imath}\beta_p r\,\Delta x} \xi^s. \tag{8.2.1.4}$$

It is apparent from (8.2.1.4) that the error will remain bounded as time increases (s increases) only if

$$|\xi| \leqslant 1. \tag{8.2.1.5}$$

To demonstrate the practical application of this constraint, consider the following difference representation of the heat flow equation

$$u_{r,s+1} - u_{r,s} = \rho(u_{r+1,s} - 2u_{r,s} + u_{r-1,s}). \tag{8.2.1.6}$$

Rearrangement of (8.2.1.6) yields

$$u_{r,s+1} = \rho u_{r+1,s} + (1 - 2\rho)u_{r,s} + \rho u_{r-1,s}. \qquad (8.2.1.7)$$

Substitution of $E^{p,q}(r\,\Delta x, s\,\Delta t)$ into (8.2.1.7) yields (dropping the subscript on β)

$$\xi^{s+1}e^{\hat{\imath}\beta r\,\Delta x} = \rho\xi^s e^{\hat{\imath}\beta(r+1)\,\Delta x} + (1 - 2\rho)\xi^s e^{\hat{\imath}\beta r\,\Delta x} + \rho\xi^s e^{\hat{\imath}\beta(r-1)\,\Delta x}. \qquad (8.2.1.8)$$

Rearranging and simplifying this equation, one obtains

$$\xi = (1 - 2\rho) + \rho[e^{\hat{\imath}\beta\,\Delta x} + e^{-\hat{\imath}\beta\,\Delta x}]. \qquad (8.2.1.9)$$

We now employ the following relations to make (8.2.1.9) more amenable to further analysis:

$$e^{\hat{\imath}\beta\,\Delta x} + e^{-\hat{\imath}\beta\,\Delta x} = 2\cos\beta\,\Delta x, \qquad (8.2.1.10a)$$

$$1 - \cos\beta\,\Delta x = 2\sin^2(\beta\,\Delta x/2). \qquad (8.2.1.10b)$$

Substituting (8.2.1.10) into (8.2.1.9), we obtain

$$\xi = 1 - 2\rho(1 - \cos\beta\,\Delta x) = 1 - 4\rho\sin^2(\beta\,\Delta x/2). \qquad (8.2.1.11)$$

From (8.2.1.5) we can write directly for a stable algorithm

$$|\xi| = |1 - 4\rho\sin^2(\beta\,\Delta x/2)| \leqslant 1$$

or

$$-1 \leqslant 1 - 4\rho\sin^2(\beta\,\Delta x/2) \leqslant 1.$$

The only relevant inequality not automatically satisfied is

$$-1 \leqslant 1 - 4\rho\sin^2(\beta\,\Delta x/2),$$

or, upon rearranging,

$$\rho \leqslant \frac{1}{2\sin^2(\beta\,\Delta x/2)}.$$

We must assume that all possible values of β may be encountered during the calculations. Consequently, we must consider the worst instance. This occurs when

$$\sin^2(\beta\,\Delta x/2) = 1,$$

which leads directly to the constraint

$$\rho \leqslant \tfrac{1}{2}. \qquad (8.2.1.12)$$

The scheme presented as (8.2.1.6) is the classic explicit approximation and (8.2.1.12) is the well-known conditional stability bound.

The determination of the functional form for ξ is not always as algebraically simple as we described for the classic explicit approximation. Often it is necessary to solve rather complicated expressions for ξ. It is interesting, however, that multistep approximations, which might intuitively seem rather difficult to analyze, are quite straightforward. The overall amplification factor is simply the product of those factors obtained for each step.

In the remainder of this book, we shall use only unconditionally stable finite difference approximations unless specified otherwise.

8.3 Consistency and Convergence of Finite Difference Approximations

Consistency concerns the ability of a numerical approximation to represent the partial differential equation desired, rather than some other equation, as the discretization is refined. In other words, a numerical approximation is consistent if the truncation error vanishes as Δx_i, $\Delta t \rightarrow 0$. Convergence, on the other hand, considers the behavior of the deviation between the numerical and analytical solutions as the mesh is refined. This can be written more explicitly. A finite difference approximation is said to be convergent if

$$\|\bar{u}(r\,\Delta x,\, s\,\Delta t) - u_{r,s}\| \rightarrow 0$$

as Δx, $\Delta t \rightarrow 0$, where $\|\ \|$ is a suitable norm and $\bar{u}(r\,\Delta x,\, s\,\Delta t)$ is the exact solution at the point $(r\,\Delta x,\, s\,\Delta t)$.

Convergence is, in general, much more difficult to establish than consistency and stability. Thus, it is indeed fortunate that these latter two properties can be used to establish convergence via Lax's equivalence theorem, which reads, "Given a properly posed initial-value problem and a finite-difference approximation to it that satisfies the consistency condition, stability is the necessary and sufficient condition for convergence" (Richtmyer and Morton, 1967, p. 45).

Whereas most finite difference approximations found in the literature are consistent, not all are unconditionally consistent. A famous example of a conditionally consistent approximation is that attributed to DuFort and Frankel (1953) for the numerical solution of the heat flow equation

$$\partial u/\partial t = \partial^2 u/\partial x^2. \tag{8.3.1.1}$$

The finite difference approximation is

$$\frac{u_{r,s+1} - u_{r,s-1}}{\Delta t} = \frac{u_{r+1,s} - u_{r,s+1} - u_{r,s-1} + u_{r-1,s}}{(\Delta x)^2} \tag{8.3.1.2}$$

with a truncation error of

$$\left[\frac{(\Delta t)^2}{6}\frac{\partial^3 u}{\partial t^3} - \frac{(\Delta x)^2}{12}\frac{\partial^4 u}{\partial x^4} + \frac{(\Delta t)^2}{(\Delta x)^2}\frac{\partial^2 u}{\partial t^2}\right]_{r\,\Delta x, s\,\Delta t} + \cdots .$$

When Δt and Δx tend to zero such that $\Delta t/\Delta x$ also tends to zero, this equation is consistent with (8.3.1.1). If, however, Δt and Δx tend to zero such that $\Delta t/\Delta x$ approaches a constant c, then the second-order time derivative does not vanish and the numerical approximation is consistent with the hyperbolic equation

$$\frac{\partial u}{\partial t} + c^2\frac{\partial^2 u}{\partial t^2} = \frac{\partial^2 u}{\partial x^2}. \tag{8.3.1.3}$$

Thus, (8.3.1.2) is conditionally consistent with (8.3.1.1).

In this brief introduction to finite difference theory we have only touched the surface of a vast literature on the subject. We have not considered nonlinear approximations or the myriad of clever algorithms that have been developed to solve efficiently particular problem types. These matters will be dealt with as the need arises in the forthcoming discussions of particular applications.

References

Ames, W. F. (1977). "Numerical Methods for Partial Differential Equations," 2nd ed. Academic Press, New York.

DuFort, E. C., and Frankel, S. P. (1953). Stability Conditions in the Numerical Treatment of Parabolic Differential Equations. *Math. Tables and Other Aids to Computation,* **7**, 135–152.

Lapidus, L. and Pinder, G. F. (1982). "Numerical Solution of Partial Differential Equations in Science and Engineering." Wiley, New York.

Mitchell, A. R. and Griffiths, D. F. (1980). "The Finite Difference Method in Partial Differential Equations." Wiley, New York.

Richtmyer, R. D. and Morton, K. W. (1967). "Difference Methods for Initial-Value Problems." Interscience, New York.

Smith, G. D. (1978). "Numerical Solution of Partial Differential Equations: Finite Difference Methods," 2nd ed. Clarendon Press, Oxford.

von Rosenberg, D. U. (1969). "Methods for the Numerical Solution of Partial Differential Equations." Elsevier, New York.

9

Finite Difference Simulation of Single and Multiphase Isothermal Fluid Flow and Solute Transport

9.0 Introduction

In this chapter we investigate the application of finite difference methods to problems involving single- and multiphase flow and solute transport. Finite difference methods were the first numerical procedures used to attack these problems and remain the most popular today. They are often combined with the "method of characteristics" to solve solute transport problems. We first investigate the single-phase flow problem and subsequently focus on solute transport phenomena.

9.1 Simulation of Single-Phase Flow

9.1.1 Pressure Formulation

The three-dimensional flow of groundwater or "liquid" oil is described by Eqs. (4.2.1.1) and (4.2.1.2). Combination of these equations yields

$$\frac{\partial}{\partial x_i}\left[\frac{k_{ij}}{\mu}\left(\frac{\partial p}{\partial x_j} + \rho g \frac{\partial z}{\partial x_j}\right)\right] = (\alpha + \phi\beta)\frac{\partial p}{\partial t}, \qquad i = 1, 2, 3, \qquad (9.1.1.1a)$$

where k_{ij} is the permeability, μ the fluid viscosity, ρ the fluid density, α the formation compressibility, ϕ the porosity, β the fluid compressibility, z the elevation, and p the fluid pressure. This form of the equation contains fluid pressure as the primitive variable. Therefore, it can readily accommodate the case of a fluid whose density depends upon temperature and solute concentration as well as fluid pressure.

The finite difference representation of (9.1.1.1a) is obtained by applying the approximation theory of the preceding chapter. Using the operator notation introduced in Table 8.1 we can write directly

$$\frac{1}{(\Delta x_i)^2}\delta_{x_i}\left[\frac{k_{ij}}{\mu}\left(\delta_{x_j}p + \rho g \delta_{x_j} z\right)\right]_{s+1} = \left(\frac{\alpha + \phi\beta}{\Delta t}\right)\nabla_t p_{s+1}, \qquad (9.1.1.1b)$$

where $s + 1$ denotes the current time level, and for convenience it is assumed that Δx_1, Δx_2, and Δx_3 of the chosen mesh are constant.

When a finite difference subscript is omitted, as is the case for (9.1.1.1b), the difference is taken at $(r_1\Delta x_1,\ r_2\Delta x_2,\ r_3\Delta x_3,\ s\ \Delta t)$; otherwise the subscript will be written explicitly. The subscript r_i is defined through the relation $x_{ir_i} = r_i\,\Delta\,x_i$.

The structure of the algebraic equations arising from (9.1.1.1b) lends itself to efficient equation-solving algorithms. We begin with the two-dimensional form of (9.1.1.1b) and then extend these concepts to three dimensions.

Alternating Direction Implicit (ADI) Method of Peaceman and Rachford

The alternating direction implicit (ADI) procedure is one of the earliest and most popular implicit algorithms for the solution of (9.1.1.1) in two dimensions. It was introduced by Peaceman and Rachford in their classic 1955 paper. To simplify notation we assume the spatial dependence of ρ is weak, so that (9.1.1.1b) can be written as

$$\frac{1}{(\Delta x_i)^2}\,\delta_{x_i}\left(\frac{k_{ij}}{\mu}\,\delta_{x_j}\,p\right)_{s+1} = \left(\frac{\alpha + \phi\beta}{\Delta t}\right)\nabla_t p_{s+1}. \tag{9.1.1.2}$$

If the finite difference net consists of an $N \times M$ array of nodes, then this equation would require the simultaneous solution of $N \times M$ equations to obtain the nodal values of p. Alternatively one can rewrite (9.1.1.2), for the case of $k_{12} = k_{21} = 0$ using a two-step procedure:

$$\frac{1}{(\Delta x_1)^2}\,\delta_{x_1}\left(\frac{k_{11}}{\mu}\delta_{x_1}p\right)_{s+1/2} = -\frac{1}{(\Delta x_2)^2}\,\delta_{x_2}\left(\frac{k_{22}}{\mu}\delta_{x_2}p\right)_s + \frac{\alpha + \phi\beta}{\Delta t/2}(p_{s+1/2} - p_s) \qquad [x_1 \text{ sweep}], \tag{9.1.1.3a}$$

$$\frac{1}{(\Delta x_2)^2}\,\delta_{x_2}\left(\frac{k_{22}}{\mu}\,\delta_{x_2}p\right)_{s+1} = -\frac{1}{(\Delta x_1)^2}\,\delta_{x_1}\left(\frac{k_{11}}{\mu}\,\delta_{x_1}p\right)_{s+1/2} + \frac{\alpha + \phi\beta}{\Delta t/2}(p_{s+1} - p_{s+1/2}) \qquad [x_2 \text{ sweep}]. \tag{9.1.1.3b}$$

By splitting the operator in this manner, one generates two equations, each of which is computationally one dimensional (see Fig. 9.1). During the x_1 sweep all unknowns at the $s + 1/2$ time level are aligned along the x_1 coordinate, the information pertaining to the x_2 segment of the operator is lagged one-half time step. This generates M smaller matrix

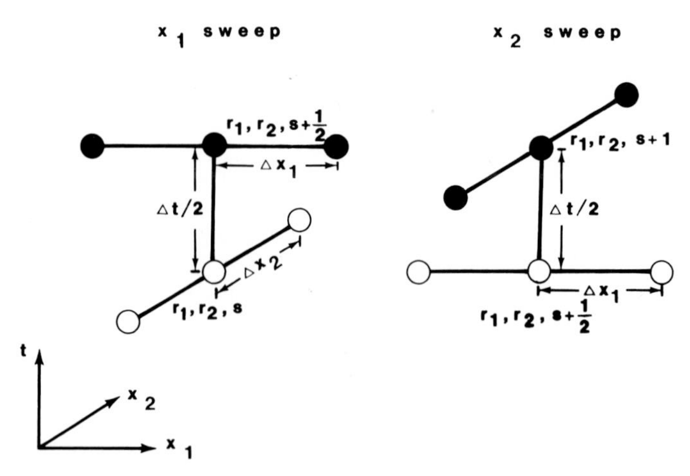

Fig. 9.1. Computational molecule for alternating direction implicit procedure. Shaded nodes are unknown and open nodes are known during calculation. For iterative alternating direction implicit procedure the time stepping is replaced with iteration increments.

problems, one for each x_1 line of nodes. Each such matrix problem requires the solution of N equations. The same procedure is used during the x_2 sweep in generating N matrix equations, each with M unknowns. Note that the intermediate values, i.e., those obtained during the x_1 sweep, are not to be considered a solution; they are quite often inaccurate. One is also constrained to use the same Δt on each two-sweep step.

Because during each sweep one solves for unknown nodal values along either one row or one column of the finite difference net, the resulting algorithm is remarkably efficient. During each row or column calculation a tridiagonal matrix is generated. This is evident from (9.1.1.3), wherein only three unknown nodal values are involved in generating each equation. A tridiagonal matrix equation of this kind is readily solved using the highly efficient Thomas algorithm. A detailed description of this algorithm can be found in a number of books on matrix algebra and numerical analysis; we outline it briefly below.

Examination of the amplification factors for (9.1.1.3a) and (9.1.1.3b) reveals that each step is conditionally stable. However, the amplification factor for the combined procedure shows that the overall process is unconditionally stable. One can show [see Peaceman (1977), p. 59] that this two-step procedure is equivalent to a Crank–Nicolson centered in time approximation with a perturbation term. The resulting approximation is second-order accurate in space and time; that is, the error is $O((\Delta x_1)^2 + (\Delta x_2)^2 + (\Delta t)^2)$.

Thomas Algorithm

Many efficient numerical schemes generate systems of equations with tridiagonal matrices. A typical set of equations can be written

$$[A]\{u\} = \{d\}, \tag{9.1.1.4}$$

where

$$[A] = \begin{bmatrix} b_1 & c_1 & & & \\ a_2 & b_2 & c_2 & & \\ & \ddots & \ddots & \ddots & \\ & & & a_N & b_N \end{bmatrix},$$

$$\{u\} = \begin{Bmatrix} u_1 \\ u_2 \\ \vdots \\ u_N \end{Bmatrix}, \qquad \{d\} = \begin{Bmatrix} d_1 \\ d_2 \\ \vdots \\ d_N \end{Bmatrix}.$$

The coefficient matrix $[A]$ contains information on the discretization strategy and parametric values. The vector $\{d\}$ is made up of known information and the vector $\{u\}$ consists of the sought-after values. The Thomas algorithm is derived by decomposing $[A]$ into a product of lower and upper triangular matrices:

$$[A] = [L][U],$$

where $[L]$ and $[U]$ are defined as

$$[L] = \begin{bmatrix} \beta_1 & & & \\ a_2 & \beta_2 & & \\ & \ddots & \ddots & \\ & & a_N & \beta_N \end{bmatrix},$$

$$[U] = \begin{bmatrix} 1 & \gamma_1 & & & \\ & 1 & \gamma_2 & & \\ & & \ddots & \ddots & \\ & & & 1 & \gamma_{N-1} \\ & & & & 1 \end{bmatrix}.$$

Multiplying $[L]$ by $[U]$ and equating the elements of $[L][U]$ to the elements of $[A]$, term by term, we obtain the following relations:

$$\beta_1 = b_1,$$

$$\gamma_i = c_i/\beta_i, \qquad i = 1, 2, \ldots, N-1,$$

$$\beta_i = b_i - a_i\gamma_{i-1}, \qquad i = 2, 3, \ldots, N.$$

Equation (9.1.1.4) is written as

$$[L][U]\{u\} = \{d\},$$

or

$$[L]\{s\} = \{d\},$$

where $\{s\}$ is an intermediate unknown vector defined as

$$\{s\} = [U]\{u\}.$$

The elements of $\{s\}$ can be easily determined by performing a forward solution of $[L]\{s\} = \{d\}$. This leads to

$$s_1 = d_1/\beta_1,$$

$$s_i = (d_i - a_i s_{i-1})/\beta_i, \qquad i = 2, 3, \ldots, N.$$

Knowing $\{s\}$, one can determine $\{u\}$ by performing a backward solution of $[U]\{u\} = \{s\}$. This yields

$$u_N = s_N,$$

$$u_i = s_i - \gamma_i u_{i+1}, \qquad i = N-1, N-2, \ldots, 1.$$

In summary, the computational procedure consists of two steps. The first step is to create arrays for β_i and s_i using the forward recursion relations, and the second step is to determine the desired unknowns u_i using the backward recursion relations.

The Thomas algorithm presented above is highly efficient. It requires only five multiplications (or divisions) and three subtractions per grid point. The computational labor is directly proportional to the number of equations N, compared with $N^3/2$, which is the labor required to solve a full system of N simultaneous equations. Finally, the simple Thomas algorithm can be directly extended to deal with a system of block tridiagonal equations. A derivation of this block Thomas algorithm can be found in Remson *et al.* (1971).

Alternating Direction Implicit Method of Douglas and Rachford

It would seem a straightforward matter to extend the Peaceman–Rachford ADI algorithm to three dimensions. Unfortunately, the obvious formulation does not lead to an unconditionally stable approximation. Douglas and Rachford (1956) generated the first unconditionally stable ADI algorithm for three space dimensions. Assuming that the off-diagonal components of the permeability tensor are zero, their algorithm reads

$$\frac{1}{(\Delta x_1)^2}\delta_{x_1}\left(\frac{k_{11}}{\mu}\delta_{x_1}p\right)_{s+1/3} + \frac{1}{(\Delta x_2)^2}\delta_{x_2}\left(\frac{k_{22}}{\mu}\delta_{x_2}p\right)_s + \frac{1}{(\Delta x_3)^2}\delta_{x_3}\left(\frac{k_{33}}{\mu}\delta_{x_3}p\right)_s$$
$$= \frac{\alpha + \phi\beta}{\Delta t}(p_{s+1/3} - p_s) \qquad [x_1 \text{ sweep}], \tag{9.1.1.5a}$$

$$\frac{1}{(\Delta x_1)^2}\delta_{x_1}\left(\frac{k_{11}}{\mu}\delta_{x_1}p\right)_{s+1/3} + \frac{1}{(\Delta x_2)^2}\delta_{x_2}\left(\frac{k_{22}}{\mu}\delta_{x_2}p\right)_{s+2/3} + \frac{1}{(\Delta x_3)^2}\delta_{x_3}\left(\frac{k_{33}}{\mu}\delta_{x_3}p\right)_s$$
$$= \frac{\alpha + \phi\beta}{\Delta t}(p_{s+2/3} - p_s) \qquad [x_2 \text{ sweep}], \tag{9.1.1.5b}$$

$$\frac{1}{(\Delta x_1)^2}\delta_{x_1}\left(\frac{k_{11}}{\mu}\delta_{x_1}p\right)_{s+1/3} + \frac{1}{(\Delta x_2)^2}\delta_{x_2}\left(\frac{k_{22}}{\mu}\delta_{x_2}p\right)_{s+2/3} + \frac{1}{(\Delta x_3)^2}\delta_{x_3}\left(\frac{k_{33}}{\mu}\delta_{x_3}p\right)_{s+1}$$
$$= \frac{\alpha + \phi\beta}{\Delta t}(p_{s+1} - p_s) \qquad [x_3 \text{ sweep}]. \tag{9.1.1.5c}$$

Note that the equations corresponding to the x_2 sweep and x_3 sweep can also be written in the form

$$\frac{1}{(\Delta x_2)^2}\delta_{x_2}\left(\frac{k_{22}}{\mu}\delta_{x_2}p\right)_{s+2/3}$$
$$= \frac{1}{(\Delta x_2)^2}\delta_{x_2}\left(\frac{k_{22}}{\mu}\delta_{x_2}p\right)_s + \frac{\alpha + \phi\beta}{\Delta t}(p_{s+2/3} - p_{s+1/3}) \; [x_2 \text{ sweep}], \tag{9.1.1.5d}$$

$$\frac{1}{(\Delta x_3)^2}\delta_{x_3}\left(\frac{k_{33}}{\mu}\delta_{x_3}p\right)_{s+1}$$
$$= \frac{1}{(\Delta x_3)^2}\delta_{x_3}\left(\frac{k_{33}}{\mu}\delta_{x_3}p\right)_s + \frac{\alpha + \phi\beta}{\Delta t}(p_{s+1} - p_{s+2/3}) \; [x_3 \text{ sweep}]. \tag{9.1.1.5e}$$

Examination of (9.1.1.5) reveals that on each sweep one row or column of nodes is being effectively considered. This is the same situation that existed in the two-dimensional case; only the structure of the known vector of information changes because now derivative approximations in two directions are being lagged. Once again a tridiagonal coefficient matrix is generated and the highly efficient Thomas algorithm may be invoked to solve these equations.

A stability analysis of (9.1.1.5) reveals that the complete algorithm is unconditionally stable. In contrast to the Peaceman–Rachford algorithm presented earlier, which can be shown to be a perturbation of the Crank–Nicolson approximation, the Douglas–Rachford algorithm is a perturbation of a backward in time difference equation. As such it is second-order accurate in space and first-order accurate in time (Peaceman, 1977, p. 61). This algorithm is also applicable to two space dimensions. Although additional ADI algorithms [such as Douglas (1962)] can be found in the literature, the algorithms given are the most often encountered. We now turn our attention to a variant on the ADI approach, known as the iterative alternating direction implicit method.

Iterative Alternating Direction Implicit (IADI) Method

The iterative alternating direction implicit method (IADI) is useful when one encounters problems wherein the ADI procedure fails to converge or when a steady-flow solution is desired. It can also be used effectively in the event weakly nonlinear coefficients are encountered. The basic idea is to proceed from one time level to another by a series of discrete steps. Each step is called an iteration and is denoted by the superscript k. A commonly encountered IADI form of (9.1.1.2) is

$$\frac{1}{(\Delta x_1)^2}\delta_{x_1}\left(\frac{k_{11}}{\mu}\delta_{x_1}p\right)_{s+1}^{k+1/2} + \frac{1}{(\Delta x_2)^2}\delta_{x_2}\left(\frac{k_{22}}{\mu}\delta_{x_2}p\right)_{s+1}^{k}$$

$$= \frac{\alpha + \phi\beta}{\Delta t}(p_{s+1}^{k+1/2} - p_s) + H_l(p_{s+1}^{k+1/2} - p_{s+1}^{k}) \quad [x_1 \text{ sweep}], \quad (9.1.1.6a)$$

$$\frac{1}{(\Delta x_1)^2}\delta_{x_1}\left(\frac{k_{11}}{\mu}\delta_{x_1}p\right)_{s+1}^{k+1/2} + \frac{1}{(\Delta x_2)^2}\delta_{x_2}\left(\frac{k_{22}}{\mu}\delta_{x_2}p\right)_{s+1}^{k+1}$$

$$= \frac{\alpha + \phi\beta}{\Delta t}(p_{s+1}^{k+1} - p_s) + H_l(p_{s+1}^{k+1} - p_{s+1}^{k+1/2}) \quad [x_2 \text{ sweep}], \quad (9.1.1.6b)$$

where H_l is an iteration parameter and l the iteration parameter index. Were all of the terms involving the unknown iteration level in each step collected on the left-hand side of (9.1.1.6), it would be evident that a tridiagonal system of equations has, once again, been generated. Thus the Thomas algorithm is applicable.

The iteration function H_l is the product of a normalizing parameter γ_l and an iteration parameter σ_l, that is,

$$H_l = \sigma_l \gamma_l. \quad (9.1.1.7)$$

The normalizing parameter is generally written

$$
\begin{aligned}
(\Delta x_1\,\Delta x_2)\gamma_l = {} & \frac{k_{11}}{\mu}\bigg|_{r_1+1/2,r_2} \frac{\Delta x_{2r_2}}{x_{1r_1+1} - x_{1r_1}} \\
& + \frac{k_{11}}{\mu}\bigg|_{r_1-1/2,r_2} \frac{\Delta x_{2r_2}}{x_{1r_1} - x_{1r_1-1}} \\
& + \frac{k_{22}}{\mu}\bigg|_{r_1,r_2+1/2} \frac{\Delta x_{1r_1}}{x_{2r_2+1} - x_{2r_2}} \\
& + \frac{k_{22}}{\mu}\bigg|_{r_1,r_2-1/2} \frac{\Delta x_{1r_1}}{x_{2r_2} - x_{2r_2-1}}.
\end{aligned}
\tag{9.1.1.8}
$$

Although a single iteration parameter may be employed, in practice, a set of parameters are normally used, e.g. $\sigma_1, \sigma_2, \sigma_3, \ldots, \sigma_K$, where $\sigma_1 < \sigma_2 < \sigma_3 \cdots < \sigma_K$. These parameters are used sequentially from smallest to largest (or vice versa) and the sequence is repeated until convergence is achieved. The parameters are often selected so that they generate an increasing geometric series, that is, $\sigma_l/\sigma_{l-1} = \alpha$, where α is a constant determined in the following manner. If we require K parameters and let the largest and smallest be $\sigma_{\max}$ and $\sigma_{\min}$, respectively, then the multiplier α is given by (Thomas, 1982)

$$
\ln(\alpha) = \frac{\ln(\sigma_{\max}/\sigma_{\min})}{K-1}. \tag{9.1.1.9}
$$

Thomas reports that the number of parameters depends upon the range of σ_l; for a small range, 0.01–2, for example, four or five parameters are appropriate, whereas for a range of σ_l of 0.0001–2, six to eight values are usually used.

To use (9.1.1.9) we must determine $\sigma_{\max}$ and $\sigma_{\min}$. The magnitude of $\sigma_{\max}$ is not critical and a value between one and two is generally used. The use of a high value of $\sigma_{\max} \leqslant 2$ is appropriate in strongly anisotropic cases where $k_{11} >> k_{22}$ or $k_{22} >> k_{11}$ (see Trescott *et al.*, 1976). The rate of convergence of this iterative method is very sensitive to $\sigma_{\min}$. While a theoretical foundation has been laid for the calculation of $\sigma_{\min}$ for idealized problems, there is not a formula available for calculating this parameter for practical situations. The formula for the idealized case is generally used for the more complicated problem and is given by (Thomas, 1982, p. 89)

$$\sigma_{\min} = \min\left\{\frac{\pi^2}{2N^2}\left[1 + \frac{k_{22}(\Delta x_1)^2}{k_{11}(\Delta x_2)^2}\right]^{-1}, \frac{\pi^2}{2M^2}\left[1 + \frac{k_{11}(\Delta x_2)^2}{k_{22}(\Delta x_1)^2}\right]^{-1}\right\}, \quad (9.1.1.10)$$

where N and M are the number of spatial increments in the x_1 and x_2 directions, respectively. In application it may be necessary to experiment with the value of $\sigma_{\min}$ because of the sensitivity of the rate of convergence to this parameter.

Solution for the Increment

In iterative methods such as presented above, we are actually trying to solve for changes rather than the actual variable value. That is, we are seeking the change in the variable over an iteration or a time step. The equations presented above are, of course, written in terms of the variable. To make more effective use of the limited number of significant figures available in the digital computer, it is advantageous to rewrite our equations in terms of increments. Let us define the incremental values

$$P^{k+1/2} \equiv p_{s+1}^{k+1/2} - p_{s+1}^{k} \quad \text{or} \quad p_{s+1}^{k+1/2} = P^{k+1/2} + p_{s+1}^{k}, \quad (9.1.1.11a)$$

$$P^{k+1} \equiv p_{s+1}^{k+1} - p_{s+1}^{k} \quad \text{or} \quad p_{s+1}^{k+1} = P^{k+1} + p_{s+1}^{k}. \quad (9.1.1.11b)$$

Substitution into Eqs. (9.1.1.6a) and (9.1.1.6b) yields

$$\frac{1}{(\Delta x_1)^2}\delta_{x_1}\left[\frac{k_{11}}{\mu}\delta_{x_1}(P^{k+1/2} + p_{s+1}^{k})\right] + \frac{1}{(\Delta x_2)^2}\delta_{x_2}\left(\frac{k_{22}}{\mu}\delta_{x_2}p\right)_{s+1}^{i}$$

$$= \frac{\alpha + \phi\beta}{\Delta t}(P^{k+1/2} + p_{s+1}^{k} - p_s) + H_l P^{k+1/2}, \quad (9.1.1.12a)$$

$$\frac{1}{(\Delta x_1)^2}\delta_{x_1}\left(\frac{k_{11}}{\mu}\delta_{x_1}p\right)_{s+1}^{k+1/2} + \frac{1}{(\Delta x_2)^2}\delta_{x_2}\left[\frac{k_{22}}{\mu}\delta_{x_2}(P^{k+1} + p_{s+1}^{k})\right]$$

$$= \frac{\alpha + \phi\beta}{\Delta t}(P^{k+1} + p_{s+1}^{k} - p_s) + H_l(P^{k+1} - P^{k+1/2}). \quad (9.1.1.12b)$$

These equations can be rewritten in the form

$$\frac{1}{(\Delta x_1)^2}\delta_{x_1}\left(\frac{k_{11}}{\mu}\delta_{x_1}P^{k+1/2}\right) - \left(\frac{\alpha + \phi\beta}{\Delta t} + H_l\right)P^{k+1/2}$$

$$= -\left[\frac{1}{(\Delta x_1)^2}\delta_{x_1}\left(\frac{k_{11}}{\mu}\delta_{x_1}\right) + \frac{1}{(\Delta x_2)^2}\delta_{x_2}\left(\frac{k_{22}}{\mu}\delta_{x_2}\right)\right]p_{s+1}^{k}$$

$$+ \frac{\alpha + \phi\beta}{\Delta t}(p_{s+1}^{k} - p_s), \quad (9.1.1.13a)$$

$$\frac{1}{(\Delta x_2)^2}\delta_{x_2}\left(\frac{k_{22}}{\mu}\delta_{x_2}P^{k+1}\right) - \left(\frac{\alpha + \phi\beta}{\Delta t} + H_l\right)P^{k+1}$$

$$= -\frac{1}{(\Delta x_1)^2}\delta_{x_1}\left(\frac{k_{11}}{\mu}\delta_{x_1}p\right)_{s+1}^{k+1/2} - \frac{1}{(\Delta x_2)^2}\delta_{x_2}\left(\frac{k_{22}}{\mu}\delta_{x_2}p\right)_{s+1}^{k}$$

$$+ \frac{\alpha + \phi\beta}{\Delta t}(p_{s+1}^k - p_s) - H_l P^{k+1/2}. \tag{9.1.1.13b}$$

It is now evident that (9.1.1.3) forms a tridiagonal set of equations in the incremental variables $P^{k+1/2}$ and P^{k+1}. These equations can be further modified for computational convenience as illustrated in Thomas (1982, p. 90).

The termination of the iterative procedure depends upon the selection of a closure criterion. There are a number of possibilities. A common choice is the norm $\|p_{s+1}^{k+1} - p_{s+1}^k\|_\infty$ combined with its normalized variant $\|(p_{s+1}^{k+1} - p_{s+1}^k)/p_{s+1}^{k+1}\|_\infty$. Iteration is terminated when

$$\|p_{s+1}^{k+1} - p_{s+1}^k\|_\infty < \varepsilon_1$$

and

$$\left\|\frac{p_{s+1}^{k+1} - p_{s+1}^k}{p_{s+1}^{k+1}}\right\|_\infty < \varepsilon_2,$$

where ε_1 and ε_2 are usually problem dependent and determined heuristically through numerical experiment. Another commonly used criterion is the satisfaction of the mass (or energy) balance. Examination of (9.1.1.13a) reveals that, as $\lim|p_{s+1}^{k+1/2} - p_{s+1}^k| \to 0$, the left-hand side of this equation should vanish. Thus substitution of the last iterate into this expression should generate a residual that is a measure of lack of convergence. Criteria can be established using the sum over all nodes of either the absolute values of these residuals or their normalized equivalents.

Line Successive Overrelaxation

Line successive overrelaxation (LSOR) is an alternative to IADI and often is more effective in treating cross-sectional problems. For information about a comparison of LSOR and IADI, the reader is referred to Bjordamman and Coats (1969). The concept of LSOR is similar to the IADI procedure. This is made more clearly evident through an examination of Figs. 9.1 and 9.2. The LSOR equivalent of the general two-dimensional

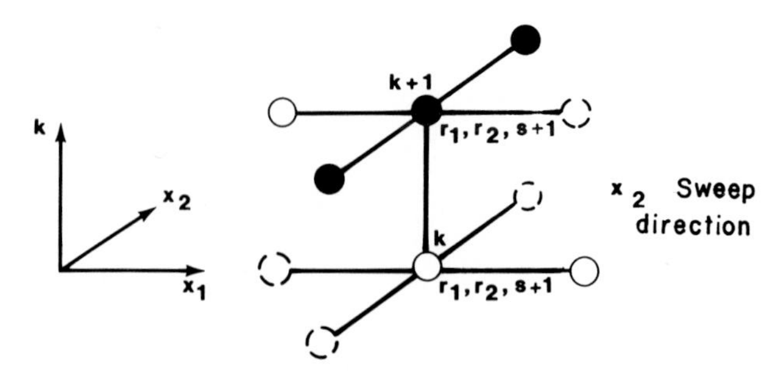

Fig. 9.2. Computational molecule for LSOR. Shaded nodes are unknown during calculation and open nodes are known. Dashed nodes are not explicitly used in equations for node $(r_1, r_2, s+1)$.

formulation (9.1.1.2) can be written for an x_2 sweep direction as

$$-\frac{1}{(\Delta x_1)^2}\frac{k_{11}}{\mu}\delta_{x_1}p^{k+1}_{r_1-1/2,r_2,s+1}+\frac{1}{(\Delta x_2)^2}\delta_{x_2}\left(\frac{k_{22}}{\mu}\delta_{x_2}p\right)^*_{r_1,r_2,s+1}$$

$$+\frac{\alpha+\phi\beta}{\Delta t}\overset{*}{p}_{r_1,r_2,s+1}=-\frac{1}{(\Delta x_1)^2}\frac{k_{11}}{\mu}\delta_{x_1}p^{k}_{r_1+1/2,r_2,s+1}$$

$$+\frac{\alpha+\phi\beta}{\Delta t}p_{r_1,r_2,s}, \qquad (9.1.1.14)$$

where the asterisk indicates an intermediate value residing at the $k+1$ level. Equation (9.1.1.14) is written and subsequently solved for all the nodes on a given r_2 line; one then repeats the same procedure for the r_2+1 line. While the left-hand side of (9.1.1.14) appears to contain four unknown values, the first term in the equation is known either from a boundary condition or from the solution just obtained during the preceding sweep of the r_1-1 line. Thus, this algorithm once again generates an efficiently solved tridiagonal system of equations. The second step in this algorithm uses the extrapolation equation

$$p^{k+1}_{r_1,r_2,s+1}=p^{k}_{r_1,r_2,s+1}+\beta(\overset{*}{p}_{r_1,r_2,s+1}-p^{k}_{r_1,r_2,s+1}), \qquad (9.1.1.15)$$

where β is an iteration parameter.

The remaining problem is the determination of the relaxation parameter β. The optimal parameter value β_{opt} is given as (Peaceman, 1977, p. 118)

$$\beta_{\text{opt}}=\frac{2}{1+[1-\mu^2(G_{LJ})]^{1/2}}, \qquad (9.1.1.16a)$$

where $\mu(G_{LJ})$ is the maximum eigenvalue of the line Jacobian iteration matrix G_{LJ} and is given by

$$\mu(G_{LJ}) = \max[(1 - 2\sin^2(\pi/2N), \{1 + (\Delta x_2/\Delta x_1)(\pi^2/2M^2)\}^{-1}], \tag{9.1.1.16b}$$

where it is assumed $\Delta x_2 \geqslant \Delta x_1$, N and $M >> 1$, with M and N being the number of Δx_1 blocks and the number of Δx_2 blocks, respectively. Various methods have been devised for updating β during the iteration process. These methods are described in Breitenbach *et al.* (1969) and Remson *et al.* (1971).

In certain problems, the rate of convergence of LSOR can be improved by applying a one- or a two-dimensional correction proposed by Watts (1971, 1973) and Settari and Aziz (1974). The modified algorithm is known as LSORC (line successive overrelaxation with additive corrections). A concise description of LSORC is presented by Peaceman (1977).

The above analysis is readily extended to the case of n lines considered simultaneously. Of course, the bandwidth of the n lines of equations will increase and the Thomas algorithm will not be applicable. The most popular forms of multiline iteration are two-line SOR and the slice successive overrelaxation (SSOR) for three-dimensional problems. There are many variants on the fundamental concepts presented above; a comprehensive yet concise discussion of these extensions can be found in Peaceman (1977) and Aziz and Settari (1979).

Strongly Implicit Procedure

The strongly implicit procedure (SIP) as introduced by Stone (1968) is one of a class of incomplete factorization methods. To understand the fundamental ideas behind this approach we first write the matrix equation arising out of a standard five-point difference approximation of a second-order equation. Consider the regular nodal arrangement of Fig. 9.3. Using the nomenclature of Peaceman (1977), we obtain the relation

$$\left[\begin{array}{ccc|ccc|ccc}
E_{11} & F_{11} & 0 & H_{11} & & & & & \\
D_{21} & E_{21} & F_{21} & 0 & H_{21} & & & & \\
0 & D_{31} & E_{31} & 0 & 0 & H_{31} & & & \\
\hline
B_{12} & 0 & 0 & E_{12} & F_{12} & 0 & H_{12} & & \\
 & B_{22} & 0 & D_{22} & E_{22} & F_{22} & 0 & H_{22} & \\
 & & B_{32} & 0 & D_{32} & E_{32} & 0 & 0 & H_{32} \\
\hline
 & & & B_{13} & 0 & 0 & E_{13} & F_{13} & 0 \\
 & & & & B_{23} & 0 & D_{23} & E_{23} & F_{23} \\
 & & & & & B_{33} & 0 & D_{33} & E_{33}
\end{array}\right] \cdot \left\{\begin{array}{c} P^*_{11} \\ P^*_{21} \\ P^*_{31} \\ \hline P^*_{12} \\ P^*_{22} \\ P^*_{32} \\ \hline P^*_{13} \\ P^*_{23} \\ P^*_{33} \end{array}\right\} = \left\{\begin{array}{c} Q_{11} \\ Q_{21} \\ Q_{31} \\ \hline Q_{12} \\ Q_{22} \\ Q_{32} \\ \hline Q_{13} \\ Q_{23} \\ Q_{33} \end{array}\right\}. \tag{9.1.1.17}$$

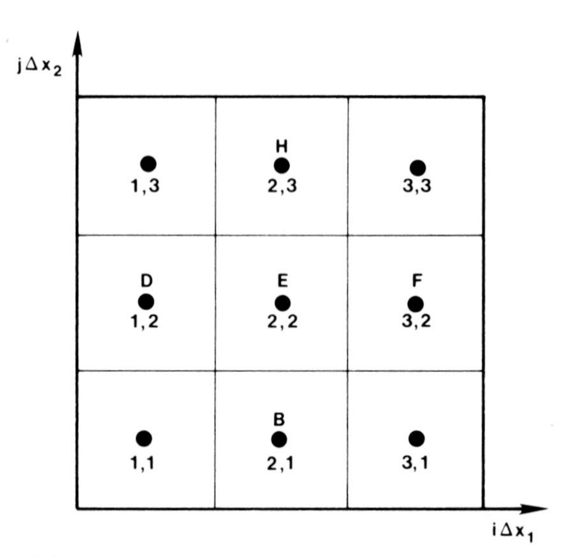

Fig. 9.3. Finite difference net used to describe the strongly implicit procedure. The symbols B, D, E, F, and H are displayed for node (2, 2).

A comparison of the expanded form of (9.1.1.2) and this matrix equation reveals that the coefficients in (9.1.1.17) for our problem are

$$B_{ij} = \left(\frac{-1}{(\Delta x_2)^2}\left\langle\frac{k_{22}}{\mu}\right\rangle\right)_{i,j-1/2}, \qquad F_{ij} = \left(\frac{-1}{(\Delta x_1)^2}\left\langle\frac{k_{11}}{\mu}\right\rangle\right)_{i+1/2,j},$$

$$D_{ij} = \left(\frac{-1}{(\Delta x_1)^2}\left\langle\frac{k_{11}}{\mu}\right\rangle\right)_{i-1/2,j}, \qquad H_{ij} = \left(\frac{-1}{(\Delta x_2)^2}\left\langle\frac{k_{22}}{\mu}\right\rangle\right)_{i,j+1/2},$$

$$E_{ij} = -(B_{ij} + D_{ij} + F_{ij} + H_{ij}) + \left(\frac{\alpha + \phi\beta}{\Delta t}\right)_{i,j},$$

$$Q_{ij} = \left(\frac{\alpha + \phi\beta}{\Delta t}\right)_{i,j}(p_{i,j,s}),$$

$$p^*_{ij} = p_{i,j,s+1},$$

where the mobility terms in angular brackets are computed by taking the harmonic mean of the mobility values of the adjacent nodes.

Let us write (9.1.1.17) as

$$[A]\{p\} = \{q\}. \tag{9.1.1.18}$$

We now assume there exists a matrix $[A']$ which is very similar to $[A]$ but is easily factored. Moreover, we shall construct $[A']$ so that the factors $[L']$ and $[U']$ are upper and lower triangular matrices containing only three elements per row. Thus $[L']$ and $[U']$ are of the following special form:

$$[L'] \equiv \begin{bmatrix} d_{11} & & & & & & & & \\ c_{21} & d_{21} & & & & & & & \\ 0 & c_{31} & d_{31} & & & & & & \\ b_{12} & 0 & 0 & d_{12} & & & & & \\ & b_{22} & 0 & c_{22} & d_{22} & & & & \\ & & b_{32} & 0 & c_{32} & d_{32} & & & \\ & & & b_{13} & 0 & 0 & d_{13} & & \\ & & & & b_{23} & 0 & c_{23} & d_{23} & \\ & & & & & b_{33} & 0 & c_{33} & d_{33} \end{bmatrix}, \quad (9.1.1.19a)$$

$$[U'] \equiv \begin{bmatrix} 1 & e_{11} & 0 & f_{11} & & & & & \\ & 1 & e_{21} & 0 & f_{21} & & & & \\ & & 1 & 0 & 0 & f_{31} & & & \\ & & & 1 & e_{12} & 0 & f_{12} & & \\ & & & & 1 & e_{22} & 0 & f_{22} & \\ & & & & & 1 & 0 & 0 & f_{32} \\ & & & & & & 1 & e_{13} & 0 \\ & & & & & & & 1 & e_{23} \\ & & & & & & & & 1 \end{bmatrix}. \quad (9.1.1.19b)$$

Multiplication of $[L']$ and $[U']$ gives

$$[A'] = \left[\begin{array}{ccc|ccc|ccc} E'_{11} & F'_{11} & 0 & H'_{11} & & & & & \\ D'_{21} & E'_{21} & F'_{21} & G'_{21} & H'_{21} & & & & \\ 0 & D'_{31} & E'_{31} & 0 & G'_{31} & H'_{31} & & & \\ \hline B'_{12} & C'_{12} & 0 & E'_{12} & F'_{12} & 0 & H'_{12} & & \\ & B'_{22} & C'_{22} & D'_{22} & E'_{22} & F'_{22} & G'_{22} & H'_{22} & \\ & & B'_{32} & 0 & D'_{32} & E'_{32} & 0 & G'_{32} & H'_{32} \\ \hline & & & B'_{13} & C'_{13} & 0 & E'_{13} & F'_{13} & 0 \\ & & & & B'_{23} & C'_{23} & D'_{23} & E'_{23} & F'_{23} \\ & & & & & B'_{33} & 0 & D'_{33} & E'_{33} \end{array}\right], \quad (9.1.1.19c)$$

where

$$B'_{ij} = b_{ij}, \qquad F'_{ij} = d_{ij}e_{ij},$$

$$C'_{ij} = b_{ij}e_{i,j-1}, \qquad G'_{ij} = c_{ij}f_{i-1,j},$$

$$D'_{ij} = c_{ij}, \qquad H'_{ij} = d_{ij}f_{ij},$$

$$E'_{ij} = b_{ij}f_{i,j-1} + c_{ij}e_{i-1,j} + d_{ij}.$$

Examination of the approximate matrix $[A']$ and the original matrix $[A]$ reveals that, although they are quite similar, they differ in one important way. The approximate matrix contains two additional diagonals, represented by the coefficients C'_{ij} and G'_{ij}. The objective now is to utilize $[A']$ in an effective iterative scheme. This is achieved by employing $[A']$ in the following way:

$$[A']\{p\}^{k+1} = [A']\{p\}^k - \beta([A]\{p\}^k - \{q\}), \tag{9.1.1.20}$$

where, as earlier, β is an iteration parameter. Earlier on in this section we introduced the concept of solving for the increment to enhance solution accuracy. By defining the increment as

$$\{P\}^{k+1} = \{p\}^{k+1} - \{p\}^k, \tag{9.1.1.21}$$

Eq. (9.1.1.20) becomes

$$[A']\{P\}^{k+1} = \beta\{r\}^k, \tag{9.1.1.22}$$

where $\{r\}^k$ is the residual defined through the relation

$$\{r\}^k = \{q\} - [A]\{p\}^k. \tag{9.1.1.23}$$

Recall, however, that we defined $[A']$ so that it factored into the upper and lower triangular matrices defined by (9.1.1.19). Thus, we rewrite (9.1.1.22) as

$$[L'][U']\{P\}^{k+1} = \beta\{r\}^k. \tag{9.1.1.24}$$

Letting $\{w\}^{k+1} = [U']\{P\}^{k+1}$, we can express (9.1.1.24) as the two-step procedure

$$[L']\{w\}^{k+1} = \beta\{r\}^k, \tag{9.1.1.25a}$$

$$[U']\{P\}^{k+1} = \{w\}^{k+1}. \tag{9.1.1.25b}$$

The simplicity and efficiency of this algorithm is derived from the structure of $[L']$ and $[U']$. Referring to (9.1.1.19) and (9.1.1.25), we see that the forward sweep of (9.1.1.25a) reduces to

$$\begin{aligned} b_{ij}w^{k+1}_{i,j-1} + c_{ij}w^{k+1}_{i-1,j} + d_{ij}w^{k+1}_{ij} \\ = \beta r^k_{ij}, \qquad 1 \leq i \leq N, \quad 1 \leq j \leq M. \end{aligned} \tag{9.1.1.26a}$$

Because b_{i1} and c_{1j} are zero, Eq. (9.1.1.26a) can be readily solved explicitly in order of increasing i and increasing j. The backward sweep yields the relation

$$\begin{aligned} P^{k+1}_{ij} + e_{ij}P^{k+1}_{i+1,j} + f_{ij}P^{k+1}_{i,j+1} \\ = w^{k+1}_{ij}, \qquad 1 \leq i \leq N, \quad 1 \leq j \leq M. \end{aligned} \tag{9.1.1.26b}$$

Because $\{w\}^{k+1}$ is known and e_{Nj} and F_{iM} are zero, we can also solve (9.1.1.26b) explicitly in order of decreasing i and decreasing j. The desired solution for $\{p\}^{k+1}$ is now obtained directly from (9.1.1.21).

The selection of the elements of $[L']$ and $[U']$ remains the unanswered question. There are several avenues of approach to this problem. Here we present the most popular one, the method of Stone (1968). According to this scheme one defines the elements of $[L']$ and $[U']$ as

$$b_{ij} = B_{ij}/(1 + \alpha_k e_{i,j-1}), \tag{9.1.1.27a}$$

$$c_{ij} = D_{ij}/(1 + \alpha_k f_{i-1,j}), \tag{9.1.1.27b}$$

$$d_{ij} = E_{ij} + \alpha_k b_{ij} e_{i,j-1} + \alpha_k c_{ij} f_{i-1,j} - b_{ij} f_{i,j-1} - c_{ij} e_{i-1,j}, \tag{9.1.1.27c}$$

$$e_{ij} = (F_{ij} - \alpha_k b_{ij} e_{i,j-1})/d_{ij}, \tag{9.1.1.27d}$$

$$f_{ij} = (H_{ij} - \alpha_k c_{ij} f_{i-1,j})/d_{ij}, \tag{9.1.1.27e}$$

where α_k is an iteration parameter.

Because B_{i1} and D_{1j} are zero, it is possible to solve (9.1.1.27a)–(9.1.1.27e) sequentially in terms of increasing i and increasing j. With this information in hand we return to (9.1.1.26) to obtain the desired solution for P_{ij}^{k+1} and so on. To obtain good convergence, Stone (1968) recommends that the iteration parameter β in (9.1.1.26a) be set to unity. The α_k iteration parameters should be cycled as iteration proceeds. Although the minimum value of α_k is not critical, the maximum value of α_k is very important. Peaceman (1977, p. 134) recommends the following strategy for the calculation of α_k:

$$\alpha_{\max} = 1 - \min_{ij}\left[\frac{\pi^2}{2N^2[1 + (AY/AX)]}, \frac{\pi^2}{2M^2[1 + (AX/AY)]}\right]$$

where AX are the coefficients D_{ij} and F_{ij} and AY are the coefficients B_{ij} and H_{ij}. To space geometrically the α_k between 1 and $1-\alpha_{\max}$ one can employ the algorithm

$$1 - \alpha_p = (1 - \alpha_{\max})^{(p-1)/(q-1)}, \qquad p = 1, 2, \ldots, q,$$

where q is the number of parameters in the cycle. In problems such as described herein $\alpha_k = 1$ should be included in the iteration sequence. Peaceman (1977) reports that Stone has markedly improved the rate of convergence of his method by alternating the order in which the equations are solved. Specifically he executes the factorization step in order of decreasing j and increasing i on every second iteration. Similarly, he proceeds in order of decreasing i and increasing j on the back substitution step of the same iteration.

The above SIP algorithm is applicable to two-dimensional flow problems. Its extension to three dimensions can be found in Weinstein *et al.* (1969).

Direct Solution Procedure

An alternative way of dealing with a system of $N \times M$ equations resulting from the two-dimensional finite difference approximation is to employ a direct solution procedure. Various solution algorithms are available for banded equations [see, for example, Jennings (1971) and Segui (1975)]. The amount of work involved in treating a banded matrix is proportional to the bandwidth squared. Consequently, one should number the nodes carefully to minimize the bandwidth. Several strategies for node numbering are described by Cuthill (1972), George (1973), and Price and Coats (1974). For a rectangular grid, Price and Coats have shown that it is highly advantageous to use the $D4$ diagonal ordering of the nodes shown in Fig. 9.4. A comparison of the work required by the $D4$ ordering and the standard ordering (where the nodes are numbered along the j axis when $M \leq N$ is presented in Fig. 9.5. Note that the maximum reduction in the work effort is obtained when $M = N$.

We now describe the algorithm associated with $D4$ ordering. Let the matrix equation $[A]\{u\} = \{d\}$ be partitioned so that

$$[A]\{u\} = \begin{bmatrix} A_1 & A_2 \\ A_3 & A_4 \end{bmatrix} \begin{Bmatrix} u_1 \\ u_2 \end{Bmatrix} = \begin{Bmatrix} d_1 \\ d_2 \end{Bmatrix}, \tag{9.1.1.28}$$

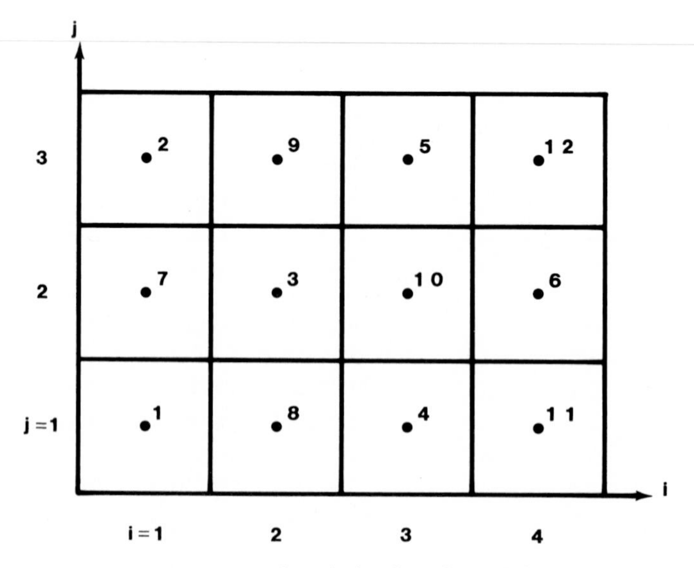

Fig. 9.4. Node ordering for scheme $D4$.

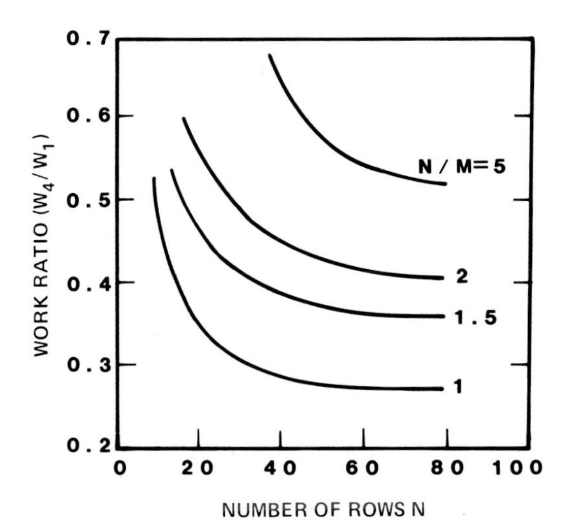

Fig. 9.5. Ratio of the work required by scheme $D4$ (w_4) to the work required by standard ordering (w_1) in two dimensions versus number of rows (N) for various values of N/M. (Adapted from Aziz and Settari, 1979).

where the structure of the partitioned matrix $[A]$ is depicted in Fig. 9.6, with $\boldsymbol{A}_1$ and $\boldsymbol{A}_4$ corresponding to the diagonal submatrices and $\boldsymbol{A}_2$ and $\boldsymbol{A}_3$ are the remaining banded submatrices.

Using (9.1.1.28), we first perform the forward elimination on the lower half of $[A]$ and transform $[A] \rightarrow [A^*]$, $\{d\} \rightarrow \{d^*\}$, where

$$[A^*] = \begin{bmatrix} \boldsymbol{A}_1 & \boldsymbol{A}_2 \\ \boldsymbol{0} & \boldsymbol{A}_4^* \end{bmatrix}, \qquad \{d^*\} = \begin{Bmatrix} \boldsymbol{d}_1 \\ \boldsymbol{d}_2^* \end{Bmatrix}.$$

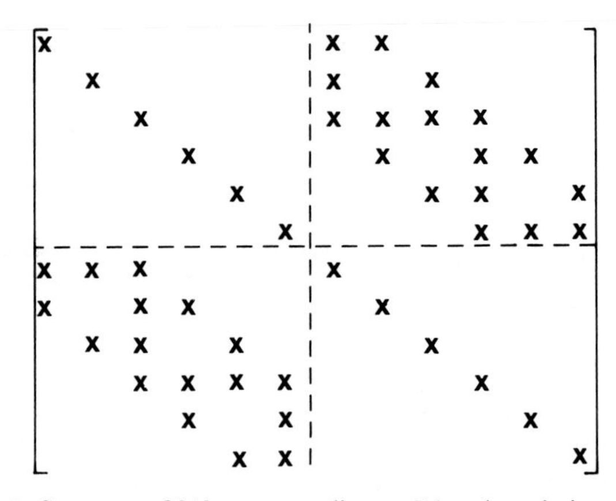

Fig. 9.6. Structure of $[A]$ corresponding to $D4$ node ordering scheme.

Note in particular that $\boldsymbol{A}_1$, $\boldsymbol{A}_2$, and $\boldsymbol{d}_1$ remain unchanged, whereas $\boldsymbol{A}_4$ now becomes a new banded submatrix $\boldsymbol{A}_4^*$. Because of the special structure of $[A^*]$, we can solve for $\boldsymbol{u}_2$, which is the lower half of $\boldsymbol{u}$, by considering the following subsystem of equations:

$$\boldsymbol{A}_4^* \boldsymbol{u}_2 = \boldsymbol{d}_2^* \tag{9.1.1.29}$$

and employing a standard band elimination scheme.

After $\boldsymbol{u}_2$ has been determined, $\boldsymbol{u}_1$ can be obtained by performing the following back substitution:

$$\boldsymbol{u}_1 = (\boldsymbol{A}_1)^{-1}(\boldsymbol{d}_1 - \boldsymbol{A}_2 \boldsymbol{u}_2), \tag{9.1.1.30}$$

where the inverse of the diagonal matrix $(\boldsymbol{A}_1)^{-1}$ is easily determined. It is apparent that the $D4$ ordering leads to a substantial reduction of computational effort because in the elimination, one only needs to deal with submatrix $\boldsymbol{A}_4$ instead of the original matrix $[A]$. In addition, the computer storage requirement is also reduced by half.

9.1.2 Potential Formulation

The groundwater flow equation is often written in terms of the fluid potential or hydraulic head. The appropriate transformation is

$$h = \frac{1}{g} \int_{p_0}^{p} \frac{d\xi}{\rho(\xi)} + z, \tag{9.1.2.1}$$

where h is the hydraulic head and p_0 is a reference pressure. The governing equation is now obtained by substitution of (9.1.2.1) into (9.1.1.1) to yield (4.2.1.9a), that is,

$$\frac{\partial}{\partial x_i}\left(K_{ij} \frac{\partial h}{\partial x_j}\right) - S_s \frac{\partial h}{\partial t} = 0 \tag{9.1.2.2}$$

where $K_{ij} \equiv k_{ij}\rho g/\mu$ and $S_s \equiv (\alpha + \phi\beta)\rho g$. The parameters K_{ij} and S_s are the hydraulic conductivity and specific storage, respectively, and normally are assumed to be functions of space only. An appropriate finite difference representation of (9.1.2.2) is obtained by representing the spatial terms with a central difference approximation and the time derivative by a weighted difference representation, that is,

$$\frac{1}{\Delta x_i \, \Delta x_j} \delta_{x_i}[K_{ij} \delta_{x_j}(h_{s+1} + (1 - \gamma)h_s)] = S_s(h_{s+1} - h_s)/\Delta t, \tag{9.1.2.3}$$

where γ is a time weighting factor, $0 \leqslant \gamma \leqslant 1$. Unconditional stability requires $\frac{1}{2} \leqslant \gamma \leqslant 1$. Minor oscillations in time can be circumvented by selecting $\gamma > \frac{1}{2}$; smooth results are achieved with $\gamma = 0.66$.

9.1.3 *Flow of a Real Gas*

The equation describing the flow of a real gas was presented in Chapter 4; it reads

$$\frac{\partial}{\partial x_i}\left(k_{ij}\frac{\partial m}{\partial x_j}\right) = \lambda\frac{\partial m}{\partial t} \tag{9.1.3.1}$$

where $\lambda \equiv \phi\beta\mu$, k_{ij} is the gas permeability, β the gas compressibility factor [see Eq. (4.22.10c), μ the gas dynamic viscosity, and m the gas pseudopressure, which is defined by (4.2.2.8a) and we rewrite here as

$$m(p) = \int_{p_o}^{p}\frac{2\xi\, d\xi}{\sigma(\xi)\mu(\xi)}, \tag{9.1.3.2}$$

where $\sigma(\xi)$ is the gas deviation factor.

While (9.1.3.1) appears relatively innocuous, it does involve the solution of a nonlinear equation because λ is a function of m (see Fig. 4.11). The nonlinearities associated with (9.1.3.1) are not severe. As a result most techniques for the solution of nonlinear partial differential equations can be employed. The simplest iterative procedure employs a Picard-type approach, wherein the coefficient λ is lagged by one iteration. Writing the obvious finite difference approximation for (9.1.3.1), we obtain the first-order convergent method:

$$\frac{1}{\Delta x_i\,\Delta x_j}\delta_{x_i}(k_{ij}\delta_{x_j})[\gamma m_{s+1}^{k+1} + (1-\gamma)m_s] = \frac{\lambda(m^k)(m_{s+1}^{k+1} - m_s)}{\Delta t}. \tag{9.1.3.3}$$

The solution procedure requires the repetitive solution of (9.1.3.3) until $\|m_{s+1}^{k+1} - m_{s+1}^{k}\| < \varepsilon$ for a specified tolerance ε. One then proceeds to the next $(s+2)$ time step and begins once again the iterative procedure.

9.1.4 *Flow in Saturated–Unsaturated Soils*

The flow of water in saturated–unsaturated soil can be described by Eq. (4.7.2.1), which reads

$$\frac{\partial}{\partial x_i}\left[K_{ij}k_{rw}\left(\frac{\partial\psi}{\partial x_j} + \frac{\partial z}{\partial x_j}\right)\right] = \phi\frac{\partial S_w}{\partial t}, \tag{9.1.4.1}$$

where K_{ij} is the saturated hydraulic conductivity, k_{rw} the relative permeability with respect to the water phase, ψ the pressure head, ϕ porosity, and S_w the saturation of water.

In addition to (9.1.4.1) the system requires boundary and initial conditions and constitutive relations. Typical constitutive relations are found in Fig. 4.13. Note that this single-equation formulation, while recognizing the existence of two phases in these constitutive equations, neglects the influence of air pressure gradients and the associated air velocity. The more general case will be considered shortly in our discussion of multiphase flow simulation.

Returning to our governing equation (9.1.4.1), we find it advantageous to expand the right-hand side as follows:

$$\phi \frac{\partial S_{\mathrm{w}}}{\partial t} = \phi \frac{dS_{\mathrm{w}}}{d\psi} \frac{\partial \psi}{\partial t} = \frac{d\theta}{d\psi} \frac{\partial \psi}{\partial t} \equiv C \frac{\partial \psi}{\partial t}, \tag{9.1.4.2}$$

where C is called the specific moisture capacity and θ the volumetric moisture content. The constitutive curves of Fig. 4.13 are prepared in terms of θ so that C is the slope of the ψ–θ curve.

Substitution of (9.1.4.2) into (9.1.4.1) yields the nonlinear equation

$$\frac{\partial}{\partial x_i} \left[K_{ij} k_{r\mathrm{w}} \left(\frac{\partial \psi}{\partial x_j} + \frac{\partial z}{\partial x_j} \right) \right] = C \frac{\partial \psi}{\partial t}. \tag{9.1.4.3}$$

The obvious central difference approximation to (9.1.4.3) generates poor solutions in many instances. The reason behind this becomes more apparent when we expand the spatial pressure term:

$$\frac{\partial}{\partial x_i} \left[K_{ij} k_{r\mathrm{w}} \frac{\partial \psi}{\partial x_j} \right] = K_{ij} k_{r\mathrm{w}} \frac{\partial^2 \psi}{\partial x_i \, \partial x_j} + \frac{\partial (K_{ij} k_{r\mathrm{w}})}{\partial x_i} \frac{\partial \psi}{\partial x_j}. \tag{9.1.4.4}$$

The first term on the right-hand side is analogous to dispersive transport, and the second resembles convective transport with $\partial(K_{ij}k_{r\mathrm{w}})/\partial x_i$ the pseudovelocity. It is well known that convective dispersive transport equations are very difficult to solve when convection dominates. Thus it is not surprising that (9.1.4.3) causes analysts distress when the relative permeability varies rapidly with space. Borrowing from our knowledge of the convective–dispersive transport equation, we would logically seek an asymmetric form for the approximation of (9.1.4.4). Employing an upstream difference approximation for the convective term in (9.1.4.4), we obtain for the x_1 direction

$$\frac{\partial}{\partial x_1} \left[K_{11} k_{r\mathrm{w}} \frac{\partial \psi}{\partial x_1} \right]$$

$$\simeq \frac{a_{r_1}}{(\Delta x_1)^2} [\psi_{r_1+1} - 2\psi_{r_1} + \psi_{r_1-1}] + \frac{a_{r_1} - a_{r_1-1}}{\Delta x_1} \frac{\psi_{r_1} - \psi_{r_1-1}}{\Delta x_1}, \tag{9.1.4.5}$$

where $a_{r_1} \equiv (K_{11}k_{r\mathrm{w}})_{r_1}$. Rearranging (9.1.4.5), we obtain

$$\frac{1}{(\Delta x_1)} \left(a_{r_1} \frac{\psi_{r_1+1} - \psi_{r_1}}{\Delta x_1} - a_{r_1-1} \frac{\psi_{r_1} - \psi_{r_1-1}}{\Delta x_1} \right), \tag{9.1.4.6}$$

which is the generally employed formulation. Assuming that $K_{12} = K_{21} = 0$ and there is negligible gravity effect, the finite difference approximation for (9.1.4.3) can now be written

$$\sum_{i=1}^{2} \frac{1}{(\Delta x_i)^2} (a_{r_i} \delta_{x_i} \psi_{r_i+1/2} - a_{r_i-1} \delta_{x_i} \psi_{r_i-1/2}) = C_{r_i,s+1} \frac{\psi_{r_i,s+1} - \psi_{r_i,s}}{\Delta t}, \quad (9.1.4.7)$$

where it is understood that $r_i - 1$ is taken in the upstream direction. The nonlinear coefficients a_{r_i} and $C_{r_i,s+1}$ are generally evaluated using simple Picard iteration (see Section 4.6.1). While there may be occasions when a Newton–Raphson formulation would be appropriate, this approach is not normally used for the single-equation description of saturated–unsaturated flow.

9.2 Simulation of Multiphase Flow

9.2.1 Pressure Formulation

Multiphase flow simulation arises in unsaturated groundwater flow, oil reservoir simulation, and geothermal reservoir simulation. The equations describing the flow of two immiscible fluids are presented in Section 4.7.3 as Eqs. (4.7.3.1). They read for the case of oil (o) and water (w)

$$\frac{\partial}{\partial x_i}\left[k_{ij}\lambda_w\left(\frac{\partial p_w}{\partial x_j} + \rho_w g \frac{\partial z}{\partial x_j}\right)\right] = \phi \frac{\partial S_w}{\partial t}, \quad (9.2.1.1a)$$

$$\frac{\partial}{\partial x_i}\left[k_{ij}\lambda_o\left(\frac{\partial p_o}{\partial x_j} + \rho_o g \frac{\partial z}{\partial x_j}\right)\right] = \phi \frac{\partial S_o}{\partial t}, \quad (9.2.1.1b)$$

where k_{ij} is the intrinsic permeability, ϕ porosity, $\lambda_\alpha \equiv k_{r\alpha}/\mu_\alpha$ the relative mobility of phase α, and S_α the saturation of phase α. Because the saturations must sum to unity, that is, $S_o + S_w = 1$, the second equation can be rewritten in the alternative form

$$\frac{\partial}{\partial x_i}\left[k_{ij}\lambda_o\left(\frac{\partial p_o}{\partial x_j} + \rho_o g \frac{\partial z}{\partial x_j}\right)\right] = -\phi \frac{\partial S_w}{\partial t}. \quad (9.2.1.1c)$$

These equations must be augmented with appropriate initial and boundary conditions as well as constitutive relations before they can be solved. The constitutive relations relate the mobility and saturation to pressure, that is,

$$k_{r\alpha} = k_{r\alpha}(S_\alpha), \quad (9.2.1.2a)$$

$$S_\alpha = S_\alpha(p_{\alpha/\beta}), \quad (9.2.1.2b)$$

where $p_{\alpha/\beta}$ is the capillary pressure defined as $p_{\alpha/\beta} = p_\alpha - p_\beta$, $\beta \neq \alpha$.

In the case of air–water flow in the unsaturated zone, Eqs. (9.2.1.1) can be attacked in their current form [see, for example, Green *et al.* (1971)]. The equations generally used for immiscible flow in oil reservoirs, however, are modifications of (9.2.1.1). We examined these in Section 4.5.2, where we described in some detail the "formation volume factor" that accounts for volume changes due to different fluid pressures in the reservoir and at the surface. When this factor is introduced into (9.2.1.1), one obtains the following expressions for an oil–water system:

$$\frac{\partial}{\partial x_i}\left[\frac{k_{ij}\lambda_{\mathrm{w}}}{B_{\mathrm{w}}}\left(\frac{\partial p_{\mathrm{w}}}{\partial x_j} + \rho_{\mathrm{w}} g \frac{\partial z}{\partial x_j}\right)\right] = \frac{\partial}{\partial t}\left(\frac{\phi S_{\mathrm{w}}}{B_{\mathrm{w}}}\right), \qquad (9.2.1.3\mathrm{a})$$

$$\frac{\partial}{\partial x_i}\left[\frac{k_{ij}\lambda_{\mathrm{o}}}{B_{\mathrm{o}}}\left(\frac{\partial p_{\mathrm{o}}}{\partial x_j} + \rho_{\mathrm{o}} g \frac{\partial z}{\partial x_j}\right)\right] = \frac{\partial}{\partial t}\left(\frac{\phi S_{\mathrm{o}}}{B_{\mathrm{o}}}\right), \qquad (9.2.1.3\mathrm{b})$$

where B_α is the formation volume factor for fluid phase α, defined as

$$B_\alpha = \rho_{\alpha S}/\rho_\alpha,$$

where ρ_α is the mass density of phase α and the subscript S denotes surface conditions. Note that when a source term Q_α appears in (9.2.1.3), it is related to the surface flow through the relation $Q_\alpha = B_\alpha Q_{\alpha S}$.

9.2.2 *The Simultaneous Solution Method (SS) (Douglas et al.*, 1959)

Let us reduce (9.2.1.3) to its areal counterpart in two space dimensions. Vertical averaging yields

$$\frac{\partial}{\partial x_i}\left[l\frac{k_{ij}\lambda_\alpha}{B_\alpha}\left(\frac{\partial p_\alpha}{\partial x_j} + \rho_\alpha g \frac{\partial z}{\partial x_j}\right)\right] = l\frac{\partial}{\partial t}\left(\phi\frac{S_\alpha}{B_\alpha}\right)$$

$$\alpha = \mathrm{o}, \mathrm{w}, \quad i = 1, 2, \qquad (9.2.2.1)$$

where it is understood that the formation coefficients and the dependent variables are vertically averaged over the thickness l; for example,

$$S(x_1, x_2) = (1/l)\int_0^l S(\mathbf{x})\, dx_3. \qquad (9.2.2.2)$$

In many applications complete vertical mixing is assumed. In some cases, however, there is a vertical segregation of fluids and the averaged parameters must be adjusted using "pseudofunctions," which accommodate the segregation phenomenon. We shall consider these functions separately later in Section 9.2.6.

Assuming that $k_{12} = k_{21} = 0$, the finite difference form of (9.2.2.1) can be written

$$\frac{1}{(\Delta x_i)^2} \delta_{x_i} \left[\frac{lk_{ii}\lambda_\alpha}{B_\alpha} (\delta_{x_i} p_\alpha + \rho_\alpha g \delta_{x_i} z) \right] = \frac{l}{\Delta t} \nabla_t \left(\phi \frac{S_\alpha}{B_\alpha} \right),$$

$$\alpha = \mathrm{o, w}, \quad i = 1, 2. \tag{9.2.2.3}$$

This formulation is a backward difference approximation in time wherein all the terms on the left-hand side of (9.2.2.3) are at the $s + 1$ time level. The simultaneous solution algorithm is generated by first expanding the right-hand side (9.2.2.3) to yield

$$\nabla_t \left(\phi \frac{S_\alpha}{B_\alpha} \right) = S_{\alpha s} \nabla_t \left(\frac{\phi}{\beta_\alpha} \right) + \left(\frac{\phi}{B_\alpha} \right)_{s+1} \nabla_t S_\alpha$$

$$= S_{\alpha s} \phi_s \nabla_t \left(\frac{1}{B_\alpha} \right) + \left(\frac{\phi}{B_\alpha} \right)_{s+1} \nabla_t S_\alpha$$

$$+ S_{\alpha s} \frac{1}{B_{\alpha s+1}} \nabla_t \phi, \qquad \alpha = \mathrm{o, w}. \tag{9.2.2.4}$$

Let us now apply the chain rule to the terms on the right-hand side of (9.2.2.4). We assume the following functional relations:

$$B_\alpha = B(p_\alpha), \qquad S_\alpha = S_\alpha(p_{\alpha/\beta}), \qquad \phi = \phi(p) \simeq \phi[\tfrac{1}{2}(p_\alpha + p_\beta)].$$

Equation (9.2.2.4) now becomes

$$\nabla_t \left(\phi \frac{S_\alpha}{B_\alpha} \right) = S_{\alpha s} \phi_s \frac{d(1/\mathrm{B}_\alpha)}{dp_\alpha} \nabla_t p_\alpha + \left(\frac{\phi}{B_\alpha} \right)_{s+1} \frac{dS_\alpha}{dp_{\mathrm{n/W}}} (\nabla_t p_{\mathrm{n}} - \nabla_t p_{\mathrm{W}})$$

$$+ \frac{1}{2} \frac{S_{\alpha s}}{B_{\alpha s+1}} \frac{d\phi}{dp} (\nabla_t p_{\mathrm{n}} + \nabla_t p_{\mathrm{W}}), \qquad \alpha = \mathrm{o, w}, \tag{9.2.2.5}$$

where n is the nonwetting phase and W is the wetting phase. In this case, $n \equiv \mathrm{o}$ and $W \equiv \mathrm{w}$. The derivatives appearing in (9.2.2.5) are approximated using a chord slope, that is,

$$\frac{dS_\alpha}{dp_{\alpha/\beta}} = \frac{S_{\alpha s+1} - S_{\alpha s}}{p_{\alpha/\beta s+1} - p_{\alpha/\beta s}}. \tag{9.2.2.6}$$

Substitution of (9.2.2.5) into (9.2.2.3) yields the desired difference expressions:

$$\frac{1}{(\Delta x_i)^2}\delta_{x_i}\left[\frac{lk_{ii}\lambda_\alpha}{B_\alpha}(\delta_{x_i}p_\alpha + \rho_\alpha g\delta_{x_i}z)\right]$$

$$= \frac{l}{\Delta t}\left[S_{\alpha s}\phi_s\frac{d(1/B_\alpha)}{dp_\alpha}\nabla_t p_\alpha + \left(\frac{\phi}{B_\alpha}\right)_{s+1}\frac{dS_\alpha}{dp_{\mathrm{n/w}}}(\nabla_t p_\mathrm{n} - \nabla_t p_\mathrm{w})\right.$$

$$\left. + \frac{1}{2}\frac{S_{\alpha s}}{B_{\alpha s+1}}\frac{d\phi}{dp}(\nabla_t p_\mathrm{n} + \nabla_t p_\mathrm{w})\right], \qquad \alpha = \mathrm{o, w}, \quad i = 1, 2 \qquad (9.2.2.7)$$

The two equations represented by (9.2.2.7) are generated for each node in the finite difference net. The resulting algebraic equations can be written

$$\delta_{x_i}[a_{\mathrm{w}i}(\delta_{x_i}p_\mathrm{w} + \rho_\mathrm{w}g\delta_{x_i}z)] = b_{\mathrm{w1}}\nabla_t p_\mathrm{w} + b_{\mathrm{w2}}\nabla_t p_\mathrm{o}, \qquad i = 1, 2, \quad (9.2.2.8\mathrm{a})$$

$$\delta_{x_i}[a_{\mathrm{o}i}(\delta_{x_i}p_\mathrm{o} + \rho_\mathrm{o}g\delta_{x_i}z)] = b_{\mathrm{o1}}\nabla_t p_\mathrm{o} + b_{\mathrm{o2}}\nabla_t p_\mathrm{w}, \qquad i = 1, 2, \quad (9.2.2.8\mathrm{b})$$

where

$$a_{\mathrm{w}i} = \frac{lk_{ii}\lambda_\mathrm{w}}{(\Delta x_i)^2 B_\mathrm{w}}, \qquad i = 1, 2,$$

$$b_{\mathrm{w1}} = \frac{l}{\Delta t}\left[S_{\mathrm{w}s}\phi_s\frac{d(1/B_\mathrm{w})}{dp_\mathrm{w}} - \left(\frac{\phi}{B_\mathrm{w}}\right)_{s+1}\frac{dS_\mathrm{w}}{dp_{\mathrm{o/w}}}\right.$$

$$\left. + \frac{1}{2}\frac{S_{\mathrm{w}s}}{B_{\mathrm{w}s+1}}\frac{d\phi}{dp}\right],$$

$$b_{\mathrm{w2}} = \frac{l}{\Delta t}\left[\left(\frac{\phi}{B_\mathrm{w}}\right)_{s+1}\frac{dS_\mathrm{w}}{dp_{\mathrm{o/w}}} + \frac{1}{2}\frac{S_{\mathrm{w}s}}{B_{\mathrm{w}s+1}}\frac{d\phi}{dp}\right],$$

$$a_{\mathrm{o}i} = \frac{lk_{ii}\lambda_\mathrm{o}}{(\Delta x_i)^2 B_\mathrm{o}}, \qquad i = 1, 2,$$

$$b_{\mathrm{o1}} = \frac{l}{\Delta t}\left[S_{\mathrm{o}s}\phi_s\frac{d(1/B_\mathrm{o})}{dp_o} + \left(\frac{\phi}{B_\mathrm{o}}\right)_{s+1}\frac{dS_\mathrm{o}}{dp_{\mathrm{o/w}}}\right.$$

$$\left. + \frac{1}{2}\frac{S_{\mathrm{o}s}}{B_{\mathrm{o}s+1}}\frac{d\phi}{dp}\right],$$

$$b_{\mathrm{o2}} = \frac{l}{\Delta t}\left[-\left(\frac{\phi}{B_\mathrm{o}}\right)_{s+1}\frac{dS_\mathrm{o}}{dp_{\mathrm{o/w}}} + \frac{1}{2}\frac{S_{\mathrm{o}s}}{B_{\mathrm{o}s+1}}\frac{d\phi}{dp}\right].$$

The set of equations (9.2.2.8) are generally numbered so that block matrices are generated, that is, $p_{w1}, p_{o1}, p_{w2}, p_{o2} \ldots$. Thus many of the efficient matrix solving routines (e.g., the block Thomas algorithm) can be used directly. One simply replaces arithmetic operations with matrix operations. Because (9.2.2.8) represents a set of nonlinear algebraic equations, these equations must, in general, be solved iteratively. Let us now consider this aspect of the problem in more detail.

The approach to use in evaluating the nonlinear terms in (9.2.2.8) depends upon the degree of nonlinearity involved. For weakly nonlinear terms, such as those involving B_α, μ_α, and ρ_α, a suitable estimate for the $s + 1$ time step is one based on the dependent variable calculation of the s time step. In fact, in the work of Green *et al.* (1971), this approach was also used in the calculation of the mobility terms in the saturated–unsaturated flow equations.

In most instances the mobility and the $dS_\alpha/dp_{\alpha/\beta}$ coefficient in the time derivative are considered strong nonlinearities and must be handled carefully. There are a number of ways of treating these coefficients. We begin with the Newton–Raphson approximation.

Newton–Raphson Method

Let us rearrange (9.2.2.8) and introduce the iteration index k so as to yield

$$A_{\rm w}^k p_{{\rm w}s+1}^k + B_{\rm w}^k(p_{{\rm w}s+1}^k - p_{{\rm w}s}) + C_{\rm w}^k(p_{{\rm o}s+1}^k - p_{{\rm o}s}) + G_{\rm w}^k = R_{\rm w}^k, \quad (9.2.2.9a)$$

$$A_{\rm o}^k p_{{\rm o}s+1}^k + B_{\rm o}^k(p_{{\rm o}s+1}^k - p_{{\rm o}s}) + C_{\rm o}^k(p_{{\rm w}s+1}^k - p_{{\rm w}s}) + G_{\rm o}^k = R_{\rm o}^k, \quad (9.2.2.9b)$$

where

$$A_\alpha^k = \delta_{x_i} a_{\alpha i} \delta_{x_i}(\cdot), \qquad i = 1, 2,$$

$$B_\alpha^k \equiv -b_{\alpha 1}^k,$$

$$C_\alpha^k \equiv -b_{\alpha 2}^k,$$

$$G_\alpha^k \equiv \delta_{x_i} a_{\alpha i} \rho_\alpha g \delta_{x_1} z.$$

Here R_α^k is the residual of the equation for phase α. Before we pursue the Newton–Raphson development further, let us pause momentarily to revisit the complexity inherent in (9.2.2.9). Because the coefficients A_α^k involve the spatial difference operators $\delta_{x_i} a_{\alpha i} \delta_{x_i}(\cdot)$, Eq. (9.2.2.9a) contains five values of $p_{{\rm w}s+1}^k$, one for each position in the five-point star associated with the location $(r_1\,\Delta x_1, r_2\,\Delta x_2)$, and (9.2.2.9b) contains five values of $p_{{\rm o}s+1}^k$. Let us represent these five values using the notation $p_{\alpha j,s+1}$, $j = 0, 1, 2, 3, 4$, as illustrated in Fig. 9.7.

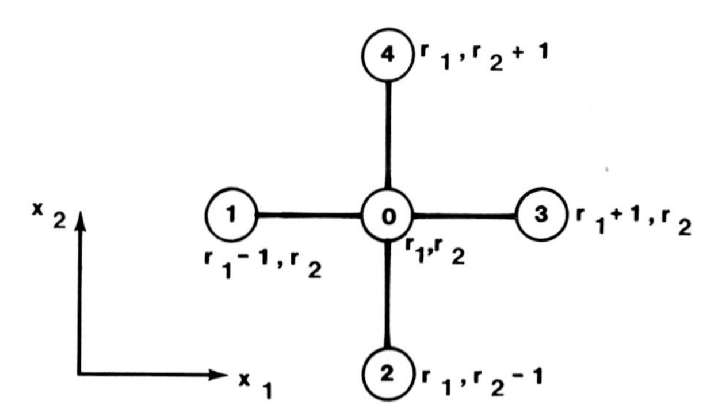

Fig. 9.7. Representation of the five-point star using a single index running from 0 to 4.

Using this nodal location convention, one can write the Newton–Raphson scheme in the following way:

$$R_\alpha^{k+1} = R_\alpha^k + \sum_{j=0}^{4} \left[\frac{\partial R_\alpha^k}{\partial p_{\mathrm{o}j,s+1}} \Delta p_{\mathrm{o}j,s+1} + \frac{\partial R_\alpha^k}{\partial p_{\mathrm{w}j,s+1}} \Delta p_{\mathrm{w}j,s+1} \right]$$

$$= 0, \qquad \alpha = \mathrm{o, w}, \tag{9.2.2.10}$$

where $\Delta p_{\alpha j,s+1} \equiv p_{\alpha j,s+1}^{k+1} - p_{\alpha j,s+1}^{k}$. For problems involving other than five-point stars, j would range over a different nodal sequence. If we were to write our equations in matrix form, the summation indicated in (9.2.2.10) would be over the nonzero elements in the row representing the difference equation. Rewriting (9.2.2.9) so that known information is collected into a single term, we have

$$D_{\mathrm{w}}^k p_{\mathrm{w}s+1}^k + C_{\mathrm{w}}^k p_{\mathrm{o}s+1}^k + f_{\mathrm{w}}^k = R_{\mathrm{w}}^k, \tag{9.2.2.11a}$$

$$D_{\mathrm{o}}^k p_{\mathrm{o}s+1}^k + C_{\mathrm{o}}^k p_{\mathrm{w}s+1}^k + f_{\mathrm{o}}^k = R_{\mathrm{o}}^k, \tag{9.2.2.11b}$$

where

$$f_{\mathrm{w}}^k = -B_{\mathrm{w}}^k p_{\mathrm{w}s} - C_{\mathrm{w}}^k p_{\mathrm{o}s} + G_{\mathrm{w}}^k,$$

$$f_{\mathrm{o}}^k = -B_{\mathrm{o}}^k p_{\mathrm{o}s} - C_{\mathrm{o}}^k p_{\mathrm{w}s} + G_{\mathrm{o}}^k,$$

$$D_{\mathrm{w}}^k = A_{\mathrm{w}}^k + B_{\mathrm{w}}^k,$$

$$D_{\mathrm{o}}^k = A_{\mathrm{o}}^k + B_{\mathrm{o}}^k.$$

Upon substitution of (9.2.2.11) into (9.2.2.10) and setting $R_\alpha^{k+1} = 0$, the final equations become

$$\sum_{j=0}^{4}\left\{\left[C_{wj}^{k}+\left(p_{o}^{k}\frac{\partial C_{w}^{k}}{\partial p_{o}}\right)_{j}+\frac{\partial f_{w}^{k}}{\partial p_{oj}}+\left(p_{w}^{k}\frac{\partial D_{w}^{k}}{\partial p_{o}}\right)_{j}\right]\Delta p_{oj}\right.$$

$$\left.+\left[D_{wj}^{k}+\left(p_{w}^{k}\frac{\partial D_{w}^{k}}{\partial p_{w}}\right)_{j}+\frac{\partial f_{w}^{k}}{\partial p_{wj}}\right]\Delta p_{wj}+\left(p_{o}^{k}\frac{\partial C_{w}^{k}}{\partial p_{w}}\right)_{j}\right\}=-R_{w}^{k}, \tag{9.2.2.12a}$$

$$\sum_{j=0}^{4}\left\{\left[C_{oj}^{k}+\left(p_{w}^{k}\frac{\partial C_{o}^{k}}{\partial p_{w}}\right)_{j}+\frac{\partial f_{o}^{k}}{\partial p_{wj}}+\left(p_{o}^{k}\frac{\partial D_{o}^{k}}{\partial p_{w}}\right)_{j}\right]\Delta p_{wj}\right.$$

$$\left.+\left[D_{oj}^{k}+\left(p_{o}^{k}\frac{\partial D_{o}^{k}}{\partial p_{o}}\right)_{j}+\frac{\partial f_{o}^{k}}{\partial p_{oj}}\right]\Delta p_{oj}+\left(p_{w}^{k}\frac{\partial C_{o}^{k}}{\partial p_{o}}\right)_{j}\right\}=-R_{o}^{k}. \tag{9.2.2.12b}$$

The various procedures that can be employed to determine the parameter derivatives in (9.2.2.12) have been detailed in Section 4.6.2.

To test the accuracy of the various nonlinear solution schemes, Settari and Aziz (1975) solved the Buckley–Leverett problem (two-phase flow with negligible capillarity gradients) using several methods. The problem was set up to use the following parameters: L, the domain length, is 1000 ft with a cross-sectional area of 10,000 ft^2; $B_w = B_o = 1$; $\mu_w = \mu_o = 1$ cp; $k = 300$ md; and $\phi = 0.2$. The oil is produced at $x = L$ at the rate of 426.5 ft^3/day and water is reinjected at the same rate at $x = 0$. The reservoir has an initial saturation of $S_w = 0.16$. Because the SS method presented required $p_{o/w}$ functions, a linear function defined by $p_{o/w} = 0.6$ when $S_w = 0.1$ and $p_{o/w} = 0$ at $S_w = 0.8$ is used. The results are not influenced by the form of this capillary pressure function. Solutions obtained using the Newton–Raphson method and the relative permeability functions of Fig. 9.8 are presented in Fig. 9.9. One should keep in mind that upstream weighting is to be used in calculating the mobility coefficient of each phase. The commonly used upstream weighting scheme is defined by

$$(k_{r\alpha}/\mu_{\alpha})_{i+1/2,j} = \eta(k_{r\alpha}/\mu)_{i,j} + (1-\eta)(k_{r\alpha}/\mu)_{i+1,j},$$

where $\eta = 1$ if flow is from node i to $i+1$ and $\eta = 0$ if flow is from node $i+1$ to i.

In certain instances the time derivative coefficients B_α and C_α cause difficulties with the Newton–Raphson formulation. To accommodate these situations, a modified Newton–Raphson scheme for these terms was suggested by Peaceman (1967) and is described in detail in Settari and Aziz (1975). The method employs the most recent values of pressure to

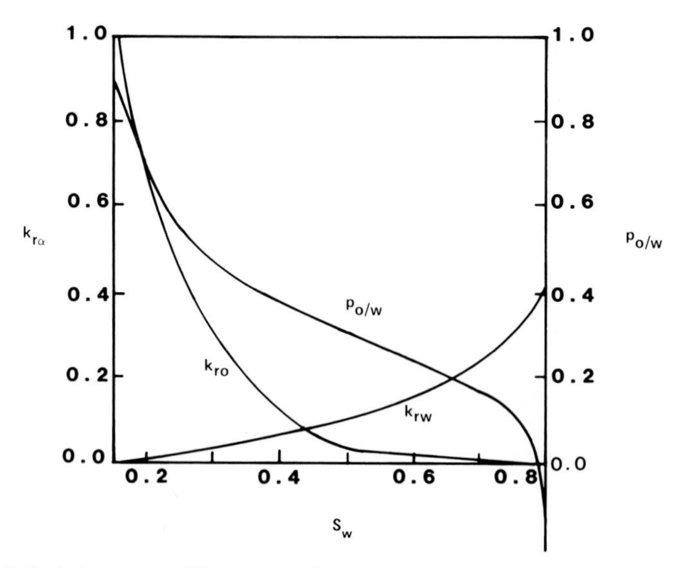

Fig. 9.8. Relative permeability and capillary pressure functions employed by Settari and Aziz (1975) in their computation of the Buckley–Leverett problem.

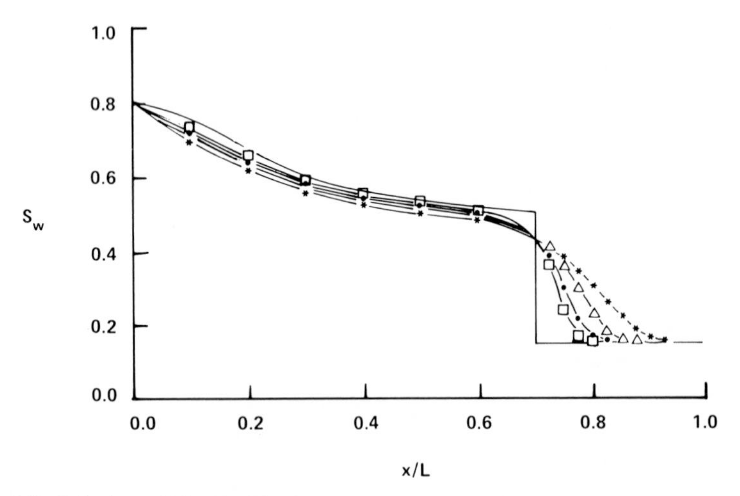

Fig. 9.9. Solutions of the Buckley-Leverett problem obtained by Settari and Aziz (1975) using the Newton–Raphson scheme. Legend: □, Δt = 25 days; •, Δt = 50 days; △, Δt = 100 days; *, Δt = 188 days.

compute the capillary pressure, that is,

$$p^*_{\alpha/\beta} = p^{k+1}_\alpha - p^{k+1}_\beta, \qquad \alpha \neq \beta. \tag{9.2.2.13}$$

One then uses this capillary pressure in the saturation projection, that is,

$$S^{k+1}_{\alpha r_1,r_2,s+1} = S^k_{\alpha r_1,r_2,s+1} + \left.\frac{dS_{\alpha r_1,r_2,s+1}}{dp_{\alpha/\beta}}\right|_k (p^*_{\alpha/\beta} - p^k_{\alpha/\beta})_{r_1,r_2,s+1}. \tag{9.2.2.14}$$

The capillary pressure is now adjusted through the constitutive relation

$$p^{k+1}_{\alpha/\beta} = f(S^{k+1}_{\alpha r_1,r_2,s+1}) \tag{9.2.2.15}$$

and one of the phase pressures adjusted to accommodate this new capillary pressure.

Linearized Method

This method uses a linearized form of one of the terms in the Newton–Raphson algorithm, that is [see (9.2.2.12)],

$$\sum_{j=0}^{4}\left[D_{\alpha j,s+1} + \left(p_\alpha \frac{dD^k_\alpha}{dp_\alpha}\right)_{j,s+1}\right]\Delta p_{\alpha j,s+1}$$
$$\simeq \sum_{j=0}^{4}\left[D_{\alpha j,s} + \left(p_\alpha \frac{dD_\alpha}{dp_\alpha}\right)_{j,s}\right]\Delta p_{\alpha j,s+1}, \qquad \alpha = \text{o, w}. \tag{9.2.2.16}$$

Aziz and Settari (1979, p. 159) show that the use of (9.2.2.16) in the SS method is equivalent to the first iteration in a Newton–Raphson approximation of the mobility term. The stability bound associated with this method is evident in Fig. 9.10, which reproduces the solutions presented by Settari and Aziz (1975).

Semiimplicit Method of Nolen and Berry (1972)

Nolen and Berry introduced a variant on the linearized method that retained more of the inherent nonlinearity in the convective terms. Their approach employs the following form of the nonlinear convective term:

$$\sum_{j=0}^{4}\left[D_{\alpha j,s+1} + \left(p_\alpha \frac{dD^k_\alpha}{dp_\alpha}\right)_{j,s+1}\right]\Delta p_{\alpha j,s+1}$$
$$\simeq \sum_{j=0}^{4}\left[D_{\alpha j,s} + \left(p_\alpha \frac{dD_\alpha}{dp_\alpha}\right)_{j,s+1}\right]\Delta p_{\alpha j,s+1}. \tag{9.2.2.17}$$

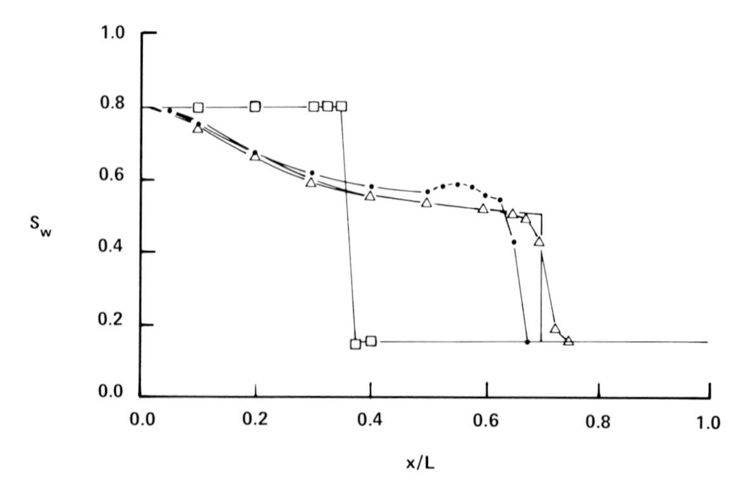

Fig. 9.10. Solutions of the Buckley–Leverett problem obtained by Settari and Aziz (1975) using explicit transmissivity calculation scheme. Legend: □, Δt = 25 days; •, Δt = 50 days; △, Δt = 100 days.

The nonlinearity appearing on the right-hand side of (9.2.2.17) was accommodated using Newton–Raphson iteration but with a chord slope representation of the $(dD_\alpha/dp_\alpha)\,|_{j,s+1}$ term. In actual practice, this differentiation must be chained out such that the slope of the relative permeability–saturation curve and that of the saturation–capillary pressure curve are evaluated as

$$\frac{dD_\alpha}{dS_\alpha}\frac{dS_\alpha}{dp_{\alpha/\beta}}\frac{dp_{\alpha/\beta}}{dp_\alpha} \simeq \frac{D_\alpha(S^k_{\alpha,s+1}) - D_\alpha(S_{\alpha,s})}{S^k_{\alpha,s+1} - S_{\alpha,s}}\frac{dS_\alpha}{dp_{\alpha/\beta}}\frac{dp_{\alpha/\beta}}{dp_\alpha}, \tag{9.2.2.18}$$

where

$$dp_{\alpha/\beta}/dp_\alpha = 1 \qquad \text{and} \qquad dp_{\alpha/\beta}/dp_\beta = -1.$$

In Fig. 9.11 we have presented the results of Settari and Aziz (1975) obtained for the test problem using the semiimplicit method with a chord slope. Notice that this method provides more accurate solutions than does the fully implicit Newton–Raphson approach.

9.2.3 The Implicit Pressure–Explicit Saturation Method (IMPES)

The IMPES method apparently originated with the papers of Sheldon *et al.* (1959) and Stone and Garder (1961). Although the method is normally identified with three-phase simulation, we shall consider herein the oil–

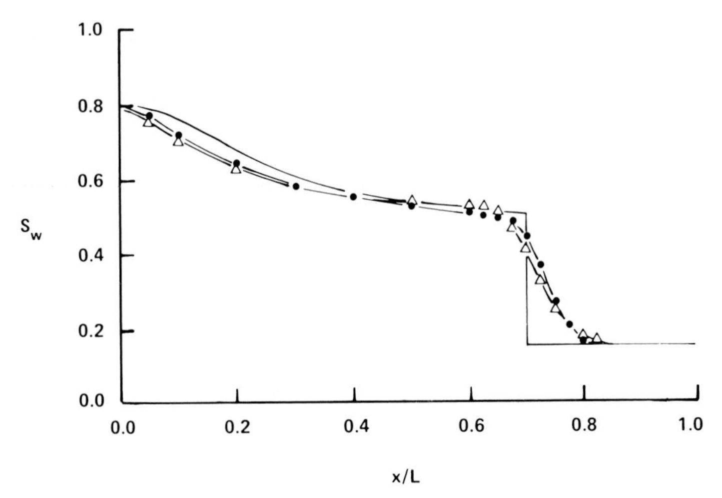

Fig. 9.11. Solution of the Buckley–Leverett problem obtained by Settari and Aziz (1975) using the semiimplicit scheme of Nolen and Berry (1972). Legend: •, $\Delta t = 100$ days; Δ, $\Delta t = 188$ days.

water example presented in earlier sections of this chapter. The idea of the IMPES method is to eliminate the saturation terms from the flow equations to obtain an equation that involves only one dependent variable (pressure or potential). The combined equation can then be solved and saturation computed by referring back to the original flow equations.

Let us assume that the capillary pressure $p_{o/w} = p_o - p_w$ associated with the spatial derivative does not change over a time step. This does not mean the pressures themselves are invariant but rather the difference between them is. This assumption yields the relation $\nabla_t p_w = \nabla_t p_o$, which, in combination with (9.2.2.3) and the definition $p \equiv p_o$, yields

$$\delta_{x_i}[a_{wi}(\delta_{x_i}p - \delta_{x_i}p_{o/w} + \rho_w g\delta_{x_i}z)] = \frac{d_{w1}}{\Delta t}\nabla_t p + \frac{d_{w2}}{\Delta t}\nabla_t S_w, \quad (9.2.3.1a)$$

$$\delta_{x_i}[a_{oi}(\delta_{x_i} p + \rho_o g\delta_{x_i}z)] = \frac{d_{o1}}{\Delta t}\nabla_t p + \frac{d_{o2}}{\Delta t}\nabla_t S_o, \quad (9.2.3.1b)$$

where

$$a_{\alpha i} = lk_{ii}\lambda_\alpha/(\Delta x_i)^2 B_\alpha, \qquad i = 1, 2,$$

$$d_{\alpha 1} = \frac{l}{\Delta t}\left[(S_\alpha\phi)_s \frac{d(1/B_\alpha)}{dp_\alpha} + \frac{S_{\alpha s}}{B_{\alpha s+1}}\frac{d\phi}{dp}\right],$$

$$d_{\alpha 2} = (\phi/B_\alpha)_{s+1}.$$

The objective now is to reduce (9.2.3.1) to one equation in pressure, taking advantage of the fact that the sum of the phase saturations is invariant. This is achieved by multiplying (9.2.3.1a) by d_{o2} and (9.2.3.1b) by d_{w2} and adding; there results

$$d_{o2}\delta_{x_i}[a_{wi}(\delta_{x_i} p - \delta_{x_i} p_{o/w} + \rho_w g \delta^{x_i} z)]$$

$$+ d_{w2}\delta_{x_i}[a_{oi}(\delta_{x_i} p + \rho_o g \delta_{x_i} z)] = \left(\frac{d_{o2}d_{w1}}{\Delta t} + \frac{d_{o1}d_{w2}}{\Delta t}\right) \nabla_t p. \quad (9.2.3.2)$$

This equation can be rewritten in a form similar to that encountered earlier for single-phase flow:

$$\delta_{x_i}[a_{wi}(\delta_{x_i} p + \rho_w g \delta_{x_i} z)] + \frac{d_{w2}}{d_{o2}} \delta_{x_i}[a_{oi}(\delta_{x_i} p + \rho_o g \delta_{x_i} z)]$$

$$= \frac{1}{\Delta t}\left(d_{w1} + \frac{d_{o1}d_{w2}}{d_{o2}}\right) \nabla_t p + \delta_{x_i}(a_{wi}\delta_{x_i} p_{o/w,s}). \quad (9.2.3.3)$$

Notice that we have employed the capillary pressure from the last time step in evaluating the right-hand side of (9.2.3.3).

Although (9.2.3.3) provides pressures, one must obtain estimates of the capillary pressure and saturation to proceed. The saturations are obtained by substitution of p into (9.2.3.1b). Having obtained $S_{\alpha s+1}$, one can obtain the updated capillary pressures from constitutive relations. The coefficients in (9.2.3.3) are nonlinear in this formulation and an iterative scheme is necessary to obtain a solution. Other formulations of the IMPES procedure are possible; some use different forms for the starting equations, and others consider different variables at the new time level.

There are a number of additional schemes that have been developed for the solution of the multiphase flow equations. For a discussion of the sequential solution method (SEQ) the reader is referred to Aziz and Settari (1979, p. 171) and for a presentation of the leapfrog method one may consult either Peaceman (1977, p. 154) or Thomas (1982, p. 127).

9.2.4 *Stability of Multiphase Algorithms*

It is evident from Fig. 9.10, which presents the linearized method solution for the problem, that conditional stability bounds can be expected in any of the explicit methods. Thus, the linearized method, the semiimplicit method, and the IMPES method should exhibit stability constraints. A fully nonlinear stability analysis is not possible at this time, and simplifying assumptions must be utilized. A discussion of these analyses can be found in Aziz and Settari (1979, Chapter 5) and in Peaceman (1977, Chapter 6). Even the Newton–Raphson SS method may not be convergent for a poor initial guess. Considering the complexity of most problems encountered in field applications, a heuristic stability analysis is often the only resort. Because of the uncertainty in the stability behavior of the nonlinear algorithms, there is a general tendency to employ the more computationally burdensome but more stable implicit algorithms.

9.2.5 Production Terms

To this point we have neglected the production terms in our governing equations. Mathematically, they are easily accommodated by adding the appropriate terms to the flow equations. Difficulty arises, however, when one must determine the relative production of each phase in a multiphase system. There are several approaches to this problem.

Explicit Production Method

Let us assume, as is often the case, that the total production rate is given as q_T. Because q_T is known, we can identify it with the new time level $s+1$, that is $q_{Ts+1} = q_T$. To determine the water and oil production rates we employ the mobility terms for the cell containing the well. Thus,

$$q_{\mathrm{w}s+1} = \left[\left(\frac{k_{\mathrm{rw}}}{\mu_{\mathrm{w}} B_{\mathrm{w}}}\right) \Big/ \left(\frac{k_{\mathrm{rw}}}{\mu_{\mathrm{w}} B_{\mathrm{w}}} + \frac{k_{\mathrm{ro}}}{\mu_{\mathrm{o}} B_{\mathrm{o}}}\right)\right]_s q_{\mathrm{T}s+1}, \tag{9.2.5.1a}$$

$$q_{\mathrm{o}s+1} = \left[\left(\frac{k_{\mathrm{ro}}}{\mu_{\mathrm{o}} B_{\mathrm{o}}}\right) \Big/ \left(\frac{k_{r\mathrm{w}}}{\mu_{\mathrm{w}} B_{\mathrm{w}}} + \frac{k_{r\mathrm{o}}}{\mu_{\mathrm{o}} B_{\mathrm{o}}}\right)\right]_s q_{\mathrm{T}s+1}. \tag{9.2.5.1b}$$

For many problems this explicit formulation is adequate. When the stability of the scheme is strained because small space increments or large time steps are employed, an oscillatory solution may be encountered. To circumvent this problem one can employ an implicit production scheme, which we shall now consider.

Implicit Production Method

We assume once again that the total production rate is specified. The water production can be determined from

$$q_{\mathrm{w}s+1} = \left[\left(\frac{k_{r\mathrm{w}}}{\mu_{\mathrm{w}} B_{\mathrm{w}}}\right) \Big/ \left(\frac{k_{r\mathrm{w}}}{\mu_{\mathrm{w}} B_{\mathrm{w}}} + \frac{k_{r\mathrm{o}}}{\mu_{\mathrm{o}} B_{\mathrm{o}}}\right)\right]_{s+1/2} q_{\mathrm{T}}. \tag{9.2.5.2}$$

Because the mobilities are unknown, an extrapolation procedure is used to provide an approximation of them. If we define the mobility as

$$M_\alpha = \left(\frac{k_{r\alpha}}{\mu_\alpha B_\alpha}\right) \Big/ \left(\frac{k_{rw}}{\mu_w B_w} + \frac{k_{ro}}{\mu_o B_o}\right),$$

we can utilize the relation

$$M_{w,s+1/2} = M_{w,s} + \tfrac{1}{2}\, \partial M_w / \partial S_w \big|_s (S_{ws+1} - S_{ws}). \qquad (9.2.5.3)$$

The term $\partial M_w / \partial S_w$ evaluated at time level s using a chord slope will normally suffice. Difficult problems, however, may require an iterative update. Note that this formulation involves the unknown saturations and therefore affects the structure of the difference approximation more than the explicit production method.

Allocation of Total Production Rates

In the previous formulation, we dealt with the case involving a well that penetrates only one grid block. This case is encountered if an areal simulation of the reservoir system is performed. However, if a vertical cross section or a three-dimensional simulation is performed, a single well may penetrate several grid blocks. In this case, it is necessary to employ a procedure for allocating the total well production rate to the individual grid blocks. Several procedures can be found in the literature. Perhaps the simplest and most common procedure is to allocate the total flow rate according to the total block mobility. For instance, if the well penetrates n grid blocks, the total flow rate from block r is given by

$$q_{Tr} = Q_T (\lambda_T)_r \Big/ \sum_{r=1}^{n} (\lambda_T)_r, \qquad r = 1, 2, \ldots, n \qquad (9.2.5.4)$$

where Q_T is the total well production rate, and $(\lambda_T)_r$ is the total mobility of grid block r. Note that this equation is based on the assumption of uniform grid blocks and isotropic permeability. When this assumption is not valid, a more complicated equation for computing q_{Tr} is required; this is given in Nolen and Berry (1972).

Allowance for Wellbore Boundary Conditions

Another practical aspect that has not been dealt with in our preceding discussion is the treatment of local boundary conditions in the wellbore. Such treatment is necessary if one is concerned with the well operation and performance of the well. For a description of techniques for incorporating the bottom hole pressure in the well into the finite difference computation, the reader is referred to Thomas (1982, pp. 153–159).

9.2.6 Pseudofunctions

Simulation of a three-dimensional problem using an algorithm for two space dimensions is an often-encountered problem. This procedure is motivated by the significant computational savings achieved through a reduction in dimensionality. To eliminate a dimension one formally integrates over it and redefines the equation parameters in terms of these integrals. The procedure was outlined in some detail in Section 4.2.1.

The most common and physically interesting problems are associated with vertical averaging. Implicit in this procedure is a loss of information concerning vertical variation in both the equation coefficients and the dependent variables. In an effort to capture some of this information, reservoir engineers have attempted to introduce known physical information into the vertical integration procedure. In the ensuing development we follow the vertical equilibrium modeling approach used by Coats *et al.* (1971) and Aziz and Settari (1979, p. 377).

Typical integrals encountered in the vertical integration procedure are

$$I_1 = \int_0^l k_{r\alpha}(S_\alpha)k_{ij}(x_i)\,dx_3, \tag{9.2.6.1a}$$

$$I_2 = \int_0^l S_\alpha(x_i)\phi(x_i)\,dx_3. \tag{9.2.6.1b}$$

To accommodate these integrals one first defines vertically averaged quantities

$$\overline{k}_{ij} = (1/l)\int_0^l k_{ij}(x_i)\,dx_3 \tag{9.2.6.2a}$$

and

$$\overline{\phi} = (1/l)\int_0^l \phi(x_i)\,dx_3. \tag{9.2.6.2b}$$

These functions are now modified to obtain a permeability-weighted vertically averaged relative permeability

$$\overline{k}_{r\alpha} = \int_0^l k_{ij}(x_i)k_{r\alpha}(S_\alpha)\,dx_3 \bigg/ \int_0^l k_{ij}(x_i)\,dx_3 \tag{9.2.6.3a}$$

and a porosity-weighted vertically averaged saturation

$$\overline{S}_{\alpha} = \int_0^l \phi(x_i)S(x_i)\,dx_3 \Bigg/ \int_0^l \phi(x_i)\,dx_3. \qquad (9.2.6.3b)$$

Let us consider now the equilibrium distribution of water and oil saturation wherein capillary forces have been neglected. The distribution of oil and water is given in Fig. 9.12a, where $S_{\alpha c}$ is the irreducible saturation of phase α. The initial vertically averaged saturation of the water phase is given by

$$\overline{S}_{wi} = \frac{[x_3(i) - 0] + (S_{wc})[l - x_3(i)]}{l}. \qquad (9.2.6.4)$$

When the water displaces the oil in a column of soil and the interface rises (Fig. 9.12b) the new vertically averaged equilibrium saturation is given by

$$\overline{S}_{wI} = \{[x_3(i) - 0] + (1 - S_{oc}) \cdot [x_3(I) - x_3(i)] + (S_{wc})[l - x_3(I)]\}/l. \qquad (9.2.6.5)$$

The corresponding relation for the drainage situation (Fig. 9.12c) is

$$\overline{S}_{wD} = \{(S_{wc})[l - x_3(D)] + x_3(D)\}/l. \qquad (9.2.6.6)$$

The next step is to utilize the constraint of pressure continuity at the interface. The pressure in each phase can be calculated relative to the interface pressure as follows:

$$p_{\alpha}(x_3) = p_{\alpha}(x_3^*) + \gamma_{\alpha}(x_3 - x_3^*), \qquad \alpha = \text{o, w}, \qquad (9.2.6.7)$$

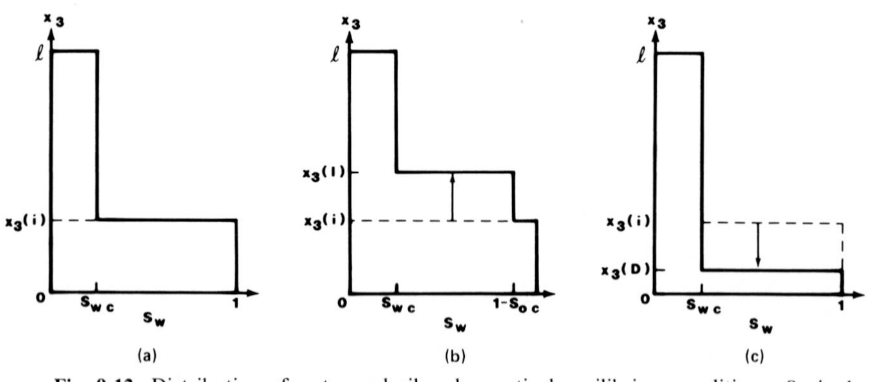

Fig. 9.12. Distribution of water and oil under vertical equilibrium conditions. $S_{\alpha c}$ is the irreducible saturation of phase α. The initial interface location is indicated by $x_3 = i$. The locations $x_3(I)$ and $x_3(D)$ indicate two new equilibrium states. (Adapted from Aziz and Settari, 1979.)

where x_3^* is the interface location. Substitution of α = o, w and subtraction of the result, keeping in mind that $p_o(x_3^*) = p_w(x_3^*)$, we obtain

$$p_o(x_3) - p_w(x_3) = \gamma_o(x_3 - x_3^*) - \gamma_w(x_3 - x_3^*). \qquad (9.2.6.8)$$

If we choose the top of the reservoir, that is, $x_3 = l$, as the reference frame for the areal calculations we have for that location

$$p_o(l) - p_w(l) = (\gamma_w - \gamma_o)[l - x_3^*]. \qquad (9.2.6.9)$$

Following Coats *et al.* (1971), we define

$$\bar{p}_{o/w} \equiv p_o(l) - p_w(l) = (\gamma_w - \gamma_o)[l - x_3^*], \qquad (9.2.6.10)$$

where $\bar{p}_{o/w}$ is called the "pseudocapillary pressure," although it is not related to capillarity whatsoever.

To obtain the final relations we substitute (9.2.6.10) into (9.2.6.5) and (9.2.6.6) with x_3^* set equal to $x_3(I)$ and $x_3(D)$, respectively. The desired equations take the form

$$\bar{S}_{wI} = -\frac{(1 - S_{oc} - S_{wc})\bar{p}_{o/w}}{l(\gamma_w - \gamma_o)} + 1 - S_{oc}\left[1 - \frac{x_3(i)}{l}\right] \qquad (9.2.6.11a)$$

and

$$\bar{S}_{wD} = 1 - [(1 - S_{wc})\bar{p}_{o/w}/l(\gamma_w - \gamma_o)]. \qquad (9.2.6.11b)$$

These saturation relations are used to supplement the vertically averaged (flow) equations.

The vertically averaged relative permeabilities are obtained by again utilizing Fig. 9.12 as the frame of reference. The appropriate relations are given by

$$\bar{k}_{rwI} = \{k_{rwro}[x_3(I) - x_3(i)] + [x_3(i)]\}/l, \qquad (9.2.6.12a)$$

$$\bar{k}_{rwD} = (1)[x_3(D)]/l, \qquad (9.2.6.12b)$$

where k_{rwro} is the relative permeability of water at residual oil saturation. The equations describing the vertically averaged relative permeabilities of oil can be established using a similar argument. The final expressions are obtained by substitution of (9.2.6.10) into these equations, which describe the vertically averaged relative permeabilities $\bar{k}_{r\alpha I}$, $\bar{k}_{r\alpha D}$.

Pseudofunction relations are also used for purposes other than the areal simulation of multiphase reservoirs. One of the common applications is in the simulation of well performance in an areal model. In this case one can employ a vertical model to generate pseudofunction relations for the relative permeability functions.

9.2.7 Potential Formulation

When the fluid density is considered to be a function only of pressure, it is possible to recast the preceding equations governing two-phase flow in terms of a fluid potential. The concept and procedure is analogous to that presented earlier for groundwater flow. One defines a potential function for each phase, that is,

$$\Phi_\alpha = \int_{p_{0\alpha}}^{p_\alpha} \frac{d\xi}{\rho(\xi)} + gz. \tag{9.2.7.1}$$

This function can be differentiated using Leibniz's rule to yield, for example,

$$\frac{\partial}{\partial x_i}\Phi_\alpha = \frac{1}{\rho}\frac{\partial}{\partial x_i}p_\alpha + \frac{\partial}{\partial x_i}gz. \tag{9.2.7.2}$$

A similar relation is obtained for the time derivative. These expressions can now be substituted in the governing equations, that is, (9.2.1.3), and a relation devoid of the gravity term results. Thereafter one can employ the development described in the preceding sections to develop a simulator.

9.2.8 Grid Orientation Effect

A numerical difficulty known as "grid orientation effect" is often encountered in using upstream weighting in conjunction with a five-point difference approximation to simulate fluid displacement in a confined reservoir via a five-spot well pattern. As illustrated in Fig. 9.13, the resulting two-dimensional flow problem can be solved using either a diagonal or a parallel grid. Unfortunately, experience indicates that under certain conditions, the two grids yield solutions that are quite different from each other, and that both of these solutions are grossly in error even though the grids may be very refined. Such a grid orientation phenomenon has been reported by many researchers, including Coats *et al.* (1974), Todd *et al.* (1972), and Yanosik and McCracken (1976). Generally, the solution sensitivity to grid orientation increases with increasing value of the unfavorable mobility ratio of the two fluid phases. It is most pronounced for a piston-type displacement. Increasing capillary pressure and gravity terms in the flow equations tend to decrease the grid orientation effect for some immiscible as well as miscible displacement problems. Several researchers have attempted to seek an improved solution using other types of numerical approximations. Yanosik and McCracken (1976) have introduced a nine-point finite difference approximation with upstream weighting and obtained a realistic prediction of unfavorable mobility ratio displacements. On the other hand, Settari *et al.* (1977) and Spivak *et al.* (1977) have obtained an accurate solution to similar types of displacement

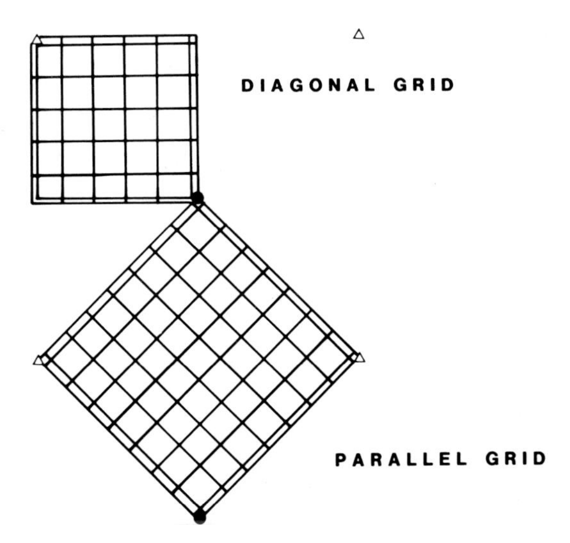

Fig. 9.13. Diagonal and parallel grids for a five-spot well flow problem. (Circles denote injection wells and triangles denote production wells.)

problems using a Galerkin finite element approximation with Hermitian basis functions. Another way to overcome the grid orientation effect is to use the upstream weighted finite element schemes that we have described in Chapter 4.

9.2.9 *Simulation of Three-Phase Flow*

In the previous sections, we focused on the simulation of two-phase flow problems. More specifically, we chose to examine the oil–water system as a special case. The mathematical model which incorporates the three phases–water, oil, and dissolved gas–with fluid properties determined by pressure is known as the "black oil simulator." The governing equations are an extension of those presented in (9.2.1.3), that is,

$$\frac{\partial}{\partial x_i}\left[\frac{k_{ij}\lambda_\alpha}{B_\alpha}\left(\frac{\partial p_\alpha}{\partial x_j} + \rho_\alpha g \frac{\partial z}{\partial x_j}\right)\right] = \frac{\partial}{\partial t}\left(\frac{\phi S_\alpha}{B_\alpha}\right), \qquad \alpha = \text{o, w}, \tag{9.2.9.1}$$

$$\frac{\partial}{\partial x_i}\left[\frac{k_{ij}\lambda_o R_s}{B_o}\left(\frac{\partial p_o}{\partial x_j} + \rho_o g \frac{\partial z}{\partial x_j}\right)\right] + \frac{\partial}{\partial x_i}\left[\frac{k_{ij}\lambda_g}{B_g}\left(\frac{\partial p_g}{\partial x_j} + \rho_g g \frac{\partial z}{\partial x_j}\right)\right] = \phi \frac{\partial}{\partial t}\left(\frac{S_g}{B_g} + \frac{S_o R_s}{B_o}\right), \tag{9.2.9.2}$$

where the subscript g designates the gas phase, $R_s = \rho_o/\rho_g\,|_s$ is the solution gas–oil ratio, and the subscript s indicates surface conditions. Although these equations are somewhat different than those we have already encountered, their solution does not entail any new concepts. Thus, we shall not detail the three-phase flow formulation but rather refer the reader to more specialized oil reservoir engineering texts, such as Aziz and Settari (1979), Crichlow (1977), Peaceman (1977), and Thomas (1982).

9.3 Simulation of Single-Phase Species Transport

9.3.1 Equation Formulation and Approximation

The species transport equations were developed in detail in Chapter 5. In this section, we examine the use of the finite difference method to solve these equations for practical field problems. To avoid excessive repetition we shall not consider all of the problems presented in Chapter 5. Rather we shall select the specific problem of species transport in a concentration dependent flow field.

Recall from Section 5.6.2 that the density dependent transport and flow equations can be written in the form

$$\frac{\partial}{\partial x_i}\left(\phi\rho D_{ij}\frac{\partial u}{\partial x_j}\right) + \frac{\rho k_{ij}}{\mu}\left(\frac{\partial p}{\partial x_j} + \rho g_j\right)\frac{\partial u}{\partial x_i} = \phi\rho\frac{\partial u}{\partial t} \tag{9.3.1.1a}$$

and

$$\frac{\partial}{\partial x_i}\left[\frac{\rho k_{ij}}{\mu}\left(\frac{\partial p}{\partial x_j} + \rho g_j\right)\right] = \frac{\partial}{\partial t}(\phi\rho), \tag{9.3.1.1b}$$

where u is the mass fraction of a specified species, often chlorides in the case of density-dependent flow. In writing (9.3.1.1) we have neglected source terms, radioactive decay, and adsorption for clarity of presentation. The components of the dispersion coefficient are given in Eq. (5.2.5.1).

The first step in numerical approximation is to expand the right-hand side of Eq. (9.3.1.1b) in terms of the dependent variables u and p. Let us assume the following functional dependence:

$$\phi = \phi(p), \tag{9.3.1.2a}$$

$$\rho = \rho(u, p). \tag{9.3.1.2b}$$

The right-hand side of (9.3.1.1b) becomes

$$\frac{\partial}{\partial t}(\phi\rho) = \rho\frac{\partial\phi}{\partial t} + \phi\frac{\partial\rho}{\partial t} = \rho\frac{d\phi}{dp}\frac{\partial p}{\partial t} + \phi\left(\frac{\partial\rho}{\partial p}\frac{\partial p}{\partial t} + \frac{\partial\rho}{\partial u}\frac{\partial u}{\partial t}\right). \tag{9.3.1.3}$$

In the selection of difference approximations to (9.3.1.3) one wishes to preserve the following consistency relation [see INTERCOMP (1976)]

$$\nabla_t(ab) \equiv (ab)_{s+1} - (ab)_s. \tag{9.3.1.4}$$

This is achieved using the approximation

$$\nabla_t(ab) = a_{s+1}\nabla_t b + b_{s+1}\nabla_t a, \tag{9.3.1.5}$$

which by convention can be written

$$\nabla_t(ab) = a\nabla_t b + b\nabla_t a. \tag{9.3.1.6}$$

Equations (9.3.1.3) can now be approximated using

$$\frac{\partial}{\partial t}(\phi\rho) = \rho\frac{d\phi}{dp}\nabla_t p + \phi\frac{\partial\rho}{\partial p}\nabla_t p + \frac{\partial\rho}{\partial u}\nabla_t u. \tag{9.3.1.7}$$

The backward difference formula of (9.3.1.7) is now combined with a central difference representation of the spatial terms in (9.3.1.1) to give

$$\frac{1}{\Delta x_i}\delta_{x_i}\left(\phi\rho D_{ij}\frac{\delta_{x_j}}{\Delta x_j}u\right) + \frac{\rho k_{ij}}{\mu}\left(\frac{\delta_{x_i}}{\Delta x_j}p + \rho g_j\right)\mu_{x_i}\frac{\delta_{x_i}}{\Delta x_i}u = \left(\frac{\phi\rho}{\Delta t}\right)\nabla_t u \tag{9.3.1.8a}$$

$$\frac{1}{\Delta x_i}\delta_{x_i}\left[\frac{\rho k_{ij}}{\mu}\left(\frac{\delta_{x_j}}{\Delta x_j} + \rho g_j\right)\right] = \left(\rho\frac{d\phi}{dp} + \phi\frac{\partial\rho}{\partial p}\right)\nabla_t p + \frac{\partial\rho}{\partial u}\nabla_t u, \tag{9.3.1.8b}$$

where μ_{x_i} is the averaging operator of Table 8.1.

Equations (9.3.1.8) represent $2N$ nonlinear equations in $2N$ unknowns, where N is the number of nodal points. Because the nonlinearities in these equations are relatively weak, they are easily accommodated using a Picard-type iteration procedure. It is advantageous, however, to recast the difference equations in terms of the incremental change over a time step to minimize round-off error during the calculations. The standard relation $\{p_{s+1}\} = \{p_s\} + \{\delta p\}$, $\{u_{s+1}\} = \{u_s\} + \{\delta u\}$ is substituted into (9.3.1.8) to obtain the desired algebraic expressions. The constitutive coefficients, for example, $\partial\rho/\partial p$, are assumed known from experimentally based relations. One can solve these equations either simultaneously or sequentially. Inasmuch as the nonlinearities in (9.3.1.8) generally require the use of an iterative procedure, a sequential solution approach, wherein one dependent variable is obtained per solution step, is normally more desirable than obtaining both $\{u\}$ and $\{p\}$ in one step.

The most popular finite difference algorithms are based upon the five-point molecule in two space dimensions and the seven-point molecule in three space dimensions. While these molecules are very well suited to the approximation of many partial differential equations encountered in mathematical physics, they are unable to represent adequately the dispersive term in (9.3.1.8a). This is readily seen if we expand the dispersive term over the j subscript as follows:

$$\frac{1}{\Delta x_i} \delta_{x_i}\left(\phi \rho D_{ij} \frac{\delta_{x_j}}{\Delta x_j} u \right) = \frac{1}{\Delta x_i} \delta_{x_i}\left(\phi \rho D_{i1} \frac{\delta_{x_1}}{\Delta x_1} u \right) + \frac{1}{\Delta x_i} \delta_{x_i}\left(\phi \rho D_{i2} \frac{\delta_{x_2}}{\Delta x_2} u \right). \tag{9.3.1.9}$$

Evidently, the cross derivative exists when the components D_{12} and D_{21} of the dispersion tensor are nonzero. Although it is often possible in the flow equation to eliminate this term through a judicious choice of coordinate directions, this is not generally possible for the dispersive term. Moreover, as we demonstrated in Section 8.1.3, this term cannot be represented by the same nodes used for the approximation of the other terms in (9.3.1.9). To include this term formally results in an increased number of nonzero off-diagonals in the coefficient matrix of the algebraic equations. This not only increases the computational effort required to obtain a solution, but also renders many of the more efficient equation solving schemes either less effective or inapplicable altogether. The obvious choice here is to lag the cross-derivative terms either one time increment or one iteration so that they become known quantities that can be transferred to the right-hand side of the algebraic equations. If the iterative alternative is chosen, Eq. (9.3.1.9) might read for $i = 1$

$$\frac{1}{\Delta x_1} \delta_{x_1} \left(\phi \rho D_{1j} \frac{\delta_{x_j}}{\Delta x_1} \right)^k u^{k+1} = \frac{1}{(\Delta x_1)^2} \delta_{x_1}\left(\phi \rho D_{11}\, \delta_{x_1} \right)^k u^{k+1}$$
$$+ \frac{1}{\Delta x_1} \delta_{x_1}\left(\phi \rho D_{12} \frac{\delta_{x_2}}{\Delta x_2} \right)^k u^k. \tag{9.3.1.10}$$

A similar equation can be derived for $i = 2$.

The approximation of the convective term in (9.3.1.1a) also warrants additional consideration. The central difference representation of Eq. (9.3.1.8a) is attractive because it minimizes numerical diffusion. On the other hand, numerical behavior that leads to solution oscillations is often a problem. This oscillatory behavior is eliminated when one selects a spatial increment which satisfies the equation

$$V_i\, \Delta x_i / D_{ii} \leqslant 2. \tag{9.3.1.11}$$

This constraint can prove inconvenient when large velocities or small dispersion coefficients are encountered. An alternative to satisfying (9.3.1.11) is to represent the first-order term using an upstream difference approximation as described earlier for the multiphase flow equations.

This approach would result in the following:

$$\frac{\rho k_{ij}}{\mu}\left(\frac{\partial p}{\partial x_j} + \rho g_j\right)\frac{\partial u}{\partial x_i} \simeq \left[\frac{\rho k_{ij}}{\mu}\left(\frac{\delta_{x_j}}{\Delta x_j} p + \rho g_j\right)\right]_{r_i - 1} \frac{\nabla_{x_i}}{\Delta x_i} u.$$

Although this approximation will eliminate numerical oscillations, it will generate numerical diffusion. An attractive alternative is to utilize a weighted difference (combined upstream and central) approximation to $\partial u/\partial x_i$ so that there is just enough asymmetry to damp oscillations but not so much that numerical diffusion becomes a serious problem.

9.4 Simulation of Multiphase Species Transport

The simulation of multiphase species transport wherein species can move from one phase into another is generally regarded as one of the most challenging oil–reservoir compositional modeling problems. Enhanced oil recovery using a carbon dioxide flood is one example of an important oil reservoir engineering problem described by the multiphase species transport equations. In groundwater hydrology, chlorinated hydrocarbons introduced accidentally into the subsurface may either move in low concentrations as dissolved constituents in the water phase or as a separate organic phase of low solubility or both. While this particular problem has yet to be simulated, it also is described by the multiphase species transport equations.

9.4.1 Governing Equations

The point of departure in developing the multiphase species transport equations is (5.2.1.1). Let us write this equation for species k in a three-phase gas–oil–water (g–o–w) system. Following established practice in the oil industry we neglect dispersive effects and sum the equations for each phase: there results

$$-\partial(n^{\mathrm{o}}\rho^{\mathrm{o}}\omega_k^{\mathrm{o}}v_i^{\mathrm{o}} + n^{\mathrm{g}}\rho^{\mathrm{g}}\omega_k^{\mathrm{g}}v_i^{\mathrm{g}} + n^{\mathrm{w}}\rho^{\mathrm{w}}\omega_k^{\mathrm{w}}v_i^{\mathrm{w}})/\partial x_i$$
$$= \partial(n^{\mathrm{o}}\rho^{\mathrm{o}}\omega_k^{\mathrm{o}} + n^{\mathrm{g}}\rho^{\mathrm{g}}\omega_k^{\mathrm{g}} + n^{\mathrm{w}}\rho^{\mathrm{w}}\omega_k^{\mathrm{w}})/\partial t, \qquad (9.4.1.1)$$

where ω_k^α is the mass fraction of species k in phase α, α = g, o, w. One can now introduce Darcy's law written for each phase velocity v_i^α, that is,

$$v_i^\alpha = -\frac{k_{ij}k_r^\alpha}{n^\alpha\mu^\alpha}\left(\frac{\partial p^\alpha}{\partial x_j} + \rho^\alpha g_j\right). \qquad (9.4.1.2)$$

Substituting (9.4.1.2) into (9.4.1.1) and ϕS^α for n^α, we obtain

$$\frac{\partial}{\partial x_i}\left\{\frac{k_{ij}}{\mu}\left[\rho^{\mathrm{o}}\omega_k^{\mathrm{o}}k_{\mathrm{r}}^{\mathrm{o}}\left(\frac{\partial p^{\mathrm{o}}}{\partial x_j}+\rho^{\mathrm{o}}g_j\right)+\rho^{\mathrm{g}}\omega_k^{\mathrm{g}}k_{\mathrm{r}}^{\mathrm{g}}\left(\frac{\partial p^{\mathrm{g}}}{\partial x_j}+\rho^{\mathrm{o}}g_j\right)\right.\right.$$
$$\left.\left.+\rho^{\mathrm{w}}\omega_k^{\mathrm{w}}k_{\mathrm{r}}^{\mathrm{w}}\left(\frac{\partial p^{\mathrm{w}}}{\partial x_j}+\rho^{\mathrm{w}}g_j\right)\right]\right\} \tag{9.4.1.3}$$
$$=\frac{\partial}{\partial t}[\phi(S^{\mathrm{o}}\rho^{\mathrm{o}}\omega_k^{\mathrm{o}}+S^{\mathrm{g}}\rho^{\mathrm{g}}\omega_k^{\mathrm{g}}+S^{\mathrm{w}}\rho^{\mathrm{w}}\omega_k^{\mathrm{w}})].$$

At this point we have, for η components, the unknown variables listed in Table 9.1. Also listed in this table are the available constitutive relations. To close the system of equations one must impose the following additional constraints:

$$\sum_k \omega_k^\alpha = 1, \qquad \alpha = \mathrm{o, w, g} \tag{9.4.1.4}$$

$$f_i^{\mathrm{o}}(\omega_k^{\mathrm{o}}, p^{\mathrm{o}}) = f_i^{\mathrm{w}}(\omega_k^{\mathrm{w}}, p^{\mathrm{w}}), \qquad i = 1, \ldots, \eta, \quad 1 \le k \le \eta, \tag{9.4.1.5a}$$

$$f_i^{\mathrm{w}}(\omega_k^{\mathrm{w}}, p^{\mathrm{w}}) = f_i^{\mathrm{g}}(\omega_k^{\mathrm{g}}, p^{\mathrm{g}}), \qquad i = 1, \ldots, \eta, \quad 1 \le k \le \eta, \tag{9.4.1.5b}$$

where f_i^α is the fugacity of species i in phase α. The 2η equations generated by (9.4.1.5) describe the instantaneous thermodynamic phase equilibrium of all species between all phases. A tally shows that we now have $3\eta + 15$ equations in $3\eta + 15$ unknowns.

Although the equation-of-state method for computing phase equilibria appears to be the method of choice at this time [see, for example, Coats (1980), Fussell (1979), Fussell and Fussell (1979), and Mehra *et al.* (1982)], an alternative formulation based on equilibrium constants is possible. In this approach, the phase equilibrium is described by experimentally de-

Table 9.1
Unknown Variables and Constitutive Relation for Compositional Simulator

Variable	*Number of unknowns*	*Constitutive relation*	*Number of equations*
ω_k^α	3η	Governing equations	η
p^α	3		
S^α	3	$S^\alpha = S^\alpha(p^\alpha, p^\beta)$	3
ρ^α	3	$\rho^\alpha = \rho^\alpha(p^\alpha, \omega_k^\alpha)$	3
μ^α	3	$\mu^\alpha = \mu^\alpha(p^\alpha, \omega_k^\alpha)$	3
k_r^α	3	$k_r^\alpha = k_r^\alpha(S^\beta)$, $\beta = o, g, w$	3
	$3\eta + 15$		$\eta + 12$

termined equilibrium constants. Typical relations used in lieu of (9.4.1.5) are

$$\omega_k^{g}/\omega_k^{o} = K_k^{go}(p_g, p_o, \omega_k^{g}, \omega_k^{o}) \quad (9.4.1.6a)$$

and

$$\omega_k^{g}/\omega_k^{w} = K_k^{gw}(p_g, p_o, \omega_k^{g}, \omega_k^{w}), \quad (9.4.1.6b)$$

where K_k^{go} and K_k^{gw} are equilibrium constants.

The numerical solution of (9.4.1.3)–(9.4.1.5) borrows heavily from our earlier discussion of multiphase flow simulation. The approximation methods developed therein are employed and augmented by methods designed to solve the phase equilibrium and state equations. The details of this procedure are beyond the scope of this work and the interested reader is referred to Coats (1980). To date, problems involving species transport in multiphase flow appear to have been simulated only using finite difference methods, although the use of Galerkin and higher-order accurate collocation methods may hold considerable promise as a practical alternative.

9.5 Summary

Problems of mass and energy transport in porous media are among the most difficult, challenging, and important encountered by the reservoir engineer or hydrologist. Because of their inherent mathematical complexity, they generally have been treated using relatively simple approaches based on finite difference theory. Beyond the scope of this book is the exceedingly interesting class of problems involving chemical reactions. Particularly challenging are those that require a detailed representation of reaction kinetics. More complex physical systems and more sophisticated numerical methodology will continue to provide significant challenges to the analyst for many years to come.

References

Aziz, K. and Settari, A. (1979). "Petroleum Reservoir Simulation." Applied Science Publishers, London.

Bjordamman, J. and Coats, K. H. (1969). Comparison of alternating-direction and successive over-relaxation techniques in simulation of reservoir fluid flow. *Soc. Pet. Eng. J.* **9,** 47–58.

Breitenback, E. A., Thurnau, D. H., and van Poolen H. K. (1969). "Solution of immiscible fluid flow simulation equations," *Soc. Pet. Eng. J.* **9,** 155–169.

Coats, K. H., Nielsen, R. L., Terhund, M. H., and Weber, A. G. (1967). Simulation of three-dimensional, two phase flow in oil and gas reservoirs. *Soc. Pet. Eng. J.* **7,** 377–388.

Coats, K. H., Dempsey, J. R., and Henderson, J. H. (1971). The use of vertical equilibrium in two-dimensional simulation of three-dimension reservoir performance. *Trans. Soc. Pet. Eng.* **251,** 63–71.

Coats, K. H., George, W. D., and Marcum, B. E. (1974). Three-dimensional simulation of steamflooding. *Soc. Pet. Eng. J.* **15,** 573–592.

Coats, K. H. (1980). An equation of state compositional model. *Soc. Pet. Eng. J.* **20,** 363–376.

Crichlow, H. B. (1977). "Modern Reservoir Engineering, A Simulation Approach." Prentice-Hall, Englewood Cliffs, New Jersey.

Cuthill, E. (1972). Several strategies for reducing the band-width of matrices. *In.* "Sparse Matrices and Their Applications," D. J. Rose and R. A. Willoughby, eds., pp. 157–166. Plenum Press, New York.

Douglas, J. (1962). Alternating direction methods for three space variables. *Numer. Math.* **4,** 41–63.

Douglas, J., and Rachford, H. H. (1956). On the numerical solution of heat conduction problems in two and three space variables. *Trans. Am. Math. Soc.* **82,** 421–439.

Douglas, J., Peaceman, D. W., and Rachford, H. H. (1959). A method for calculating multi-dimensional immiscible displacement. *Trans. Am Inst. Min. Metall. Pet. Eng.* **216,** 297–308.

Fussell, L. T. (1979). A technique for calculating multiphase equilibria. *Soc. Pet. Eng. J.* **19** (4), 203–210.

Fussell, L. T., and Fussell, D. D. (1979). An iterative technique for compositional reservoir models, *SPEJ AIME* **19** (4), 211–220.

George, A. (1973). Nested dissection of a regular finite element mesh. *SIAM J. Numer. Anal.* **10** (2), 345–363.

Green, D. W., Dabiri, H., Weinaug, C. F., and Prill, R. (1971). Numerical modeling of unsaturated groundwater flow and comparison of the model to a field experiment. *Water Resour. Res.* **6** (3), 862–874.

INTERCOMP Resources Development and Engineering, Inc. (1976). A model for calculating effects of liquid waste disposal in deep saline aquifers. U.S. Geological Survey, Water Resources Investigations 76–61.

Jennings, A. (1971). Solution of variable bandwidth positive definite simultaneous equations. *Comput. J.* **14** (4).

Mehra, R. K., Heidemann, R. A., and Aziz, K. (1982). Computation of Multiphase Equilibrium for Compositional Simulation. *SPEJ AIME* **221,** 61–68.

Nolen, J. S., and Berry, D. W. (1972). Tests of the stability and time-step sensitivity of semi-implicit reservoir simulation techniques. *Trans. SPE AIME* **253,** 253–266.

Peaceman, D. W. (1967). Numerical solution of the nonlinear equations for two-phase flow through porous media. *In* "Nonlinear Partial Differential Equations—Methods of Solutions," W. F. Ames, ed., 79–91. Academic Press, New York.

Peaceman, D. W. (1977). "Fundamentals of Numerical Reservoir Simulation." Elsevier, Amsterdam.

Peaceman, D. W., and Rachford, H. H. (1955). The Numerical Solution of Parabolic and Elliptic Differential Equations. *SIAM J. Appl. Math* **3,** 28–41.

Price, H. S., and Coats, K. H. (1974). Direct Methods in Reservoir Simulation. *Soc. Pet. Eng. J.* **14,** 295–308.

Remson, I., Hornberger, G. M., and Moltz, F. J. (1971). "Numerical Methods in Subsurface Hydrology." Wiley-Interscience, New York.

Segui, N. T. (1975). Computer programs for the solutions of systems of linear algebraic equations. Contractor Report No. NASA CR-2173, U.S. National Aeronautics and Space Administration.

Settari, A., and Aziz, K. (1974). A generalization of the additive correction methods for the iterative solution of matrix equations. *SIAM J. Numer. Anal.* **14,** 221–236.

Settari, A., and Aziz, K. (1975). Treatment of nonlinear terms in the numerical solution of partial differential equations for multiphase flow in porous media. *Int. J. Multiphase Flow,* **1,** 817–844.

Settari, A. Price, H. S., and Dupont, T. (1977). Development and application of variational methods for simulation of miscible displacement in porous media. *Soc. Pet. Eng. J.* **17,** 228–246.

Sheldon, J. W., Zondek, B., and Cardwell, W. T. (1959). One-dimensional incompressible non-capillary, two-phase fluid flow in a porous medium. *Trans. Soc. Pet. Eng. AIME* **216,** 290–296.

Spivak, A., Price, H. S., and Settari, A. (1977). Solution of the equations for multi-dimensional two-phase immiscible flow by variational methods. *Soc. Pet. Eng. J.* **17** (1), 27–41.

Stone, H. L. (1968). Iterative solution of implicit approximations of multi-dimensional partial differential equations. *SIAM J. Numer. Anal.* **5,** 530–558.

Stone, H. L., and Garder, A. O. Jr. (1961). Analysis of gas-cap or dissolved-gas reservoirs. *Trans. Soc. Pet. Eng. AIME* **222,** 92–104.

Thomas, G. W. (1982). "Principles of Hydrocarbon Reservoir Simulation." International Human Resources Development Corp., Boston.

Todd, M. R., O'Dell, P. M., and Hirasaki, G. J. (1972). Methods for increased accuracy in numerical reservoir simulators. *Soc. Pet. Eng. J.* **12,** 515–530.

Trescott, P. C., Pinder, G. F., and Larson, S. P. (1976). "Finite-Difference Model for Aquifer Simulation in Two Dimensions with Results of Numerical Experiments." Book 7, ch. C1. U.S. Geological Survey.

Watts, J. W. (1971). An iterative matrix solution method suitable for anisotropic problems. *Soc. Pet. Eng. J.* **11,** 47–51.

Watts, J. W. (1973). A method for improving line successive overrelaxation in anisotropic problems—A theoretical analysis. *Soc. Pet. Eng. J.* **13,** 105–118.

Weinstein, H. G., Stone, H. L., and Kwan, T. V. (1969). An iterative procedure for solution of systems of parabolic and elliptic equations in three dimensions. *Ind. Eng. Chem. Fundam.* **8,** 281–287.

Yanosik, J. L., and McCracken, T. A. (1976). A nine-point finite difference reservoir simulator for realistic prediction of unfavorable mobility ratio displacements. 4th Symposium on Numerical Simulation of Reservoir Performance, Paper No. SPE 5734, 209–230.

10

Alternative Finite Difference Methods

10.1 Introduction

In this chapter, we present alternative finite difference methods for solving specific subsurface flow problems. These methods are divided into two groups. In the first group some type of transformation of variables is employed to reduce governing partial differential equations or their solution regions into a form suitable for efficient solution by standard finite difference methods. In the second group, the method of characteristics is used as the means to convert time dependent hyperbolic or parabolic equations to simple differential equations, which are then solved by explicit time-stepping schemes. Although the methods to be described in the following sections have a limited scope of application, they are usually more cost effective than the conventional finite difference and finite element methods. For some problems the solutions produced by the latter are not only more expensive but substantially less accurate than those obtained by the proposed alternative approach.

10.2 Finite Difference Methods by Transformations

Three types of transformation are presented. These include the similarity, Kirchhoff, and conformal transformations. The first two are used in solving unsaturated water flow problems. The conformal transformation is used in solving steady-state free surface groundwater flow problems.

10.2.1 Similarity Transformation

Let us consider the problem of one-dimensional unsaturated flow in soils. An analytical solution to this problem is made difficult by the nonlinear form of the material coefficients. In the case of horizontal flow (adsorption), the governing equation can be converted to a simpler nonlinear ordinary differential equation by using the Boltzmann similarity transformation. Numerical solution of the resulting ordinary differential equation reduces to a straightforward step-by-step integration.

In the case of vertical flow (infiltration), the governing equation contains a gravity term and must be treated by a combined application of perturbation and similarity approximations.

Although the discussion presented in this subsection is limited to one-dimensional flow problems, it should be realized that the solution approach can be readily extended to handle radial flow and some simple cases of two-dimensional and three-dimensional flow. Additional solution examples can be found in Drake *et al.* (1969) and Philip (1969).

(1) Horizontal Unsaturated Flow

The governing equation for this case was presented in Chapter 4. It may be written in the form

$$\frac{\partial}{\partial x}\left(D\frac{\partial\theta}{\partial x}\right) = \frac{\partial\theta}{\partial t}, \tag{10.2.1.1}$$

where θ is moisture content and D is the diffusion coefficient. The coefficient D is a nonlinear function of θ and is defined as

$$D = Kk_r\, d\psi/d\theta,$$

where K is the saturated hydraulic conductivity, k_r the relative permeability, and ψ the capillary pressure head. Equation (10.2.1.1) describes moisture movement in a tube of soil as is depicted in Fig. 10.1. The initial and

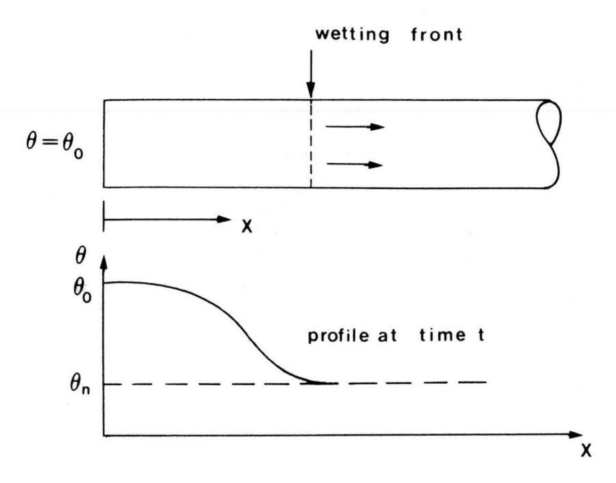

Fig. 10.1. One-dimensional adsorption in a soil tube.

boundary conditions are given by

$$\theta(x, 0) = \theta_n, \tag{10.2.1.2a}$$

$$\theta(0, t) = \theta_o, \tag{10.2.1.2b}$$

$$\theta(\infty, t) = \theta_n. \tag{10.2.1.2c}$$

Philip (1955) solved the problem by introducing the Boltzmann transformation of the form

$$\phi = x/t^{1/2}. \tag{10.2.1.3}$$

Using the chain rule, the derivatives in Eq. (10.2.1.1) can be converted as follows:

$$\frac{\partial\theta}{\partial t} = \frac{d\theta}{d\phi}\frac{\partial\phi}{\partial t} = -\frac{xt^{-3/2}}{2}\frac{d\theta}{d\phi}, \tag{10.2.1.4a}$$

$$\frac{\partial\theta}{\partial x} = \frac{d\theta}{d\phi}\frac{\partial\phi}{\partial x} = t^{-1/2}\frac{d\theta}{d\phi}, \tag{10.2.1.4b}$$

$$\frac{\partial^2\theta}{\partial x^2} = \frac{d}{d\phi}\left(t^{-1/2}\frac{d\theta}{d\phi}\right)\frac{\partial\phi}{\partial x} = t^{-1}\frac{d^2\theta}{d\phi^2}, \tag{10.2.1.4c}$$

$$\frac{\partial D}{\partial x} = \frac{dD}{d\phi}\frac{\partial\phi}{\partial x} = t^{-1/2}\frac{dD}{d\phi}. \tag{10.2.1.4d}$$

Substitution of (10.2.1.4a)–(10.2.1.4d) into (10.2.1.1) yields

$$Dt^{-1}\frac{d^2\theta}{d\phi^2} + t^{-1}\frac{d\theta}{d\phi}\frac{dD}{d\phi} = -\tfrac{1}{2}\phi t^{-1}\frac{d\theta}{d\phi},$$

or

$$\frac{d}{d\phi}\left(D\frac{d\theta}{d\phi}\right) = -\frac{\phi}{2}\frac{d\theta}{d\phi}. \tag{10.2.1.5}$$

Before solving Eq. (10.2.1.5), we must obtain the corresponding initial and boundary conditions from Eqs. (10.2.1.2a)–(10.2.1.2c). These conditions are as follows:

$$\theta(\phi = \infty) = \theta_n \quad \text{or} \quad d\theta\,(\phi = \infty)/d\phi = 0, \tag{10.2.1.6a}$$

$$\theta(\phi = 0) = \theta_o. \tag{10.2.1.6b}$$

A suitable form of Eq. (10.2.1.5) for numerical approximation is obtained by integrating once with respect to ϕ and applying the condition

$$d\theta\,(\phi = \infty)/d\phi = 0,$$

$$\int_{\theta_n}^{\theta} \phi \, d\theta = -2D \, d\theta/d\phi, \qquad (10.2.1.7a)$$

which, when the upper limit of integration is set equal to $\theta_{i+1/2}$, becomes

$$\int_{\theta_n}^{\theta_{i+1/2}} \phi \, d\theta = -2D_{i+1/2} \, d\theta/d\phi. \qquad (10.2.1.7b)$$

Eq. (10.2.1.7) is a first-order integrodifferential equation that we now solve subject to the conditions (10.2.1.6a) and (10.2.1.6b). First we develop its finite difference approximation by subdividing the interval θ_o to θ_n into n equal increments as depicted in Fig. 10.2. Thus, $\theta_i = \theta_o - i\,\Delta\theta$ for $0 \leqslant i \leqslant n$, where $\Delta\theta = (\theta_o - \theta_n)/n$. The left-hand term in Eq. (10.2.1.7b) is simply the area bounded by the $\phi - \theta$ curve and the lines $\phi = \theta_n$ and $\theta = \theta_{i+1/2}$. This area can be approximated by

$$\int_{\theta_n}^{\theta_{i+1/2}} \phi \, d\theta = \int_{\theta_n}^{\theta_i} \phi \, d\theta + \int_{\theta_i}^{\theta_{i+1/2}} \phi \, d\theta \simeq \int_{\theta_n}^{\theta_i} \phi \, d\theta - \phi_i \, \Delta\theta/2. \quad (10.2.1.8)$$

From Eqs. (10.2.1.7b) and (10.2.1.8) it follows that

$$\begin{aligned} \int_{\theta_n}^{\theta_i} \phi \, d\theta - \phi_i \, \Delta\theta/2 &= -2D_{i+1/2} \frac{d\theta}{d\theta} \\ &\simeq 2D_{i+1/2} \, \Delta\theta/(\phi_{i+1} - \phi_i), \end{aligned} \qquad (10.2.1.9)$$

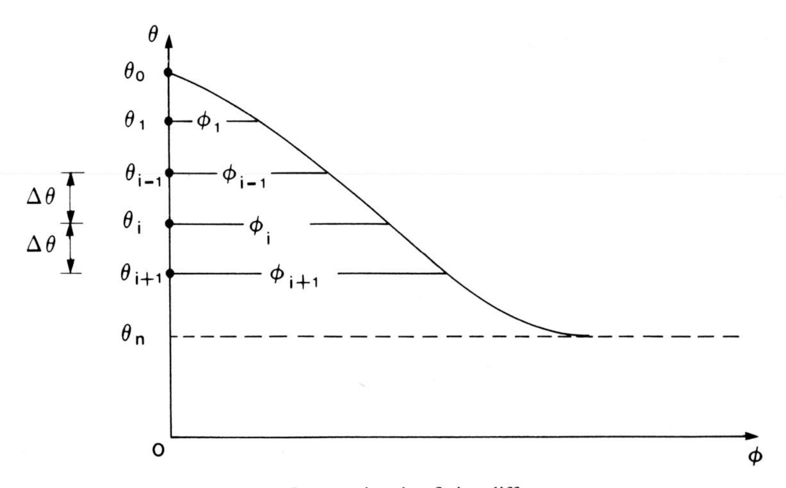

Fig. 10.2. Integration by finite differences.

where

$$D_{i+1/2} \simeq \tfrac{1}{2}(D_i + D_{i+1}). \tag{10.2.1.10}$$

Equation (10.2.1.9) may be rearranged to give

$$\phi_{i+1} - \phi_i = \frac{2D_{i+1/2}}{(1/\Delta\theta)\left(\int_{\theta_n}^{\theta_i} \phi\, d\theta\right) - (\phi_i/2)}. \tag{10.2.1.11}$$

If we change i to $i - 1$, then Eq. (10.2.1.11) becomes

$$\phi_i - \phi_{i-1} = \frac{2D_{i-1/2}}{(1/\Delta\theta)\left(\int_{\theta_n}^{\theta_{i-1}} \phi\, d\theta\right) - (\phi_{i-1}/2)}, \tag{10.2.1.12}$$

where the integral can be approximated by

$$\int_{\theta_n}^{\theta_{i-1}} \phi\, d\theta = \int_{\theta_n}^{\theta_i} \phi\, d\theta + \int_{\theta_i}^{\theta_{i-1}} \phi\, d\theta \simeq \int_{\theta_n}^{\theta_i} \phi\, d\theta + (\phi_i + \phi_{i-1})\, \Delta\theta/2. \tag{10.2.1.13}$$

Substituting Eq. (10.2.1.13) into (10.2.1.12), we obtain

$$\phi_i - \phi_{i-1} = \frac{2D_{i-1/2}}{(1/\Delta\theta)\left(\int_{\theta_n}^{\theta_i} \phi\, d\theta\right) + (\phi_i/2)}. \tag{10.2.1.14}$$

Equations (10.2.1.11) and (10.2.1.14) may be written in the form

$$\phi_{i+1} - \phi_i = 2D_{i+1/2}/I_{i+1/2}, \tag{10.2.1.15}$$

$$\phi_i - \phi_{i-1} = 2D_{i-1/2}/I_{i-1/2}, \tag{10.2.1.16}$$

where

$$I_{i+1/2} = \frac{1}{\Delta\theta}\int_{\theta_n}^{\theta_i} \phi\, d\theta - \frac{\phi_i}{2}, \tag{10.2.1.17a}$$

$$I_{i-1/2} = \frac{1}{\Delta\theta}\int_{\theta_n}^{\theta_i} \phi\, d\theta + \frac{\phi_i}{2}. \tag{10.2.1.17b}$$

From Eqs. (10.2.1.17a) and (10.2.1.17b), we obtain the following recurrence relation:

$$I_{i+1/2} = I_{i-1/2} - \phi_i. \tag{10.2.1.18}$$

For computational purposes, we rewrite Eq. (10.2.1.15) as

$$\phi_{i+1} = \phi_i + 2D_{i+1/2}/I_{i+1/2}. \tag{10.2.1.19}$$

Based on Eqs. (10.2.1.18) and (10.2.1.19), a general iterative solution procedure may be described as

(i) Tabulate $D_{i+1/2}$ from the known D versus θ function.
(ii) Make an initial estimate of the value of $I_{1/2} = (1/\Delta\theta) \int_{\theta_n}^{\theta_o} \phi \, d\theta$ and use Eq. (10.2.1.19) and the initial condition $\phi_o = 0$ to compute ϕ_1.
(iii) Knowing $I_{1/2}$ we can compute $I_{3/2}$ from Eq. (10.2.1.18).
(iv) Continue the alternating use of Eqs. (10.2.1.19) and (10.2.1.18) to evaluate ϕ_i and $I_{i+1/2}$ for $2 < i < n - 1$.
(v) Compare the value of $I_{n-1/2}$ given by this procedure with the analytical value $\hat{I}_{n-1/2}$ determined as described later.
(vi) Repeat steps (ii)–(v) using improved estimates of $I_{1/2}$ until $|\hat{I}_{n-1/2} - I_{n-1/2}|$ is sufficiently small that final adjustments can be made by interpolation.

To obtain the analytical expression for $\hat{I}_{n-1/2}$ referred to in (v), we assume that θ_{n-1} is sufficiently close to θ_n, so that D in this range may be regarded as constant. If this assumption is valid, then an analytical solution for θ_{n-1} may be obtained by solving the equation

$$D_{n-1/2}\, \partial^2\theta/\partial x^2 = \partial\theta/\partial t \tag{10.2.1.20}$$

subject to conditions

$$\theta(x, t) = \theta_n \qquad \text{for} \quad x > 0, \quad t = 0, \tag{10.2.1.21a}$$

$$\lim_{x\to\infty} \theta(x, t) = \theta_n, \qquad \text{for} \quad t \geqslant 0, \tag{10.2.1.21b}$$

$$\theta(x, t) = \theta_{n-1}, \qquad \text{for} \quad x = \phi_{n-1}t^{1/2}, \quad t > 0. \tag{10.2.1.21c}$$

The solution is given by (see Kirkham and Powers, 1972)

$$\theta = \frac{\Delta\theta}{\operatorname{erfc}\{\phi_{n-1}/[2(D_{n-1/2})^{1/2}]\}} \operatorname{erfc}\left[\frac{x}{2(tD_{n-1/2})^{1/2}}\right] + \theta_n.$$

This can be expressed in the form

$$[(\theta - \theta_n)/\Delta\theta] \operatorname{erfc}\{\phi_{n-1}/[2(D_{n-1/2})^{1/2}]\} = \operatorname{erfc}\{\phi/[2(D_{n-1/2})^{1/2}]\}$$

or

$$\phi = 2(D_{n-1/2})^{1/2} \operatorname{inverfc}\left\{[(\theta - \theta_n)/\Delta\theta] \operatorname{erfc}[\phi_{n-1}/2(D_{n-1/2})^{1/2}]\right\}, \tag{10.2.1.21d}$$

where the notation inverfc is used to denote the inverse of the complementary error function. To obtain $\hat{I}_{n-1/2}$, we employ Eq. (10.2.1.17a) and replace i by $n - 1$. Thus

$$\hat{I}_{n-1/2} = \frac{1}{\Delta\theta} \int_{\theta_n}^{\theta_{n-1}} \phi \, d\theta - \frac{\phi_{n-1}}{2}. \tag{10.2.1.22a}$$

It can be shown (see Kirkham and Powers, 1971) that combination of Eqs. (10.2.1.22a) and (10.2.1.21d) leads to

$$\hat{I}_{n-1/2} = \frac{\phi_{n-1}}{2} + \frac{2D_{n-1/2}}{\phi_{n-1}} A\left[\frac{\phi_{n-1}}{2(D_{n-1/2})^{1/2}}\right], \tag{10.2.1.22b}$$

where

$$A(x) = [2\pi^{-1/2} x \exp(-x^2)/\operatorname{erfc}(x)] - 2x^2.$$

The function $A(x)$ can either be computed numerically or obtained from tables. Typical values are listed in Table 10.1. For $x > 3$, $A(x)$ can be computed from the following asymptotic expansion:

$$A = 1 - \frac{2}{\lambda} + \frac{10}{\lambda^2} - \frac{74}{\lambda^3} + \frac{706}{\lambda^4} - \frac{8162}{\lambda^5} + \frac{108830}{\lambda^6} - \cdots,$$

where $\lambda = 2x^2$.

To speed up convergence of the nonlinear iteration, it is desirable to obtain a good initial estimate of $I_{1/2}$. This is achieved by solving Eq. (10.2.1.1) subject to Eqs. (10.2.1.2a)–(10.2.1.2c) and using a constant

Table 10.1
Typical Values of Function $A(x)$

x	A	x	A	x	A
0.0	0.000	1.5	0.763	5.0	0.964
0.2	0.199	2.0	0.836	6.0	0.974
0.4	0.353	2.5	0.881	7.0	0.981
0.6	0.472	3.0	0.911	8.0	0.985
0.8	0.566	3.5	0.930	9.0	0.988
1.0	0.639	4.0	0.946	10.0	0.990

value of diffusivity given by

$$\tilde{D} = 2\int_{\theta_n}^{\theta_o} D(\theta - \theta_n)\, d\theta/(\theta_o - \theta_n)^2.$$

The solution takes the form

$$\theta - \theta_n = (\theta_o - \theta_n)\, \mathrm{erfc}[\phi/(2\tilde{D}^{1/2})]. \tag{10.2.1.23}$$

Using Eq. (10.2.1.17a) with $i = 0$ and noting that $\phi_o = 0$, we obtain

$$I_{1/2} = \frac{1}{\Delta\theta}\int_{\theta_n}^{\theta_o} \phi\, d\theta = \frac{1}{\Delta\theta}\int_0^{\infty} (\theta - \theta_n)\, d\phi. \tag{10.2.1.24a}$$

Substitution of Eq. (10.2.1.23) into (10.2.1.24a) yields the following first estimate of $I_{1/2}$:

$$(I_{1/2})_1 = 2n(\tilde{D}/\pi)^{1/2}. \tag{10.2.1.24b}$$

The second estimate of $I_{1/2}$ is given in terms of the first estimate and the discrepancy between the computed and analytical values of $I_{n-1/2}$. Let Δ be defined as

$$\Delta = I^*_{n-1/2} - \hat{I}_{n-1/2}, \tag{10.2.1.25}$$

where $\hat{I}_{n-1/2}$ and $I^*_{n-1/2}$ are the analytical value and the value of $I_{n-1/2}$ given by the numerical solution procedure, respectively. We adopt the following as the second estimate of $I_{1/2}$:

$$(I_{1/2})_2 = (I_{1/2})_1 - (\Delta/2)_1. \tag{10.2.1.26}$$

Once the first and second estimates of $I_{1/2}$ have been determined, the subsequent estimates can be computed from the following interpolation formula (see Fig. 10.3):

$$\frac{(I_{1/2})_{i-1} - (I_{1/2})_i}{(I_{1/2})_{i-1} - (I_{1/2})_{i-2}} = \frac{\Delta_{i-1}}{\Delta_{i-1} + \Delta_{i-2}}$$

or

$$(I_{1/2})_i = (I_{1/2})_{i-1} - \left(\frac{\Delta_{i-1}}{\Delta_{i-1} + \Delta_{i-2}}\right)[(I_{1/2})_{i-1} - (I_{1/2})_{i-2}]$$

$$\text{for} \quad i = 3, 4, 5, \ldots. \tag{10.2.1.27}$$

The solution just developed gives further insight into the nature of a similarity solution of a partial differential equation. This solution is a

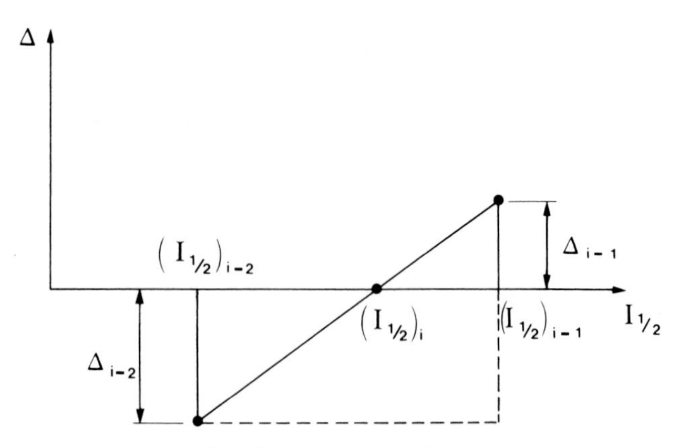

Fig. 10.3. Linear interpolation of $I_{1/2}$.

unique set of values of θ vs. ϕ. For a particular time value t_1, the similarity transformation becomes $x = \phi\sqrt{t_1}$ and is thus similar in shape to θ vs. x; this may be viewed as one reason for the use of the name "similarity solution" (see Fig. 10.4.).

(2) Horizontal Saturated–Unsaturated Flow

A more complicated case of one-dimensional horizontal flow is illustrated in Fig. 10.5. Water is applied under constant pressure head, $\psi > 0$, at the upstream end. Consequently, there exist both saturated and unsaturated flow regions. The interface of these two regions is a moving boundary located at $x = x_f$. The solution of the problem, given in Remson *et al.* (1971), consists of two parts. The first part is to determine $x_f(t)$ by solving

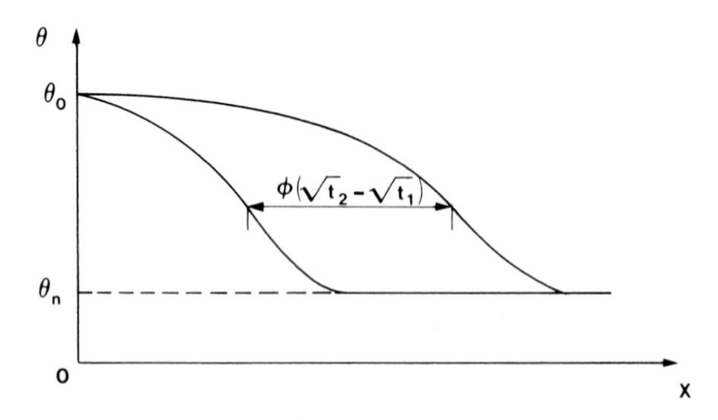

Fig. 10.4. Feature of the similarity solution.

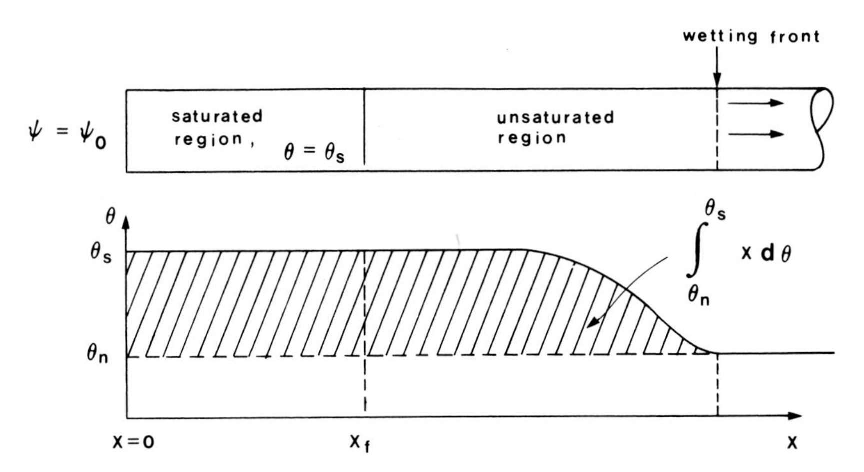

Fig. 10.5. One-dimensional saturated–unsaturated flow without gravity.

the saturated Darcy flow equation, and the second part is to solve the unsaturated flow equation using the similarity transformation and the expression of x_f derived from the first part.

To locate x_f, we first note that the Darcy flow velocity in the saturated region is equal to the rate of change of the cumulative moisture volume per unit area, at time t:

$$\mathbf{V} = \frac{d}{dt}\int_{\theta_n}^{\theta_s} x\, d\theta, \tag{10.2.1.28}$$

where θ_s is the saturated moisture content.

Application of Darcy's law to the saturated region yields

$$\mathbf{V} = K_s(\psi_o - 0)/x_f, \tag{10.2.1.29}$$

where K_s is the saturated hydraulic conductivity. Combination of Eqs. (10.2.1.28) nd (10.2.1.29) yields

$$x_f = \frac{K_s\psi_o}{(d/dt)\displaystyle\int_{\theta_n}^{\theta_s} x\, d\theta}. \tag{10.2.1.30}$$

Next, we transform the unsaturated flow equation and its initial and boundary conditions at $x = x_s$ and $x = \infty$. The result is

$$\frac{d}{d\phi}\left(D\frac{d\theta}{d\phi}\right) = -\frac{\phi}{2}\frac{d\theta}{d\phi} \tag{10.2.1.31}$$

with the conditions

$$\theta(\phi_f) = \theta_s, \tag{10.2.1.32a}$$

$$\theta(\infty) = \theta_n, \tag{10.2.1.32b}$$

where ϕ_f is given by

$$\phi_f = \frac{x_f}{t^{1/2}} = K_s\psi_o \bigg/ \left[t^{1/2} \frac{d}{dt} \int_\theta^{\theta_s} \phi t^{1/2} \, d\theta \right]. \tag{10.2.1.33}$$

Since the relationship between ϕ and θ is independent of time, the integral in Eq. (10.2.1.33) may be written as

$$\frac{d}{dt} \int_{\theta_n}^{\theta_s} \phi t^{1/2} \, d\theta = \frac{1}{2} t^{-1/2} \int_{\theta_n}^{\theta_s} \phi \, d\theta + t^{1/2} \frac{d}{dt} \int_{\theta_n}^{\theta_s} \phi \, d\theta. \tag{10.2.1.34a}$$

Since $\int_{\theta_n}^{\theta_s} \phi \, d\theta$ represents the total area under the $\phi - \theta$ curve, it is constant, and thus Eq. (10.2.1.34a) becomes

$$\frac{d}{dt} \int_{\theta_n}^{\theta_s} \phi t^{1/2} \, d\theta = \frac{1}{2} t^{-1/2} \int_{\theta_n}^{\theta_s} \phi \, d\theta. \tag{10.2.1.34b}$$

Substitution of Eq. (10.2.1.34b) into (10.2.1.33) gives

$$\phi_f = 2K_s\psi_o \bigg/ \int_{\theta_n}^{\theta_s} \phi \, d\theta. \tag{10.2.1.35}$$

The numerical solution of Eq. (10.2.1.31) is the same as before except that now $\phi_o = \phi_f$ and $\theta(\phi_o) = \theta_s$ instead of $\phi_o = 0$ and $\theta(\phi_o) = \theta_o$. The only change required is that after estimating $I_{1/2}$ so that

$$I_{1/2} = (1/\Delta\theta) \int_{\theta_n}^{\theta_s} \phi \, d\theta$$

we must compute $\phi_o = \phi_f$ from

$$\phi_o = 2K_s\psi_o/\Delta\theta \, I_{1/2}. \tag{10.2.1.36}$$

(3) Unsaturated Infiltration

The numerical procedure for solving the absorption case can be extended to deal with the unsaturated vertical infiltration case. The governing equation now contains a gravity term that does not permit a direct application of a similarity transformation. This difficulty can be overcome by the use of a perturbation technique (Philip, 1969). To describe the procedure involved, we consider the flow situation depicted in Fig. 10.6

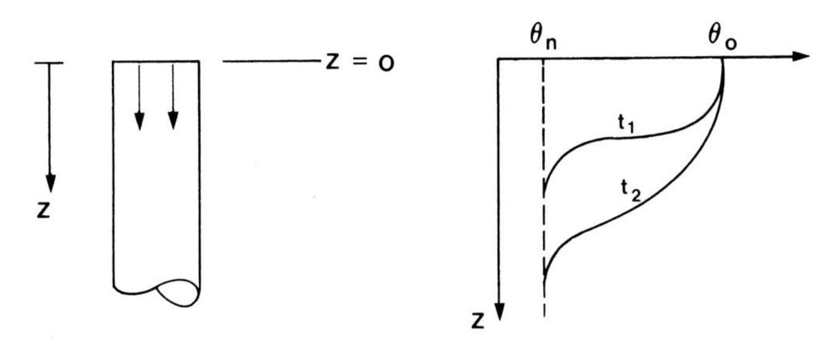

Fig. 10.6. Infiltration in a soil column.

and write the governing equation in the form

$$\frac{\partial}{\partial z}\left(D\frac{\partial \theta}{\partial z}\right) - \frac{dK}{d\theta}\frac{\partial \theta}{\partial z} = \frac{\partial \theta}{\partial t}. \tag{10.2.1.37}$$

The initial and boundary conditions considered are given by

$$\theta(x, 0) = \theta_n, \tag{10.2.1.38a}$$

$$\theta(0, t) = \theta_o, \tag{10.2.1.38b}$$

$$\theta(\infty, t) = \theta_n. \tag{10.2.1.38c}$$

First, we consider z as a function of θ and t, and write

$$\frac{dz}{dt} = \left(\frac{\partial z}{\partial \theta}\right)_t \frac{d\theta}{dt} + \left(\frac{\partial z}{\partial t}\right)_\theta. \tag{10.2.1.39}$$

From Eq. (10.2.1.39), it follows that (Philip, 1969)

$$\left(\frac{\partial \theta}{\partial t}\right)_z \left(\frac{\partial z}{\partial \theta}\right)_t + \left(\frac{\partial z}{\partial t}\right)_\theta = 0. \tag{10.2.1.40}$$

Next, we express Eq. (10.2.1.37) in the form

$$\left(\frac{\partial \theta}{\partial t}\right)_z \left(\frac{\partial z}{\partial \theta}\right)_t = \frac{\partial}{\partial z}\left(D\frac{\partial \theta}{\partial z}\right)\left(\frac{\partial z}{\partial \theta}\right)_t - \frac{dK}{d\theta}\left(\frac{\partial \theta}{\partial z}\right)\left(\frac{\partial z}{\partial \theta}\right)_t. \tag{10.2.1.41}$$

Combination of Eqs. (10.2.1.41) and (10.2.1.40) yields

$$-\frac{\partial z}{\partial t} = \frac{\partial}{\partial \theta}\left(D\frac{\partial \theta}{\partial z}\right) - \frac{dK}{d\theta}, \tag{10.2.1.42}$$

where, for brevity, the subscript θ has been dropped from $\partial z/\partial t$.

Integrating Eq. (10.2.1.42) with respect to θ, we obtain

$$-\int_{\theta_n}^{\theta} \frac{\partial z}{\partial t}\, d\theta = D\frac{\partial \theta}{\partial z} - \left(D\frac{\partial \theta}{\partial z}\right)_{\theta_n} - (K - K_n). \qquad (10.2.1.43)$$

Because the boundary condition (10.2.1.38c) implies that $\lim_{\theta \to \theta_n} [D\,(\partial\theta/\partial z)] = 0$, Eq. (10.2.1.43) reduces to

$$-\int_{\theta_n}^{\theta} \frac{\partial z}{\partial t}\, d\theta = D\frac{\partial \theta}{\partial z} - (K - K_n). \qquad (10.2.1.44)$$

Next, we seek a series solution of the form

$$z(\theta, t) = \phi_1 t^{1/2} + \phi_2 t + \phi_3 t^{3/2} + \phi_4 t^2 + \cdots. \qquad (10.2.1.45)$$

Insertion of Eq. (10.2.1.45) into Eq. (10.2.1.44) yields

$$-\int_{\theta_n}^{\theta} [\tfrac{1}{2}\phi_1 t^{-1/2} + \phi_2 + \tfrac{3}{2}\phi_3 t^{1/2} + 2\phi_4 t + \cdots]\, d\theta$$
$$= D(\phi_1' t^{1/2} + \phi_2' t + \phi_3' t^{3/2}$$
$$+ \phi_4' t^2 + \cdots)^{-1} + K - K_n, \qquad (10.2.1.46)$$

where the prime denotes differentiation with respect to θ. Now

$$(\phi_1' t^{1/2} + \phi_2' t + \phi_3' t^{3/2} + \phi_4' t^2 + \cdots)^{-1}$$
$$= \frac{1}{\phi_1' t^{1/2}}\left\{1 - \frac{1}{\phi_1'}(\phi_2' t^{1/2} + \phi_3' t + \phi_4' t^{3/2} + \cdots)\right.$$
$$\left. + \frac{1}{(\phi_1')^2}[(\phi_2' t^{1/2})^2 + \phi_2'\phi_3' t^{3/2} + \cdots] + \cdots\right\}. \qquad (10.2.1.47)$$

Substituting Eq. (10.2.1.47) into Eq. (10.2.1.46), and assuming that the expansion in $t^{1/2}$ is convergent, we may equate coefficients of powers of $t^{1/2}$ and obtain the following set of ordinary integrodifferential equations:

$$\int_{\theta_1}^{\theta} \phi_1\, d\theta = \frac{-2D}{\phi_1'}, \qquad (10.2.1.48a)$$

$$\int_{\theta_n}^{\theta} \phi_2\, d\theta = \frac{D\phi_2'}{(\phi_1')^2} + (K - K_n), \qquad (10.2.1.48b)$$

$$\int_{\theta_n}^{\theta} \phi_3\, d\theta = \frac{2D}{3}\left[\frac{\phi_3'}{(\phi_1')^2} - \frac{(\phi_2')^2}{(\phi_1')^3}\right], \qquad (10.2.1.48c)$$

$$\int_{\theta_n}^{\theta} \phi_4\, d\theta = \frac{D}{2}\left[\frac{\phi_4'}{(\phi_1')^2} - \frac{(\phi_2')^2}{(\phi_1')^3}\left\{\frac{2\phi_3'}{\phi_2'} - \frac{\phi_2'}{\phi_1'}\right\}\right], \qquad (10.2.1.48d)$$

and so on, the equation for ϕ_n ($n \geqslant 3$) being of the form

$$\int_{\theta_n}^{\theta} \phi_j \, d\theta = \frac{2D}{n} \left[\frac{\phi'_n}{(\phi'_1)^2} - R_n(\theta) \right], \qquad (10.2.1.48e)$$

where R_n may be determined from $\phi_1, \ldots, \phi_{n-1}$. We have to solve Eqs. (10.2.1.48e) subject to the condition

$$\phi_j \big|_{\theta=\theta_o} = 0, \qquad j = 1, 2, 3, 4, \ldots . \qquad (10.2.1.49)$$

This condition is directly obtainable from the boundary condition given by Eq. (10.2.1.38b). It is evident that the leading term, ϕ_1, of the series solution for one-dimensional infiltration is exactly the solution for the one-dimensional adsorption equation. Thus, the procedure for computing ϕ_1 is identical to that described previously. With ϕ_1 known, we may solve Eq. (10.2.1.48b) for ϕ_2 and then Eq. (10.2.1.48c) for ϕ_3 and so on. The various Eqs. (10.2.1.48b)–(10.2.1.48e) are linear and may be efficiently solved by the use of appropriate finite difference forms and forward integration initiated using an assumed value of $I_{1/2} = \int_{\theta_n}^{\theta_o} \phi \, d\theta$. Due to the linear form of the resulting finite difference equations, we only need to perform the calculation for two arbitrary values of $I_{1/2}$, and then use linear interpolation to obtain the final solution. Each of the linear equations (10.2.1.48b)–(10.2.1.48e) may be written in the form

$$\int_{\theta_n}^{\theta} f \, d\theta = \alpha \frac{df}{d\theta} - \beta, \qquad (10.2.1.50)$$

where f and $df/d\theta$ denote the functions ϕ_j and ϕ'_j, and α and β are coefficients which are known functions of f and θ.

To solve Eq. (10.2.1.50), we employ the following recursion formulas:

$$f_{i+1} = f_i - \left(I_{i+1/2} + \frac{\beta_{i+1/2}}{\Delta\theta} \right) \Big/ \left[\frac{\alpha_{r+1/2}}{(\Delta\theta)^2} - \frac{1}{8} \right], \qquad (10.2.1.51a)$$

$$I_{i+1/2} = I_{i-1/2} - (f_i/2) \qquad (10.2.1.51b)$$

for grid points $i = 0, 1, 2, 3, \ldots, n$, where

$$I_{i+1/2} = (1/\Delta\theta) \int_{\theta_n}^{\theta_i} f \, d\theta - (f_i/2). \qquad (10.2.1.51c)$$

By the alternating use of Eqs. (10.2.1.51a) and (10.2.1.51b), we can compute f for various values of θ, starting from the boundary condition $f = 0$ for $\theta = \theta_o$.

A rigorous treatment of the convergence of the series solution of the transformed infiltration equation (10.2.1.43) seems rather difficult. Empirical experience suggests that a suitable criterion for "practical convergence" is (Philip, 1969)

$$t \leqslant t_{\text{grav}} = [S/(K_o - K_n)]^2,$$

where t_{grav} is a characteristic time of the infiltration process and S is the sorptivity, defined as

$$S = \int_{\theta_n}^{\theta_o} \phi_1 \, d\theta.$$

When $t > t_{\text{grav}}$, the perturbation solution is not satisfactory because the series in Eq. (10.2.1.45) usually converge too slowly or fail to converge. We now present Philip's asymptotic solution valid for large t. This can be obtained by writing z in the form

$$z(\theta, t) = u(t - t_n) + \zeta(\theta). \tag{10.2.1.52a}$$

Thus

$$\partial\theta/\partial z = d\theta/d\zeta. \tag{10.2.1.52b}$$

Substituting Eqs. (10.2.1.52a) and (10.2.1.52b) into (10.2.1.43), we obtain

$$u(\theta - \theta_n) = (K - K_n) - D(d\theta/d\zeta). \tag{10.2.1.53}$$

Now

$$\lim_{t\to\infty}\left[\lim_{\theta\to\theta_o}\left(\frac{\partial\theta}{\partial z}\right)\right] = 0, \tag{10.2.1.54}$$

which implies that

$$\lim_{\theta\to\theta_o}\left(\frac{d\theta}{d\zeta}\right) = 0. \tag{10.2.1.55}$$

From Eqs. (10.2.1.52a) and (10.2.1.55), we may deduce

$$u = (K_o - K_n)/(\theta_o - \theta_n). \tag{10.2.1.56}$$

Substitution of Eq. (10.2.1.56) into (10.2.1.53) yields

$$[(K_o - K_n)/(\theta_o - \theta_n)](\theta - \theta_n) = (K - K_n) - D(d\theta/d\zeta),$$

or

$$d\zeta = -D\,d\theta \Big/ \left[-(K - K_n) + \frac{K_o - K_n}{\theta_o - \theta_n}(\theta - \theta_n)\right]. \tag{10.2.1.57}$$

Upon integration, Eq. (10.2.1.57) becomes

$$\zeta = (\theta_o - \theta_n) \int_{\theta}^{\hat{\theta}} \frac{D\, d\theta}{(K_o - K_n)(\theta - \theta_n) - (K - K_n)(\theta_o - \theta_n)}, \qquad (10.2.1.58)$$

where the zero of ζ is taken at $\hat{\theta} = \theta_o - \Delta\theta$, $\Delta\theta$ being a small increment. It is usually necessary to take the zero of ζ at θ values other than θ_o and θ_n because, in general, ζ has singularities for both of these θ values. The integral in Eq. (10.2.1.58) can be evaluated simply by numerical integration.

Having obtained the moisture content distribution, we can then proceed to evaluate the rate and quantity of infiltration. The latter is just the cumulative volume of water that has infiltrated into the soil. This is equal to the volume under the moisture profile plus the volume of water that passes out at the bottom of the profile. Thus

$$I = \int_{\theta_n}^{\theta_o} z\, d\theta + K_n t, \qquad (10.2.1.59)$$

where I is the cumulative infiltration at time t. Substituting Eq. (10.2.1.45) into (10.2.1.59), we obtain

$$I = St^{1/2} + (A_2 + K_n)t + A_3 t^{3/2} + A_4 t^2 + \cdots, \qquad (10.2.1.60)$$

where

$$S = \int_{\theta_n}^{\theta_o} \phi_1\, d\theta \qquad (10.2.1.61a)$$

as before and

$$A_j = \int_{\theta_n}^{\theta_o} \phi_j\, d\theta, \qquad j = 2, 3, 4, \ldots . \qquad (10.2.1.61b)$$

The infiltration rate may be obtained by differentiating Eq. (10.2.1.60) with respect to t:

$$q = \tfrac{1}{2} St^{-1/2} + (A_2 + K_n) + \tfrac{3}{2} A_3 t^{1/2} + 2A_4 t + \cdots . \qquad (10.2.1.62)$$

(4) Illustrative Examples

Two example problems solved by Philip (1969) are presented here. These problems concern the horizontal adsorption and vertical infiltration in "Yolo" clay, whose $K(\theta)$, $\psi(\phi)$, and $D(\phi)$ functions are depicted in Fig. 10.7. We assume that $\theta_o = 0.495$ and $\theta_n = 0.238$ in both examples. The result of the adsorption problem is depicted in Fig. 10.8a, where ϕ

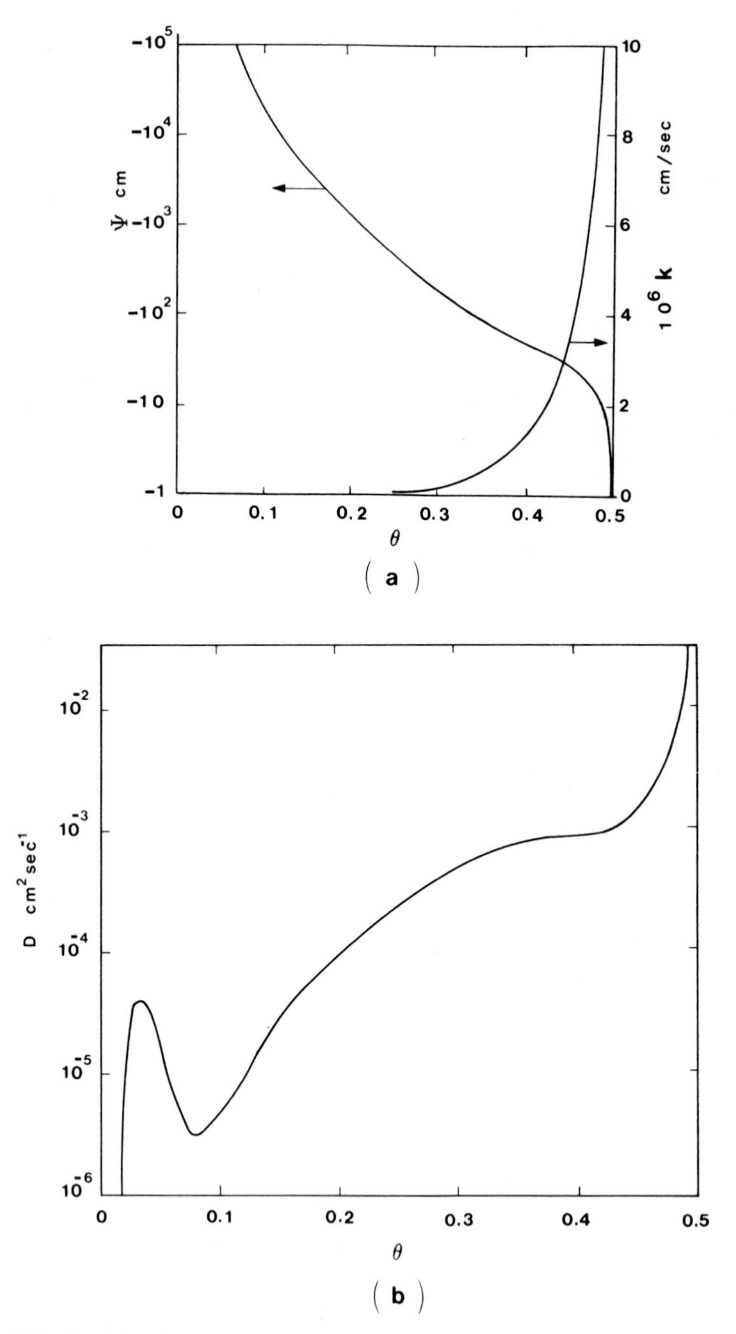

Fig. 10.7. Constitutive relations for Yolo light clay: (a) $\psi - \theta$ and $K - \theta$ curves; (b) $D - \theta$ curve. (After Philip, 1969.)

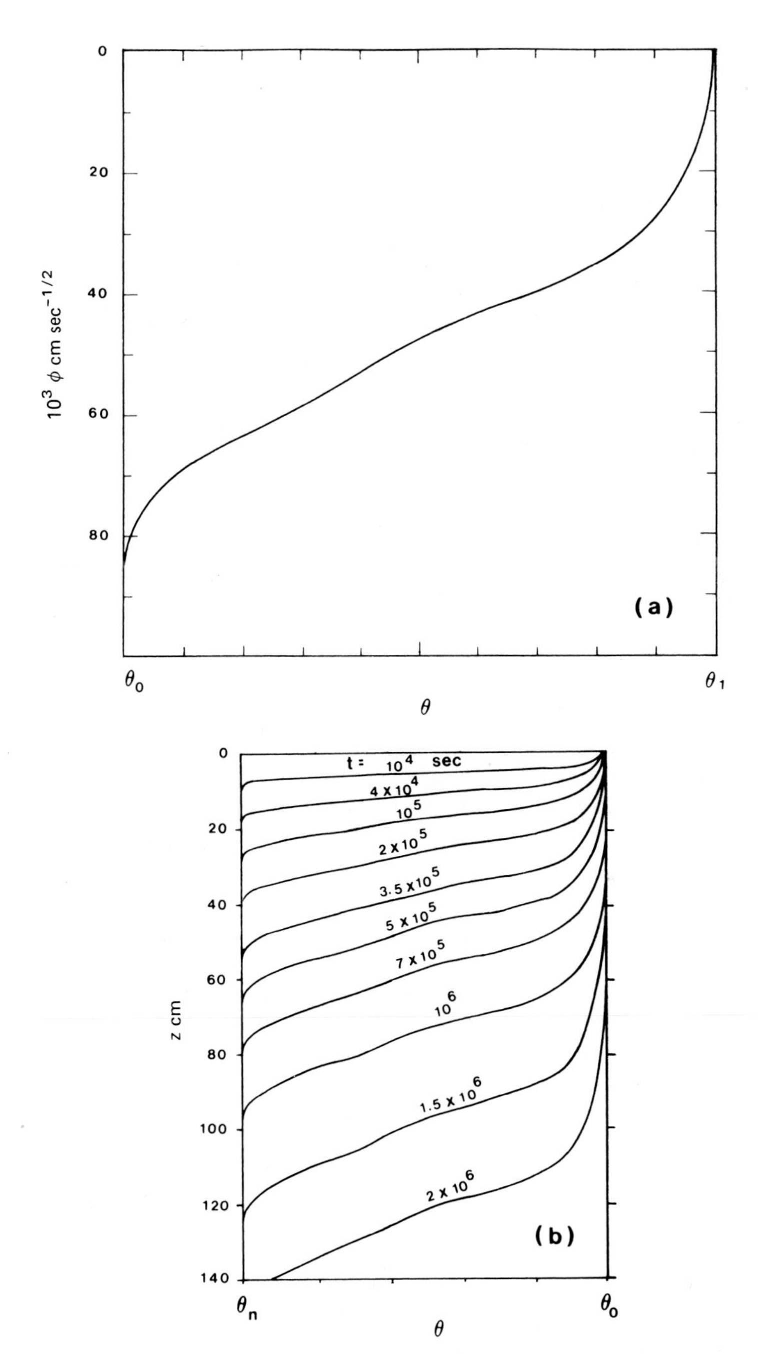

Fig. 10.8. Computed moisture profiles: (a) adsorption case, ϕ vs. θ plot; (b) infiltration case, z vs. θ plot. $\theta_0 = 0.495$, $\theta_n = 0.238$. (After Philip, 1969.)

is plotted as a function of θ. The result of the infiltration problem is shown in Fig. 10.8b as a sequence of moisture profiles calculated using the first four terms of the series in Eq. (10.2.1.45).

10.2.2 Kirchhoff's Transformation

Another type of transformation that is useful in solving unsaturated flow problems is the "Kirchhoff transformation." We demonstrate its application by referring to a specific problem of two dimensional flow in an isotropic soil slab shown in Fig. 10.9. The governing equation for this problem takes the form

$$\frac{\partial}{\partial x}\left(Kk_r \frac{\partial \psi}{\partial x}\right) + \frac{\partial}{\partial z}\left(Kk_r \frac{\partial \psi}{\partial z}\right) + K \frac{dk_r}{d\psi}\frac{\partial \psi}{\partial z} = C\frac{\partial \psi}{\partial t}. \tag{10.2.2.1}$$

The derivation of Eq. (10.2.2.1) is given in Chapter 4. As before, ψ is the capillary head, K is the saturated hydraulic conductivity, k_r is the relative permeability, and C is the specific moisture capacity, $C \equiv d\theta/d\psi$. The initial and boundary conditions are depicted in Fig. 10.9. We now define Kirchhoff's transformation as

$$F = \int_{\psi_o}^{\psi} Kk_r(\xi)\, d\xi, \tag{10.2.2.2}$$

where ψ_o is an arbitrary reference value of ψ. Using Eq. (10.2.2.2) and the curve of k_r versus ψ, we can readily obtain the curves of F versus

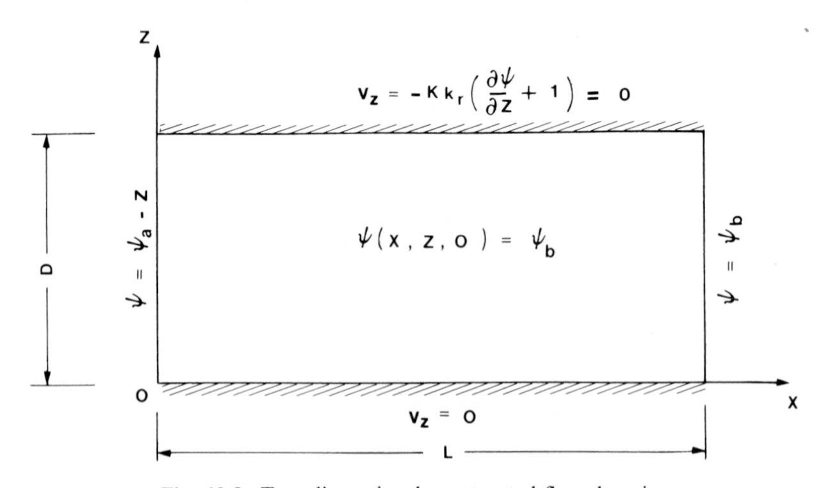

Fig. 10.9. Two-dimensional unsaturated flow domain.

ψ and F versus k_r. Differentiation of Eq. (10.2.2.2) gives

$$\frac{dF}{d\psi} = Kk_r(\psi), \qquad \frac{\partial F}{\partial t} = Kk_r \frac{\partial \psi}{\partial t}, \tag{10.2.2.3a}$$

$$\frac{\partial F}{\partial x} = Kk_r \frac{\partial \psi}{\partial x}, \qquad \frac{\partial F}{\partial z} = Kk_r \frac{\partial \psi}{\partial z}. \tag{10.2.2.3b}$$

Combination of Eqs. (10.2.2.3a), (10.2.2.3b) and (10.2.2.1) gives

$$\frac{\partial^2 F}{\partial x^2} + \frac{\partial^2 F}{\partial z^2} + \left(\frac{1}{k_r}\frac{dk_r}{d\psi}\right)\frac{\partial F}{\partial z} = \left(\frac{C}{Kk_r}\right)\frac{\partial F}{\partial t},$$

or

$$\frac{\partial^2 F}{\partial x^2} + \frac{\partial^2 F}{\partial z^2} + \alpha \frac{\partial F}{\partial z} = \beta \frac{\partial F}{\partial t}, \tag{10.2.2.4}$$

where

$$\alpha = \frac{1}{k_r}\frac{dk_r}{d\psi} = \frac{d[\ln(k_r)]}{d\psi} \tag{10.2.2.5a}$$

and

$$\beta = C/Kk_r. \tag{10.2.2.5b}$$

It is evident that the original governing equation now becomes a quasi-linear equation which is simpler to discretize by finite differences. In a similar manner, we can transform the initial and boundary conditions of the problem to the following:

$$F(x, z, 0) = F_b, \tag{10.2.2.6a}$$

$$F(0, z, t) = F_a(z), \tag{10.2.2.6b}$$

$$F(L, z, t) = F_b, \tag{10.2.2.6c}$$

$$\partial F/\partial z \Big|_{z=0} = -Kk_r = \partial F/\partial z \Big|_{z=D} \tag{10.2.2.6d}$$

The transformed governing equation can be solved by several time-stepping schemes. We elect to present only the ADI and the implicit time-stepping schemes.

Let $F_{I,J}^k$ denote the value of F at node (I, J) at time level k. With the ADI scheme (see Chapter 9 for details of the ADI procedure), linear finite difference equations are obtained and these can be written using a mesh-centered grid as follows:

(a) for the nodes on a horizontal grid line,

$$\frac{1}{(\Delta x)^2}\,\delta_x^2 F_{I,J}^{k+1/2} + \frac{1}{(\Delta z)^2}\,\delta_z^2 F_{I,J}^{k} + \frac{\alpha_{I,J}^{k}}{\Delta z}\,\mu\delta_z F_{I,J}^{k} = 2\beta_{I,J}^{k}\,(F_{I,J}^{k+1/2} - F_{I,J}^{k})/\Delta t, \tag{10.2.2.7a}$$

and (b) for the nodes on a vertical grid line,

$$\begin{aligned}\frac{1}{(\Delta x)^2}\,\delta_x^2 F_{I,J}^{k+1/2} + \frac{1}{(\Delta z)^2}\,\delta_z^2 F_{I,J}^{k+1} + \frac{\delta_{I,J}^{k+1/2}}{\Delta z}\,\mu\delta_z F_{I,J}^{k+1} \\ = 2\beta_{I,J}^{k+1/2}\,(F_{I,J}^{k+1} - F_{I,J}^{k+1/2})/\Delta t,\end{aligned} \tag{10.2.2.7b}$$

where δ_x, δ_z, and μ are the finite difference operators, defined previously in Chapter 8 (see Table 8.1).

Boundary conditions are treated in the usual manner. Once again a tridiagonal system of linear equations is obtained for each grid line, and this system is solved using the Thomas algorithm (see Chapter 9).

With the implicit scheme, the entire system of nonlinear equations is generated. For a typical node (I, J), we can write

$$\begin{aligned}\frac{1}{(\Delta x)^2}\,\delta_x^2 F_{I,J}^{k+1} + \frac{1}{(\Delta z)^2}\,\delta_z^2 F_{I,J}^{k+1} + \frac{\alpha_{I,J}^{k+1/2}}{\Delta z}\,\mu\delta_z F_{I,J}^{k+1} \\ = \beta_{I,J}^{k+1/2}\,(F_{I,J}^{k+1} - F_{I,J}^{k})/\Delta t,\end{aligned} \tag{10.2.2.8}$$

where $\alpha_{I,J}^{k+1/2}$ and $\beta_{I,J}^{k+1/2}$ are evaluated using the midtime value of F, i.e.,

$$F_{I,J}^{k+1/2} = \tfrac{1}{2}(F_{I,J}^{k} + F_{I,J}^{k+1}). \tag{10.2.2.9}$$

The resulting nonlinear equations can be solved efficiently using either the line successive overrelaxation (LSOR) or the alternating direction implicit iterative procedure (ADIPIT) (see Chapter 9 for details on the solution procedures).

Figure 10.10 shows the result for a typical problem solved by Rubin (1968) using the ADI time-stepping scheme with extrapolated values of the coefficients α and β in Eqs. (10.2.2.7a) and (10.2.2.7b). After the nodal values of F have been obtained the corresponding values of ψ and θ are obtained from the graph of F vs. ψ and ψ vs. θ, respectively.

10.2.3 Conformal Transformation

(1) General Formulation

Problems involving a free surface are rather inconvenient to solve in a physical plane by the standard (rectangular grid) finite difference method.

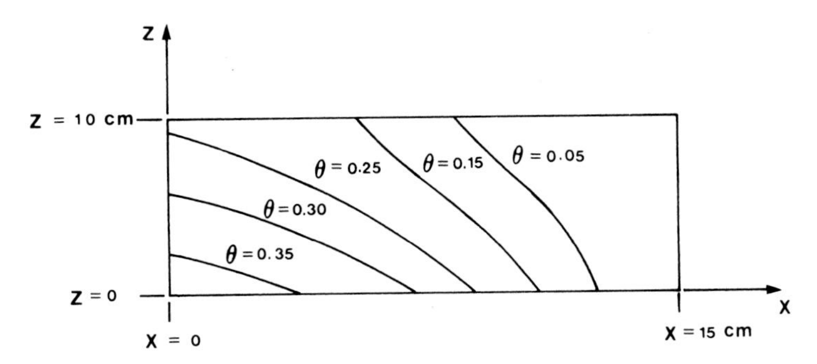

Fig. 10.10. Moisture content contours at $t = 10$ min. (Rehovolt sand, initial $\theta = 0.005$, $L = 15$ cm, $D = 10$ cm, $\psi_0 = -18$ cm). (After Rubin, 1968.)

Unlike the finite element and the boundary element methods, the finite difference method has limited flexibility in accommodating the changing free surface position. To overcome this difficulty, we now consider an alternative approach that employs a coordinate transformation to convert an irregular flow domain in the physical plane into a rectangular domain. This approach has been used by Jeppson (1968a, 1968b) to solve several steady seepage problems.

To describe the general formulation, we refer to Fig. 10.11, which illustrates the case of seepage of water from a canal through a soil medium underlain by a drainage layer. It is assumed that the total seepage rate Q is unknown and that the soil is homogeneous and isotropic. In addition, the flow domain is assumed to be symmetric with respect to the y axis. From the assumption of isotropic and homogeneous porous medium, it follows that the velocities and hydraulic head satisfy the following equations:

$$V_x = -K\,\partial h/\partial x, \tag{10.2.3.1a}$$

$$V_y = -K\,\partial h/\partial y, \tag{10.2.3.1b}$$

and

$$K\left(\frac{\partial^2 h}{\partial x^2} + \frac{\partial^2 h}{\partial y^2}\right) = 0. \tag{10.2.3.1c}$$

Let the "velocity potential" Φ be defined as $\Phi = -Kh$. Equations (10.2.3.1a) and (10.2.3.1b) thus become

$$V_x = \partial\Phi/\partial x, \qquad V_y = \partial\Phi/\partial y, \tag{10.2.3.2a}$$

and

$$\frac{\partial^2\Phi}{\partial x^2} + \frac{\partial^2\Phi}{\partial y^2} = \nabla^2\Phi = 0. \tag{10.2.3.2b}$$

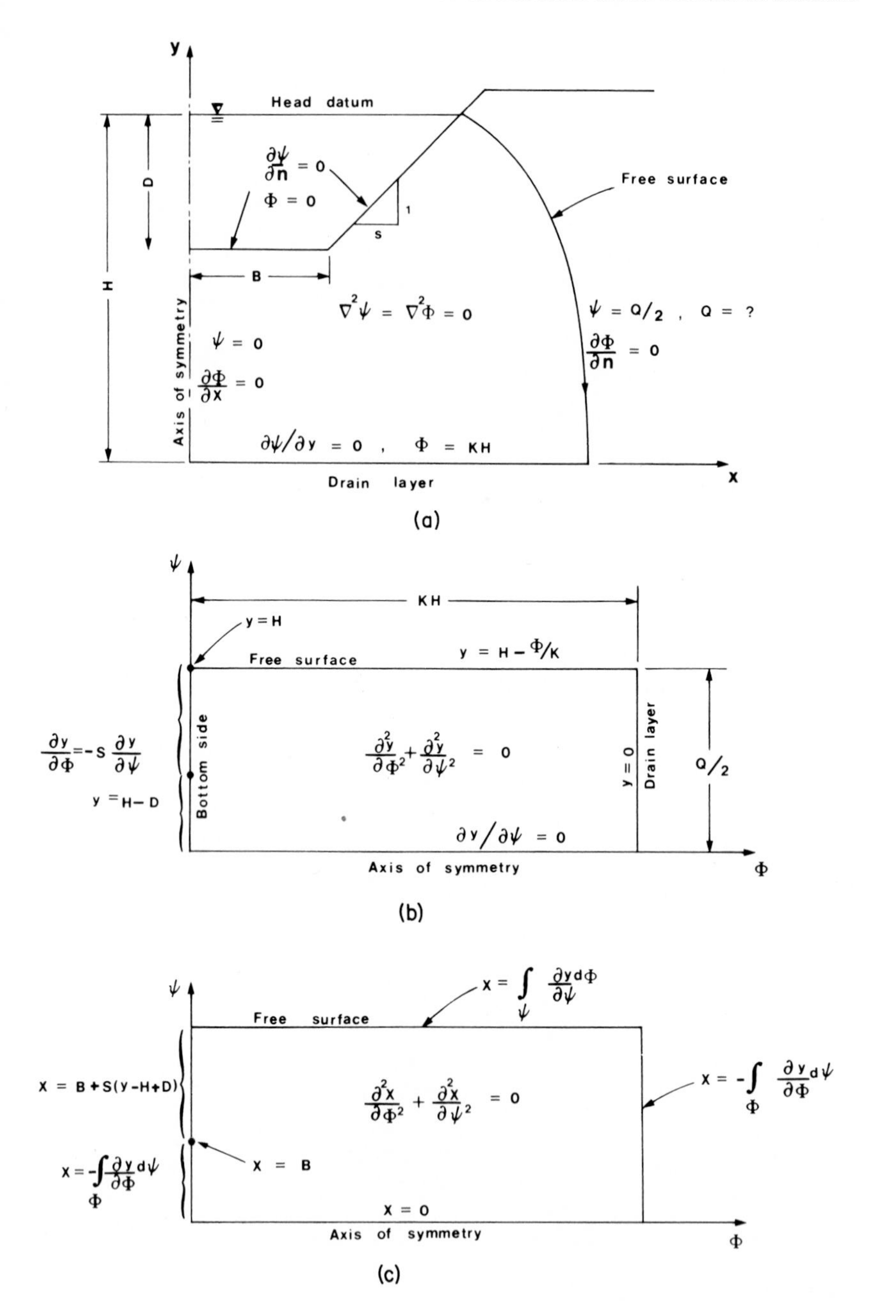

Fig. 10.11. Steady seepage from a canal. (a) formulation in the (Φ, ψ) plane, y variable; (b) formulation in the (Φ, ψ) plane, x variable; and (c) formulation in the physical plane. (After Jeppson, 1968a.)

Before performing the transformation, we need to introduce one more variable. The appropriate one to use is the streamfunction, defined as $d\psi = V_x\, dy - V_y\, dx$, or

$$V_x = \frac{\partial \psi}{\partial y}, \qquad V_y = -\frac{\partial \psi}{\partial x}. \qquad (10.2.3.3a)$$

It immediately follows that each line with a constant ψ value corresponds to a streamline because along this line $d\psi = 0$. In addition, the slope of the line lies in the direction of the velocity vector, i.e., $dy/dx = V_y/V_x$.

Next, we eliminate h by differentiating and combining Eqs. (10.2.3.1a) and (10.2.3.1b). Thus,

$$-\frac{\partial V_y}{\partial x} + \frac{\partial V_x}{\partial y} = K\frac{\partial^2 h}{\partial x\, \partial y} - K\frac{\partial^2 h}{\partial y\, \partial x} = 0 \qquad (10.2.3.3b)$$

or

$$\frac{\partial^2 \psi}{\partial x^2} + \frac{\partial^2 \psi}{\partial y^2} = \nabla^2 \psi = 0. \qquad (10.2.3.3c)$$

Equations (10.2.3.2b) and (10.2.3.3c) form a pair of governing equations with Φ and ψ as dependent variables in the physical (x, y) plane. The corresponding boundary conditions are illustrated in Fig. 10.11. These boundary conditions can be described as follows. Along the bottom and side of the canal, the streamlines are in the normal direction, and the hydraulic head is constant. The head can be set equal to zero if we adopt the top water level as the datum. These conditions are equivalent to $\partial\psi/\partial n = 0$ and $\Phi = 0$. Along the vertical axis of symmetry the velocity is vertical, thus leading to the boundary conditions $\partial\Phi/\partial x = 0$ and $\psi = 0$. Along the drain layer, the value of hydraulic head is equal to $-H$ and the velocity is in the vertical direction thus leading to $\Phi = KH$ and $\partial\psi/\partial y = 0$, respectively. The remaining boundary conditions are those on the free surface. Since the free surface is also a streamline, ψ must be constant and in this case equal to one-half of the total seepage rate. In addition, the normal derivative of the potential function Φ must be zero. To obtain the equivalent formulation in the (Φ, ψ) plane, we first transform the physical domain into the rectangular domain shown in Figs. 10.11b and 10.11c. The rectangular shape results from the fact that each of the functions, Φ and ψ, takes on two different values along the opposite segments of the flow boundary. Next, we proceed to derive the partial differential equations that describe the distributions of x and y in the domain of Φ and ψ. These equations can be obtained by applying the rules for the transformation of variables as follows: Because Φ and ψ are functions of x and y, we can write $\Phi = f(x, y)$, $\psi = g(x, y)$, and hence, in general,

$$F(\Phi, \psi, x(\Phi, \psi), y(\Phi, \psi)) = f(x, y) - \Phi = 0,$$

$$G(\Phi, \psi, x(\Phi, \psi), y(\Phi, \psi)) = g(x, y) - \psi = 0.$$

Using the chain rule of differentiation, we obtain

$$\left.\frac{\partial F}{\partial \Phi}\right|_{\psi} = \frac{\partial F}{\partial \Phi} + \frac{\partial F}{\partial x}\frac{\partial x}{\partial \Phi} + \frac{\partial F}{\partial y}\frac{\partial y}{\partial \Phi} = 0,$$

$$\left.\frac{\partial G}{\partial \Phi}\right|_{\psi} = \frac{\partial G}{\partial \Phi} + \frac{\partial G}{\partial x}\frac{\partial x}{\partial \Phi} + \frac{\partial G}{\partial y}\frac{\partial y}{\partial \Phi} = 0.$$

Solving for $\partial x/\partial \Phi$ and $\partial y/\partial \Phi$ using Cramer's rule and noting that $\partial F/\partial \Phi = -1$ and $\partial G/\partial \Phi = 0$, we obtain

$$\frac{\partial x}{\partial \Phi} = -\frac{\begin{vmatrix} -1 & \partial F/\partial y \\ 0 & \partial G/\partial y \end{vmatrix}}{\begin{vmatrix} \partial F/\partial x & \partial F/\partial y \\ \partial G/\partial x & \partial G/\partial y \end{vmatrix}}, \qquad \frac{\partial y}{\partial \Phi} = -\frac{\begin{vmatrix} \partial F/\partial x & -1 \\ \partial G/\partial x & 0 \end{vmatrix}}{\begin{vmatrix} \partial F/\partial x & \partial F/\partial y \\ \partial G/\partial x & \partial G/\partial y \end{vmatrix}},$$

Since $\partial F/\partial x = \partial f/\partial x = \partial \Phi/\partial x$, $\partial F/\partial y = \partial \Phi/\partial y$, $\partial G/\partial x = \partial \psi/\partial x$, and $\partial G/\partial y = \partial \psi/\partial y$, the above expressions become

$$\frac{\partial x}{\partial \Phi} = \frac{1}{J}\frac{\partial \psi}{\partial y} = \frac{V_x}{J}, \tag{10.2.3.4a}$$

$$\frac{\partial y}{\partial \Phi} = -\frac{1}{J}\frac{\partial \psi}{\partial x} = \frac{V_y}{J}, \tag{10.2.3.4b}$$

where J is the Jacobian given by

$$J = \begin{vmatrix} \dfrac{\partial \phi}{\partial x} & \dfrac{\partial \Phi}{\partial y} \\ \dfrac{\partial \psi}{\partial x} & \dfrac{\partial \psi}{\partial y} \end{vmatrix} = \begin{vmatrix} V_x & V_y \\ -V_y & V_x \end{vmatrix} = V_x^2 + V_y^2.$$

In a similar manner, we can show that

$$\frac{\partial x}{\partial \psi} = -\frac{1}{J}\frac{\partial \Phi}{\partial y} = \frac{-V_y}{J}, \tag{10.2.3.4c}$$

$$\frac{\partial y}{\partial \psi} = \frac{1}{J}\frac{\partial \Phi}{\partial x} = \frac{V_x}{J}. \tag{10.2.3.4d}$$

It follows from Eqs. (10.2.3.4a) and (10.2.3.4d) that

$$\partial x/\partial \Phi = \partial y/\partial \psi \tag{10.2.3.5a}$$

and from Eqs. (10.2.3.4b) and (10.2.3.4c),

$$\partial x/\partial \psi = -\partial y/\partial \Phi. \tag{10.2.3.5b}$$

Equations (10.2.3.5a) and (10.2.3.5b) correspond to the well-known Cauchy–Riemann conditions encountered in the theory of complex variables. Consequently, the transformation from the (x, y) plane to the (Φ, ψ) plane corresponds to the "conformal" transformation of complex variables.

Either x or y can be eliminated from the Cauchy–Riemann conditions by differentiation and combination. Thus differentiating Eq. (10.2.3.5a) with respect to Φ and Eq. (10.2.3.5b) with respect to ψ and adding yield

$$\frac{\partial^2 x}{\partial \Phi^2} + \frac{\partial^2 x}{\partial \psi^2} = 0. \tag{10.2.3.6a}$$

Similarly, we also obtain

$$\frac{\partial^2 y}{\partial \Phi^2} + \frac{\partial^2 y}{\partial \psi^2} = 0. \tag{10.2.3.6b}$$

Equations (10.2.3.6a) and (10.2.3.6b) are the required governing equations for the inverse functions. The boundary conditions for y are depicted in Fig. 10.11a. These conditions can be derived in the following manner. Along the axis of symmetry, the equipotential lines are horizontal, thus leading to the boundary condition $\partial y/\partial \psi = 0$. Along the drain layer it is obvious that $y = 0$. Along the horizontal bottom of the canal, $y = H - D$. Along the free surface, the pressure is atmospheric $(p = 0)$, thus leading to the boundary condition $\Phi = -K(y - H)$ or

$$y = H - \Phi/K. \tag{10.2.3.7a}$$

Finally, along the side of the canal, the streamlines are in the normal direction, thus leading to the boundary condition

$$V_x/V_y = -1/S \qquad (S = \text{side slope}),$$

which upon combination with Eqs. (10.2.3.4a)–(10.2.3.4d) becomes

$$-\frac{1}{S}\frac{\partial y}{\partial \Phi} = \frac{\partial y}{\partial \psi} = \frac{\partial x}{\partial \Phi}. \tag{10.2.3.7b}$$

The boundary conditions for x are depicted in Fig. 10.11c. These conditions can be derived as follows.

Along the axis of symmetry, it is seen from Fig. 10.11a that $x = 0$.

Along the side of the canal, the x coordinate is given by $x = B + S(y - H + D)$. The remaining boundary conditions can be obtained by integrating the Cauchy–Riemann equations (10.2.3.5a) and (10.2.3.5b) along the boundary segments which correspond to either a streamline or an equipotential line. Suppose that the distribution of y is known, we then obtain conditions along the bottom of the canal and the free surface as follows.

Along the bottom of the canal and the drain layer, Φ is constant and thus x is obtained by integrating Eq. (10.2.3.5b),

$$x = -\int_{\Phi} \partial y/\partial\Phi \, d\psi, \tag{10.2.3.7c}$$

and along the free surface, x is obtained by integrating Eq. (10.2.3.5a),

$$x = \int_{\psi} \partial y/\partial\psi \, d\Phi. \tag{10.2.3.7d}$$

SOLUTION APPROACH Once the partial differential equations and the corresponding boundary conditions have been obtained for the inverse functions x and y in the (Φ, ψ) domain, the problem can be solved in a straightforward manner using the finite difference method. Two solution techniques are presented in Jeppson's work (1968a). We elect to describe the technique that employs complex variables as an aid for solving the governing equations for x and y simultaneously. The advantage of this approach lies in its systematic nature, which makes the computer implementation simpler than the alternative approach. We initiate our discussion of the solution procedure with the introduction of the complex variable ξ, defined as

$$\xi = x + iy = F(\Phi + i\psi) = F(w), \tag{10.2.3.8}$$

where $i = \sqrt{-1}$.

The partial differential equation and boundary conditions for ξ are depicted in Fig. 10.12. These boundary conditions are obtained simply by combining the information in Figs. 10.11a and 10.11b. Note that there are several ways of applying the conditions along the boundary of the canal. For example, on the side, instead of the boundary condition depicted in Fig. 10.12, one may use the condition

$$x + i\,\partial y/\partial\Phi = B + S(y - H + D) - iS\,\partial y/\partial\psi.$$

We are now ready to solve the transformed seepage problem using finite differences. Before proceeding to perform the finite difference discretization, it is important to recognize two special features of this problem.

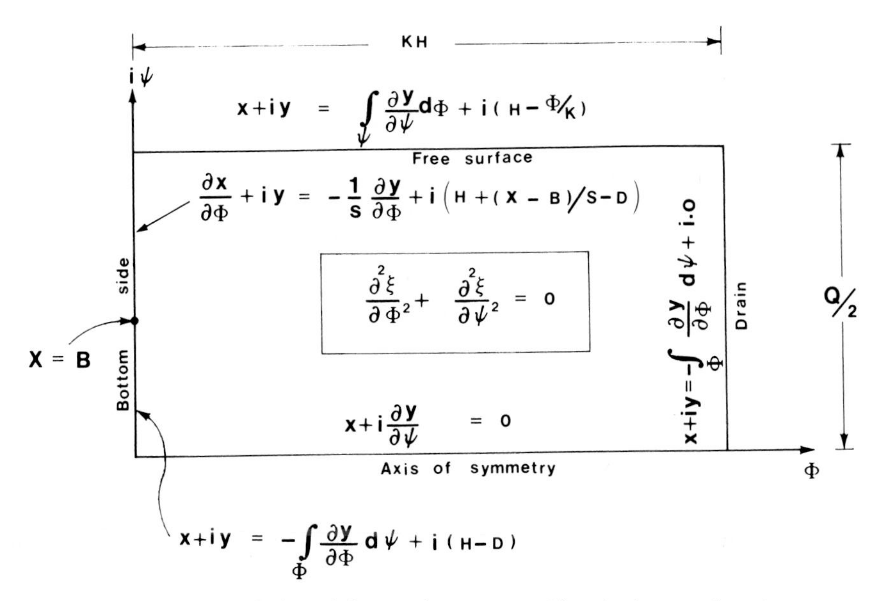

Fig. 10.12. Formulation of the canal seepage problem in the complex plane.

The first feature is the unknown location of the top boundary of the (Φ, ψ) domain, and the second feature is the unknown location of the common point of the bottom and side boundaries in the (Φ, ψ) domain (Fig. 10.12). The need for such information becomes quite apparent upon examination of finite difference operators and treatment of boundary conditions. Let us now employ a mesh-centered grid with nodal spacing of $\Delta\Phi$ and $\Delta\psi$ for constant Φ lines and constant ψ lines, respectively. A sample mesh consisting of m constant Φ lines and n constant ψ lines is depicted in Fig. 10.13. For an interior node (I, J), the five-point approximation of the Laplacian operator can be shown to take the form

$$\frac{\partial^2 \xi}{\partial \Phi^2} + \frac{\partial^2 \xi}{\partial \psi^2} = 0 \simeq \frac{1}{\Delta\Phi}\left[\frac{\xi_{I+1,J} - \xi_{I,J}}{\Delta\phi} - \frac{\xi_{I,J} - \xi_{I-1,J}}{\Delta\Phi}\right] + \frac{1}{\Delta\psi}\left[\frac{\xi_{I,J+1} - \xi_{I,J}}{\Delta\psi} - \frac{\xi_{I,J} - \xi_{I,J-1}}{\Delta\psi}\right]. \quad (10.2.3.9a)$$

Equation (10.2.3.9a) can be expressed as

$$-2(1+R^2)\xi_{I,J} + [\xi_{I+1,J} + \xi_{I-1,J} + R^2(\xi_{I,J+1} + \xi_{I,J-1})] = 0, \quad (10.2.3.9b)$$

where R is the ratio of the spacing of Φ = constant grid lines and ψ = constant grid lines, $R = \Delta\Phi/\Delta\psi$, and $\xi_{I,J}$ contains the values at node I, J of the real and imaginary parts of the complex variable $x + iy$.

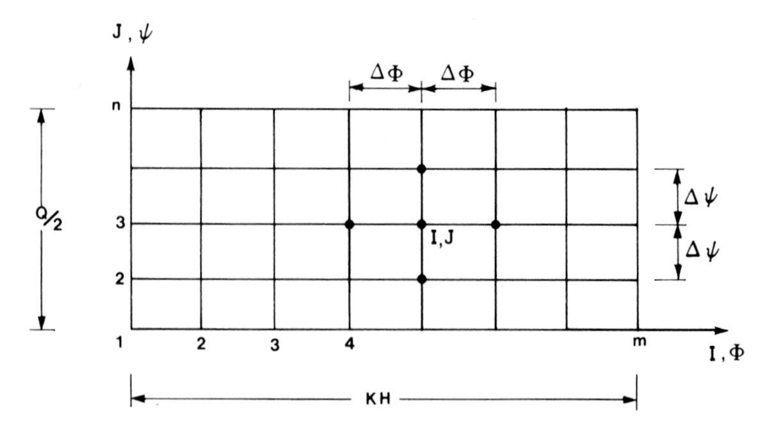

Fig. 10.13. Finite difference discretization of the canal seepage problem.

The finite difference operators for the boundary conditions can be developed analogously. For the zero normal derivative boundary condition $\partial y/\partial\psi = 0$, along the axis of symmetry, the finite difference operator is the same as Eq. (10.2.3.9b) with ξ replaced by y and with $y_{I,J-1}$ set equal to $y_{I,J+1}$. All of the integrals appearing on various boundary segments can be readily evaluated using either trapezoidal or Simpson's rule for numerical integration.

For example, the integration along the free surface (top boundary, $J = n$) can be evaluated as follows (Fig. 10.14):

$$x = \int_\psi \partial y/\partial\psi \, d\Phi = \int_\psi f \, d\Phi = x_{1,n} + \text{area under } f\text{–}\Phi \text{ curve}, \tag{10.2.3.10a}$$

which, upon application of trapezoidal rule, becomes

$$x_{I,n} = x_{1,n} + (\Delta\Phi/2)(f_1 + 2f_2 + 2f_3 + \cdots + 2f_{I-1} + f_I), \tag{10.2.3.10b}$$

where

$$f_\alpha = (\partial y/\partial\psi)_\alpha \simeq (y_{\alpha+1,n} - y_{\alpha,n})/\Delta\psi, \qquad \alpha = 1, 2, \ldots, I. \tag{10.2.3.10c}$$

The integration along the drain and the bottom of the canal can be done in a similar manner.

The solution of the algebraic equations resulting from the finite difference approximation can be performed efficiently using the point successive relaxation procedure. By this procedure, the grid points are repeatedly swept with the values of the functions (x and y) adjusted at each grid point by the finite difference operator. For convenience, we elect to sweep across from bottom to top of the network ($J = 1, 2, \ldots$ with

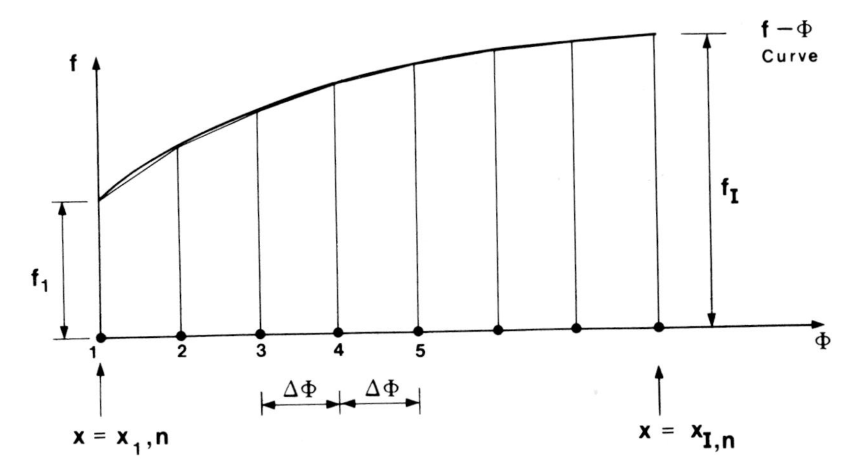

Fig. 10.14. Evaluation of the integral along the free surface using trapezoidal rule.

increasing ψ) moving from left to right ($I = 1, 2, \ldots$ with increasing Φ). There exist two equations at each node, one for the variable x and the other for the variable y.

The equations for a typical interior node I, J can be simply derived from Eq. (10.2.3.9b). These equations are expressed in matrix form as

$$[A_{I,J}]\{\xi_{I,J}^{k+1}\} = \{F_{I,J}\}, \tag{10.2.3.11}$$

where

$$[A_{I,J}] = -2(1 + R^2)\begin{bmatrix} 1 & 0 \\ 0 & 1 \end{bmatrix}, \qquad \{\xi_{I,J}^{k+1}\} = \begin{Bmatrix} x_{I,J}^{k+1} \\ y_{I,J}^{k+1} \end{Bmatrix},$$

$$\{F_{I,J}\} = -\begin{Bmatrix} x_{I+1,J}^{k} + x_{I-1,J}^{k+1} + R^2(x_{I,J+1}^{k} + x_{I,J-1}^{k+1}) \\ y_{I+1,J}^{k} + y_{I-1,J}^{k+1} + R^2(y_{I,J+1}^{k} + y_{I,J-1}^{k+1}) \end{Bmatrix},$$

and the superscripts $k+1$ and k denote the current and previous iterations, respectively. The equations for a typical node (I, J) on the side boundary can be derived by referring to Fig. 10.12 and replacing normal derivatives $\partial x/\partial\Phi$ and $\partial y/\partial\Phi$ by central difference operators. These equations can be written as

$$(x_{2,J} - x_{0,J})/(2\Delta\Phi) = -(1/S)\,(y_{2,J} - y_{0,J})/(2\Delta\Phi), \tag{10.2.3.12a}$$

$$y_{1,J} = H + (x_{1,J} - B)/S - D. \tag{10.2.3.12b}$$

The values $x_{0,J}$ and $y_{0,J}$ at the fictitious node $(0, J)$ can be eliminated in the usual manner by applying the five-point Laplacian operator at node

(1, J). Thus,

$$x_{0,J} = 2(1 + R^2)x_{1,J} - [x_{2,J} + R^2(x_{1,J+1} + x_{1,J-1})], \qquad (10.2.3.13a)$$

$$y_{0,J} = 2(1 + R^2)y_{1,J} - [y_{2,J} + R^2(y_{1,J+1} + y_{1,J-1})]. \qquad (10.2.3.13b)$$

Substituting Eqs. (10.2.3.13a) and (10.2.3.13b) into Eq. (10.2.3.12a) and rearranging both Eqs. (10.2.3.12a) and (10.2.3.12b), we obtain the final result, which can be written in matrix form as

$$[A_{1,J}]\{\xi_{1,J}^{k+1}\} = \{F_{1,J}\}, \qquad (10.2.3.14)$$

where

$$[A_{1,J}] = \begin{bmatrix} 2(1 + R^2) & (2/S)(1 + R^2) \\ -1/S & 1 \end{bmatrix}, \qquad \{\xi_{1,J}^{k+1}\} = \begin{Bmatrix} x_{1,J}^{k+1} \\ y_{1,J}^{k+1} \end{Bmatrix},$$

$$\{F_{1,J}\} = \begin{Bmatrix} R^2(x_{1,J+1}^{k} + x_{1,J-1}^{k+1}) + (R^2/S)(y_{1,J+1}^{k} + y_{1,J-1}^{k+1}) + 2x_{2,J} + (2/S)y_{2,J} \\ H - (B/S) - \mathrm{D} \end{Bmatrix}.$$

The equations for other boundary nodes can be derived in a similar manner.

The final solution of the problem is obtained by a cycle of solutions, each subsequent one of which is more nearly correct. The complete solution procedure consists of the following steps:

(1) Because the value of R is unknown a priori, we have to start by assuming it. Using the assumed value of R, we generate and solve simultaneously the equations for x and y values for each node, moving from bottom to top along the first vertical line of the discretized (Φ, ψ) domain. For each node, the most recent values of x and y are obtained by the use of an overrelaxation factor. As the solution proceeds, we compare the computed value of x with the half-bottom width B to determine whether the current node belongs to the bottom or side boundary of the canal. The common point of the bottom and side boundaries is determined when $x = B$.

(2) Next we make a series of sweeps from the second through the last vertical lines. After sweeping the last vertical line, we run a check for convergence of the x and y values. If necessary, iteration is performed until satisfactory convergence is achieved.

(3) After the completion of step 2, we check to see whether the assumed value of R agrees with that computed using the most current nodal values of x and y. The computed value of R is obtained by taking the arithmetic average of the values for individual cells of the flow net in the physical plane. Thus,

$$R_{\text{computed}} = (1/N)\sum_{C=1}^{N} R_C, \qquad N = (m - 1)(n - 1),$$

in which N is the total number of cells in the flow region and R_C is the cell value of R. For a typical cell shown in Fig. 10.15, R_C is determined from

$$R_C = \bar{d}_\Phi / \bar{d}_\psi,$$

where

$$\bar{d}_\Phi = \tfrac{1}{2}(d_{\Phi,1} + d_{\Phi,2}), \qquad \bar{d}_\psi = \tfrac{1}{2}(d_{\psi,1} + d_{\psi,2}),$$

$$d_{\Phi,1} = [(y_{I+1,J} - y_{I,J})^2 + (x_{I+1,J} - x_{I,J})^2]^{1/2},$$

$$d_{\Phi,2} = [(y_{I+1,J+1} - y_{I,J+1})^2 + (x_{I+1,J+1} - x_{I,J+1})^2]^{1/2},$$

$$d_{\psi,1} = [(y_{I,J+1} - y_{I,J})^2 + (x_{I,J+1} - x_{I,J})^2]^{1/2},$$

$$d_{\psi,2} = [(y_{I+1,J+1} - y_{I+1,J})^2 + (x_{I+1,J+1} - x_{I+1,J})^2]^{1/2}.$$

It follows from the properties of the conformal transformation and flow net that if the computed solution for x and y is correct, the assumed value of R should agree closely with the computed value of R. When this is not the case, then it is necessary to repeat the solution procedure

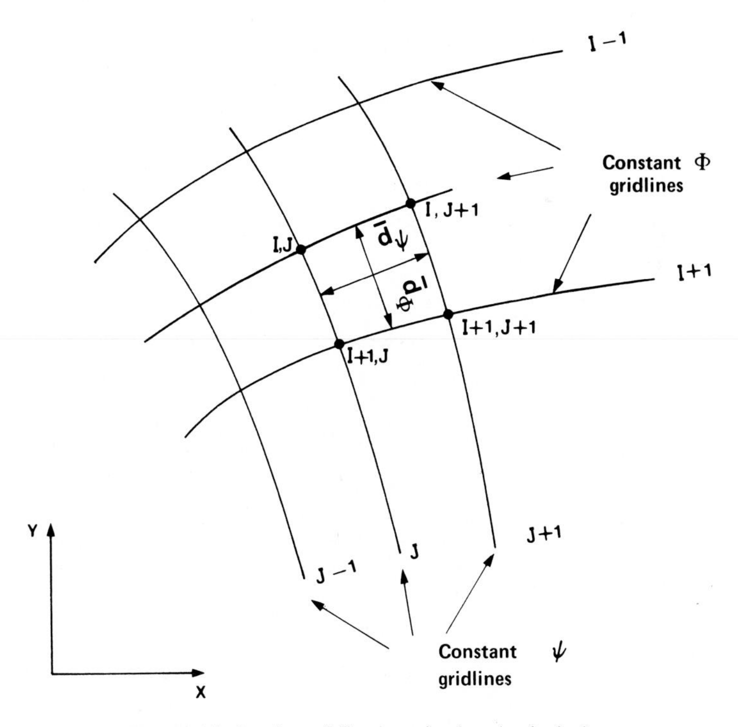

Fig. 10.15. Portion of flow net in the physical plane.

in steps 1 and 2 using a revised value of R. Iteration must then be performed until satisfactory convergence is obtained.

Illustrative Examples Solutions of problems involving two-dimensional seepage from three geometric canals—trapezoidal, triangular, and rectangular—are presented by Jeppson (1968a). The results for the rectangular and trapezoidal canals are depicted in Fig. 10.16. As illustrated, an advantage of the conformal transformation finite difference approach to

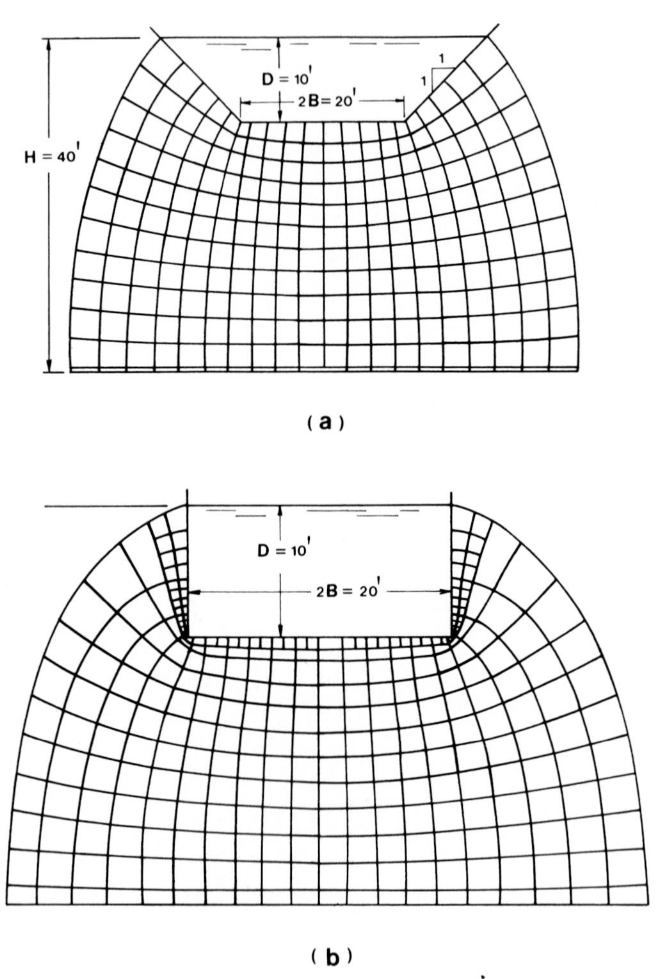

Fig. 10.16. Computed flow nets: (a) trapezoidal canal; (b) rectangular canal. (After Jeppson, 1968a).

free surface problems is the ease with which a flow net can be plotted and velocities and flow rate determined from the final solution, which consists of nodal values of $x(\Phi, \psi)$ and $y(\Phi, \psi)$. Although we do not give a detailed account, problems involving anisotropic material and axisymmetric problems can also be readily treated by extension of the formulation presented (see Jeppson, 1968b).

10.3 Method of Characteristics

10.3.1 General Theory

The method of characteristics has received widespread acceptance in the analysis of transient fluid flow governed by partial differential equations of hyperbolic type. Before dealing with some specific subsurface problems, it is appropriate to describe the general procedure for deriving the "characteristic" equations and to present solution algorithms of the first- and second-order equations.

First-Order Hyperbolic Equations

Consider the quasi-linear equation of the form

$$a \frac{\partial u}{\partial x} + b \frac{\partial u}{\partial t} = c, \tag{10.3.1.1a}$$

where x and t denote the space and time variables and a, b, and c are, in general, functions of x, t, and u. If $b \neq 0$, Eq. (10.3.1.1) can be written as

$$\frac{a}{b} \frac{\partial u}{\partial x} + \frac{\partial u}{\partial t} - \frac{c}{b} = 0. \tag{10.3.1.1b}$$

Now, from chain rule of differentiation, it follows that

$$\frac{du}{dt} = \frac{\partial u}{\partial t} + \frac{dx}{dt} \frac{\partial u}{\partial x}$$

or

$$\frac{dx}{dt} \frac{\partial u}{\partial x} + \frac{\partial u}{\partial t} - \frac{du}{dt} = 0. \tag{10.3.1.2}$$

Subtracting Eq. (10.3.1.2) from (10.3.1.1b), we obtain

$$\left(\frac{a}{b} - \frac{dx}{dt}\right) \frac{\partial u}{\partial x} - \left(\frac{c}{b} - \frac{du}{dt}\right) = 0. \tag{10.3.1.3}$$

It is apparent that Eq. (10.3.1.3) reduces to an ordinary differential equation along a curve defined by

$$(a/b) - (dx/dt) = 0$$

or

$$dx/dt = a/b. \tag{10.3.1.4}$$

Such a curve is called a "characteristic" curve or simply a characteristic. Along its path, Eq. (10.3.1.3) reduces to

$$du/dt = c/b, \tag{10.3.1.5}$$

which can be solved by simple analytical or numerical integration. For convenience, we write Eqs. (10.3.1.4) and (10.3.1.5) together as

$$dx/a = dt/b = du/c. \tag{10.3.1.6}$$

As an example, suppose that the function u satisfies the equation

$$x^{1/2}\frac{\partial u}{\partial x} + u\frac{\partial u}{\partial t} + u^2 = 0 \tag{10.3.1.7}$$

and the initial condition

$$u(x, 0) = 1, \qquad 0 < x < \infty.$$

In this case, $a = x^{1/2}$, $b = u$, $c = -u^2$, and Eq. (10.3.1.6) becomes

$$dx/x^{1/2} = dt/u = du/-u^2. \tag{10.3.1.8}$$

Thus the characteristic curve through a point $x = x_I$ and $t = 0$ in the (x, t) plane is given by

$$\int_{x_I}^{x} dx/x^{1/2} = \int_0^t dt/u,$$

$$2(x^{1/2} - x_I^{1/2}) = \int_0^t dt/u. \tag{10.3.1.9}$$

To evaluate the integral on the left-hand side we need to express u in terms of t. This is achieved by integrating the equation

$$dt/u = du/-u^2.$$

Thus

$$\int_0^t dt = -\int_{u_o}^{u} du/u. \tag{10.3.1.10}$$

Because the initial condition is $u_o = 1$,

$$t = \ln(1/u)$$

or

$$1/u = e^t. \tag{10.3.1.11}$$

Substituting Eq. (10.3.1.11) into (10.3.1.9) and integrating, we obtain the characteristic equation

$$t = \ln(2x^{1/2} + 1 - 2x_I^{1/2}). \tag{10.3.1.12}$$

Along this characteristic the solution of Eq. (10.3.1.7) is given by

$$u = e^{-t} = 1/(2x^{1/2} + 1 - 2x_I^{1/2}). \tag{10.3.1.13}$$

Second-Order Hyperbolic Equations

The procedure for deriving characteristics of first-order equations can be extended to second-order equations. Consider the general hyperbolic equation of the form

$$a\frac{\partial^2 u}{\partial x^2} + b\frac{\partial^2 u}{\partial x\,\partial t} + c\frac{\partial^2 u}{\partial t^2} + e = 0, \tag{10.3.1.14}$$

where a, b, c, and e are, in general, functions of x, t, and the first derivatives of u. Let us define p and q as

$$p = \partial u/\partial x, \qquad q = \partial u/\partial t.$$

Thus

$$dp = \frac{\partial^2 u}{\partial x^2}\,dx + \frac{\partial^2 u}{\partial t\,\partial x}\,dt, \tag{10.3.1.15a}$$

$$dq = \frac{\partial^2 u}{\partial x\,\partial t}\,dx + \frac{\partial^2 u}{\partial t^2}\,dt. \tag{10.3.1.15b}$$

Solving the last two equations, we obtain

$$\frac{\partial^2 u}{\partial x^2} = \frac{dp}{dx} - \frac{\partial^2 u}{\partial x\,\partial t}\frac{dt}{dx}, \tag{10.3.1.16a}$$

$$\frac{\partial^2 u}{\partial t^2} = \frac{dq}{dt} - \frac{\partial^2 u}{\partial x\,\partial t}\frac{dx}{dt}. \tag{10.3.1.16b}$$

Substitution of Eqs. (10.3.1.16a) and (10.3.1.16b) into (10.3.1.14) yields

$$\left(\frac{\partial^2 u}{\partial x\,\partial t}\right)\left(-a\frac{dt}{dx} - c\frac{dx}{dt} + b\right) + \left(e + a\frac{dp}{dx} + c\frac{dq}{dt}\right) = 0, \tag{10.3.1.17}$$

which upon multiplication by $-dt/dx$ becomes

$$\frac{\partial^2 u}{\partial x\,\partial t}\left[a\left(\frac{dt}{dx}\right)^2 + c - b\frac{dt}{dx}\right] - \left[a\frac{dp}{dx}\frac{dt}{dx} + c\frac{dq}{dt}\frac{dt}{dx} + e\frac{dt}{dx}\right] = 0. \tag{10.3.1.18}$$

Suppose, in the (x, t) plane, we define "characteristic" curves so that the first bracketed expression is zero. On such curves, the original equation (10.3.1.14) then reduces to

$$a\frac{dp}{dx}\frac{dt}{dx} + c\frac{dq}{dx} + e\frac{dt}{dx} = 0,$$

or

$$a\,(dt/dx)\,dp + c\,dq + e\,dt = 0. \tag{10.3.1.19}$$

The characteristic curves are defined by the equation

$$a(dt/dx)^2 - b(dt/dx) + c = 0,$$

or

$$(dt/dx)_{\pm} = b \pm \sqrt{b^2 - 4ac}/2a. \tag{10.3.1.20}$$

For convenience, we let $(dt/dx)_+ = f$ and $(dt/dx)_- = g$. In the simplest case where Eq. (10.3.1.14) is a linear equation with constant coefficients, there exist two characteristic curves that are striaght lines having the slopes $(dt/dx)_+$ and $(dt/dx)_-$, which are given by Eq. (10.3.1.20). We now describe the procedure for solving Eq. (10.3.1.14) by numerical integration along the two characteristic lines. First, we note that on the (x, t) plane the x axis corresponds to the initial line on which the initial conditions are specified. We also note that on the t axis and on the vertical line $x = L$, the boundary conditions are specified. Next, we consider the first two points P and Q on the initial line. By moving along the characteristic lines passing through these points, the point R can be located as shown in Fig. 10.17.

Next, we determine the complete information at point R by the following procedure:

(1) The coordinates of point R in the (x, t) plane are determined by solving the equations

$$t_R - t_P = f_P(x_R - x_P), \tag{10.3.1.21a}$$

$$t_R - t_Q = g_Q(x_R - x_Q), \tag{10.3.1.21b}$$

where $f = (dt/dx)_+$ and $g = (dt/dx)_-$.

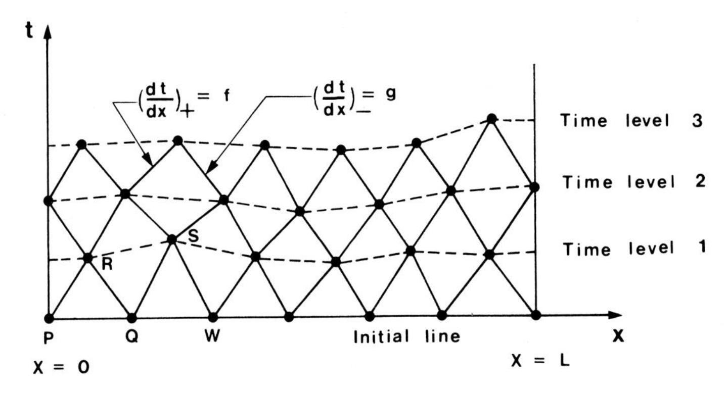

Fig. 10.17. Grid pattern for numerical solution by the method of characteristics.

Solving for x_R and t_R we obtain

$$x_R = \frac{1}{f_P - g_Q}(f_P x_P - g_Q x_Q + t_Q - t_P), \qquad (10.3.1.22a)$$

$$t_R = t_P + f_P(x_R - x_P). \qquad (10.3.1.22b)$$

(2) Next, Eq. (10.3.1.19) is written in finite difference form and applied over the line segments PR and QR. This leads to

$$a_P f_P(p_R - p_P) + c_P(q_R - q_P) + e_P(t_R - t_P) = 0, \qquad (10.3.1.23a)$$

$$a_Q g_Q(p_R - p_Q) + c_Q(q_R - q_Q) + e_Q(t_R - t_Q) = 0. \qquad (10.3.1.23b)$$

Solving for p_R and q_R, we obtain

$$p_R = \frac{1}{a_P f_P/c_P - a_Q f_Q/c_Q}[a_P f_P p_p/c_P - a_Q f_Q p_Q/c_Q + q_P - q_Q$$
$$+ (e_P/c_P - e_Q/c_Q)t_R + e_P t_P/c_P - e_Q t_Q/c_Q], \qquad (10.3.1.24a)$$

$$q_R = e_P(t_P - t_Q)/c_P + a_P f_P(p_P - p_R)/c_P + q_P. \qquad (10.3.1.24b)$$

(3) The value of u at point R can now be determined from the following differential equation:

$$du = p\,dx + q\,dt. \qquad (10.3.1.25a)$$

Writing Eq. (10.3.1.25a) in finite difference form, we can derive the following approximation for u_R:

$$u_R = \tfrac{1}{2}[u_P + \tfrac{1}{2}(p_R + p_P)(x_R - x_P) + \tfrac{1}{2}(q_R + q_P)(t_R - t_P)$$
$$+ u_Q + \tfrac{1}{2}(p_R + p_Q)(x_R - x_Q) + \tfrac{1}{2}(q_R + q_Q)(t_R - t_Q)]. \qquad (10.3.1.25b)$$

If the partial differential equation (10.3.1.4) is linear, the calculation of the unknowns at point R is considered completed. However, if Eq. (10.3.1.4) is quasi-linear, iteration should be performed. Improved values of x_R and t_R can be obtained by solving

$$t_R - t_P = \tfrac{1}{2}(f_P + f_R)(x_R - x_P) \qquad (10.3.1.26a)$$

and

$$t_R - t_Q = \tfrac{1}{2}(f_P + f_Q)(x_R - x_Q). \qquad (10.3.1.26b)$$

Improved values of p_R and q_R can be obtained by solving

$$\bar{a}_P\bar{f}_P(p_R - p_P) + \bar{c}_P(q_R - q_P) + \bar{e}_P(t_R - t_P) = 0 \qquad (10.3.1.27a)$$

and

$$\bar{a}_Q\bar{g}_Q(p_R - p_Q) + \bar{c}_Q(q_R - q_Q) + \bar{e}_Q(t_R - t_Q) = 0, \qquad (10.3.1.27b)$$

where

$$\bar{a}_P = (a_P + a_R)/2, \qquad \bar{c}_P = (c_P + c_R)/2, \qquad \bar{e}_P = (e_P + e_R)/2,$$

$$\bar{a}_Q = (a_Q + a_R)/2, \qquad \bar{c}_Q = (c_Q + c_R)/2, \qquad \bar{e}_Q = (e_Q + p_R)/2.$$

Provided Q is close to P, the number of iterations will usually be small.

The calculation in steps 1–3 can be repeated for a second point S, which is the intersection of characteristic lines through Q and another initial point W. The solution procedure can be continued in this manner to evaluate the function u throughout the region in the (x, t) plane as desired (see Fig. 10.17). Note that along the vertical boundaries of the (x, t) solution domain, either the value of u or of p is given as the boundary condition. Consequently, we only need to use two equations, one along one of the characteristics and the other derived from Eq. (10.3.1.25a).

10.3.2 Transient Groundwater Flow

The method of characteristics can be applied to transient confined groundwater flow problems governed by a pair of first-order hyperbolic equations, provided the inertial term is included in Darcy's law. [Such an application is described in Wiley (1976). Other applications of the method of characteristics are described in Streeter and Wiley (1967) and in Ames (1977).] For illustrative purposes, we consider a simple case of axisymmetric flow in a homogeneous and isotropic aquifer overlain by an aquitard. For this flow situation, Darcy's law and the continuity equation can be expressed as

$$\frac{1}{g\phi}\frac{\partial V}{\partial t} + \frac{V}{K} = -\frac{\partial h}{\partial r}, \qquad (10.3.2.1a)$$

$$\frac{\partial V}{\partial r} + \frac{V}{r} - \frac{q'}{b} = -S_s\frac{\partial h}{\partial t}, \qquad (10.3.2.1b)$$

where V is the Darcy velocity; h is the hydraulic head; g is the gravitational acceleration; K, ϕ, S_s, and b are the hydraulic conductivity, porosity, specific storage, and thickness of the aquifer, respectively; and q' is the leakage rate from the aquitard to the aquifer. The leakage rate q' may be expressed as $q' = K'(h_o - h)/b'$, in which K' and b' are the hydraulic conductivity and thickness of the aquitard, respectively.

In practice, the inertial term $(\partial V/\partial t)/g\phi$ of Eq. (10.3.2.1a) is negligible when compared with the other two terms. This makes it convenient to combine Eqs. (10.3.2.1a) and (10.3.2.1b) and obtain the well-known second-order parabolic equation

$$K\left(\frac{\partial^2 h}{\partial r^2} + \frac{1}{r}\frac{\partial h}{\partial r}\right) - \frac{q'}{b} = S_s\frac{\partial h}{\partial t}. \qquad (10.3.2.2)$$

In solving the problem by the characteristic method, it is advantageous to use Eqs. (10.3.2.1a) and (10.3.2.1b) rather than Eq. (10.3.2.2) in the formulation. We thus rewrite Eqs. (10.3.2.1a) and (10.3.2.1b) as

$$\frac{\alpha^2}{g\phi}\frac{\partial V}{\partial t} + \frac{\partial h}{\partial r} + \frac{V}{K} = 0, \qquad (10.3.2.3a)$$

$$\frac{1}{S_s}\left[\frac{\partial V}{\partial r} + \frac{V}{r} - \frac{q'}{b'}\right] + \frac{\partial h}{\partial t} = 0, \qquad (10.3.2.3b)$$

in which α denotes a scalar multiplier. A value of $\alpha = 1$ corresponds to the true representation of the physical flow behavior. To derive the characteristic equations, it is convenient to combine Eqs. (10.3.2.3a) and (10.3.2.3b) and obtain the following second-order equation:

$$\frac{\alpha^2}{g\phi}\frac{\partial^2 V}{\partial t^2} + \frac{1}{K}\frac{\partial V}{\partial t} - \frac{1}{S_s}\left[\frac{\partial^2 V}{\partial r^2} + \frac{\partial}{\partial r}\left(\frac{V}{r}\right) - \frac{\partial}{\partial r}\left(\frac{q'}{b}\right)\right] = 0,$$

or

$$a\frac{\partial^2 V}{\partial r^2} + c\frac{\partial^2 V}{\partial t^2} + e = 0, \qquad (10.3.2.4)$$

where

$$a = -1/S_s, \qquad c = \alpha^2/g\phi, \qquad q' = (K'/b')(h_o - h), \qquad (10.3.2.5a)$$

and

$$e = \frac{1}{K}\frac{\partial V}{\partial t} - \frac{1}{S_s}\left[\frac{\partial}{\partial r}\left(\frac{V}{r}\right) - \frac{\partial}{\partial r}\left(\frac{q'}{b'}\right)\right]. \tag{10.3.2.5b}$$

Using the formula in Eq. (10.3.1.20) for characteristics of the general second-order hyperbolic equation, we obtain

$$(dt/dr)_{\pm} = \pm\alpha\,(S_s/g\phi)^{1/2},$$

or

$$(dr/dt)_{\pm} = \pm\xi/\alpha, \qquad \text{where } \xi = (g\phi\,/S_s)^{1/2}. \tag{10.3.2.6}$$

Next, we perform the following linear combination of Eqs. (10.3.2.3a) and (10.3.2.3b):

$$\frac{\partial V}{\partial t} + \frac{g\phi}{\alpha^2}\left(\frac{\partial h}{\partial r} + \frac{V}{K}\right) + \frac{1}{\alpha}\left(\frac{g\phi}{S_s}\right)^{1/2}\left[\frac{\partial V}{\partial r} + \frac{V}{r} - \frac{q'}{b}\right] + \frac{(g\phi S_s)^{1/2}}{\alpha}\frac{\partial h}{\partial t} = 0,$$

which can be arranged in the form

$$\left(\frac{\partial V}{\partial t} + \frac{\xi}{\alpha}\frac{\partial V}{\partial r}\right) + \frac{(g\phi S_s)^{1/2}}{\alpha}\left(\frac{\partial h}{\partial t} + \frac{\xi}{\alpha}\frac{\partial h}{\partial r}\right)$$

$$+ \frac{g\phi}{\alpha^2}\frac{V}{k} + \frac{\xi}{\alpha}\left(\frac{V}{r} - \frac{q'}{b}\right) = 0. \tag{10.3.2.7}$$

Equation (10.3.2.7) represents the equation along the characteristic $(dr/dt)_{+} = \xi/\alpha$. Thus, we can write

$$\frac{dV}{dt} + \frac{(g\phi S_s)^{1/2}}{\alpha}\frac{dh}{dt} + \frac{g\phi}{\alpha^2}\frac{V}{K} + \frac{\xi}{\alpha}\left(\frac{V}{r} - \frac{q'}{b}\right) = 0,$$

or

$$\frac{\alpha}{\xi}\frac{dV}{dt} + S_s\frac{dh}{dt} + S_s\frac{\xi V}{\alpha K} + \frac{V}{r} + \frac{K'(h - h_o)}{bb'} = 0. \tag{10.3.2.8a}$$

Similarly, the equation along the characteristic $(dr/dt)_{-} = -\xi/\alpha$ can be obtained as

$$-\frac{\alpha}{\xi}\frac{dV}{dt} + S_s\frac{dh}{dt} - S_s\frac{\xi V}{\alpha K} + \frac{V}{r} + \frac{K'(h - h_o)}{bb'} = 0. \tag{10.3.2.8b}$$

Once the characteristics and the associated differential equations have been derived, the numerical solution follows in a similar manner to that described for second-order hyperbolic equations. For the axisymmetric well flow problem, it is advantageous to adopt a grid with variable spatial and time increments. A typical grid is depicted in Fig. 10.18. For convenience, a constant value of Δt_i is used between two successive time levels, and the same spatial increments are used in all time levels. Thus, the locations of all grid points are fixed a priori. This is achieved by varying the value of the parameter α so that Eqs. (10.3.2.6) are satisfied. (Note that α becomes a nonphysical parameter when its value is not equal to 1.) However, as long as the value of α is held at a sufficiently low level so that the inertial term in the equation of motion is insignificant as compared to the friction term, the solution obtained should represent the true physical behavior as accurately as the solution of Eq. (10.3.2.2). The solution algorithm for an interior point can be described as follows:

(1) First, we approximate the characteristic equations using finite differences and solve for the values of α that correspond to the prescribed location of the required interior point. For point P in Fig. 10.18, the finite difference approximation is given by

$$R_P - R_A = (\xi/\alpha^+)(t_P - t_A) \tag{10.3.2.9a}$$

and

$$R_P - R_B = (\xi/\alpha^-)(t_P - t_B), \tag{10.3.2.9b}$$

where α^+ and α^- are the values of the characteristic lines AP and BP

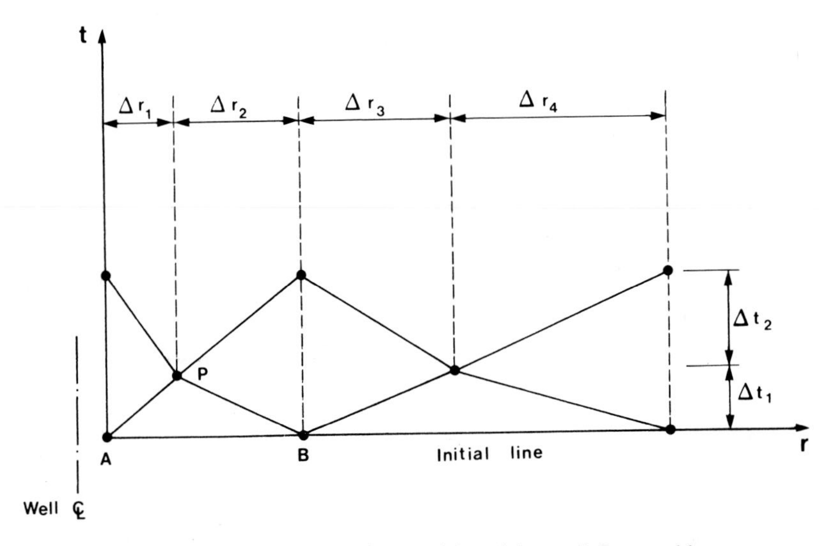

Fig. 10.18. Typical grid portion used in solving well flow problem.

and R_P, R_A and R_B are values of the radial coordinate for points P, A, and B, respectively.

(2) We then approximate Eqs. (10.3.2.8a) and (10.3.2.8b) and solve for the values of h and V. For point P, the finite difference approximation of Eqs. (10.3.2.8a) and (10.3.2.8b) can be shown to take the form

$$h_P + c_1 V_P = c_2 \tag{10.3.2.10a}$$

and

$$h_P + c_3 V_P = c_4, \tag{10.3.2.10b}$$

where (referring to Fig. 10.18)

$$c_1 = \left[\frac{\alpha^+ \xi}{2g\phi}\left(1 + \frac{R_P}{R_A}\right) + \frac{R_P}{2K}\ln\frac{R_P}{R_A}\right] \Big/ \left(1 + w\right),$$

$$c_2 = \left\{h_A\left(1 - w\right) + 2wh_o + V_A\left[\frac{\alpha^+ \xi}{2g\phi}\left(1 + \frac{R_A}{R_P}\right) - \frac{R_A}{2K}\ln\frac{R_P}{R_A}\right]\right\} \Big/ \left(1 + w\right),$$

$$c_3 = \left[\frac{\alpha^- \xi}{2g\phi}\left(1 + \frac{R_P}{R_B}\right) + \frac{R_P}{2K}\ln\frac{R_B}{R_P}\right] \Big/ \left(1 + w\right),$$

$$c_4 = \left\{h_B\left(1 - w\right) + 2wh_o - V_B\left[\frac{\alpha^- \xi}{2g\phi}\left(1 + \frac{R_B}{R_P}\right) - \frac{R_B}{2K}\ln\frac{R_B}{R_P}\right]\right\} \Big/ \left(1 + w\right),$$

$$w = K'\,\Delta t_1/(2S_s bb'), \qquad \text{and} \qquad \Delta t_1 = t_P - t_A = t_P - t_B.$$

The calculation in steps 1 and 2 can be repeated for other interior points in a particular time level. For the boundary points along the t axis, only one characteristic needs to be used because one of the unknown variables can be determined from each of the prescribed boundary conditions at the well boundary and the radius of influence.

In order to ensure that the inertial term of the flow equation is small compared to the remaining terms, it is necessary to set an upper limit on the value of the parameter α. For a typical characteristic line BP, the value of α used should satisfy the following criterion:

$$\alpha < \frac{\varepsilon g\phi}{\xi K}\,\frac{R_P R_B}{R_P + R_B}\,\frac{(\,|Q_P| + |Q_B|\,)}{|Q_P - Q_B|}\ln\frac{R_B}{R_B}, \tag{10.3.2.11}$$

where Q_P and Q_B are the discharges at sections corresponding to R_P and R_B, respectively, and ε is a parameter that provides a measure of the ratio of the inertial to the friction term in the Darcy equation of motion. (Note that Q and V are related by $Q = 2\pi rbV$.)

The inequality (10.3.2.11) is derived by simply comparing the magnitude of the inertial and friction terms in Eq. (10.3.2.10b). Wiley (1976) indicates that a value of ε between 0.02 and 0.10 often leads to a fairly accurate solution.

A typical example problem solved by Wiley is depicted in Fig. 10.19a. For this problem, a value of $\varepsilon = 0.01$ is used. The results obtained are compared in Fig. 10.19b with the analytical solution. Excellent agreement between the numerical and analytical values of drawdown can be observed.

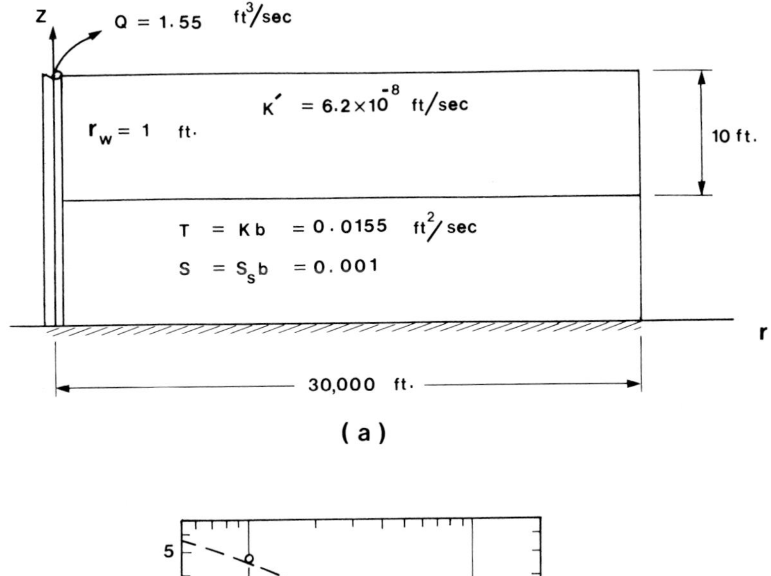

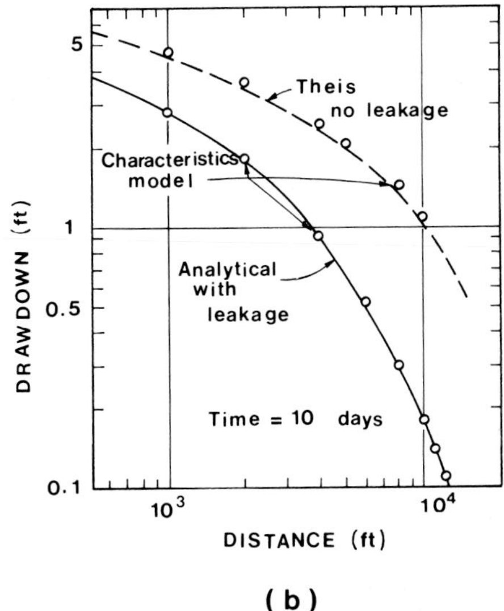

Fig. 10.19. Solution of the well flow problem using the method of characteristics: (a) problem data; (b) drawdown–distance plot. (After Wiley, 1976.)

10.3.3 Two-Phase Flow

Next, we demonstrate the application of the characteristic method to two-phase flow problems. The cases of negligible and significant capillary effects are considered.

Buckley–Leverett Problem

A simple case of horizontal flow of two immiscible fluids, oil and water, in a core of uniform porous medium is illustrated in Fig. 10.20. Initially, the porous medium is saturated with oil, and then water is injected at a constant rate to displace oil. It is assumed that both fluids are incompressible and that capillary effects are negligible (i.e., $p_w = p_o = p$). Such a flow problem is commonly referred to as the Buckley–Leverett problem (Buckley and Leverett, 1942). It can be described by two governing equations, one for pressure and the other for water saturation. These equations are simply Darcy's law for both phases and the continuity equation for the water phase. They may be written in the form

$$V = V_w + V_o = -k(\lambda_w + \lambda_o)(\partial p/\partial x) \tag{10.3.3.1a}$$

and

$$-\partial V_w/\partial x = \partial S_w/\partial t, \tag{10.3.3.1b}$$

where V_f (f = o, w) denotes the velocity of phase f, k is the intrinsic permeability, ϕ is porosity, S_w is saturation of water, and λ_f is the mobility of phase f, i.e., $\lambda_f = k_{rf}/\mu_f$, where k_{rf} and μ_f are the relative permeability and dynamic viscosity of phase f, respectively. It can be seen that the two equations are uncoupled and that if we know the water saturation, the mobilities can be evaluated and the pressure obtained from Eq. (10.3.3.1a) by a simple integration. Thus, the main task is to solve the

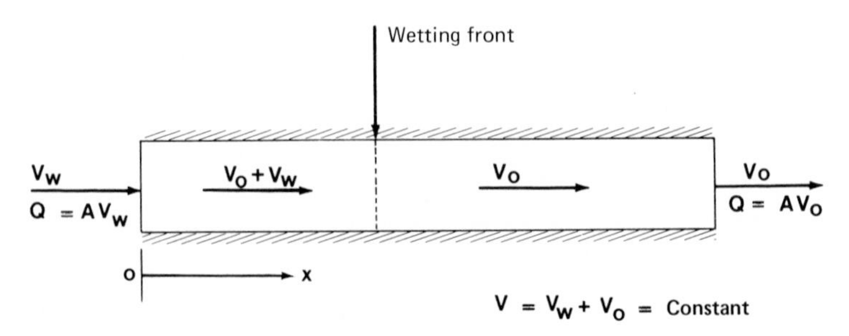

Fig. 10.20. Buckley–Leverett problem.

saturation equation, which is a first-order hyperbolic equation. To solve Eq. (10.3.3.2), we first define a fractional flow function f_w as

$$f_w = V_w/V = \frac{-k\lambda_w \, \partial p/\partial x}{-k(\lambda_w + \lambda_o)\, \partial p/\partial x}$$

$$= \frac{1}{1 + \lambda_o/\lambda_w} = \frac{1}{1 + (k_{ro}\mu_w)/(\mu_o k_{rw})}. \qquad (10.3.3.2)$$

Because the relative permeabilities are nonlinear functions of S_w, it follows that f_w is also a nonlinear function of S_w. Figure 10.21 illustrates the plot of f_w and its derivative, f'_w. We can now rewrite Eq. (10.3.3.2) in the form

$$\phi \frac{\partial S_w}{\partial t} + V \frac{\partial f_w}{\partial x} = 0,$$

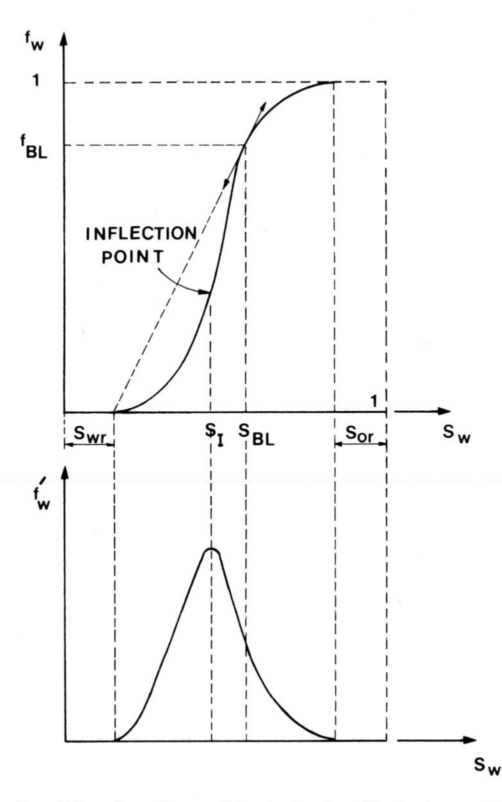

Fig. 10.21. Fractional flow function and its derivative (S_{wr} is the residual water saturation and S_{or} the residual oil saturation). (After Morel-Seytoux, 1969.)

or, by applying the chain rule,

$$\frac{\partial S_w}{\partial t} + \frac{V}{\phi} f'_w \frac{\partial S_w}{\partial x} = 0, \tag{10.3.3.3}$$

which is known as the Buckley–Leverett equation.

Because Eq. (10.3.3.3) is a first-order hyperbolic equation, it has one characteristic equation, and this may be derived simply by matching with the following equation, which is the definition of the total derivative:

$$\frac{dS_w}{dt} = \frac{\partial S_w}{\partial t} + \frac{\partial S_w}{\partial x}\left(\frac{dx}{dt}\right)_{S_w = \text{const.}} = 0. \tag{10.3.3.4}$$

Combining Eqs. (10.3.3.3) and (10.3.3.4), we obtain

$$\left(\frac{dx}{dt}\right)_{S_w} = \frac{V}{\phi} f'_w, \tag{10.3.3.5a}$$

which may be integrated to give the x coordinate corresponding to any prescribed value of S_w:

$$x\Big|_{S_w} = \frac{Vt}{\phi} f'_w + x_0, \tag{10.3.3.5b}$$

where x_0 is the initial position (at $t = 0$) of the prescribed S_w value. It

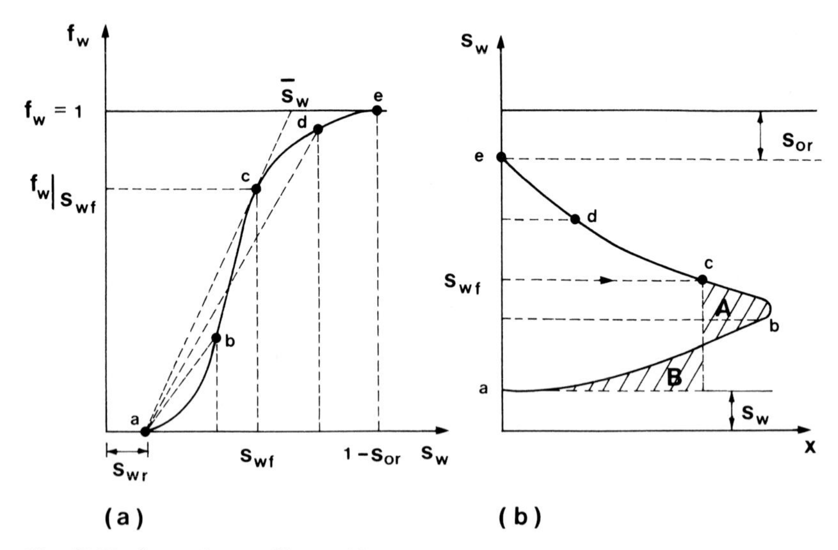

Fig. 10.22. Saturation profile resulting from a given f_w curve: (a) f_w vs. S_w; (b) S_w vs. x. (After Morel-Seytoux, 1969.)

is evident that the characteristic equation (10.3.3.5b) can be used to construct the water saturation profile at any time t. Fig. 10.22 shows the shape of a typical profile obtained. As can be expected this profile (the solid line) takes the same shape as that of f'_w. Consequently, the saturation is not uniquely defined but has triple values. Such a profile is physically impossible, which means that the straight application of Eq. (10.3.3.5b) is unacceptable. The physical reasoning is that the intermediate values of the water saturation, which have greater velocities (see Fig. 10.21), will tend to overtake the lower saturations, resulting in the formation of a saturation discontinuity. Because Eq. (10.3.3.3) is derived under the assumption that the function S_w is a continuous and differentiable function of x and t, it is invalid in the region where the saturation profile or its slope undergoes a discontinuity or abrupt front. Behind the front, where $S_{wf} < S_w < 1 - S_{or}$, the characteristic flow equation is applicable. Thus, to draw the correct saturation profile, we need to locate the position of the front such that the continuity of the mass is satisfied, i.e., the shaded area A and B, shown in Fig. 10.22b, are equal.

Alternatively, one can use the more elegant method presented by Welge (1952). To describe this method, consider the saturation profile depicted in Fig. 10.23. Let the saturation front S_{wf} be located at $x_2 = x_f$ and let $\overline{S}_w$ be the average saturation behind the front. Application of a simple material balance leads to

$$A\phi x_2(\overline{S}_w - S_{wr}) = Qt = AVt, \qquad (10.3.3.6a)$$

or

$$\overline{S}_w - S_{wr} = Vt/x_2\phi, \qquad (10.3.3.6b)$$

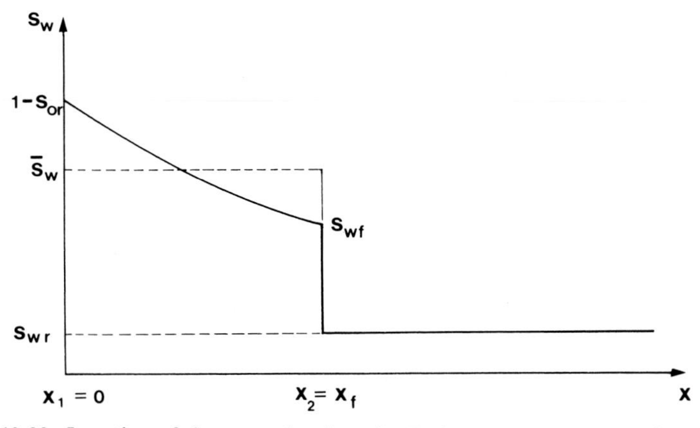

Fig. 10.23. Location of the saturation front by Welge's method. $\bar{s}_w = [\int_{x1}^{x2} s_w\, dx]/x_2 = (1/f'_w|_{Swf}) + S_{wr}$.

where A is the flow area in the plane normal to x. Now, from the characteristic equations (10.3.3.5b), $x_2 = (Vt/\phi) f'_w|_{S_{wf}}$. Thus, Eq. (10.3.3.6b) becomes

$$f'_w|_{S_{wf}} = 1/(\overline{S}_w - S_{wr}). \tag{10.3.3.7}$$

By definition, $\overline{S}_w$ can also be computed from direct integration of the saturation profile (Fig. 10.23). Thus,

$$x_2\overline{S}_w = \int_{x_1}^{x_2} S_w\, dx, \tag{10.3.3.8a}$$

or

$$\overline{S}_w = \int_0^{x_f} S_w\, dx/x_f. \tag{10.3.3.8b}$$

Noting that the integration is to be performed for a fixed value of t and that x is proportional to f'_w, Eq. (10.3.3.8b) becomes

$$\overline{S}_w = \left[\int_{1-S_{or}}^{S_{wf}} S_w\, df'_w\right] \Big/ f'_w\Big|_{S_{wf}}. \tag{10.3.3.9}$$

Integrating the numerator by parts yields

$$\overline{S}_w = \left[S_w f'_w\Big|_{1-S_{or}}^{S_{wf}} - f_w\Big|_{1-S_{or}}^{S_{wf}}\right] \Big/ f'_w\Big|_{S_{wf}}$$

$$= S_{wf} - \left(f_w|_{S_{wf}} - 1\right)/f'_w|_{S_{wf}}. \tag{10.3.3.10}$$

Solving for $f'_w|_{S_{wf}}$ and equating the result with Eq. (10.3.3.7), we obtain

$$f'_w|_{S_{wf}} = \frac{(1 - f_w|_{S_{wf}})}{\overline{S}_w - S_{wf}} = \frac{1}{\overline{S}_w - S_{wr}}. \tag{10.3.3.11}$$

As illustrated in Fig. 10.22a, Eq. (10.3.3.11) indicates that at the saturation value equal to S_{wf}, the tangent slope of the fractional flow curve is the same as the slope of the chord *ac*. (Note that the extrapolated tangent intercepts the line $f_w = 1$ at the point where $S_w = \overline{S}_w$.) Thus, we can deduce that the Welge method of locating the position of the front simply consists of plotting the fractional flow curve and locating S_{wf} as shown in Fig. 10.22a. Once S_{wf} is obtained, x_f can be located, as shown in Fig. 10.22b.

General Flow Problem with Capillarity

The procedure for solving the Buckley–Leverett problem can be extended to include the effect of capillarity. In this case, the oil and water pressures are related by

$$p_o = p_w + p_c, \tag{10.3.3.12}$$

where p_c is the capillary pressure, which is another nonlinear function of saturation. The governing equations of the flow problem now take the form

$$V = V_w + V_o = -k\left(\lambda_w \frac{\partial p_w}{\partial x} + \lambda_o \frac{\partial p_o}{\partial x}\right) \tag{10.3.3.13}$$

and

$$\phi \frac{\partial S_w}{\partial t} + \frac{\partial V_w}{\partial x} = \phi \frac{\partial S_w}{\partial t} + V \frac{\partial F_w}{\partial x} = 0, \tag{10.3.3.14}$$

where F_w denotes a fractional flow function that includes the capillary term. The relationship between F_w and f_w (the fractional flow function without capillary term) can be obtained as

$$F_w = \frac{V_w}{V} = \frac{-k\lambda_w \, \partial p_w/\partial x}{V}\left(\frac{\lambda_w + \lambda_o}{\lambda_w + \lambda_o}\right). \tag{10.3.3.15}$$

Because $f_w = \lambda_w/(\lambda_o + \lambda_w)$, Eq. (10.3.3.15) becomes

$$F_w = -f_w \frac{k}{V}(\lambda_o + \lambda_w)\frac{\partial p_w}{\partial x}. \tag{10.3.3.16}$$

Combination of Eqs. (10.3.3.16), (10.3.3.12), and (10.3.3.13) leads to

$$\begin{aligned} F_w &= f_w\left[-\frac{k}{V}\left(\lambda_w \frac{\partial p_w}{\partial x} + \lambda_o \frac{\partial p_o}{\partial x}\right) + \frac{k\lambda_o}{V}\frac{\partial p_c}{\partial x}\right] \\ &= f_w\left[1 + \frac{k\lambda_o}{V}\frac{\partial p_c}{\partial x}\right]. \end{aligned} \tag{10.3.3.17}$$

As in the Buckley–Leverett flow problem, the main task is to solve the governing equation for water saturation (10.3.3.14). Once the water saturation is known, p_w and p_o can be obtained by a simple integration of Darcy's equations. To solve Eq. (10.3.3.14), we first introduce the dimensionless Leverett J function and a function E as follows:

$$J(S_w) = (k/\phi)^{1/2}(1/\sigma)p_c(S_w) \tag{10.3.3.18}$$

and

$$E(S_w) = k_{ro}(S_w)f_w(S_w)J'(S_w), \tag{10.3.3.19}$$

where σ is the surface tension between two fluids and $J' = dJ/dS_w$. It follows that the expression for F_w and Eq. (10.3.3.14) now take the form

$$F_w = f_w\left[1 + \frac{k\lambda_o}{V}\sigma\left(\frac{\phi}{k}\right)^{1/2} J' \frac{\partial S_w}{\partial x}\right]$$

$$= f_w + \varepsilon LE\frac{\partial S_w}{\partial x} = f_w(1 + N_c) \tag{10.3.3.20}$$

and

$$\phi\frac{\partial S_w}{\partial t} + Vf'_w\frac{\partial S_w}{\partial x} + \varepsilon LV\frac{\partial}{\partial x}\left(E\frac{\partial S_w}{\partial x}\right) = 0, \tag{10.3.3.21}$$

where ε is a dimensionless parameter defined as

$$\varepsilon = \sigma(k\phi)^{1/2}/(LV\mu_o), \tag{10.3.3.22a}$$

L is a characteristic length, and N_c is a dimensionless number defined as

$$N_c = \frac{\sigma(k\phi)^{1/2}}{V\mu_o} k_{ro} J' \frac{\partial S_w}{\partial x}. \tag{10.3.3.22b}$$

It is evident that the effect of capillarity will be small if the capillary number $N_c << 1$. In addition, the governing equation for water saturation, Eq. (10.3.3.21), is a second-order nonlinear equation for which no exact analytical solution is known. If the capillary effect is negligible, the term involving ε in this equation can be dropped, and we then obtain the Buckley–Leverett equation. However, in the process the order of the differential operator is reduced, and the qualitative structure of the solution may be lost. This situation corresponds to that of a so-called "singular perturbation" problem, which is well known in the theory of fluid dynamics [see Van Dyke, (1975)]. The parameter ε thus behaves like a singular perturbation parameter.

To solve the flow problem with capillarity, it is logical to adopt a procedure similar to that for solving a singular perturbation problem. First, we obtain the "asymptotic" solution for the limiting case governed by the Buckley–Leverett equation. We expect the solution obtained to be invalid in the region near the saturation front where the values of $\partial S_w/\partial x$ and hence the capillary numbers N_c are large. In this region, the solution is determined by solving the second-order equation (10.3.3.21), which includes the capillary term. The composite saturation profile is finally obtained by combining the two solutions by an appropriate "matching" procedure.

To illustrate the complete solution procedure, let χ be a dimensionless variable defined as

$$\chi = [x - \xi(t)]/\varepsilon L. \tag{10.3.3.23a}$$

By the use of the chain rule, we obtain

$$\left.\frac{\partial S_w}{\partial x}\right|_t = \left.\frac{\partial S_w}{\partial \chi}\right|_t \left(\frac{\partial \chi}{\partial x}\right) = \frac{1}{\varepsilon L}\left.\frac{\partial S_w}{\partial \chi}\right|_t, \tag{10.3.3.23b}$$

$$\left.\frac{\partial S_w}{\partial t}\right|_x = \left.\frac{\partial S_w}{\partial \chi}\right|_t \left(\frac{\partial \chi}{\partial t}\right) + \left.\frac{\partial S_w}{\partial t}\right|_\chi = \frac{1}{\varepsilon L}\left(-\frac{\partial S_w}{\partial \chi}\frac{d\xi}{dt}\right) + \left.\frac{\partial S_w}{\partial t}\right|_\chi. \tag{10.3.3.23c}$$

Substitution of (10.3.3.23b) and (10.3.3.23c) into (10.3.3.21) yields

$$\frac{\partial}{\partial \chi}\left(E\frac{\partial S_w}{\partial \chi}\right) + \left(f'_w - \frac{\phi}{V}\frac{d\xi}{dt}\right)\frac{\partial S_w}{\partial \chi} + \frac{\varepsilon \phi L}{V}\frac{\partial S_w}{\partial t} = 0. \tag{10.3.3.24}$$

Because ε now multiplies a lower-order term, this term can be neglected. Equation (10.3.3.24) reduces to

$$\frac{\partial}{\partial \chi}\left(E\frac{\partial S_w}{\partial \chi}\right) + \left(f'_w - \frac{\phi}{V}\frac{d\xi}{dt}\right)\frac{\partial S_w}{\partial \chi} = 0, \tag{10.3.3.25}$$

which may be integrated to give

$$E\frac{\partial S_w}{\partial \chi} + f_w - \frac{\phi}{V}\frac{d\xi}{dt}S_w = C(t), \tag{10.3.3.26}$$

where $C(t)$ is an arbitrary function of t to account for the temporal behavior neglected in (10.3.3.25). To determine $C(t)$, we simply apply Eq. (10.3.3.26) at $S_w = S_{wr}$. The first and second left-hand terms drop out because $\partial S_w/\partial \chi$ and f_w are equal to zero at $S_w = S_{wr}$. The result thus becomes

$$C(t) = -S_{wr}\frac{\phi}{V}\frac{d\xi}{dt}. \tag{10.3.3.27}$$

Substitution of Eq. (10.3.3.27) into (10.3.3.26) and rearranging gives

$$E\frac{\partial S_w}{\partial \chi} = \left[(S_w - S_{wr})\frac{\phi}{V}\frac{d\xi}{dt} - f_w\right],$$

or (10.3.3.28)

$$E\frac{\partial S_w}{\partial \chi} = (S_w - S_{wr})f'_w(S_\xi) - f_w,$$

where $f'_w(S_\xi)$ is defined as

$$f'_w(S\xi) = \frac{\phi}{V}\frac{d\xi}{dt}.$$

Equation (10.3.3.28) still contains the unknown, $f'_w(S_\xi)$. This unknown can be determined by evaluating Eq. (10.3.3.28) at a match point, where $x = x_m$ and $S_w = S_{wm}$. At this match point the slope $\partial S_w/\partial \chi$ determined from Eq. (10.3.3.28) is equated to the slope of the Buckley–Leverett saturation profile. Thus, we obtain

$$f'_w(S_\xi) = [E(S_{wm}) \tan^{-1} \theta_m + f_w(S_{wm})]/(S_{wm} - S_{wr}), \tag{10.3.3.29a}$$

where θ_m is the angle between the slope of the saturation profile at $x = x_m$ and the horizontal. It remains to determine the location of the match point. This is achieved by making use of the material balance condition. As illustrated in Fig. 10.24, the areas must be the same beneath the Buckley–Leverett profile and the profile with capillary. Let $\hat{S}$ represent the Buckley–Leverett solution. It follows that

$$\int_{\chi_m}^{\chi_{wr}} (S_w - S_{wr})\, d\chi = \int_{\chi_m}^{\chi_f} (\hat{S}_w - S_{wr})\, d\chi. \tag{10.3.3.29b}$$

Now from Eqs. (10.3.3.23a) and (10.3.3.28),

$$d\chi = dx/\varepsilon L \tag{10.3.3.30a}$$

and

$$d\chi = E\, dS_w/[(S_w - S_{wr})f'_w(S_\xi) - f_w]. \tag{10.3.3.30b}$$

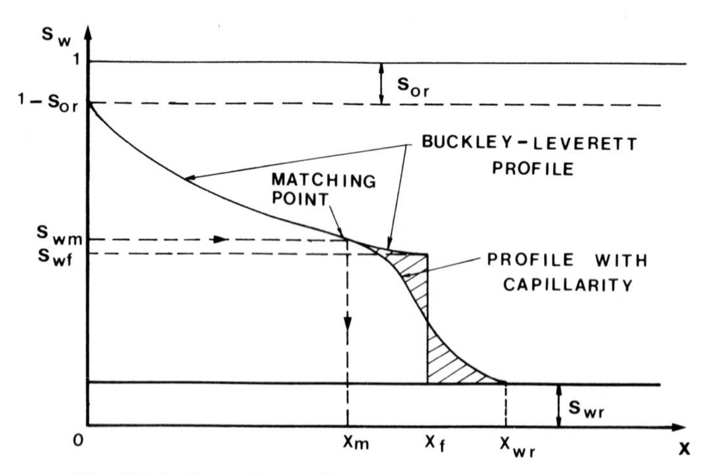

Fig. 10.24. Saturation profiles with and without capillarity.

Substituting Eqs. (10.3.3.30a) and (10.3.3.30b) into (10.3.3.29b) and making an appropriate change of the limits of integration, we obtain

$$\int_{S_{wm}}^{S_{wr}} \frac{(S_w - S_{wr})E(S_w)\, dS_w}{G(S_w, S_{wm}, t)} = \frac{1}{\varepsilon L}\int_{x_m}^{x_f} (\hat{S}_w - S_{wr})\, dx,$$

which upon substitution for εL becomes

$$\int_{x_m}^{x_f} (\hat{S}_w - S_{wr})\, dx = \frac{\sigma(k\phi)^{1/2}}{V\mu_o} \int_{S_{wm}}^{S_{wr}} \frac{(S_w - S_{wr})E(S_w)\, dS_w}{G(S_w, S_{wm}, t)}, \qquad (10.3.3.31)$$

where the function G is defined as

$$G(S_w, S_{wm}, t) = (\dot{S}_w - S_{wr})f'_w(S_\xi) - f_w(S_w)$$

$$= \frac{(S_w - S_{wr})}{(S_{wm} - S_{wr})}[E(S_{wm}) \tan^{-1}\theta_m + f_w(S_{wm})] - f_w(S_w). \qquad (10.3.3.32)$$

Equation (10.3.3.31) thus enables us to solve for S_{wm} iteratively. Once S_{wm} and x_m have been determined, the x coordinate of the saturation profile with capillarity can be computed by integrating Eq. (10.3.3.30b) as

$$\int_{x_m}^{x} dx = \varepsilon L \int_{\chi_m}^{\chi} d\chi = \varepsilon L \int_{S_{wm}}^{S_w} \frac{E(S_w)\, dS_w}{G(S_w, S_{wm}, t)},$$

or

$$x - x_m = \frac{\sigma(k\phi)^{1/2}}{V\mu_o} \int_{S_{wm}}^{S_w} \frac{E(S_w)\, dS_w}{G(S_w, S_{wm}, t)}. \qquad (10.3.3.33)$$

In summary, the complete procedure for solving the one-dimensional, two-phase flow problem with capillarity consists of the following steps:

(1) Determine the Buckley–Leverett saturation profile using the technique described for the Buckley–Leverett problem.
(2) Locate the match point (x_m, S_{wm}) by solving Eq. (10.3.3.31) iteratively.
(3) Compute the remaining part of the profile, $S_{wm} < S_w < S_{wr}$ from Eq. (10.3.3.33).

The iterative procedure used for solving Eq. (10.3.3.31) is quite simple. First an initial guess of $x_m = x_m^0$ is made, where $x_m^0 < x_f$. This initial value is then substituted into Eq. (10.3.3.31). Let I_L and I_R denote the integrals on the left and right sides of Eq. (10.3.3.31), respectively. If

$I_L < I_R$, then the assumed value of x_{m} is too small. For the first iteration, x_m^1 should be computed from

$$x_m^1 = x_m^0 + (\Delta I_0/I_L)\, x_m^0,$$

where $\Delta I^0 = (I_L - I_R)^0$. We then insert x_m^1 into Eq. (10.3.3.31) and compute the integrals and hence $(\Delta I)^1$. For a subsequent iteration $r + 1$, the value of x_m^{r+1} is computed from a linear interpolation illustrated in Fig. 10.25:

$$x_{\mathrm{m}}^{r+1} = x_{\mathrm{m}}^r + \frac{(\Delta I)^{r-1}}{(\Delta I)^r + (\Delta I)^{r-1}}\,(x_{\mathrm{m}}^{r-1} - x_{\mathrm{m}}^r).$$

The iterative procedure is repeated until satisfactory convergence is obtained. Note that all integrals should be evaluated using either Simpson's rule or the trapezoidal rule.

In retrospect, the problem of one-dimensional two-phase flow with capillarity is mathematically similar to the problem of one-dimensional infiltration of water in unsaturated soils. In general terms, both problems can be classified as singular perturbation problems. In the problem of unsaturated infiltration, the gravity term plays a significant role and hence overshadows the second-order dispersion term at large time values. The order of the governing differential equation is then reduced from second order to first order.

10.3.4 *Solute Transport*

Although the method of characteristics is normally suitable for hyperbolic equations, it can also be adopted to solve a parabolic transport equation.

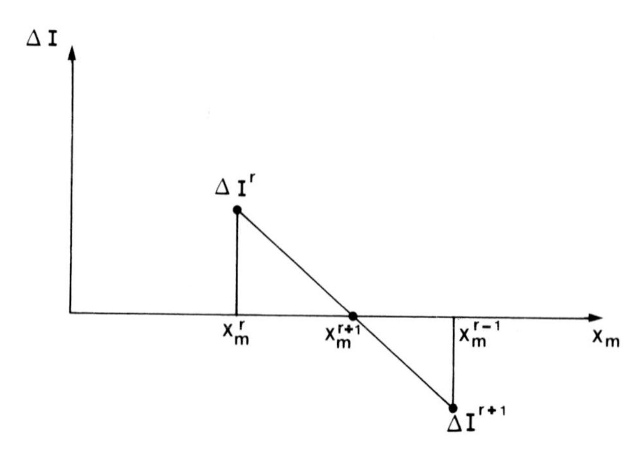

Fig. 10.25. Linear interpolation of ΔI values.

It is particularly effective when transport is dominated by convective terms. Such an application has been presented by several researchers, e.g., Garder *et al.* (1964), Pinder and Cooper (1970), and Konikow and Bredehoeft (1978).

Consider a situation involving two-dimensional transport of a solute in an aquifer. The appropriate governing partial differential equation can be written in the form

$$\frac{\partial}{\partial x_i}\left(bD_{ij}\frac{\partial c}{\partial x_j}\right) - \frac{\partial}{\partial x_i}(bV_i c) = \frac{\partial}{\partial t}(\phi bc) + c'w, \qquad (10.3.4.1a)$$

where b and ϕ are the aquifer thickness and effective porosity respectively, D_{ij} is the hydrodynamic dispersion tensor, V_i is the Darcy velocity, c is the solute concentration, w is the specific discharge, and c' is the concentration of the dissolved chemical in a source or sink fluid. If it is assumed that the groundwater flow is in a steady state, Eq. (10.3.4.1a) can then be combined with the continuity of mass equation to give (see Chapter 5)

$$\frac{1}{b}\frac{\partial}{\partial x_i}\left(bD_{ij}\frac{\partial c}{\partial x_j}\right) - V_i\frac{\partial c}{\partial x_i} = \phi\left(\frac{\partial c}{\partial t} + \frac{c}{b}\frac{\partial b}{\partial t}\right) + \frac{w(c'-c)}{b},$$

or

$$\frac{\partial c}{\partial t} + \frac{V_i}{\phi}\frac{\partial c}{\partial x_i} = \frac{1}{\phi b}\frac{\partial}{\partial x_i}\left(bD_{ij}\frac{\partial c}{\partial x_j}\right) + F, \qquad (10.3.4.1b)$$

where $F = -[(c/b)(\partial b/\partial t) + (w/b\phi)(c' - c)]$. Equation (10.3.4.1b) possesses one characteristic and this is defined by the equations

$$\frac{dx_i}{dt} = \frac{V_i}{\phi}, \qquad i = 1, 2. \qquad (10.3.4.2)$$

Along this characteristic, Eq. (10.3.4.1b) reduces to

$$\frac{dc}{dt} = \frac{1}{\phi b}\frac{\partial}{\partial x_i}\left(bD_{ij}\frac{\partial c}{\partial x_j}\right) + F,$$

which upon replacing x_1 by x and x_2 by y becomes

$$\frac{dc}{dt} = \frac{1}{\phi b}\left[\frac{\partial}{\partial x}\left(bD_{xx}\frac{\partial c}{\partial x} + bD_{xy}\frac{\partial c}{\partial y}\right) + \frac{\partial}{\partial y}\left(bD_{yy}\frac{\partial c}{\partial y} + bD_{yx}\frac{\partial c}{\partial x}\right)\right] - \left(\frac{c}{b}\frac{\partial b}{\partial t} + \frac{w(c'-c)}{b\phi}\right). \qquad (10.3.4.3)$$

The solution of the transport equation thus reduces to solving the system of equations represented by Eqs. (10.3.4.2) and (10.3.4.3). This is accomplished by adopting a block-centered finite difference grid as shown in Fig. 10.26a. In essence, Eqs. (10.3.4.2) are solved by first placing a number of particles or points in each cell of the finite difference grid and then tracking the movement of these particles (Fig. 10.26b). Because each particle has a position and concentration associated with it, we can use this information to calculate the concentration at the grid nodes resulting from the convection of the particles. Following this calculation, Eq. (10.3.4.3) is used to evaluate changes in the nodal concentrations owing to hydrodynamic dispersion, changes in saturated thickness, and the sink or source term.

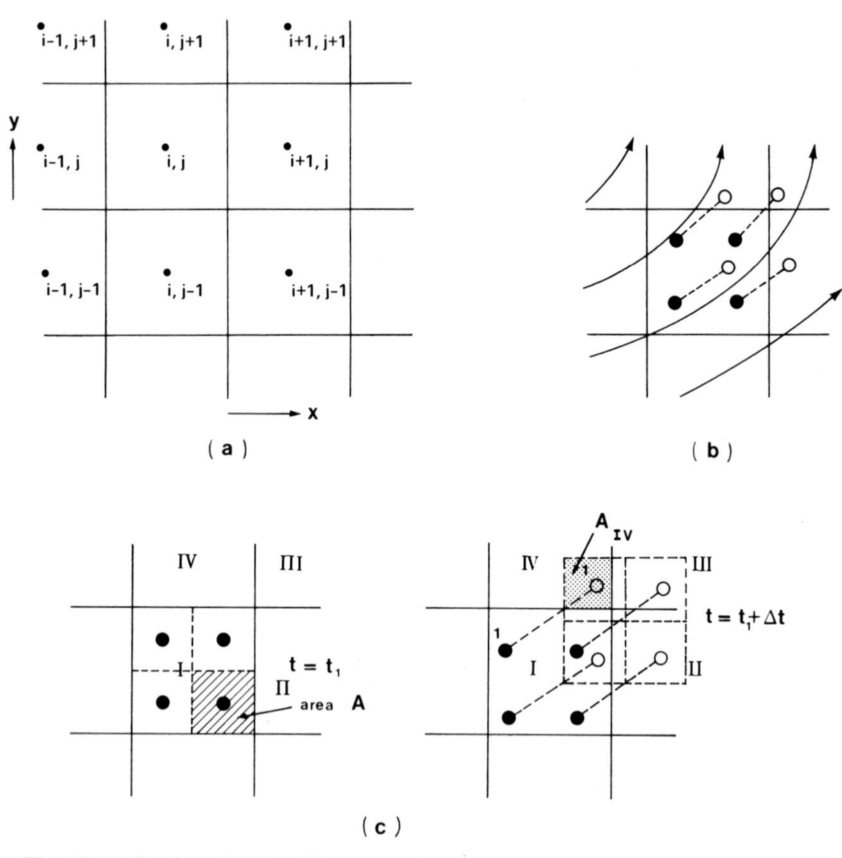

Fig. 10.26. Portion of finite difference grid, showing (a) nodal point identification, (b) particle tracking (after Konikow and Bredehoeft, 1978), and (c) application of a weighted averaging scheme. Legend for (b): ●, initial location of particle; ○, new location of particle; →, flow line and direction of flow; ----, computed path of particle.

Particle Tracking

Figure 10.26b shows the displacement of four particles initially placed in a cell. The initial concentration of these particles is set equal to the initial concentration of the node in the same cell. The new position of a particle p is computed from the following finite difference approximation of Eq. (10.3.4.2):

$$x_{p,k} = x_{p,k-1} + \frac{\Delta t}{\phi} V_{x,p,k}, \tag{10.3.4.4a}$$

and

$$y_{p,k} = y_{p,k-1} + \frac{\Delta t}{\phi} V_{y,p,k}, \tag{10.3.4.4b}$$

where $V_{x,p,k}$ and $V_{y,p,k}$ are the x and y velocities of the particle p for time level k. These velocities are calculated through bilinear interpolation of the velocities of adjacent nodes and cell boundaries.

After all particles have been displaced, the concentration at each node in the grid is computed by taking the average of the concentrations of those particles located within the area of that cell. For convenience, this average nodal concentration is denoted by $c^*_{i,j,k}$. The asterisk is used to indicate that the concentration value obtained at this stage represents the new time level only with respect to convective transport.

The algorithm employed in averaging the particle concentrations in a given cell markedly affects the behavior of the concentration solution. A strict arithmetic average generates jumps in the concentration solution as a particle moves across a cell boundary. A superior scheme involves the allocation of an area (a volume in three dimensions) to each particle. As a particle moves its area is used to assign concentration to neighboring cells. In Fig. 10.26c we demonstrate this procedure. When particle 1 moves to its new location in cell IV, the original resident cell (I) does not ignore the existence of this particle in averaging. The mass of this particle is allocated to cell I, II, III, or IV in proportion to the particle area resident in that cell. Cell IV, for example, would be allocated A_{IV}/A of the mass of particle 1 in the averaging process. This procedure, although more complicated to program, provides a more accurate solution and is worth the additional effort.

Computation of Final Nodal and Particle Concentrations

The total concentration at time level k must include changes in concentration caused by hydrodynamic dispersion and other factors apart from convective transport. These additional changes are calculated using an explicit finite difference approximation of Eq. (10.3.4.3). The approximation can be written in the form

$$\begin{aligned}\Delta c_{i,j,k} = {} & (\Delta t/\phi b)\, \{\tfrac{1}{2}[\delta_x(bD_{xx}\delta_x c_{k-1} + bD_{xy}\delta_y c_{k-1}) \\ & + \delta_y(bD_{yy}\delta_y c_{k-1} + bD_{yx}\delta_x c_{k-1}) + \delta_x(bD_{xx}\delta_x c_k^* \\ & + bD_{xy}\delta_y c_k^*) + \delta_y(bD_{yy}\delta_y c_k^* + bD_{yx}\delta_x c_k^*)] \\ & - \Delta t\left[\frac{w(c' - c)}{b\phi} + \frac{\Delta b}{2b\,\Delta t}(c_{k-1} + c_k^*)\right],\end{aligned} \qquad (10.3.4.5)$$

where δ_x and δ_y denote the central difference operators corresponding to $\partial/\partial x$ and $\partial/\partial y$, respectively.

After $\Delta c_{i,j,k}$ has been obtained, the final nodal concentrations at time level k are computed from

$$c_{i,j,k} = c_{i,j,k}^* + \Delta c_{i,j,k}. \qquad (10.3.4.6)$$

At the end of the time step, the particle concentrations are updated using the equation

$$c_{p,k} = c_{p,k-1} + \Delta c_{p,k}, \qquad (10.3.4.7)$$

where $\Delta c_{p,k}$ denotes the concentration change of particle p owing to the right-hand term of Eq. (10.3.4.3), i.e. $\Delta c_{i,j,k}$, for its resident cell. Because the concentrations of particles in a cell vary about the concentration of the node, caution must be exercised in computing $c_{p,k}$. To avoid obtaining negative values of $c_{p,k}$, the incremental change in concentration $\Delta c_{p,k}$ is determined as follows.

If the change in concentration at the node $\Delta c_{i,j,k}$ is positive, $\Delta c_{p,k}$ is obtained by simply setting it equal to $\Delta c_{i,j,k}$. On the other hand, if $\Delta c_{i,j,k}$ is negative, $\Delta c_{p,k}$ is obtained so that the ratio $\Delta c_{p,k}/c_{p,k-1}$ is equal to $\Delta c_{i,j,k}/c_{i,j,k}^*$.

To achieve numerical stability using the preceding explicit time-stepping scheme, the size of the time increment Δt must be sufficiently small. The value of Δt should be chosen to satisfy certain stability criteria.

The first stability check corresponds to the stability criterion normally used in the explicit finite difference solution of the two-dimensional groundwater flow equation. According to this criterion, Δt must be chosen so that

$$\Delta t \leqslant \min_{\text{o.g.}} \{0.5/[D_{xx}/(\Delta x)^2 + D_{yy}/(\Delta y)^2]\}, \qquad (10.3.4.8)$$

where o.g. means over grid. The second stability check is related to the movement of particles computed by Eqs. (10.3.4.4a) and (10.3.4.4b). The criterion used corresponds to the well-known Courant–Friedrichs condition for the finite difference solution of the pure convective transport equation. According to this criteiron, Δt must be chosen so that

$$(\Delta t\, V_{x,p})/(\Delta x\, \phi) \leqslant 1 \qquad (10.3.4.9a)$$

and

$$(\Delta t\, V_{y,p})/(\Delta y\, \phi) \leqslant 1. \qquad (10.3.4.9b)$$

Because $V_{x,p}$ and $V_{y,p}$ are less than or equal to the maximum x and y Darcy velocity components of the entire grid, the preceding inequalities become

$$\Delta t \leqslant \phi\, \Delta x/(V_x)_{\max} \qquad (10.3.4.10a)$$

and

$$\Delta t \leqslant \phi\, \Delta y/(V_y)_{\max}. \qquad (10.3.4.10b)$$

The third stability check takes into account the effects of mixing groundwater of one concentration with injected or recharge water of a different concentration. According to this criterion, Δt must be chosen so that

$$\Delta t \leqslant \min_{\text{o.g.}} [\phi b_{i,j,k}/w_{i,j,k}]. \qquad (10.3.4.11)$$

Treatment of No-Flow Boundaries, Sources, and Sinks

A number of problems arise in particle tracking owing to the presence of impermeable boundaries and fluid sources and sinks. Under certain conditions, the application of interpolated velocities and Eqs. (10.3.4.4a) and (10.3.4.4b) to a particle near an impermeable boundary may result in the convection of the particle into the zone of zero transmissivity. Figure 10.27 illustrates such a possible situation, which is caused by the deviation between the curvilinear flow line and the linearly projected particle path. To meet the condition required at the no-flow boundary, the particle is brought into the aquifer zone by reflection across this boundary. Fluid sources and sinks also require special treatment as they represent singularities in the flow field. To achieve radial flow around a source or sink, its nodal velocity is determined from adjacent velocities at the boundary of the cell. This nodal velocity is then used in the computation of particle velocities.

Because the velocity direction is radially outward from the source node, particles will continually move out of their cells, and eventually the cells could be void of particles. To help maintain a continuous generation of fluid and solute mass at the source, a new particle is introduced into the source cell to replace the one that has just left.

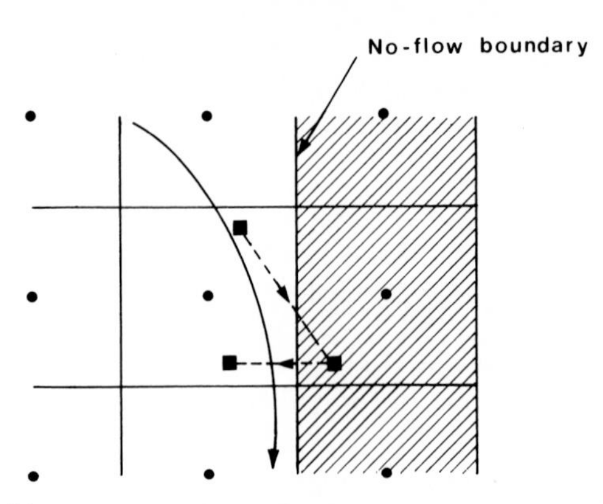

Fig. 10.27. Possible movement of particles near an impermeable boundary. Legend: ●, node of finite difference grid; ▨, zero transmissivity zone; →, actual flow path; ----, computed flow path.

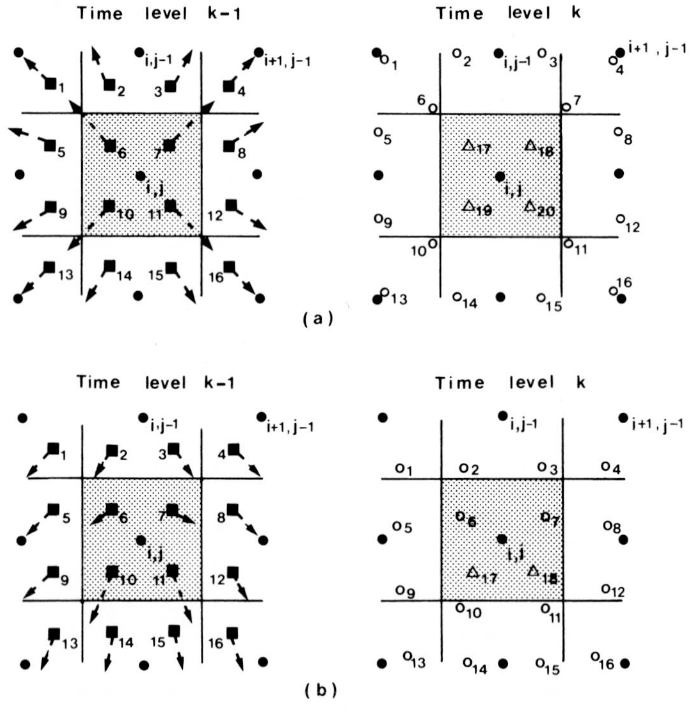

Fig. 10.28. Replacement of particles in a source cell not adjacent to an impermeable boundary (a) for relatively weak regional flow and (b) for relatively strong regional flow. Legend: ●, node of finite difference grid; ■p, initial location of particle p; ○p, new location of particle p; △p, location of replacement particle p; ⊡, fluid source. (After Konikow and Bredehoeft, 1978.)

The procedure used to replace particles in a source cell that is not adjacent to an impermeable boundary is shown in Fig. 10.28. As depicted, a uniformly spaced stream of particles is maintained by generating a new particle at the original location of the particle that left the source cell. Note that when a source is a relatively strong regional flow field, particles from outside may enter and then leave the source cell. These particles are not replaced because they did not originate at the source.

The procedure used to replace particles in a source cell adjacent to an impermeable boundary is illustrated in Fig. 10.29. It can be seen that a steady, uniformly spaced stream of particles is maintained by generating a new particle at the same relative position in the source cell as the new position in the adjacent cell of the point that left the source cell. As an example, point 7 was convected from cell $(i-1, j)$ to (i, j). Hence the replacement point 22 was placed at a location within cell $(i-1, j)$ that is identical to the new location of point 7 within cell (i, j).

For a cell containing a hydraulic sink, particles will continually move

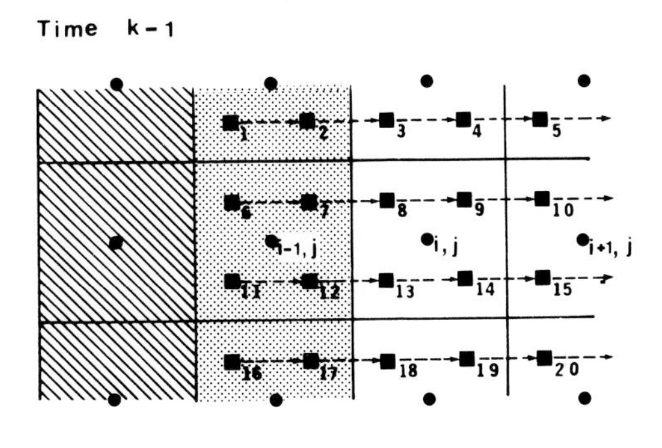

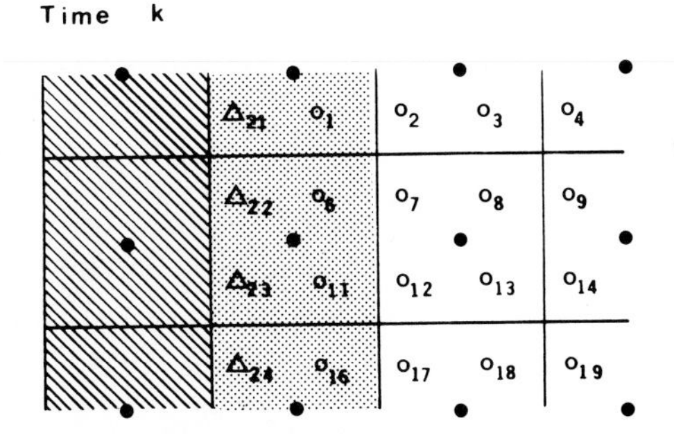

Fig. 10.29. Replacement of particles in source cells adjacent to an impermeable boundary. Legend: ●, node of finite difference grid; ■p, initial location of particle p; ○p, new location of particle p; △p, location of replacement particle p; ▩, constant head source; ▧, zero transmissivity (or no-flow boundary) zone. (After Konikow and Bredehoeft, 1978.)

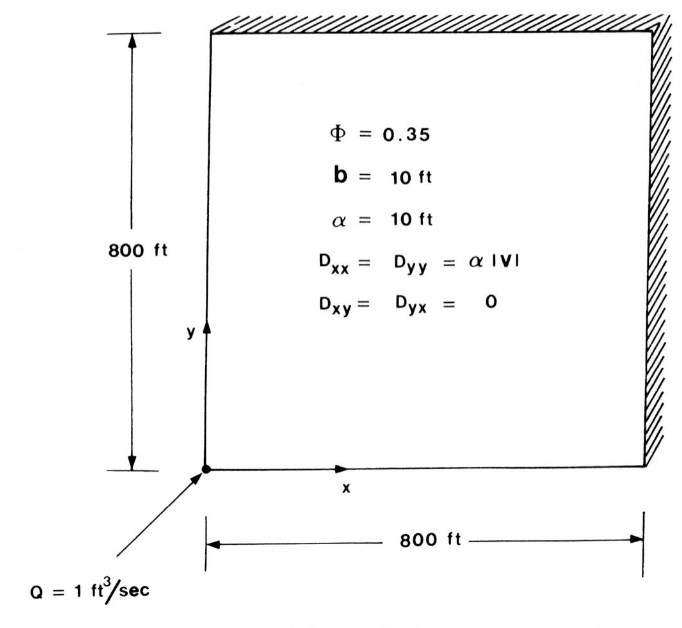

Fig. 10.30. Injection well flow and solute transport problem.

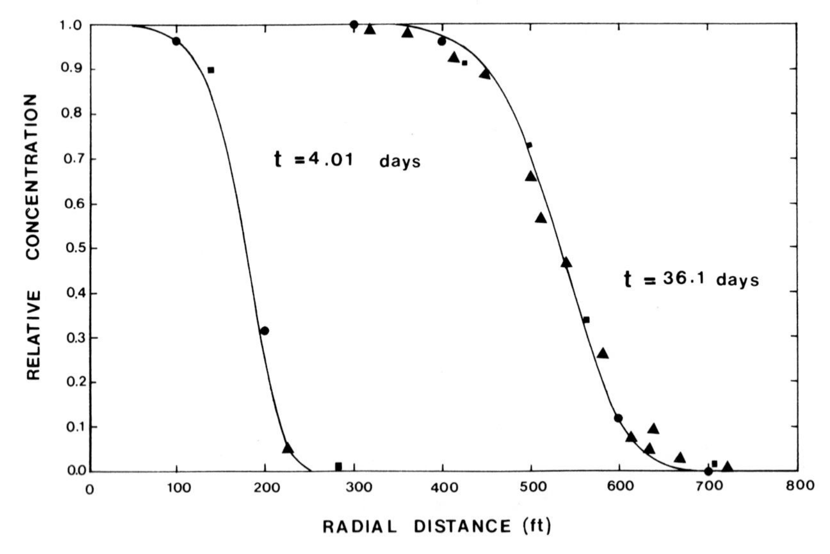

Fig. 10.31. Comparison between analytical and numerical solutions. Legend: —, analytical solution/numerical solution; ■, nodes on radii 45° to grid; ●, nodes on radii parallel to grid; ▲, nodes on intermediate radii. (After Konikow and Bredehoeft, 1978.)

into this cell, but few or none will move out of it. To avoid crowding of particles, any particle moving into a sink cell is removed from the flow field after all calculations for the time increment have been completed.

ILLUSTRATIVE EXAMPLE An example problem solved by Konikow and Bredehoeft (1978) is shown in Fig. 10.30. The problem is that of radial flow and solute transport from an injection well. The numerical results obtained are compared in Fig. 10.31 with an analytical solution given by Bear (1972, p. 638). It can be seen that there is generally good agreement between the analytical and numerical solutions for relatively short and long times. The greatest errors occur at nodes on radii from the injection well that are neither parallel to nor 45° from the x and y axes.

References

Ames, W. F. (1977). "Numerical Methods for Partial Differential Equations," 2nd ed. Academic Press, New York.

Bear, J. (1979). "Hydraulics of Groundwater." McGraw-Hill, New York.

Buckley, S. E., and Leverett, M. C. (1942). Mechanism of fluid displacement in sands. *AIME* **146,** 107–116.

Drake, R. L., Molz, F. J., Remson, I., and Fungaroli, A. A. (1969). Similarity approximation for the radial subsurface flow problem. *Water Resour. Res.* **5** (3), 673–684.

Garder, A. O., Peaceman, D. W., and Pozzi, A. L. Jr. (1964). Numerical calculation of multidimensional miscible displacement by the method of characteristics. *Soc. Pet. Eng. J.* **4** (1), 26–36.

Jeppson, R. W. (1968a). Seepage from ditches—Solution by finite differences. *J. Hydraul. Div. ASCE* **94** (HY1), 259–283.

Jeppson, R. W. (1968b). Axisymmetric seepage through homogeneous and non-homogeneous porous mediums. *Water Resour. Res.* **4** (6), 1277–1288.

Kirkham, D. and Powers, W. L. (1972). "Advanced Soil Physics," ch. 7. Wiley–Interscience, New York.

Konikow, L. F. and Bredehoeft, J. D. (1978). Computer Model of Two-dimensional Solute Transport and Dispersion in Ground Water. Book 7, ch. C2. U.S. Geological Survey.

Morel-Seytoux, H. J. (1969). Introduction to flow of immiscible liquids in porous media. *In* "Flow through Porous Media," R. J. M. De Wiest, ed., ch. 11, Academic Press, New York.

Philip, J. R. (1955). Numerical solution of equations of the diffusion type with diffusivity concentration-dependent, *Trans. Faraday Soc.* **51,** 885–892.

Philip, J. R. (1969). Theory of infiltration. *In* "Advances in Hydroscience," Vol. 5, V. T. Chow, ed. 216–291. Academic Press, New York.

Pinder, G. F. and Cooper, H. H., Jr. (1970). A numerical technique for calculating the transient position of the salt water front. *Water Resour. Res.* **6** (3), 875–882.

Remson, I., Hornberger, G. M., and Molz, F. J. (1971). "Numerical Methods in Subsurface Hydrology, with An Introduction to The Finite Element method," ch. 6. Wiley-Interscience, New York.

Rubin, J. (1968). Theoretical analysis of two-dimensional transient flow of water in unsaturated and partly unsaturated soils. *Soil Sci. Soc. Am. Proc.* **32,** 607–615.

Streeter, V. L., and Wiley, E. B. (1967). "Hydraulic Transients." McGraw-Hill, New York.

Van Dyke, M. (1975). "Perturbation Methods in Fluid Mechanics," 2nd ed. Parabolic Press, Stanford, California.

Welge, H. J. (1952). A simplified method for computing oil recovery by gas or water drive. *Trans. AIME* **195,** 91–98.

Wiley, E. B. (1976). Transient aquifer flows by characteristics method. *Am. Soc. Civ. Eng.* **102** (HY3), 293–305.

Index

G

H

I

T

U

V

W

Y